FE Reference Handbook

9.5 Version for Computer-Based Testing

978-1-932613-67-4

PO Box 1686
Clemson, SC 29633
800-250-3196
www.ncees.org

ISBN 978-1-932613-67-4

Printed in the United States of America
Fifth printing June 2018
Edition 9.5

CONTENTS

UNITS

The FE exam and this handbook use both the metric system of units and the U.S. Customary System (USCS). In the USCS system of units, both force and mass are called pounds. Therefore, one must distinguish the pound-force (lbf) from the pound-mass (lbm).

The pound-force is that force which accelerates one pound-mass at 32.174 ft/sec². Thus, 1 lbf = 32.174 lbm-ft/sec². The expression 32.174 lbm-ft/(lbf-sec²) is designated as g_c and is used to resolve expressions involving both mass and force expressed as pounds. For instance, in writing Newton's second law, the equation would be written as $F = ma/g_c$, where F is in lbf, m in lbm, and a is in ft/sec².

Similar expressions exist for other quantities. Kinetic Energy, $KE = mv^2/2g_c$, with KE in (ft-lbf); Potential Energy, $PE = mgh/g_c$, with PE in (ft-lbf); Fluid Pressure, $p = \rho gh/g_c$, with p in (lbf/ft²); Specific Weight, $SW = \rho g/g_c$, in (lbf/ft³); Shear Stress, $\tau = (\mu/g_c)(dv/dy)$, with shear stress in (lbf/ft²). In all these examples, g_c should be regarded as a force unit conversion factor. It is frequently not written explicitly in engineering equations. However, its use is required to produce a consistent set of units.

Note that the force unit conversion factor g_c [lbm-ft/(lbf-sec²)] should not be confused with the local acceleration of gravity g, which has different units (m/s² or ft/sec²) and may be either its standard value (9.807 m/s² or 32.174 ft/sec²) or some other local value.

If the problem is presented in USCS units, it may be necessary to use the constant g_c in the equation to have a consistent set of units.

METRIC PREFIXES		
Multiple	Prefix	Symbol
10^{-18}	atto	a
10^{-15}	femto	f
10^{-12}	pico	p
10^{-9}	nano	n
10^{-6}	micro	μ
10^{-3}	milli	m
10^{-2}	centi	c
10^{-1}	deci	d
10^{1}	deka	da
10^{2}	hecto	h
10^{3}	kilo	k
10^{6}	mega	M
10^{9}	giga	G
10^{12}	tera	T
10^{15}	peta	P
10^{18}	exa	E

COMMONLY USED EQUIVALENTS	
1 gallon of water weighs	8.34 lbf
1 cubic foot of water weighs	62.4 lbf
1 cubic inch of mercury weighs	0.491 lbf
The mass of 1 cubic meter of water is	1,000 kilograms
1 mg/L is	8.34 lbf/Mgal

TEMPERATURE CONVERSIONS

°F = 1.8 (°C) + 32
°C = (°F – 32)/1.8
°R = °F + 459.69
K = °C + 273.15

IDEAL GAS CONSTANTS

The universal gas constant, designated as $\bar{R}$ in the table below, relates pressure, volume, temperature, and number of moles of an ideal gas. When that universal constant, $\bar{R}$, is divided by the molecular weight of the gas, the result, often designated as R, has units of energy per degree per unit mass [kJ/(kg·K) or ft-lbf/(lbm-°R)] and becomes characteristic of the particular gas. Some disciplines, notably chemical engineering, often use the symbol R to refer to the universal gas constant $\bar{R}$.

FUNDAMENTAL CONSTANTS

Quantity		Symbol	Value	Units
electron charge		e	1.6022×10^{-19}	C (coulombs)
Faraday constant		F	96,485	coulombs/(mol)
gas constant	metric	$\bar{R}$	8,314	J/(kmol·K)
gas constant	metric	$\bar{R}$	8.314	kPa·m³/(kmol·K)
gas constant	USCS	$\bar{R}$	1,545	ft-lbf/(lb mole-°R)
		$\bar{R}$	0.08206	L·atm/(mole·K)
gravitation–Newtonian constant		G	6.673×10^{-11}	m³/(kg·s²)
gravitation–Newtonian constant		G	6.673×10^{-11}	N·m²/kg²
gravity acceleration (standard)	metric	g	9.807	m/s²
gravity acceleration (standard)	USCS	g	32.174	ft/sec²
molar volume (ideal gas), $T = 273.15\text{K}$, $p = 101.3$ kPa		V_m	22,414	L/kmol
speed of light in vacuum		c	299,792,000	m/s
Stefan-Boltzmann constant		σ	5.67×10^{-8}	W/(m²·K⁴)

CONVERSION FACTORS

Multiply	By	To Obtain
acre	43,560	square feet (ft^2)
ampere-hr (A-hr)	3,600	coulomb (C)
ångström (Å)	1×10^{-10}	meter (m)
atmosphere (atm)	76.0	cm, mercury (Hg)
atm, std	29.92	in., mercury (Hg)
atm, std	14.70	lbf/in^2 abs (psia)
atm, std	33.90	ft, water
atm, std	1.013×10^5	pascal (Pa)
bar	1×10^5	Pa
bar	0.987	atm
barrels–oil	42	gallons–oil
Btu	1,055	joule (J)
Btu	2.928×10^{-4}	kilowatt-hr (kWh)
Btu	778	ft-lbf
Btu/hr	3.930×10^{-4}	horsepower (hp)
Btu/hr	0.293	watt (W)
Btu/hr	0.216	ft-lbf/sec
calorie (g-cal)	3.968×10^{-3}	Btu
cal	1.560×10^{-6}	hp-hr
cal	4.186	joule (J)
cal/sec	4.184	watt (W)
centimeter (cm)	3.281×10^{-2}	foot (ft)
cm	0.394	inch (in)
centipoise (cP)	0.001	pascal•sec (Pa•s)
centipoise (cP)	1	g/(m•s)
centipoise (cP)	2.419	lbm/hr-ft
centistoke (cSt)	1×10^{-6}	m^2/sec (m^2/s)
cubic feet/second (cfs)	0.646317	million gallons/day (MGD)
cubic foot (ft^3)	7.481	gallon
cubic meters (m^3)	1,000	liters
electronvolt (eV)	1.602×10^{-19}	joule (J)
foot (ft)	30.48	cm
ft	0.3048	meter (m)
ft-pound (ft-lbf)	1.285×10^{-3}	Btu
ft-lbf	3.766×10^{-7}	kilowatt-hr (kWh)
ft-lbf	0.324	calorie (g-cal)
ft-lbf	1.356	joule (J)
ft-lbm	2	slug-ft/s^2
ft-lbf/sec	1.818×10^{-3}	horsepower (hp)
gallon (U.S. Liq)	3.785	liter (L)
gallon (U.S. Liq)	0.134	ft^3
gallons of water	8.3453	pounds of water
gamma (γ, Γ)	1×10^{-9}	tesla (T)
gauss	1×10^{-4}	T
gram (g)	2.205×10^{-3}	pound (lbm)
hectare	1×10^4	square meters (m^2)
hectare	2.47104	acres
horsepower (hp)	42.4	Btu/min
hp	745.7	watt (W)
hp	33,000	(ft-lbf)/min
hp	550	(ft-lbf)/sec
hp-hr	2,545	Btu
hp-hr	1.98×10^6	ft-lbf
hp-hr	2.68×10^6	joule (J)
hp-hr	0.746	kWh
inch (in.)	2.540	centimeter (cm)
in. of Hg	0.0334	atm
in. of Hg	13.60	in. of H_2O
in. of H_2O	0.0361	lbf/in^2 (psi)
in. of H_2O	0.002458	atm

Multiply	By	To Obtain
joule (J)	9.478×10^{-4}	Btu
J	0.7376	ft-lbf
J	1	newton•m (N•m)
J/s	1	watt (W)
kilogram (kg)	2.205	pound-mass (lbm)
kgf	9.8066	newton (N)
kilometer (km)	3,281	feet (ft)
km/hr	0.621	mph
kilopascal (kPa)	0.145	lbf/in^2 (psi)
kilowatt (kW)	1.341	horsepower (hp)
kW	3,413	Btu/hr
kW	737.6	(ft-lbf)/sec
kW-hour (kWh)	3,413	Btu
kWh	1.341	hp-hr
kWh	3.6×10^6	joule (J)
kip (K)	1,000	lbf
K	4,448	newton (N)
liter (L)	61.02	in^3
L	0.264	gal (U.S. Liq)
L	10^{-3}	m^3
L/second (L/s)	2.119	ft^3/min (cfm)
L/s	15.85	gal (U.S.)/min (gpm)
meter (m)	3.281	feet (ft)
m	1.094	yard
m/second (m/s)	196.8	feet/min (ft/min)
mile (statute)	5,280	feet (ft)
mile (statute)	1.609	kilometer (km)
mile/hour (mph)	88.0	ft/min (fpm)
mph	1.609	km/h
mm of Hg	1.316×10^{-3}	atm
mm of H_2O	9.678×10^{-5}	atm
newton (N)	0.225	lbf
newton (N)	1	kg•m/s^2
N•m	0.7376	ft-lbf
N•m	1	joule (J)
pascal (Pa)	9.869×10^{-6}	atmosphere (atm)
Pa	1	newton/m^2 (N/m^2)
Pa•sec (Pa•s)	10	poise (P)
pound (lbm, avdp)	0.454	kilogram (kg)
lbf	4.448	N
lbf-ft	1.356	N•m
lbf/in^2 (psi)	0.068	atm
psi	2.307	ft of H_2O
psi	2.036	in. of Hg
psi	6,895	Pa
radian	180/π	degree
slug	32.174	pound-mass (lbm)
stokes	1×10^{-4}	m^2/s
therm	1×10^5	Btu
ton (metric)	1,000	kilogram (kg)
ton (short)	2,000	pound (lb)
watt (W)	3.413	Btu/hr
W	1.341×10^{-3}	horsepower (hp)
W	1	joule/s (J/s)
weber/m^2 (Wb/m^2)	10,000	gauss

ETHICS

CODES OF ETHICS

Engineering is considered to be a "profession" rather than an "occupation" because of several important characteristics shared with other recognized learned professions, law, medicine, and theology: special knowledge, special privileges, and special responsibilities. Professions are based on a large knowledge base requiring extensive training. Professional skills are important to the well-being of society. Professions are self-regulating, in that they control the training and evaluation processes that admit new persons to the field. Professionals have autonomy in the workplace; they are expected to utilize their independent judgment in carrying out their professional responsibilities. Finally, professions are regulated by ethical standards. (Harris, C.E., M.S. Pritchard, & M.J. Rabins, *Engineering Ethics: Concepts and Cases,* Wadsworth Publishing company, pages 27–28, 1995.)

The expertise possessed by engineers is vitally important to societal welfare. In order to serve society effectively, engineers must maintain a high level of technical competence. However, a high level of technical expertise without adherence to ethical guidelines is as much a threat to public welfare as is professional incompetence. Therefore, engineers must also be guided by ethical principles.

The ethical principles governing the engineering profession are embodied in codes of ethics. Such codes have been adopted by state boards of registration, professional engineering societies, and even by some private industries. An example of one such code is the NCEES Rules of Professional Conduct, found in Section 240 of the *Model Rules* and presented here. As part of his/her responsibility to the public, an engineer is responsible for knowing and abiding by the code. Additional rules of conduct are also included in the *Model Rules*.

The three major sections of the *Model Rules* address (1) Licensee's Obligation to the Public, (2) Licensee's Obligation to Employers and Clients, and (3) Licensee's Obligation to Other Licensees. The principles amplified in these sections are important guides to appropriate behavior of professional engineers.

Application of the code in many situations is not controversial. However, there may be situations in which applying the code may raise more difficult issues. In particular, there may be circumstances in which terminology in the code is not clearly defined, or in which two sections of the code may be in conflict. For example, what constitutes "valuable consideration" or "adequate" knowledge may be interpreted differently by qualified professionals. These types of questions are called *conceptual issues,* in which definitions of terms may be in dispute. In other situations, *factual issues* may also affect ethical dilemmas. Many decisions regarding engineering design may be based upon interpretation of disputed or incomplete information. In addition, *tradeoffs* revolving around competing issues of risk vs. benefit, or safety vs. economics may require judgments that are not fully addressed simply by application of the code.

No code can give immediate and mechanical answers to all ethical and professional problems that an engineer may face. Creative problem solving is often called for in ethics, just as it is in other areas of engineering.

Model Rules, Section 240.15, Rules of Professional Conduct

To safeguard the health, safety, and welfare of the public and to maintain integrity and high standards of skill and practice in the engineering and surveying professions, the rules of professional conduct provided in this section shall be binding upon every licensee and on all firms authorized to offer or perform engineering or surveying services in this jurisdiction.

A. Licensee's Obligation to the Public

1. Licensees shall be cognizant that their first and foremost responsibility is to safeguard the health, safety, and welfare of the public when performing services for clients and employers.
2. Licensees shall sign and seal only those plans, surveys, and other documents that conform to accepted engineering and surveying standards and that safeguard the health, safety, and welfare of the public.
3. Licensees shall notify their employer or client and such other authority as may be appropriate when their professional judgment is overruled under circumstances in which the health, safety, or welfare of the public is endangered.
4. Licensees shall, to the best of their knowledge, include all relevant and pertinent information in an objective and truthful manner within all professional documents, statements, and testimony.
5. Licensees shall express a professional opinion publicly only when it is founded upon an adequate knowledge of the facts and a competent evaluation of the subject matter.
6. Licensees shall issue no statements, criticisms, or arguments on engineering and surveying matters that are inspired or paid for by interested parties, unless they explicitly identify the interested parties on whose behalf they are speaking and reveal any interest they have in the matters.
7. Licensees shall not partner, practice, or offer to practice with any person or firm that they know is engaged in fraudulent or dishonest business or professional practices.
8. Licensees who have knowledge or reason to believe that any person or firm has violated any rules or laws applying to the practice of engineering or surveying shall report it to the board, may report it to appropriate legal authorities, and shall cooperate with the board and those authorities as may be requested.
9. Licensees shall not knowingly provide false or incomplete information regarding an applicant in obtaining licensure.

10. Licensees shall comply with the licensing laws and rules governing their professional practice in each of the jurisdictions in which they practice.

B. Licensee's Obligation to Employer and Clients

1. Licensees shall undertake assignments only when qualified by education or experience in the specific technical fields of engineering or surveying involved.
2. Licensees shall not affix their signatures or seals to any plans or documents dealing with subject matter in which they lack competence, nor to any such plan or document not prepared under their responsible charge.
3. Licensees may accept assignments and assume responsibility for coordination of an entire project, provided that each technical segment is signed and sealed by the licensee responsible for preparation of that technical segment.
4. Licensees shall not reveal facts, data, or information obtained in a professional capacity without the prior consent of the client, employer, or public body on which they serve except as authorized or required by law or rules.
5. Licensees shall not solicit or accept gratuities, directly or indirectly, from contractors, their agents, or other parties in connection with work for employers or clients.
6. Licensees shall disclose to their employers or clients all known or potential conflicts of interest or other circumstances that could influence or appear to influence their judgment or the quality of their professional service or engagement.
7. Licensees shall not accept compensation, financial or otherwise, from more than one party for services pertaining to the same project, unless the circumstances are fully disclosed and agreed to in writing by all interested parties.
8. Licensees shall not solicit or accept a professional contract from a governmental body on which a principal or officer of their organization serves as a member. Conversely, licensees serving as members, advisors, or employees of a government body or department, who are the principals or employees of a private concern, shall not participate in decisions with respect to professional services offered or provided by said concern to the governmental body that they serve. (Section 150, Disciplinary Action, NCEES *Model Law*)
9. Licensees shall not use confidential information received in the course of their assignments as a means of making personal profit without the consent of the party from whom the information was obtained.

C. Licensee's Obligation to Other Licensees

1. Licensees shall not falsify or permit misrepresentation of their, or their associates', academic or professional qualifications. They shall not misrepresent or exaggerate their degree of responsibility in prior assignments nor the complexity of said assignments. Presentations incidental to the solicitation of employment or business shall not misrepresent pertinent facts concerning employers, employees, associates, joint ventures, or past accomplishments.
2. Licensees shall not offer, give, solicit, or receive, either directly or indirectly, any commission, or gift, or other valuable consideration in order to secure work, and shall not make any political contribution with the intent to influence the award of a contract by public authority.
3. Licensees shall not injure or attempt to injure, maliciously or falsely, directly or indirectly, the professional reputation, prospects, practice, or employment of other licensees, nor indiscriminately criticize other licensees' work.
4. Licensees shall make a reasonable effort to inform another licensee whose work is believed to contain a material discrepancy, error, or omission that may impact the health, safety, or welfare of the public, unless such reporting is legally prohibited.

Model Law, Section 110.20, Definitions

A. Engineer

1. Engineer—The term "Engineer," within the intent of this Act, shall mean an individual who is qualified to practice engineering by reason of engineering education, training, and experience in the application of engineering principles and the interpretation of engineering data.
2. Professional Engineer—The term "Professional Engineer," as used in this Act, shall mean an individual who has been duly licensed as a professional engineer by the board. The board may designate a professional engineer, on the basis of education, experience, and examination, as being licensed in a specific discipline or branch of engineering signifying the area in which the engineer has demonstrated competence.
3. Professional Engineer, Retired—The term "Professional Engineer, Retired," as used in this Act, shall mean an individual who has been duly licensed as a professional engineer by the board and who chooses to relinquish or not to renew a license and who applies to and is approved by the board to be granted the use of the title "Professional Engineer, Retired."
4. Engineer Intern—The term "Engineer Intern," as used in this Act, shall mean an individual who has been duly certified as an engineer intern by the board.
5. Practice of Engineering—The term "Practice of Engineering," as used in this Act, shall mean any service or creative work requiring engineering education, training, and experience in the application of engineering principles and the interpretation

of engineering data to engineering activities that potentially impact the health, safety, and welfare of the public.

The services may include, but not be limited to, providing planning, studies, designs, design coordination, drawings, specifications, and other technical submissions; teaching engineering design courses; performing surveying that is incidental to the practice of engineering; and reviewing construction or other design products for the purposes of monitoring compliance with drawings and specifications related to engineered works.

Surveying incidental to the practice of engineering excludes the surveying of real property for the establishment of land boundaries, rights of way, easements, and the dependent or independent surveys or resurveys of the public land survey system.

An individual shall be construed to practice engineering, within the meaning and intent of this Act, if he or she does any of the following:

a. Practices any discipline of the profession of engineering or holds himself or herself out as able and entitled to practice any discipline of engineering
b. Represents himself or herself to be a professional engineer by verbal claim, sign, advertisement, letterhead, or card or in any other way
c. Through the use of some other title, implies that he or she is a professional engineer or licensed under this Act

6. Inactive Status—Licensees who are not engaged in engineering practice that requires licensure in this jurisdiction may be granted inactive status. No licensee granted inactive status may practice or offer to practice engineering in this jurisdiction unless otherwise exempted in this Act. Licensees granted inactive status are exempt from continuing education requirements.

B. Professional Surveyor (Professional Land Surveyor, Professional Surveyor and Mapper, Geomatics Professional, or equivalent term)

1. Professional Surveyor—The term "Professional Surveyor," as used in this Act, shall mean an individual who has been duly licensed as a professional surveyor by the board established under this Act and who is a professional specialist in the technique of measuring land, educated in the basic principles of mathematics, the related physical and applied sciences, and the relevant requirements of law for adequate evidence and all requisite to surveying of real property, and engaged in the practice of surveying as herein defined.
2. Professional Surveyor, Retired—The term "Professional Surveyor, Retired," as used in this Act, shall mean an individual who has been duly licensed as a professional surveyor by the board and who chooses to relinquish or not to renew a license and who applies to and is approved by the board to be granted the use of the title "Professional Surveyor, Retired."
3. Surveyor Intern—The term "Surveyor Intern," as used in this Act, shall mean an individual who has been duly certified as a surveyor intern by the board.
4. Practice of Surveying—The term "Practice of Surveying," as used in this Act, shall mean providing, or offering to provide, professional services using such sciences as mathematics, geodesy, and photogrammetry, and involving both (1) the making of geometric measurements and gathering related information pertaining to the physical or legal features of the earth, improvements on the earth, the space above, on, or below the earth and (2) providing, utilizing, or developing the same into survey products such as graphics, data, maps, plans, reports, descriptions, or projects. Professional services include acts of consultation, investigation, testimony evaluation, expert technical testimony, planning, mapping, assembling, and interpreting gathered measurements and information related to any one or more of the following:
   a. Determining by measurement the configuration or contour of the earth's surface or the position of fixed objects thereon
   b. Determining by performing geodetic surveys the size and shape of the earth or the position of any point on the earth
   c. Locating, relocating, establishing, reestablishing, or retracing property lines or boundaries of any tract of land, road, right of way, or easement
   d. Making any survey for the division, subdivision, or consolidation of any tract(s) of land
   e. Locating or laying out alignments, positions, or elevations for the construction of fixed works
   f. Determining, by the use of principles of surveying, the position for any survey monument (boundary or nonboundary) or reference point; establishing or replacing any such monument or reference point
   g. Creating, preparing, or modifying electronic, computerized, or other data, relative to the performance of the activities in items a–f above

   An individual shall be construed to practice surveying, within the meaning and intent of this Act, if he or she does any of the following:

   a. Engages in or holds himself or herself out as able and entitled to practice surveying
   b. Represents himself or herself to be a professional surveyor by verbal claim, sign, advertisement, letterhead, or card or in any other way
   c. Through the use of some other title, implies that he or she is a professional surveyor or licensed under this act
5. Inactive Status—Licensees who are not engaged in surveying practice that requires licensure in this jurisdiction may be granted inactive status. No licensee granted inactive status may practice or offer to practice surveying in this jurisdiction unless otherwise exempted in this Act. Licensees granted inactive status are exempt from the continuing education requirements.

C. Board—The term "Board," as used in this Act, shall mean the jurisdiction board of licensure for professional engineers and professional surveyors, hereinafter provided by this Act.
D. Jurisdiction—The term "Jurisdiction," as used in this Act, shall mean a state, the District of Columbia, or any territory, commonwealth, or possession of the United States that issues licenses to practice and regulates the practice of engineering and/or surveying within its legal boundaries.
E. Responsible Charge—The term "Responsible Charge," as used in this Act, shall mean direct control and personal supervision of engineering or surveying work, as the case may be.
F. Rules of Professional Conduct—The term "Rules of Professional Conduct," as used in this Act, shall mean those rules of professional conduct, if any, promulgated by the board as authorized by this Act.
G. Firm—The term "Firm," as used in this Act, shall mean any form of business or entity other than an individual operating as a sole proprietorship under his or her own name.
H. Managing Agent—The term "Managing Agent," as used in this Act, shall mean an individual who is licensed under this Act and who has been designated pursuant to Section 160.20 of this Act by the firm.
I. Rules—The term "Rules," as used in this Act, shall mean those rules and regulations adopted pursuant to Section 120.60 A, Board Powers, of this Act.
J. Signature—The term "Signature," as used in this Act, shall be in accordance with the Rules.
K. Seal—The term "Seal," as used in this Act, shall mean a symbol, image, or list of information in accordance with the Rules.
L. Licensee—The term "Licensee," as used in this Act, shall mean a professional engineer or a professional surveyor.
M. Person—The term "Person," as used in this Act, shall mean an individual or firm.
N. Authoritative—The term "Authoritative," as used in this Act or Rules promulgated under this Act, shall mean being presented as trustworthy and competent when used to describe products, processes, applications, or data resulting from the practice of surveying.
O. Disciplinary Action—The term "Disciplinary Action," as used in this Act, shall mean any final written decision or settlement taken against an individual or firm by a licensing board based upon a violation of the board's laws and rules.

Model Law, Section 130.10, General Requirements for Licensure

Education, experience, and examinations are required for licensure as a professional engineer or professional surveyor.

A. Eligibility for Licensure

To be eligible for licensure as a professional engineer or professional surveyor, an individual must meet all of the following requirements:

1. Be of good character and reputation
2. Satisfy the education criteria set forth below
3. Satisfy the experience criteria set forth below
4. Pass the applicable examinations set forth below
5. Submit five references acceptable to the board

B. Engineering

1. Certification or Enrollment as an Engineer Intern
   The following shall be considered as minimum evidence that the applicant is qualified for certification as an engineer intern.
   a. Graduating from an engineering program of 4 years or more accredited by the Engineering Accreditation Commission of ABET (EAC/ABET), graduating from an engineering master's program accredited by EAC/ABET, or meeting the requirements of the NCEES Engineering Education Standard
   b. Passing the NCEES Fundamentals of Engineering (FE) examination
2. Licensure as a Professional Engineer
   a. Initial Licensure as a Professional Engineer
      An applicant who presents evidence of meeting the applicable education, examination, and experience requirements as described below shall be eligible for licensure as a professional engineer.
      (1) Education Requirements
          An individual seeking licensure as a professional engineer shall possess one or more of the following education qualifications:
          (a) A bachelor's degree in engineering from an EAC/ABET-accredited program
          (b) A master's degree in engineering from an institution that offers EAC/ABET-accredited programs
          (c) A master's degree in engineering from an EAC/M-ABET-accredited program
          (d) An earned doctoral degree in engineering acceptable to the board
      (2) Examination Requirements
          An individual seeking licensure as a professional engineer shall take and pass the NCEES Fundamentals of Engineering (FE) examination and the NCEES Principles and Practice of Engineering (PE) examination as described below.
          (a) The FE examination may be taken by a college senior or graduate of an engineering program of 4 years or more accredited by EAC/ABET, of a program that meets the requirements of the NCEES Education Standard, or of an engineering master's program accredited by EAC/ABET.
          (b) The PE examination may be taken by an engineer intern.
      (3) Experience Requirements
          An individual seeking licensure as a professional engineer shall present evidence of a specific record of progressive engineering experience satisfying one of the following described below. This experience should be of

a grade and character that indicate to the board that the applicant may be competent to practice engineering.

(a) An individual with a bachelor's degree in engineering per (1)(a) above: 4 years of experience after the bachelor's degree is conferred

(b) An individual with a master's degree in engineering per (1)(b) or (1)(c) above: 3 years of experience

(c) An individual with an earned doctoral degree in engineering acceptable to the board and who has passed the FE exam: 2 years of experience

(d) An individual with an earned doctoral degree in engineering acceptable to the board and who has elected not to take the FE exam: 4 years of experience

b. Licensure by Comity for a Professional Engineer[1,2]
The following shall be considered as minimum evidence satisfactory to the board that the applicant is qualified for licensure by comity as a professional engineer:

(1) An individual holding a certificate of licensure to engage in the practice of engineering issued by a proper authority of any jurisdiction or any foreign country, based on requirements that do not conflict with the provisions of this Act and possessing credentials that are, in the judgment of the board, of a standard that provides proof of minimal competency and is comparable to the applicable licensure act in effect in this jurisdiction at the time such certificate was issued may, upon application, be licensed without further examination except as required to examine the applicant's knowledge of statutes, rules, and other requirements unique to this jurisdiction; or

(2) An individual holding an active Council Record with NCEES, whose qualifications as evidenced by the Council Record meet the requirements of this Act, may, upon application, be licensed without further examination except as required to examine the applicant's knowledge of statutes, rules, and other requirements unique to this jurisdiction.

C. Surveying

1. Certification or Enrollment as a Surveyor Intern
The following shall be considered as minimum evidence that the applicant is qualified for certification or enrollment as a surveyor intern.

a. Graduating from a surveying program of 4 years or more accredited the Engineering Accreditation Commission of ABET (EAC/ABET), the Engineering Technology Accreditation Commission of ABET (ETAC/ABET), the Applied and Natural Science Accreditation Commission of ABET (ANSAC/ABET), or meeting the requirements of the NCEES Surveying Education Standard

b. Graduating from a program related to surveying of 4 years or more as approved by the board and with a specific record of 2 years of progressive experience in surveying

c. Graduating from a program of 4 years or more as approved by the board and with a specific record of 4 years of progressive experience in surveying

In addition to satisfying one of the above requirements, the applicant shall pass the NCEES Fundamentals of Surveying (FS) examination.

2. Licensure as a Professional Surveyor

a. Initial Licensure as a Professional Surveyor
A surveyor intern with a specific record of 4 years or more of combined office and progressive field experience satisfactory to the board in surveying, of which a minimum of 3 years of progressive field experience satisfactory on surveying projects under the supervision of a professional surveyor, shall be admitted to the NCEES Principles and Practice of Surveying examination and any required state-specific examinations. Upon passing these examinations, the applicant shall be licensed as a professional surveyor, if otherwise qualified.

b. Licensure by Comity for a Professional Surveyor
The following shall be considered as minimum evidence satisfactory to the board that the applicant is qualified for licensure by comity as a professional surveyor:

(1) An individual holding a certificate of licensure to engage in the practice of surveying issued by a proper authority of any jurisdiction or any foreign country, based on requirements that do not conflict with the provisions of this Act and possessing credentials that are, in the judgment of the board, of a standard not lower than that specified in the applicable licensure act in effect in this jurisdiction at the time such certificate was issued may, upon application be licensed without further examination except as required to examine the applicant's knowledge of statutes, rules, and other requirements unique to this jurisdiction; or

(2) An individual holding an active Council Record with NCEES, whose qualifications as evidenced by the Council Record meet the requirements of this Act, may, upon application, be licensed without further

1 Jurisdictions (boards) that do not license by discipline may license an individual as a professional engineer.
2 Jurisdictions (boards) that license by discipline may license an individual in any discipline in which the individual can verify his or her competency.

examination except as required to examine the applicant's knowledge of statutes, rules, and other requirements unique to this jurisdiction.

3. Grandfathering of Photogrammetrists—In the event that the board chooses to license photogrammetrists as professional surveyors and a photogrammetrist does not qualify under the sections above, the board may license the photogrammetrist as a professional surveyor using the following requirements and procedure.
   a. The individual was practicing surveying using photogrammetric technologies in this jurisdiction as of *[insert date]* and has at least 8 years' experience in the profession, 2 or more of which shall have been in responsible charge of photogrammetric surveying and/or mapping projects meeting ASPRS Aerial Photography and Mapping Standards or U.S. National Mapping Standards.
   b. The applicant files an application with the board by *[insert date]*. Thereafter, no photogrammetrist shall be licensed without meeting the requirements for licensure as a professional surveyor set forth by the board for all other applicants.
   c. The applicant submits certified proof of graduation from high school, high school equivalency, or a higher degree; or certified proof of a bachelor's degree in surveying or a related field of study approved by the board, which may be substituted for four of the above required years of experience; or certified proof of a master's degree in surveying or a related field of study approved by the board, which may be substituted for a maximum of five of the above required years of experience.
   d. The applicant submits proof of employment in responsible charge of photogrammetric surveying and/or mapping projects, practicing within any jurisdiction, including itemized reports detailing methods, procedures, amount of the applicant's personal involvement, and the name, address, and telephone numbers of the client for five projects completed under the supervision of the applicant within the United States. A final map for each of the five projects shall also be submitted.
   e. The applicant submits five references as to the applicant's character and quality of work, all of which shall be from professional surveyors or professional engineers currently practicing within the scope of their license in an area of surveying.

Model Law, 150.10, Grounds for Disciplinary Action—Licensees and Interns

A. The board shall have the power to suspend, revoke, place on probation, fine, recover costs, and/or reprimand, or to refuse to issue, restore, or renew a license or intern certification to any licensee or intern that is found guilty of:
   1. Any fraud or deceit in obtaining or attempting to obtain or renew a certificate of licensure
   2. Any negligence, incompetence, or misconduct in the practice of engineering or surveying
   3. Conviction of or entry of a plea of guilty or nolo contendere to any crime that is a felony, whether or not related to the practice of engineering or surveying; and conviction of or entry of a plea of guilty or nolo contendere to any crime, whether a felony, misdemeanor, or otherwise, an essential element of which is dishonesty or which is directly related to the practice of engineering or surveying
   4. Failure to comply with any of the provisions of this Act or any of the rules or regulations of the board
   5. Discipline (including voluntary surrender of a professional engineer's or professional surveyor's license in order to avoid disciplinary action) by another jurisdiction, foreign country, or the United States government, if at least one of the grounds for discipline is the same or substantially equivalent to those contained in this Act
   6. Failure to provide information requested by the board as a result of a formal or informal complaint to the board that alleges a violation of this Act
   7. Knowingly making false statements or signing false statements, certifications, or affidavits in connection with the practice of engineering or surveying
   8. Aiding or assisting another person in violating any provision of this Act or the rules or regulations of the board
   9. Violating any terms of any Order imposed or agreed to by the board or using a seal or practicing engineering or surveying while the licensee's license is inactive or restricted
   10. Signing, affixing, or permitting the licensee's seal or signature to be affixed to any specifications, reports, drawings, plans, plats, design information, construction documents or calculations, surveys, or revisions thereof which have not been prepared by the licensee or under the licensee's responsible charge
   11. Engaging in dishonorable, unethical, or unprofessional conduct of a character likely to deceive, defraud, or harm the public
   12. Providing false testimony or information to the board
   13. Habitual intoxication or addiction to the use of drugs or alcohol
   14. Providing engineering or surveying services outside any of the licensee's areas of competence

B. In addition to or in lieu of any other sanction provided in this section, any licensee or intern that violates a provision of this Act or any rule or regulation of the board may be assessed a fine in an amount determined by the board of not more than *[insert amount]* dollars for each offense
   1. Each day of continued violation may constitute a separate offense.
   2. In determining the amount of fine to be assessed pursuant to this section, the board may consider such factors as the following:
      a. Whether the amount imposed will be a substantial economic deterrent to the violation

b. The circumstances leading to the violation
c. The severity of the violation and the risk of harm to the public
d. The economic benefits gained by the violator as a result of noncompliance
e. The interest of the public
f. Consistency of the fine with past fines for similar offenses, or justification for the fine amount

Model Law, 150.30, Grounds for Disciplinary Action—Unlicensed Individuals

A. In addition to any other provisions of law, the board shall have the power to fine and recover costs from any unlicensed individual who is found guilty of:
   1. Engaging in the practice or offer to practice of engineering or surveying in this jurisdiction without being licensed in accordance with the provisions of this Act
   2. Using or employing the words "engineer," "engineering," "surveyor," "surveying," or any modification or derivative thereof in his or her name or form of business activity except as licensed in this Act
   3. Presenting or attempting to use the certificate of licensure or seal of a professional engineer or professional surveyor
   4. Engaging in any fraud or deceit in obtaining or attempting to obtain a certificate of licensure or intern certification
   5. Impersonating any professional engineer or professional surveyor
   6. Using or attempting to use an expired, suspended, revoked, inactive, retired, or nonexistent certificate of licensure
B. A fine assessed under this section may not exceed *[insert amount]* dollars for each offense.
C. Each day of continued violation may constitute a separate offense.
D. In determining the amount of fine to be assessed pursuant to this section, the board may consider such factors as the following:
   1. Whether the amount imposed will be a substantial economic deterrent to the violation
   2. The circumstances leading to the violation
   3. The severity of the violation and the risk of harm to the public
   4. The economic benefits gained by the violator as a result of noncompliance
   5. The interest of the public
   6. Consistency of the fine with past fines for similar offenses, or justification for the fine amount

Model Law, 160.10, General Requirements for Certificates of Authorization

A. A firm that practices or offers to practice engineering or surveying is required to obtain a certificate of authorization by the board in accordance with the Rules.
B. This section shall not require a certificate of authorization for a firm performing engineering or surveying for the firm itself or for a parent or subsidiary of said firm.
C. The secretary of state of this jurisdiction shall not accept organizational papers nor issue a certificate of incorporation, organization, licensure, or authorization to any firm which includes among the objectives for which it is established or within its name, any of the words "engineer," "engineering," "surveyor," "surveying," or any modification or derivation thereof unless the board has issued for said applicant a certificate of authorization or a letter indicating the eligibility of such applicant to receive such a certificate. The firm applying shall supply such certificate or letter from the board with its application for incorporation, organization, licensure, or authorization.
D. The secretary of state of this jurisdiction shall decline to authorize any trade name, trademark, or service mark that includes therein such words as set forth in the previous subsection, or any modifications or derivatives thereof, except licensees and those firms holding certificates of authorization issued under the provisions of this section.

Model Law, 160.70, Grounds for Disciplinary Action—Firms Holding a Certificate of Authorization

A. The board shall have the power to suspend, revoke, place on probation, fine, recover costs, and/or reprimand, or to refuse to issue, restore, or renew a certificate of authorization to any firm holding a certificate of authorization that is found guilty of:
   1. Any fraud or deceit in obtaining or attempting to obtain or renew a certificate of authorization
   2. Any negligence, incompetence, or misconduct in the practice of engineering or surveying
   3. Conviction of or entry of a plea of guilty or nolo contendere to any crime that is a felony, whether or not related to the practice of engineering or surveying; and conviction of or entry of a plea of guilty or nolo contendere to any crime, whether a felony, misdemeanor, or otherwise, an essential element of which is dishonesty or which is directly related to the practice of engineering or surveying
   4. Failure to comply with any of the provisions of this Act or any of the rules or regulations of the board
   5. Discipline (including voluntary surrender of a professional engineer's or professional surveyor's license in order to avoid disciplinary action) by another jurisdiction, foreign country, or the United States government, if at least one of the grounds for discipline is the same or substantially equivalent to those contained in this Act
   6 Failure to provide information requested by the board as a result of a formal or informal complaint to the board that alleges a violation of this Act

7. Knowingly making false statements or signing false statements, certifications, or affidavits in connection with the practice of engineering or surveying
8. Aiding or assisting another person in violating any provision of this Act or the rules or regulations of the board
9. Violating any terms of any Order imposed or agreed to by the board or using a seal or practicing engineering or surveying while the firm's certificate of authorization is inactive or restricted
10. Engaging in dishonorable, unethical, or unprofessional conduct of a character likely to deceive, defraud, or harm the public
11. Providing false testimony or information to the board

B. In addition to or in lieu of any other sanction provided in this section, any firm holding a certificate of authorization that violates a provision of this Act or any rule or regulation of the board may be assessed a fine in an amount determined by the board of not more than *[insert amount]* dollars for each offense.
   1. Each day of continued violation may constitute a separate offense.
   2. In determining the amount of fine to be assessed pursuant to this section, the board may consider such factors as the following:
      a. Whether the amount imposed will be a substantial economic deterrent to the violation
      b. The circumstances leading to the violation
      c. The severity of the violation and the risk of harm to the public
      d. The economic benefits gained by the violator as a result of noncompliance
      e. The interest of the public
      f. Consistency of the fine with past fines for similar offenses, or justification for the fine amount

C. In addition to any other sanction provided in this section, the board shall have the power to sanction as follows any firm where one or more of its managing agents, officers, directors, owners, or managers have been found guilty of any conduct which would constitute a violation under the provisions of this Act or any of the rules or regulations of the board:
   1. Place on probation, fine, recover costs from, and/or reprimand
   2. Revoke, suspend, or refuse to issue, restore, or renew the certificate of authorization

Model Law, 170.30, Exemption Clause

This Act shall not be construed to prevent the following:

A. Other Professions—The practice of any other legally recognized profession

B. Contingent License—A contingent license may be issued by the board or board administrator to an applicant for licensure by comity if the applicant appears to meet the requirements for licensure by comity. Such a contingent license will be in effect from its date of issuance until such time as the board takes final action on the application for licensure by comity. If the board determines that the applicant does not meet the requirements for issuance of a license, the contingent license shall be immediately and automatically revoked upon notice to the applicant and no license will be issued.

C. Employees and Subordinates—The work of an employee or a subordinate of an individual holding a certificate of licensure under this Act, or an employee of an individual practicing lawfully under Subsection B of this section, provided such work does not include final engineering or surveying designs or decisions and is done under the responsible charge of and verified by an individual holding a certificate of licensure under this Act or an individual practicing lawfully under Subsection B of this section.

INTELLECTUAL PROPERTY

Intellectual property is the creative product of the intellect and normally includes inventions, symbols, literary works, patents, and designs.

A number of options are available to individuals who wish to protect their intellectual property from being claimed or misused by others. There are four products that are commonly used to offer varying degrees of protection to owners of different types of intellectual properties: trademarks, patents, copyrights, and industrial designs.

Trademarks

A trademark is a "brand" name for an intellectual asset. A trademark includes a word, name, symbol, or device that distinguishes and identifies the source of a good or service. Registration offers protection from misuse by unauthorized groups and provides the holder with exclusive use.

Patents

A patent grants legally protected rights to the owner of the item being patented. To be eligible for patent, the item must be new, useful, and inventive. The patent is granted to the first applicant.

Copyrights

Copyrights apply to original literary, dramatic, musical, or artistic works. The work is protected by copyright from the moment that it is created. Copyright can be registered, and the registration serves as proof of ownership. Others who wish to use work that is copyrighted must generally gain permission to use it from the copyright holder.

Industrial Designs

Industrial design rights protect the appearance of new products This includes shapes, colors, and any other visual features. Registering an industrial design protects the new design from being used or sold by others.

United States Patent and Trademark Office, uspto.gov/patents.

SAFETY

DEFINITION OF SAFETY

Safety is the condition of protecting people from threats or failures that could harm their physical, emotional, occupational, psychological, or financial well-being. Safety is also the control of known threats to attain an acceptable level of risk.

The United States relies on public codes and standards, engineering designs, and corporate policies to ensure that a structure or place does what it should do to maintain a steady state of safety—that is, long-term stability and reliability. Some *Safety/Regulatory Agencies* that develop codes and standards commonly used in the United States are shown in the table.

Acronym	Name	Jurisdiction
ANSI	American National Standards Institute	Nonprofit standards organization
CGA	Compressed Gas Association	Nonprofit trade association
CSA	Canadian Standards Association	Nonprofit standards organization
FAA	Federal Aviation Administration	Federal regulatory agency
IEC	International Electrotechnical Commission	Nonprofit standards organization
ITSNA	Intertek Testing Services NA (formerly Edison Testing Labs)	Nationally recognized testing laboratory
MSHA	Mine Safety and Health Administration	Federal regulatory agency
NFPA	National Fire Protection Association	Nonprofit trade association
OSHA	Occupational Safety and Health Administration	Federal regulatory agency
UL	Underwriters Laboratories	Nationally recognized testing laboratory
USCG	United States Coast Guard	Federal regulatory agency
USDOT	United States Department of Transportation	Federal regulatory agency
USEPA	United States Environmental Protection Agency	Federal regulatory agency

SAFETY AND PREVENTION

A traditional preventive approach to both accidents and occupational illness involves recognizing, evaluating, and controlling hazards and work conditions that may cause physical or other injuries.

Hazard is the capacity to cause harm. It is an inherent quality of a material or a condition. For example, a rotating saw blade or an uncontrolled high-pressure jet of water has the capability (hazard) to slice through flesh. A toxic chemical or a pathogen has the capability (hazard) to cause illness.

Risk is the chance or probability that a person will experience harm and is not the same as a hazard. Risk always involves both probability and severity elements. The hazard associated with a rotating saw blade or the water jet continues to exist, but the probability of causing harm, and thus the risk, can be reduced by installing a guard or by controlling the jet's path. Risk is expressed by the equation:

$$\text{Risk} = \text{Hazard} \times \text{Probability}$$

When people discuss the hazards of disease-causing agents, the term *exposure* is typically used more than *probability*. If a certain type of chemical has a toxicity hazard, the risk of illness rises with the degree to which that chemical contacts your body or enters your lungs. In that case, the equation becomes:

$$\text{Risk} = \text{Hazard} \times \text{Exposure}$$

Organizations evaluate hazards using multiple techniques and data sources.

Job Safety Analysis

Job safety analysis (JSA) is known by many names, including activity hazard analysis (AHA), or job hazard analysis (JHA). Hazard analysis helps integrate accepted safety and health principles and practices into a specific task. In a JSA, each basic step of the job is reviewed, potential hazards identified, and recommendations documented as to the safest way to do the job. JSA techniques work well when used on a task that the analysts understand well. JSA analysts look for specific types of potential accidents and ask basic questions about each step, such as these:

Can the employee strike against or otherwise make injurious contact with the object?
Can the employee be caught in, on, or between objects?
Can the employee strain muscles by pushing, pulling, or lifting?
Is exposure to toxic gases, vapors, dust, heat, electrical currents, or radiation possible?

HAZARD ASSESSMENTS

Hazard Assessment

The fire/hazard diamond below summarizes common hazard data available on the Safety Data Sheet (SDS) and is frequently shown on chemical labels.

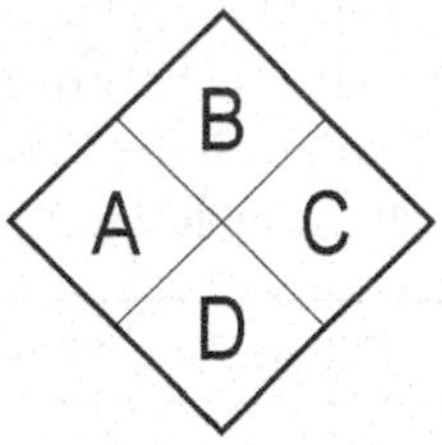

Position A – Health Hazard (Blue)
0 = normal material
1 = slightly hazardous
2 = hazardous
3 = extreme danger
4 = deadly

Position B – Flammability (Red)
0 = will not burn
1 = will ignite if preheated
2 = will ignite if moderately heated
3 = will ignite at most ambient temperature
4 = burns readily at ambient conditions

Position C – Reactivity (Yellow)
0 = stable and not reactive with water
1 = unstable if heated
2 = violent chemical change
3 = shock short may detonate
4 = may detonate

Position D – (White)
ALKALI = alkali
OXY = oxidizer
ACID = acid
Cor = corrosive
~~W~~ = use no water
☢ = radiation hazard

GHS

The *Globally Harmonized System of Classification and Labeling of Chemicals*, or GHS, is a system for standardizing and harmonizing the classification and labeling of chemicals. GHS is a comprehensive approach to:

- Defining health, physical, and environmental hazards of chemicals
- Creating classification processes that use available data on chemicals for comparison with the defined hazard criteria
- Communicating hazard information, as well as protective measures, on labels and Safety Data Sheets (SDSs), formerly called Material Safety Data Sheets (MSDSs).

GHS label elements include:

- Precautionary statements and pictograms: Measures to minimize or prevent adverse effects
- Product identifier (ingredient disclosure): Name or number used for a hazardous product on a label or in the SDS
- Supplier identification: The name, address, and telephone number of the supplier
- Supplemental information: nonharmonized information

Other label elements include symbols, signal words, and hazard statements.

GHS LABEL ELEMENTS

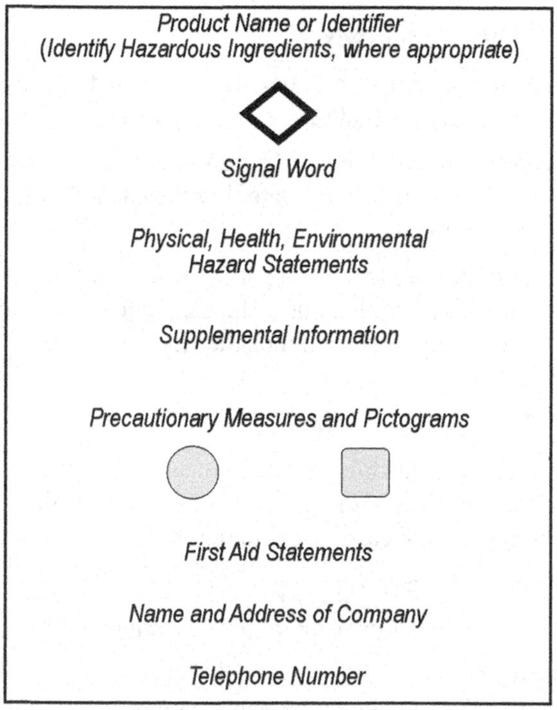
Product Name or Identifier
(Identify Hazardous Ingredients, where appropriate)
Signal Word
Physical, Health, Environmental
Hazard Statements
Supplemental Information
Precautionary Measures and Pictograms
First Aid Statements
Name and Address of Company
Telephone Number

Occupational Safety and Health Administration, *A Guide to The Globally Harmonized System of Classification and Labelling of Chemicals (GHS)*, United States Department of Labor, https://www.osha.gov/dsg/hazcom/ghsguideoct05.pdf

GHS PICTOGRAMS AND HAZARD CLASSES

- OXIDIZERS

- FLAMMABLES
- SELF-REACTIVES
- PYROPHORICS
- SELF-HEATING
- EMITS FLAMMABLE GAS
- ORGANIC PEROXIDES

- EXPLOSIVES
- SELF-REACTIVES
- ORGANIC PEROXIDES

- ACUTE TOXICITY (SEVERE)

- CORROSIVES

- GASES UNDER PRESSURE

- CARCINOGEN
- RESPIRATORY SENSITIZER
- REPRODUCTIVE TOXICITY
- TARGET ORGAN TOXICITY
- MUTAGENICITY
- ASPIRATION TOXICITY

- ENVIRONMENTAL TOXICITY

- IRRITANT
- DERMAL SENSITIZER
- ACUTE TOXICITY (HARMFUL)
- NARCOTIC EFFECTS
- RESPIRATORY TRACT IRRITATION

Occupational Safety and Health Administration, *A Guide to The Globally Harmonized System of Classification and Labelling of Chemicals (GHS)*, United States Department of Labor, https://www.osha.gov/dsg/hazcom/ghsguideoct05.pdf

TRANSPORT PICTOGRAMS

FLAMMABLE LIQUID
FLAMMABLE GAS
FLAMMABLE AEROSOL

FLAMMABLE SOLID
SELF-REACTIVE
SUBSTANCES

PYROPHORICS (SPONTANEOUSLY
COMBUSTIBLE)
SELF-HEATING SUBSTANCES

SUBSTANCES, WHICH IN
CONTACT WITH WATER,
EMIT FLAMMABLE GASES
(DANGEROUS WHEN WET)

OXIDIZING GASES
OXIDIZING LIQUIDS
OXIDIZING SOLIDS

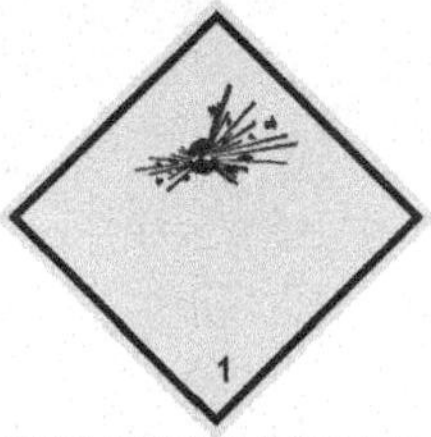

EXPLOSIVE DIVISIONS
1.1, 1.2, 1.3

EXPLOSIVE DIVISION 1.4

EXPLOSIVE DIVISION 1.5

EXPLOSIVE DIVISION 1.6

COMPRESSED GASES

ACUTE TOXICITY (POISON):
ORAL, DERMAL, INHALATION

CORROSIVE

MARINE POLLUTANT

ORGANIC PEROXIDES

Occupational Safety and Health Administration, *A Guide to The Globally Harmonized System of Classification and Labelling of Chemicals (GHS)*, United States Department of Labor, https://www.osha.gov/dsg/hazcom/ghsguideoct05.pdf

♦

ACUTE ORAL TOXICITY

	CATEGORY 1	CATEGORY 2	CATEGORY 3	CATEGORY 4	CATEGORY 5
LD_{50}	≤5 mg/kg	> 5 < 50 mg/kg	≥ 50 < 300 mg/kg	≥ 300 < 2,000 mg/kg	≥ 2,000 < 5,000 mg/kg
PICTOGRAM					NO SYMBOL
SIGNAL WORD	DANGER	DANGER	DANGER	WARNING	WARNING
HAZARD STATEMENT	FATAL IF SWALLOWED	FATAL IF SWALLOWED	TOXIC IF SWALLOWED	HARMFUL IF SWALLOWED	MAY BE HARMFUL IF SWALLOWED

♦ Safety Data Sheet (SDS)

The SDS provides comprehensive information for use in workplace chemical management. Employers and workers use the SDS as a source of information about hazards and to obtain advice on safety precautions. The SDS is product related and, usually, is not able to provide information that is specific for any given workplace where the product may be used. However, the SDS information enables the employer to develop an active program of worker protection measures, including training, which is specific to the individual workplace, and to consider any measures that may be necessary to protect the environment. Information in an SDS also provides a source of information for those involved with the transport of dangerous goods, emergency responders, poison centers, those involved with the professional use of pesticides, and consumers.

The SDS has 16 sections in a set order, and minimum information is prescribed.

♦ The Hazard Communication Standard (HCS) requires chemical manufacturers, distributors, or importers to provide SDSs to communicate the hazards of hazardous chemical products. As of June 1, 2015, the HCS requires new SDSs to be in a uniform format, and include the section numbers, the headings, and associated information under the headings below:

Section 1, Identification: Includes product identifier; manufacturer or distributor name, address, phone number; emergency phone number; recommended use; restrictions on use

Section 2, Hazard(s) identification: Includes all hazards regarding the chemical; required label elements

Section 3, Composition/information on ingredients: Includes information on chemical ingredients; trade secret claims

Section 4, First-aid measures: Includes important symptoms/effects, acute, and delayed; required treatment

Section 5, Fire-fighting measures: Lists suitable extinguishing techniques, equipment; chemical hazards from fire

Section 6, Accidental release measures: Lists emergency procedures; protective equipment; proper methods of containment and cleanup

Section 7, Handling and storage: Lists precautions for safe handling and storage, including incompatibilities

Section 8, Exposure controls/personal protection: Lists OSHA's Permissible Exposure Limits (PELs); Threshold Limit Values (TLVs); appropriate engineering controls; personal protective equipment (PPE)

Section 9, Physical and chemical properties: Lists the chemical's characteristics

Section 10, Stability and reactivity: Lists chemical stability and possibility of hazardous reactions

Section 11, Toxicological information: Includes routes of exposure; related symptoms, acute and chronic effects; numerical measures of toxicity

Section 12, Ecological information*

Section 13, Disposal considerations*

Section 14, Transport information*

Section 15, Regulatory information*

Section 16, Other information: Includes the date of preparation or last revision

*Note: Since other Agencies regulate this information, OSHA will not be enforcing Sections 12 through 15 (29 CFR 1910.1200(g)(2)).

♦ Occupational Safety and Health Administration, *A Guide to The Globally Harmonized System of Classification and Labelling of Chemicals (GHS)*, United States Department of Labor, https://www.osha.gov/dsg/hazcom/ghsguideoct05.pdf

Signal Words

The signal word found on every product's label is based on test results from various oral, dermal, and inhalation toxicity tests, as well as skin and eye corrosion assays in some cases. Signal words are placed on labels to convey a level of care that should be taken (especially personal protection) when handling and using a product, from purchase to disposal of the empty container, as demonstrated by the Pesticide Toxicity Table.

Pesticide Toxicity Categories

Signal Word on Label	Toxicity Category	Acute-Oral LD_{50} for Rats	Amount Needed to Kill an Average Size Adult	Notes
Danger–Poison	Highly Toxic	50 or less	Taste to a teaspoon	Skull and crossbones; Keep Out of Reach of Children
Warning	Moderately Toxic	50 to 500	One to six teaspoons	Keep Out of Reach of Children
Caution	Slightly Toxic	500 to 5,000	One ounce to a pint	Keep Out of Reach of Children
Caution	Relatively Nontoxic	>5,000	More than a pint	Keep Out of Reach of Children

LD_{50} - See Risk Assessment/Toxicology section on page 13.

From *Regulating Pesticides*, U.S. Environmental Protection Agency.

Flammability

Flammable describes any solid, liquid, vapor, or gas that will ignite easily and burn rapidly. A flammable liquid is defined by NFPA and USDOT as a liquid with a flash point below 100°F (38°C). Flammability is further defined with lower and upper limits:

LFL = lower flammability limit (volume % in air)
UFL = upper flammability limit (volume % in air)

The LFL is also known as the lower explosive limit (LEL). The UFL is also referred to as the upper explosive limit (UEL). There is no difference between the terms *flammable* and *explosive* as applied to the lower and upper limits of flammability.

A vapor-air mixture will only ignite and burn over the range of concentrations between LFL and UFL. Examples are:

♦

Gas/Compound	LFL	UFL
Acetone	2.6	13.0
Acetylene	2.5	100.0
Ammonia	15.0	28.0
n-butane	1.8	8.4
Carbon disulfide	1.3	50.0
Carbon monoxide	12.5	74.0
Cycloheptane	1.1	6.7
Cyclohexane	1.3	7.8
Cyclopropane	2.4	10.4
Diethyl ether	1.9	36.0
Ethane	3.0	12.4
Ethyl acetate	2.2	11.0
Ethyl alcohol	3.3	19.0
Ethyl ether	1.9	36.0
Ethyl nitrite	3.0	50.0
Ethylene	2.7	36.0
Gasoline 100/130	1.3	7.1
Gasoline 115/145	1.2	7.1
Hydrazine	4.7	100.0
Hydrogen	4.0	75.0
Hydrogen sulfide	4.0	44.0
Isobutane	1.8	8.4
Methane	5.0	15.0
Propane	2.1	9.5

♦ From *SFPE Handbook of Fire Protection Engineering*, 4 ed., Society of Fire Protection Engineers, 2008.

♦ Predicting Lower Flammable Limits of Mixtures of Flammable Gases (Le Chatelier's Rule)

Based on an empirical rule developed by Le Chatelier, the lower flammable limit of mixtures of multiple flammable gases in air can be determined. A generalization of Le Chatelier's rule is

$$\sum_{i=1}^{n}(C_i/\text{LFL}_i) \geq 1$$

where C_i is the volume percent of fuel gas, i, in the fuel/air mixture and LFL_i is the volume percent of fuel gas, i, at its lower flammable limit in air alone. If the indicated sum is greater than unity, the mixture is above the lower flammable limit. This can be restated in terms of the lower flammable limit concentration of the fuel mixture, LFL_m, as follows:

$$\text{LFL}_m = \frac{100}{\sum_{i=1}^{n}(C_{fi}/\text{LFL}_i)}$$

where C_{fi} is the volume percent of fuel gas i in the fuel gas mixture.

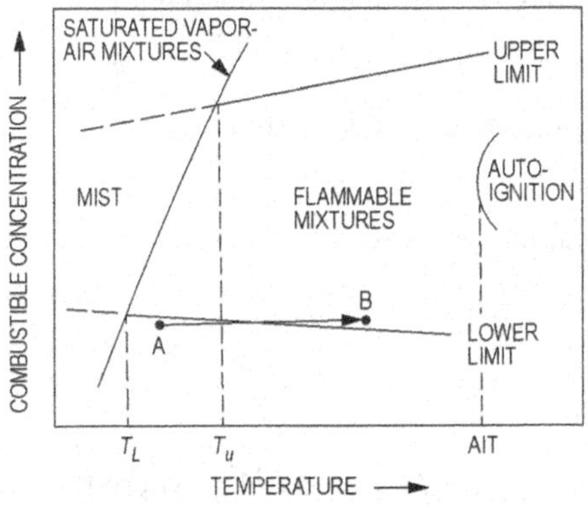

♦ *The SFPE Handbook of Fire Protection Engineering*, 1 ed., Society of Fire Protection Association, 1988. With permission.

Granular Storage and Process Safety

Some materials that are not inherently hazardous can become hazardous during storage or processing. An example is the handling of grain in grain bins. Grain bins should not be entered when the grain is being removed since grains flow to the center of the emptying bin and create suffocation hazards. Bridging may occur at the top surface due to condensation and resulting spoilage creating a crust.

Organic vapors and dusts associated with grain handling often contain toxic yeasts or molds and have low oxygen contents. These organic vapors and dusts may also be explosive.

Confined Space Safety

Many workplaces contain spaces that are considered "confined" because their configurations hinder the activities of employees who must enter, work in, and exit them. A confined space has limited or restricted means for entry or exit and is not designed for continuous employee occupancy. Confined spaces include, but are not limited to, underground vaults, tanks, storage bins, manholes, pits, silos, process vessels, and pipelines. OSHA uses the term "permit-required confined spaces" (permit space) to describe a confined space that has one or more of the following characteristics: contains or has the potential to contain a hazardous atmosphere; contains a material that has the potential to engulf an entrant; has walls that converge inward or floors that slope downward and taper into a smaller area that could trap or asphyxiate an entrant; or contains any other recognized safety or health hazard such as unguarded machinery, exposed live wires or heat stress.

OSHA has developed OSHA standards, directives (instructions for compliance officers), standard interpretations (official letters of interpretation of the standards), and national consensus standards related to confined spaces. The following gases are often present in confined spaces:

Ammonia: Irritating at 50 ppm and deadly above 1,000 ppm; sharp, cutting odor

Hydrogen sulfide: Irritating at 10 ppm and deadly at 500 ppm; accumulates at lower levels and in corners where circulation is minimal; rotten egg odor

Methane: Explosive at levels above 50,000 ppm, lighter than air, odorless

Carbon dioxide: Heavier than air, accumulates at lower levels and in corners where circulation is minimal, displaces air leading to asphyxiation

Electrical Safety

Current Level (Milliamperes)	Probable Effect on Human Body
1 mA	Perception level. Slight tingling sensation. Still dangerous under certain conditions.
5 mA	Slight shock felt; not painful but disturbing. Average individual can let go. However, strong involuntary reactions to shocks in this range may lead to injuries.
6 mA–16 mA	Painful shock, begin to lose muscular control. Commonly referred to as the freezing current or "let-go" range.
17 mA–99 mA	Extreme pain, respiratory arrest, severe muscular contractions. Individual cannot let go. Death is possible.
100 mA–2,000 mA	Ventricular fibrillation (uneven, uncoordinated pumping of the heart). Muscular contraction and nerve damage begins to occur. Death is likely.
> 2,000 mA	Cardiac arrest (stop in effective blood circulation), internal organ damage, and severe burns. Death is probable.

Worker Deaths by Electrocution; A Summary of NIOSH Surveillance and Investigative Findings, U.S. Health and Human Services, (NIOSH), 1998.

Greenwald E.K., *Electrical Hazards and Accidents–Their Cause and Prevention,* Van Nostrand Reinhold, 1991.

RISK ASSESSMENT/TOXICOLOGY

Dose-Response Curves

The dose-response curve relates toxic response (i.e., percentage of test population exhibiting a specified symptom or dying) to the logarithm of the dosage [i.e., mg/(kg•day) ingested]. A typical dose-response curve is shown below.

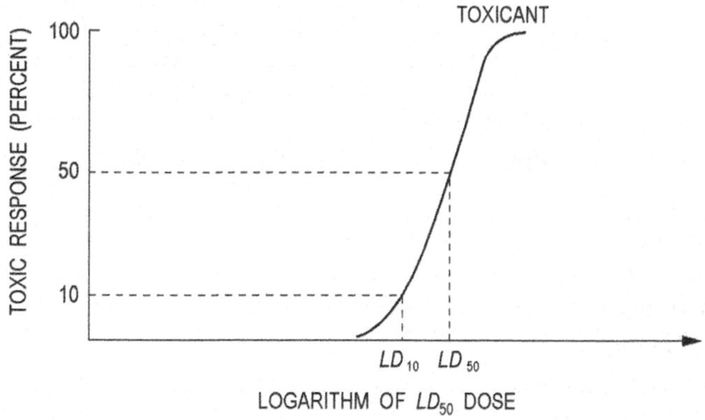

LC_{50}

Median lethal concentration in air that, based on laboratory tests, is expected to kill 50% of a group of test animals when administered as a single exposure over 1 or 4 hours.

LD_{50}

Median lethal single dose, based on laboratory tests, expected to kill 50% of a group of test animals, usually by oral or skin exposure.

Similar definitions exist for LC_{10} and LD_{10}, where the corresponding percentages are 10%.

Comparative Acutely Lethal Doses

Actual Ranking No.	LD_{50} (mg/kg)	Toxic Chemical
1	15,000	PCBs
2	10,000	Alcohol (ethanol)
3	4,000	Table salt—sodium chloride
4	1,500	Ferrous sulfate—an iron supplement
5	1,375	Malathion—pesticide
6	900	Morphine
7	150	Phenobarbital—a sedative
8	142	Tylenol (acetaminophen)
9	2	Strychnine—a rat poison
10	1	Nicotine
11	0.5	Curare—an arrow poison
12	0.001	2,3,7,8-TCDD (dioxin)
13	0.00001	Botulinum toxin (food poison)

Adapted from *Loomis's Essentials of Toxicology,* 4th ed., Loomis, T.A., and A.W. Hayes, San Diego, Academic Press, 1996.

Selected Chemical Interaction Effects

Effect	Relative toxicity (hypothetical)	Example
Additive	2 + 3 = 5	Organophosphate pesticides
Synergistic	2 + 3 = 20	Cigarette smoking + asbestos
Antagonistic	6 + 6 = 8	Toluene + benzene or caffeine + alcohol

Adapted from Williams, P.L., R.C. James, and S.M. Roberts, *Principles of Toxicology: Environmental and Industrial Applications,* 2nd ed., Wiley, 2000.

Exposure Limits for Selected Compounds

N	Allowable Workplace Exposure Level (mg/m^3)	Chemical (use)
1	0.1	Iodine
2	5	Aspirin
3	10	Vegetable oil mists (cooking oil)
4	55	1,1,2-Trichloroethane (solvent/degreaser)
5	188	Perchloroethylene (dry-cleaning fluid)
6	170	Toluene (organic solvent)
7	269	Trichloroethylene (solvent/degreaser)
8	590	Tetrahydrofuran (organic solvent)
9	890	Gasoline (fuel)
10	1,590	Naphtha (rubber solvent)
11	1,910	1,1,1-Trichloroethane (solvent/degreaser)

American Conference of Government Industrial Hygienists (ACGIH) 1996.

Carcinogens

For carcinogens, EPA considers an acceptable risk to be within the range of 10^{-4} to 10^{-6}. The added risk of cancer is calculated as follows:

Risk = dose × toxicity = $CDI \times CSF$

where

CDI = Chronic Daily Intake

CSF = Cancer Slope Factor. Slope of the dose-response curve for carcinogenic materials.

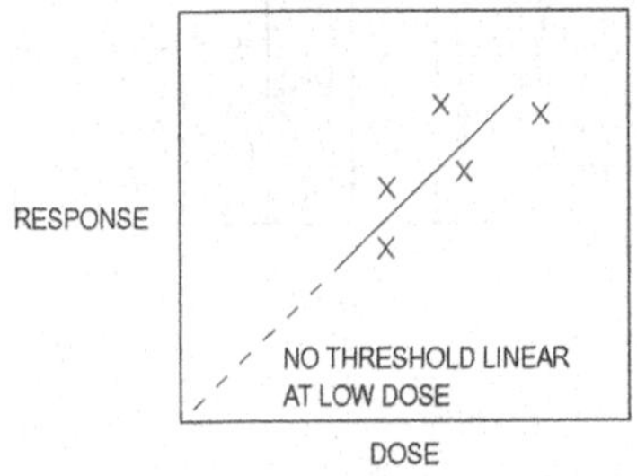

CARCINOGENIC DOSE RESPONSE CURVE

Threshold Limit Value (TLV)

TLV is the highest dose (ppm by volume in the atmosphere) the body is able to detoxify without any detectable effects.

Examples are:

Compound	TLV
Ammonia	25
Chlorine	0.5
Ethyl Chloride	1,000
Ethyl Ether	400

Noncarcinogens

For noncarcinogens, a hazard index (HI) is used to characterize risk from all pathways and exposure routes. EPA considers an $HI > 1.0$ as representing the possibility of an adverse effect occurring.

$HI = CDI_{noncarcinogen}/RfD$

$CDI_{noncarcinogen}$ = chronic daily intake of noncarcinogenic compound

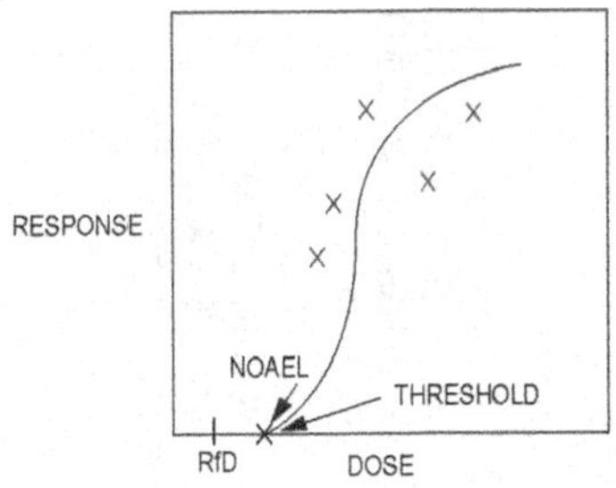

NONCARCINOGENIC DOSE RESPONSE CURVE

Dose is expressed $\left(\dfrac{\text{mass of chemical}}{\text{body weight} \cdot \text{exposure time}}\right)$

$NOAEL$ = No Observable Adverse Effect Level. The dose below which there are no harmful effects

Reference Dose

Reference dose (RfD) is determined from the Noncarcinogenic Dose-Response Curve using $NOAEL$.

RfD = lifetime (i.e., chronic) dose that a healthy person could be exposed to daily without adverse effects

$$RfD = \frac{NOAEL}{UF}$$

and

$$SHD = RfD \times W = \frac{NOAEL \times W}{UF}$$

where

SHD = safe human dose (mg/day)

$NOAEL$ = threshold dose per kg of test animal [mg/(kg•day)] from the dose-response curve

UF = the total uncertainty factor, depending on nature and reliability of the animal test data

W = the weight of the adult male

HAZARDOUS WASTE COMPATIBILITY CHART

Reactivity Group

No.	Name																																									
1	Acid, Minerals, Non-Oxidizing	1																																								
2	Acids, Minerals, Oxidizing		2																																							
3	Acids, Organic		G H	3																																						
4	Alcohols & Glycols	H	H F	H P	4																																					
5	Aldehydes	H P	H F	H P		5																																				
6	Amides	H	H GT				6																																			
7	Amines, Aliphatic & Aromatic	H	H GT	H		H		7																																		
8	Azo Compounds, Diazo Comp, Hydrazines	H G	H GT	H G	H G	H			8																																	
9	Carbamates	H G	H GT						H G	9																																
10	Caustics	H	H	H		H				H G	10																															
11	Cyanides	GT GF	GT GF	GT GF					G			11																														
12	Dithiocarbamates	H GF F	H GF F	H GF GT		GF GT		U	H G				12																													
13	Esters	H	H F						H G		H			13																												
14	Ethers	H	H F												14																											
15	Fluorides, Inorganic	GT	GT	GT												15																										
16	Hydrocarbons, Aromatic		H F														16																									
17	Halogenated Organics	H GT	H F GT					H GT	H G		H GF	H						17																								
18	Isocyanates	H G	H F GT	H G	H P			H P	H G		H P G	H G	U						18																							
19	Ketones	H	H F						H G		H	H								19																						
20	Mercaptans & Other Organic Sulfides	GT GF	H F GT						H G									H	H	H	20																					
21	Metal, Alkali & Alkaline Earth, Elemental	GF H F	GF H F	GF H F	GF H F	GF H F	GF H	GF H	GF H	GF H	GF H	GF H	GF GT H	GF H				H E	GF H	GF H	GF H	21																				
104	Oxidizing Agents, Strong	H GT		H GT	H F	H F	H F GT	F GT H	H E	H F GT		H E GT	H F GT	H F	H F		H F	H GT	H F GT	H F	H F GT	H F GE																	104			
105	Reducing Agents, Strong	H GF	H F GT	H GF	H GF F	GF H F	GF H	H GF	H G				H GT	H F				H T	GF H	GF H	GF H																		H F E	105		
106	Water & Mixtures Containing Water	H	H						G										H G			GF H																		GF GT	106	
107	Water Reactive Substances	EXTREMELY REACTIVE! Do Not Mix With Any Chemical or Waste Material																																								107
		1	2	3	4	5	6	7	8	9	10	11	12	13	14	15	16	17	18	19	20	21	22	23	24	25	26	27	28	29	30	31	32	33	34	101	102	103	104	105	106	107

KEY

REACTIVITY CODE	CONSEQUENCES
H	HEAT GENERATION
F	FIRE
G	INNOCUOUS & NON-FLAMMABLE GAS
GT	TOXIC GAS GENERATION
GF	FLAMMABLE GAS GENERATION
E	EXPLOSION
P	POLYMERIZATION
S	SOLUBILIZATION OF TOXIC MATERIAL
U	MAY BE HAZARDOUS BUT UNKNOWN

EXAMPLE:

H F GT	HEAT GENERATION, FIRE, AND TOXIC GAS GENERATION

U.S. Environmental Protection Agency, April 1980. EPA-600/2-80-076.

Exposure

Residential Exposure Equations for Various Pathways

Ingestion in drinking water

$$CDI = \frac{(CW)(IR)(EF)(ED)}{(BW)(AT)}$$

Ingestion while swimming

$$CDI = \frac{(CW)(CR)(ET)(EF)(ED)}{(BW)(AT)}$$

Dermal contact with water

$$AD = \frac{(CW)(SA)(PC)(ET)(EF)(ED)(CF)}{(BW)(AT)}$$

Ingestion of chemicals in soil

$$CDI = \frac{(CS)(IR)(CF)(FI)(EF)(ED)}{(BW)(AT)}$$

Dermal contact with soil

$$AD = \frac{(CS)(CF)(SA)(AF)(ABS)(EF)(ED)}{(BW)(AT)}$$

Inhalation of airborne (vapor phase) chemicals

$$CDI = \frac{(CA)(IR)(ET)(EF)(ED)}{(BW)(AT)}$$

Ingestion of contaminated fruits, vegetables, fish and shellfish

$$CDI = \frac{(CF)(IR)(FI)(EF)(ED)}{(BW)(AT)}$$

where ABS = absorption factor for soil contaminant (unitless)

AD = absorbed dose (mg/[kg•day])

AF = soil-to-skin adherence factor (mg/cm^2)

AT = averaging time (days)

BW = body weight (kg)

CA = contaminant concentration in air (mg/m^3)

CDI = chronic daily intake (mg/[kg•day])

CF = volumetric conversion factor for water = 1 L/1,000 cm^3

= conversion factor for soil = 10^{-6} kg/mg

CR = contact rate (L/hr)

CS = chemical concentration in soil (mg/kg)

CW = chemical concentration in water (mg/L)

ED = exposure duration (years)

EF = exposure frequency (days/yr or events/year)

ET = exposure time (hr/day or hr/event)

FI = fraction ingested (unitless)

IR = ingestion rate (L/day or mg soil/day or kg/meal)

= inhalation rate (m^3/hr)

PC = chemical-specific dermal permeability constant (cm/hr)

SA = skin surface area available for contact (cm^2)

Risk Assessment Guidance for Superfund. Volume 1, *Human Health Evaluation Manual* (part A). U.S. Environmental Protection Agency, EPA/540/1-89/002,1989.

Intake Rates—Variable Values

EPA-Recommended Values for Estimating Intake

Parameter	Standard Value
Average body weight, female adult	65.4 kg
Average body weight, male adult	78 kg
Average body weight, child[a]	
6–11 months	9 kg
1–5 years	16 kg
6–12 years	33 kg
Amount of water ingested, adult	2.3 L/day
Amount of water ingested, child	1.5 L/day
Amount of air breathed, female adult	11.3 m^3/day
Amount of air breathed, male adult	15.2 m^3/day
Amount of air breathed, child (3–5 years)	8.3 m^3/day
Amount of fish consumed, adult	6 g/day
Water swallowing rate, while swimming	50 mL/hr
Inhalation rates	
adult (6-hr day)	0.98 m^3/hr
adult (2-hr day)	1.47 m^3/hr
child	0.46 m^3/hr
Skin surface available, adult male	1.94 m^2
Skin surface available, adult female	1.69 m^2
Skin surface available, child	
3–6 years (average for male and female)	0.720 m^2
6–9 years (average for male and female)	0.925 m^2
9–12 years (average for male and female)	1.16 m^2
12–15 years (average for male and female)	1.49 m^2
15–18 years (female)	1.60 m^2
15–18 years (male)	1.75 m^2
Soil ingestion rate, child 1–6 years	>100 mg/day
Soil ingestion rate, persons > 6 years	50 mg/day
Skin adherence factor, gardener's hands	0.07 mg/cm^2
Skin adherence factor, wet soil	0.2 mg/cm^2
Exposure duration	
Lifetime (carcinogens, for noncarcinogens use actual exposure duration)	75 years
At one residence, 90th percentile	30 years
National median	5 years
Averaging time	(ED)(365 days/year)
Exposure frequency (EF)	
Swimming	7 days/year
Eating fish and shellfish	48 days/year
Oral ingestion	350 days/year
Exposure time (ET)	
Shower, 90th percentile	12 min
Shower, 50th percentile	7 min

[a] Data in this category taken from: Copeland, T., A. M. Holbrow, J. M. Otan, et al., "Use of probabilistic methods to understand the conservatism in California's approach to assessing health risks posed by air contaminants," *Journal of the Air and Waste Management Association,* vol. 44, pp. 1399-1413, 1994.

Risk Assessment Guidance for Superfund. Volume 1, *Human Health Evaluation Manual* (part A). U.S. Environmental Protection Agency, EPA/540/1-89/002, 1989.

Concentrations of Vaporized Liquids

Vaporization Rate (Q_m, mass/time) from a Liquid Surface

$$Q_m = [MKA_S P^{sat}/(R_g T_L)]$$

M = molecular weight of volatile substance

K = mass-transfer coefficient

A_S = area of liquid surface

P^{sat} = saturation vapor pressure of the pure liquid at T_L

R_g = ideal gas constant

T_L = absolute temperature of the liquid

Mass Flowrate of Liquid from a Hole in the Wall of a Process Unit

$$Q_m = A_H C_0 (2 \rho g_c P_g)^{1/2}$$

A_H = area of hole

C_0 = discharge coefficient

ρ = density of the liquid

g_c = gravitational constant

P_g = gauge pressure within the process unit

Concentration (C_{ppm}) of Vaporized Liquid in Ventilated Space

$$C_{ppm} = [Q_m R_g T \times 10^6/(kQ_V PM)]$$

T = absolute ambient temperature

k = non-ideal mixing factor

Q_V = ventilation rate

P = absolute ambient pressure

Sweep-Through Concentration Change in a Vessel

$$Q_V t = V \ln[(C_1 - C_0)/(C_2 - C_0)]$$

Q_V = volumetric flowrate

t = time

V = vessel volume

C_0 = inlet concentration

C_1 = initial concentration

C_2 = final concentration

ERGONOMICS

NIOSH Formula

Recommended Weight Limit (RWL)

RWL = 51(10/H)(1 – 0.0075|V – 30|)(0.82 + 1.8/D)(1 – 0.0032A)(FM)(CM)

where

RWL= recommended weight limit, in pounds

H = horizontal distance of the hand from the midpoint of the line joining the inner ankle bones to a point projected on the floor directly below the load center, in inches

V = vertical distance of the hands from the floor, in inches

D = vertical travel distance of the hands between the origin and destination of the lift, in inches

A = asymmetry angle, in degrees

FM = frequency multiplier (see table)

CM = coupling multiplier (see table)

Frequency Multiplier Table

	≤ 8 hr /day		≤ 2 hr/day		≤ 1 hr/day	
F, min^{-1}	V< 30 in.	V ≥ 30 in.	V < 30 in.	V ≥ 30 in.	V < 30 in.	V ≥ 30 in.
0.2	0.85		0.95		1.00	
0.5	0.81		0.92		0.97	
1	0.75		0.88		0.94	
2	0.65		0.84		0.91	
3	0.55		0.79		0.88	
4	0.45		0.72		0.84	
5	0.35		0.60		0.80	
6	0.27		0.50		0.75	
7	0.22		0.42		0.70	
8	0.18		0.35		0.60	
9		0.15	0.30		0.52	
10		0.13	0.26		0.45	
11				0.23	0.41	
12				0.21	0.37	
13			0.00			0.34
14						0.31
15						0.28

Waters, Thomas R., Ph.D., et al, *Applications Manual for the Revised NIOSH Lifting Equation*, Table 5, U.S. Department of Health and Human Services (NIOSH), January 1994.

Coupling Multiplier (CM) Table
(Function of Coupling of Hands to Load)

<table>
<tr><td colspan="3">Container</td><td colspan="2">Loose Part / Irreg. Object</td></tr>
<tr><td colspan="2">Optimal Design</td><td>Not</td><td>Comfort Grip</td><td>Not</td></tr>
<tr><td>Opt. Handles or Cut-outs</td><td>Not</td><td>POOR</td><td>GOOD</td><td></td></tr>
<tr><td>GOOD</td><td colspan="2">Flex Fingers 90 Degrees</td><td colspan="2">Not</td></tr>
<tr><td></td><td colspan="2">FAIR</td><td colspan="2">POOR</td></tr>
</table>

<table>
<tr><td>Coupling</td><td>V < 30 in. or 75 cm</td><td>V ≥ 30 in. or 75 cm</td></tr>
<tr><td>GOOD</td><td colspan="2">1.00</td></tr>
<tr><td>FAIR</td><td>0.95</td><td></td></tr>
<tr><td>POOR</td><td colspan="2">0.90</td></tr>
</table>

Waters, Thomas R., Ph.D., et al, *Applications Manual for the Revised NIOSH Lifting Equation,* Table 7, U.S. Department of Health and Human Services (NIOSH), January 1994.

Biomechanics of the Human Body

Basic Equations

$$H_x + F_x = 0$$
$$H_y + F_y = 0$$
$$H_z + W + F_z = 0$$
$$T_{Hxz} + T_{Wxz} + T_{Fxz} = 0$$
$$T_{Hyz} + T_{Wyz} + T_{Fyz} = 0$$
$$T_{Hxy} + T_{Fxy} = 0$$

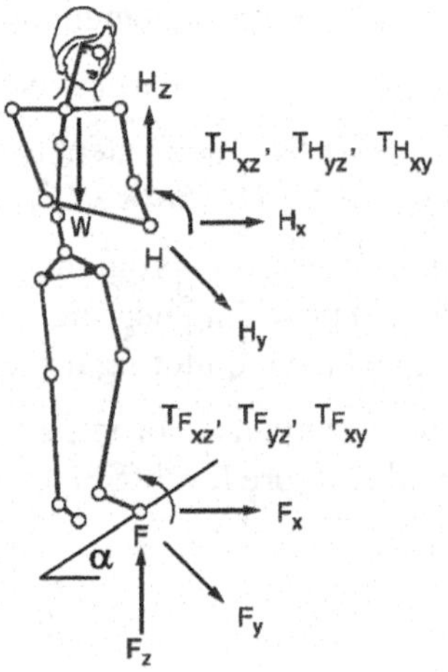

The coefficient of friction μ and the angle α at which the floor is inclined determine the equations at the foot.

$$F_x = \mu F_z$$

With the slope angle α

$$F_x = \mu F_z \cos \alpha$$

Of course, when motion must be considered, dynamic conditions come into play according to Newton's Second Law. Force transmitted with the hands is counteracted at the foot. Further, the body must also react with internal forces at all points between the hand and the foot.

Incidence Variable Values

Two concepts can be important when completing OSHA forms. These concepts are *incidence rates* and *severity rates*. On occasion it is necessary to calculate the total injury/illness incident rate of an organization in order to complete OSHA forms. This calculation must include fatalities and all injuries requiring medical treatment beyond mere first aid. The formula for determining the total injury/illness incident rate is as follows:

$$IR = N \times 200{,}000 \div T$$

IR = Total injury/illness incidence rate
N = Number of injuries, illnesses, and fatalities
T = Total hours worked by all employees during the period in question

The number 200,000 in the formula represents the number of hours 100 employees work in a year (40 hours per week × 50 weeks = 2,000 hours per year per employee). Using the same basic formula with only minor substitutions, safety managers can calculate the following types of incidence rates:

1. Injury rate
2. Illness rate
3. Fatality rate
4. Lost workday cases rate
5. Number of lost workdays rate
6. Specific hazard rate
7. Lost workday injuries rate

NOISE POLLUTION

$$\text{SPL (dB)} = 10 \log_{10}\left(P^2/P_0^2\right)$$

$$\text{SPL}_{\text{total}} = 10 \log_{10} \Sigma 10^{\text{SPL}/10}$$

Point Source Attenuation
$$\Delta\ \text{SPL (dB)} = 10 \log_{10}(r_1/r_2)^2$$

Line Source Attenuation
$$\Delta\ \text{SPL (dB)} = 10 \log_{10}(r_1/r_2)$$

where
SPL (dB) = sound pressure level, measured in decibels
P = sound pressure (Pa)
P_0 = reference sound pressure (2×10^{-5} Pa)
$\text{SPL}_{\text{total}}$ = sum of multiple sources
Δ SPL (dB) = change in sound pressure level with distance, measured in decibels
r_1 = distance from source to receptor at point 1
r_2 = distance from source to receptor at point 2

PERMISSIBLE NOISE EXPOSURE (OSHA)

Noise dose D should not exceed 100%.

$$D = 100\% \times \Sigma \frac{C_i}{T_i}$$

where C_i = time spent at specified sound pressure level, SPL, (hours)

T_i = time permitted at SPL (hours)

ΣC_i = 8 (hours)

Noise Level (dBA)	Permissible Time (hr)
80	32
85	16
90	8
95	4
100	2
105	1
110	0.5
115	0.25
120	0.125
125	0.063
130	0.031

If $D > 100\%$, noise abatement required.

If $50\% \leq D \leq 100\%$, hearing conservation program required.

Note: $D = 100\%$ is equivalent to 90 dBA time-weighted average (TWA). $D = 50\%$ equivalent to TWA of 85 dBA.

Hearing conservation program requires: (1) testing employee hearing, (2) providing hearing protection at employee's request, and (3) monitoring noise exposure.

Exposure to impulsive or impact noise should not exceed 140 dB sound pressure level (SPL).

MATHEMATICS

DISCRETE MATH

Symbols

$x \in X$	x is a member of X
$\{\ \}, \phi$	The empty (or null) set
$S \subseteq T$	S is a subset of T
$S \subset T$	S is a proper subset of T
(a,b)	Ordered pair
$P^{(s)}$	Power set of S
$(a_1, a_2, ..., a_n)$	n-tuple
$A \times B$	Cartesian product of A and B
$A \cup B$	Union of A and B
$A \cap B$	Intersection of A and B
$\forall\ x$	Universal qualification for all x; for any x; for each x
$\exists\ y$	Uniqueness qualification there exists y

A binary relation from A to B is a subset of $A \times B$.

Matrix of Relation

If $A = \{a_1, a_2, ..., a_m\}$ and $B = \{b_1, b_2, ..., b_n\}$ are finite sets containing m and n elements, respectively, then a relation R from A to B can be represented by the $m \times n$ matrix $M_R < [m_{ij}]$, which is defined by:

$$m_{ij} = \{\ 1 \text{ if } (a_i, b_j) \in R$$
$$0 \text{ if } (a_i, b_j) \notin R\}$$

Directed Graphs, or Digraphs, of Relation

A directed graph, or digraph, consists of a set V of vertices (or nodes) together with a set E of ordered pairs of elements of V called edges (or arcs). For edge (a, b), the vertex a is called the initial vertex and vertex b is called the terminal vertex. An edge of form (a, a) is called a loop.

Finite State Machine

A finite state machine consists of a finite set of states $S_i = \{s_0, s_1, ..., s_n\}$ and a finite set of inputs I; and a transition function f that assigns to each state and input pair a new state.

A state (or truth) table can be used to represent the finite state machine.

	Input			
State	i_0	i_1	i_2	i_3
S_0	S_0	S_1	S_2	S_3
S_1	S_2	S_2	S_3	S_3
S_2	S_3	S_3	S_3	S_3
S_3	S_0	S_3	S_3	S_3

Another way to represent a finite state machine is to use a state diagram, which is a directed graph with labeled edges.

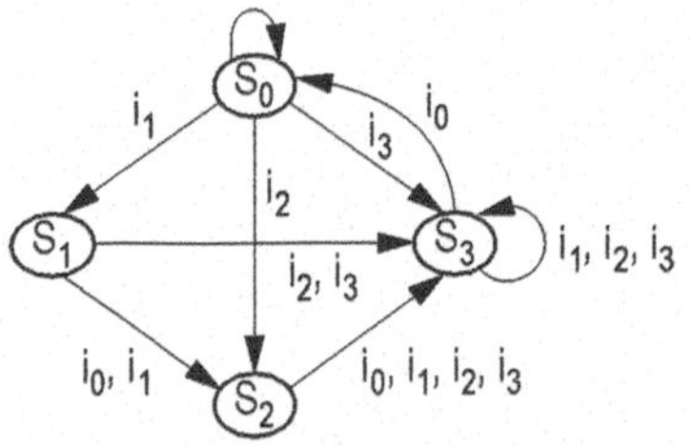

The characteristic of how a function maps one set (X) to another set (Y) may be described in terms of being either injective, surjective, or bijective.

An injective (one-to-one) relationship exists if, and only if, $\forall\ x_1, x_2 \in X$, if $f(x_1) = f(x_2)$, then $x_1 = x_2$

A surjective (onto) relationship exists when $\forall\ y \in Y, \exists\ x \in X$ such that $f(x) = y$

A bijective relationship is both injective (one-to-one) and surjective (onto).

STRAIGHT LINE

The general form of the equation is

$$Ax + By + C = 0$$

The standard form of the equation is

$$y = mx + b,$$

which is also known as the *slope-intercept* form.

The *point-slope* form is $y - y_1 = m(x - x_1)$

Given two points: slope, $m = (y_2 - y_1)/(x_2 - x_1)$

The angle between lines with slopes m_1 and m_2 is

$$\alpha = \arctan\left[(m_2 - m_1)/(1 + m_2 \cdot m_1)\right]$$

Two lines are perpendicular if $m_1 = -1/m_2$

The distance between two points is

$$d = \sqrt{(y_2 - y_1)^2 + (x_2 - x_1)^2}$$

QUADRATIC EQUATION

$$ax^2 + bx + c = 0$$

$$x = \text{Roots} = \frac{-b \pm \sqrt{b^2 - 4ac}}{2a}$$

QUADRIC SURFACE (SPHERE)

The standard form of the equation is

$$(x - h)^2 + (y - k)^2 + (z - m)^2 = r^2$$

with center at (h, k, m).

In a three-dimensional space, the distance between two points is

$$d = \sqrt{(x_2 - x_1)^2 + (y_2 - y_1)^2 + (z_2 - z_1)^2}$$

LOGARITHMS

The logarithm of x to the Base b is defined by

$$\log_b(x) = c, \text{ where } b^c = x$$

Special definitions for $b = e$ or $b = 10$ are:

$\ln x$, Base $= e$

$\log x$, Base $= 10$

To change from one Base to another:

$$\log_b x = (\log_a x)/(\log_a b)$$

e.g., $\ln x = (\log_{10} x)/(\log_{10} e) = 2.302585 (\log_{10} x)$

Identities

$$\log_b b^n = n$$

$$\log x^c = c \log x;\ x^c = \text{antilog}(c \log x)$$

$$\log xy = \log x + \log y$$

$$\log_b b = 1;\ \log 1 = 0$$

$$\log x/y = \log x - \log y$$

ALGEBRA OF COMPLEX NUMBERS

Complex numbers may be designated in rectangular form or polar form. In rectangular form, a complex number is written in terms of its real and imaginary components.

$z = a + jb$, where

a = the real component,

b = the imaginary component, and

$j = \sqrt{-1}$ (some disciplines use $i = \sqrt{-1}$)

In polar form $z = c \angle \theta$ where

$c = \sqrt{a^2 + b^2}$,

$\theta = \tan^{-1}(b/a)$,

$a = c \cos \theta$, and

$b = c \sin \theta$.

Complex numbers can be added and subtracted in rectangular form. If

$$z_1 = a_1 + jb_1 = c_1(\cos\theta_1 + j\sin\theta_1) = c_1 \angle \theta_1 \text{ and}$$

$$z_2 = a_2 + jb_2 = c_2(\cos\theta_2 + j\sin\theta_2) = c_2 \angle \theta_2, \text{ then}$$

$$z_1 + z_2 = (a_1 + a_2) + j(b_1 + b_2) \text{ and}$$

$$z_1 - z_2 = (a_1 - a_2) + j(b_1 - b_2)$$

While complex numbers can be multiplied or divided in rectangular form, it is more convenient to perform these operations in polar form.

$$z_1 \times z_2 = (c_1 \times c_2) \angle (\theta_1 + \theta_2)$$

$$z_1/z_2 = (c_1/c_2) \angle (\theta_1 - \theta_2)$$

The complex conjugate of a complex number $z_1 = (a_1 + jb_1)$ is defined as $z_1^* = (a_1 - jb_1)$. The product of a complex number and its complex conjugate is $z_1 z_1^* = a_1^2 + b_1^2$.

Polar Coordinate System

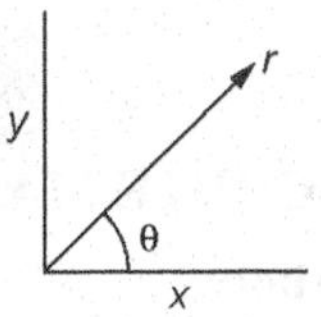

$x = r\cos\theta;\ y = r\sin\theta;\ \theta = \arctan(y/x)$

$r = |x + jy| = \sqrt{x^2 + y^2}$

$x + jy = r(\cos\theta + j\sin\theta) = re^{j\theta}$

$$[r_1(\cos\theta_1 + j\sin\theta_1)][r_2(\cos\theta_2 + j\sin\theta_2)] = r_1 r_2[\cos(\theta_1 + \theta_2) + j\sin(\theta_1 + \theta_2)]$$

$$(x + jy)^n = [r(\cos\theta + j\sin\theta)]^n = r^n(\cos n\theta + j\sin n\theta)$$

$$\frac{r_1(\cos\theta_1 + j\sin\theta_1)}{r_2(\cos\theta_2 + j\sin\theta_2)} = \frac{r_1}{r_2}\left[\cos(\theta_1 - \theta_2) + j\sin(\theta_1 - \theta_2)\right]$$

Euler's Identity

$$e^{j\theta} = \cos\theta + j\sin\theta$$

$$e^{-j\theta} = \cos\theta - j\sin\theta$$

$$\cos\theta = \frac{e^{j\theta} + e^{-j\theta}}{2},\quad \sin\theta = \frac{e^{j\theta} - e^{-j\theta}}{2j}$$

Roots

If k is any positive integer, any complex number (other than zero) has k distinct roots. The k roots of $r(\cos\theta + j\sin\theta)$ can be found by substituting successively $n = 0, 1, 2, \ldots, (k - 1)$ in the formula

$$w = \sqrt[k]{r}\left[\cos\left(\frac{\theta}{k} + n\frac{360°}{k}\right) + j\sin\left(\frac{\theta}{k} + n\frac{360°}{k}\right)\right]$$

TRIGONOMETRY

Trigonometric functions are defined using a right triangle.

$\sin\theta = y/r$, $\cos\theta = x/r$

$\tan\theta = y/x$, $\cot\theta = x/y$

$\csc\theta = r/y$, $\sec\theta = r/x$

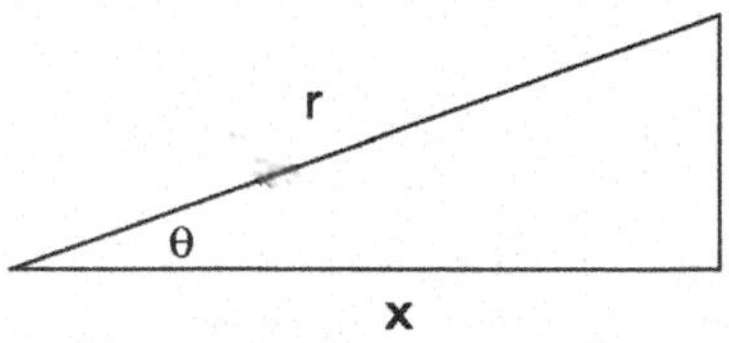

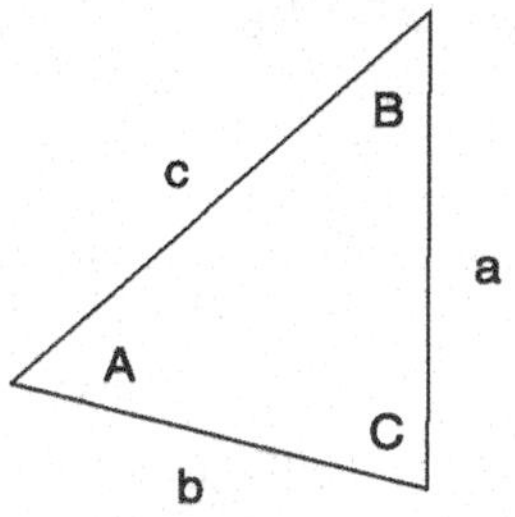

Law of Sines

$$\frac{a}{\sin A} = \frac{b}{\sin B} = \frac{c}{\sin C}$$

Law of Cosines

$$a^2 = b^2 + c^2 - 2bc\cos A$$

$$b^2 = a^2 + c^2 - 2ac\cos B$$

$$c^2 = a^2 + b^2 - 2ab\cos C$$

Brink, R.W., *A First Year of College Mathematics*, D. Appleton-Century Co., Inc., Englewood Cliffs, NJ, 1937.

Identities

$\cos\theta = \sin(\theta + \pi/2) = -\sin(\theta - \pi/2)$

$\sin\theta = \cos(\theta - \pi/2) = -\cos(\theta + \pi/2)$

$\csc\theta = 1/\sin\theta$

$\sec\theta = 1/\cos\theta$

$\tan\theta = \sin\theta/\cos\theta$

$\cot\theta = 1/\tan\theta$

$\sin^2\theta + \cos^2\theta = 1$

$\tan^2\theta + 1 = \sec^2\theta$

$\cot^2\theta + 1 = \csc^2\theta$

$\sin(\alpha + \beta) = \sin\alpha\cos\beta + \cos\alpha\sin\beta$

$\cos(\alpha + \beta) = \cos\alpha\cos\beta - \sin\alpha\sin\beta$

$\sin 2\alpha = 2\sin\alpha\cos\alpha$

$\cos 2\alpha = \cos^2\alpha - \sin^2\alpha = 1 - 2\sin^2\alpha = 2\cos^2\alpha - 1$

$\tan 2\alpha = (2\tan\alpha)/(1 - \tan^2\alpha)$

$\cot 2\alpha = (\cot^2\alpha - 1)/(2\cot\alpha)$

$\tan(\alpha + \beta) = (\tan\alpha + \tan\beta)/(1 - \tan\alpha\tan\beta)$

$\cot(\alpha + \beta) = (\cot\alpha\cot\beta - 1)/(\cot\alpha + \cot\beta)$

$\sin(\alpha - \beta) = \sin\alpha\cos\beta - \cos\alpha\sin\beta$

$\cos(\alpha - \beta) = \cos\alpha\cos\beta + \sin\alpha\sin\beta$

$\tan(\alpha - \beta) = (\tan\alpha - \tan\beta)/(1 + \tan\alpha\tan\beta)$

$\cot(\alpha - \beta) = (\cot\alpha\cot\beta + 1)/(\cot\beta - \cot\alpha)$

$\sin(\alpha/2) = \pm\sqrt{(1 - \cos\alpha)/2}$

$\cos(\alpha/2) = \pm\sqrt{(1 + \cos\alpha)/2}$

$\tan(\alpha/2) = \pm\sqrt{(1 - \cos\alpha)/(1 + \cos\alpha)}$

$\cot(\alpha/2) = \pm\sqrt{(1 + \cos\alpha)/(1 - \cos\alpha)}$

$\sin\alpha\sin\beta = (1/2)[\cos(\alpha - \beta) - \cos(\alpha + \beta)]$

$\cos\alpha\cos\beta = (1/2)[\cos(\alpha - \beta) + \cos(\alpha + \beta)]$

$\sin\alpha\cos\beta = (1/2)[\sin(\alpha + \beta) + \sin(\alpha - \beta)]$

$\sin\alpha + \sin\beta = 2\sin[(1/2)(\alpha + \beta)]\cos[(1/2)(\alpha - \beta)]$

$\sin\alpha - \sin\beta = 2\cos[(1/2)(\alpha + \beta)]\sin[(1/2)(\alpha - \beta)]$

$\cos\alpha + \cos\beta = 2\cos[(1/2)(\alpha + \beta)]\cos[(1/2)(\alpha - \beta)]$

$\cos\alpha - \cos\beta = -2\sin[(1/2)(\alpha + \beta)]\sin[(1/2)(\alpha - \beta)]$

MENSURATION OF AREAS AND VOLUMES

Nomenclature

A = total surface area

P = perimeter

V = volume

Parabola

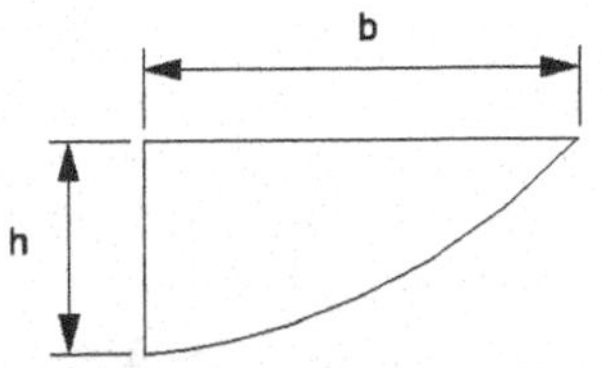

A = 2bh/3

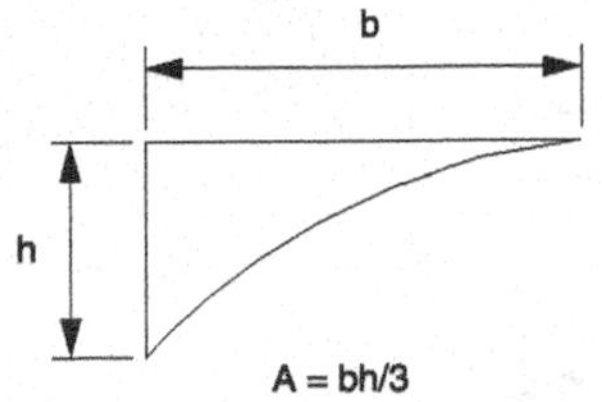

A = bh/3

Ellipse

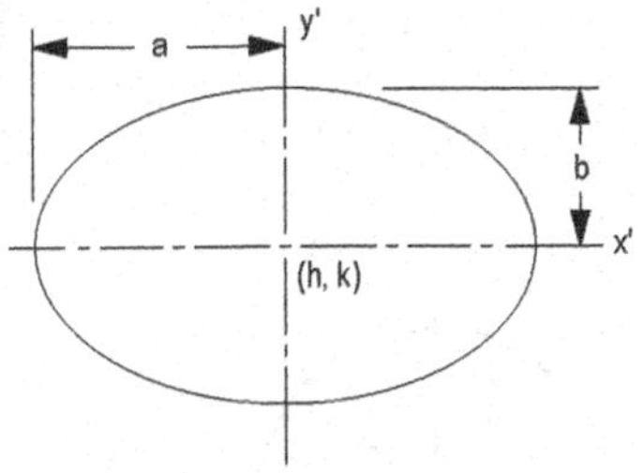

A = πab

$$P_{approx} = 2\pi\sqrt{(a^2 + b^2)/2}$$

$$P = \pi(a+b)\begin{bmatrix} 1 + (1/2)^2\lambda^2 + (1/2 \times 1/4)^2\lambda^4 \\ +(1/2 \times 1/4 \times 3/6)^2\lambda^6 + (1/2 \times 1/4 \times 3/6 \times 5/8)^2\lambda^8 \\ +(1/2 \times 1/4 \times 3/6 \times 5/8 \times 7/10)^2\lambda^{10} + \ldots \end{bmatrix},$$

where

$$\lambda = (a-b)/(a+b)$$

Gieck, K., and R. Gieck, *Engineering Formulas*, 6th ed., Gieck Publishing, 1967.

Circular Segment

♦

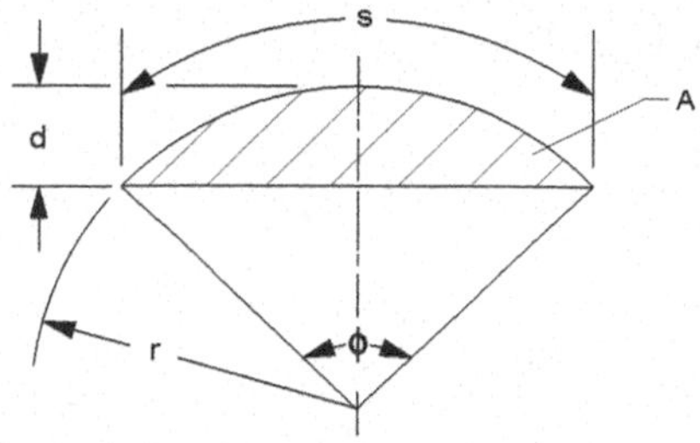

$A = \left[r^2(\phi - \sin\phi)\right]/2$

$\phi = s/r = 2\{\arccos[(r-d)/r]\}$

Circular Sector

♦

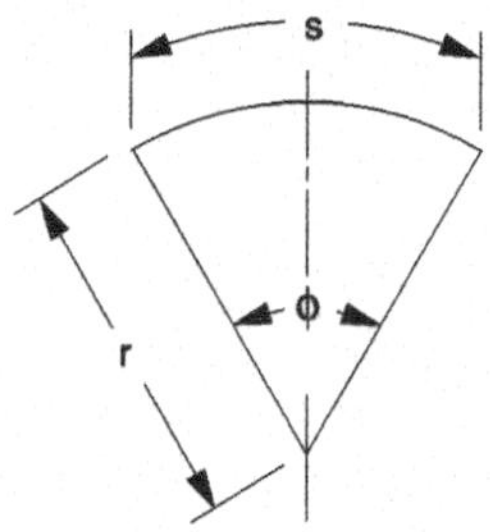

$A = \phi r^2/2 = sr/2$

$\phi = s/r$

Sphere

♦

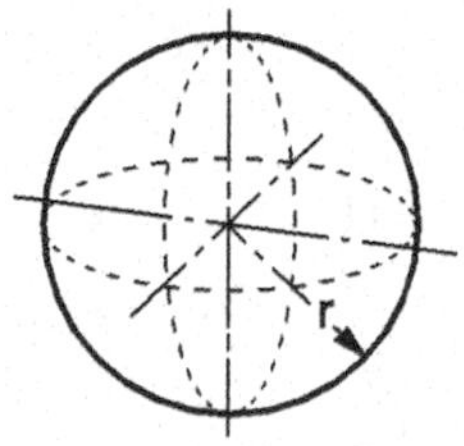

$V = 4\pi r^3/3 = \pi d^3/6$

$A = 4\pi r^2 = \pi d^2$

Parallelogram

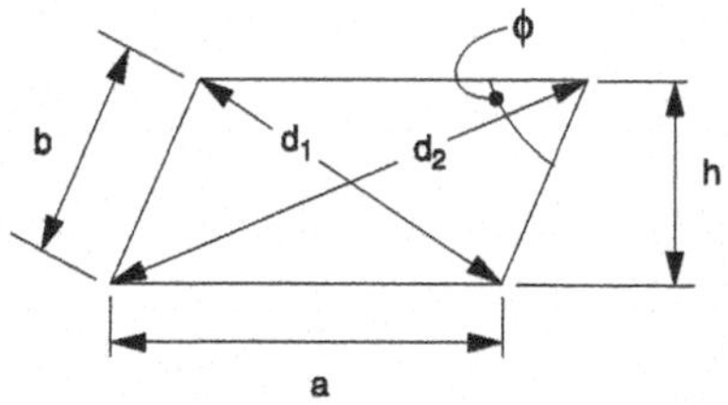

$P = 2(a + b)$

$d_1 = \sqrt{a^2 + b^2 - 2ab(\cos\phi)}$

$d_2 = \sqrt{a^2 + b^2 + 2ab(\cos\phi)}$

$d_1^2 + d_2^2 = 2(a^2 + b^2)$

$A = ah = ab(\sin\phi)$

If $a = b$, the parallelogram is a rhombus.

Regular Polygon (*n* equal sides)

♦

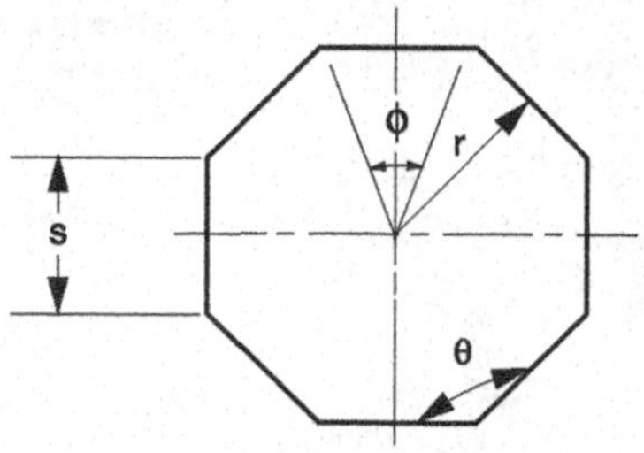

$\phi = 2\pi/n$

$\theta = \left[\frac{\pi(n-2)}{n}\right] = \pi\left(1 - \frac{2}{n}\right)$

$P = ns$

$s = 2r\left[\tan(\phi/2)\right]$

$A = (nsr)/2$

Prismoid

♦

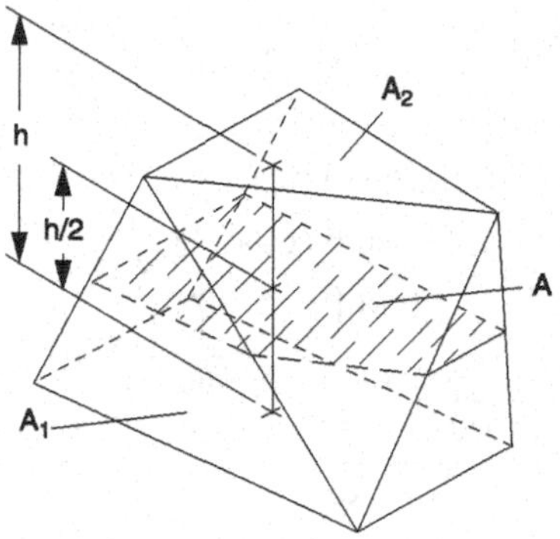

$V = (h/6)(A_1 + A_2 + 4A)$

Right Circular Cone

♦

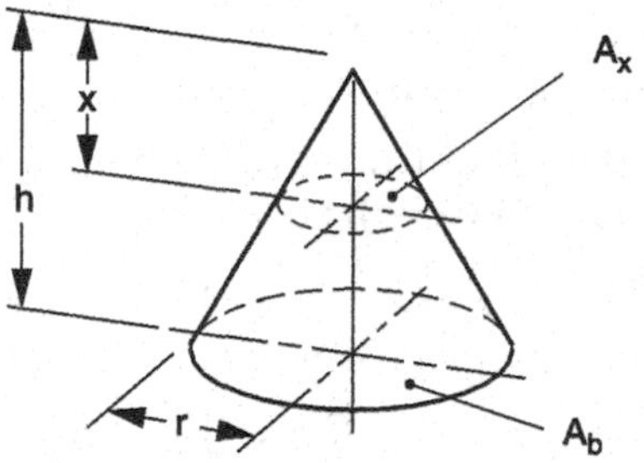

$V = (\pi r^2 h)/3$

$A =$ side area + base area

$= \pi r\left(r + \sqrt{r^2 + h^2}\right)$

$A_x : A_b = x^2 : h^2$

♦ Gieck, K., and R. Gieck, *Engineering Formulas*, 6th ed., Gieck Publishing, 1967.

Right Circular Cylinder

♦

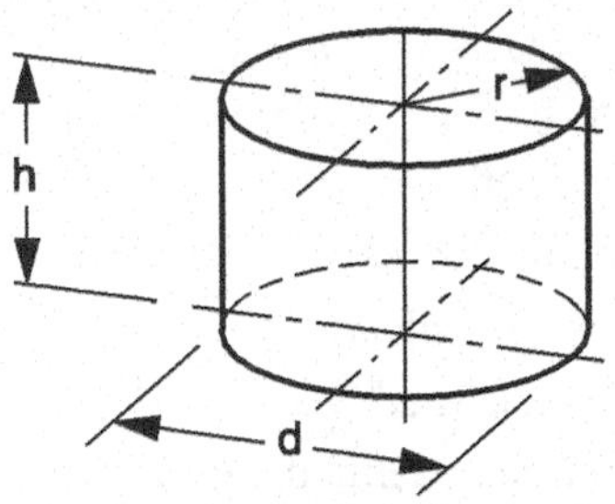

$$V = \pi r^2 h = \frac{\pi d^2 h}{4}$$

$$A = \text{side area} + \text{end areas} = 2\pi r(h + r)$$

Paraboloid of Revolution

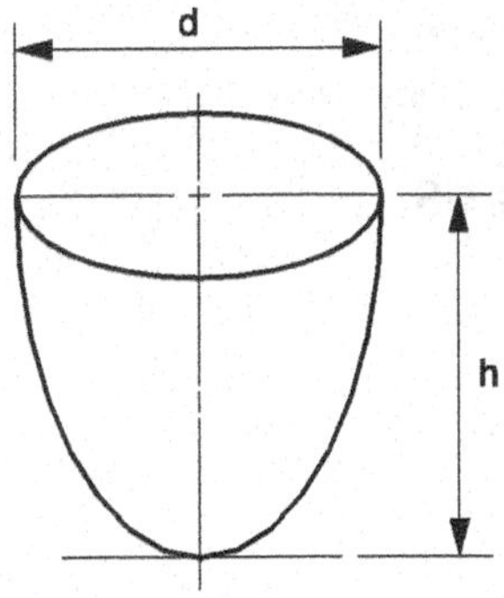

$$V = \frac{\pi d^2 h}{8}$$

CONIC SECTIONS

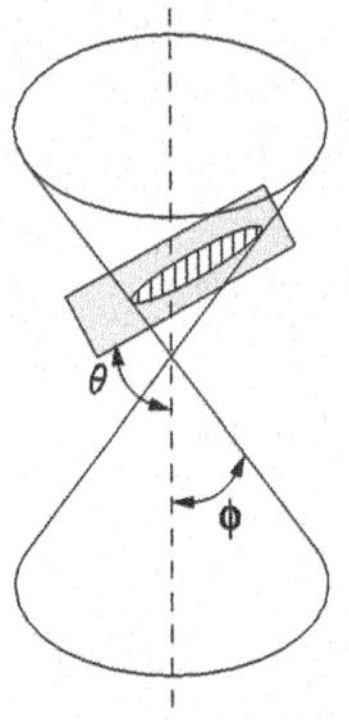

e = eccentricity = $\cos\theta/(\cos\phi)$

[Note: X' and Y', in the following cases, are translated axes.]

Case 1. Parabola $e = 1$:

•

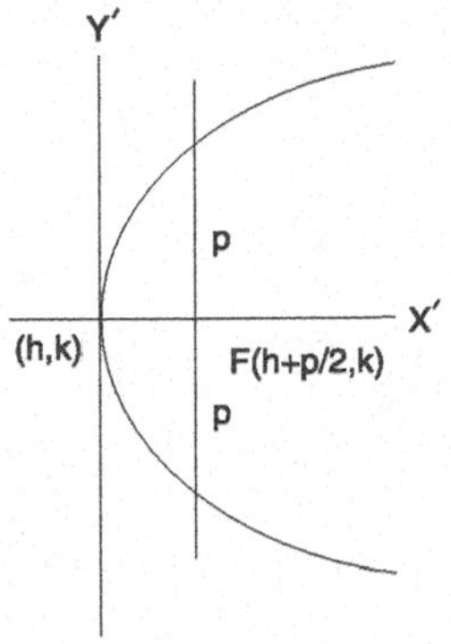

$(y - k)^2 = 2p(x - h)$; Center at (h, k)

is the standard form of the equation. When $h = k = 0$,

Focus: $(p/2, 0)$; Directrix: $x = -p/2$

Case 2. Ellipse $e < 1$:

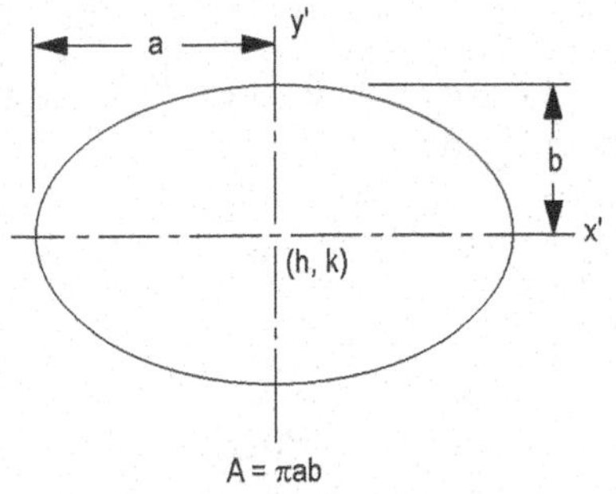

$$\frac{(x-h)^2}{a^2} + \frac{(y-k)^2}{b^2} = 1; \text{ Center at } (h, k)$$

is the standard form of the equation. When $h = k = 0$,

Eccentricity: $e = \sqrt{1 - (b^2/a^2)} = c/a$

$b = a\sqrt{1 - e^2}$;

Focus: $(\pm ae, 0)$; Directrix: $x = \pm a/e$

♦ Gieck, K., and R. Gieck, *Engineering Formulas*, 6th ed., Gieck Publishing, 1967.

• Brink, R.W., *A First Year of College Mathematics*, D. Appleton-Century Co., Inc., 1937.

Case 3. Hyperbola $e > 1$:

• 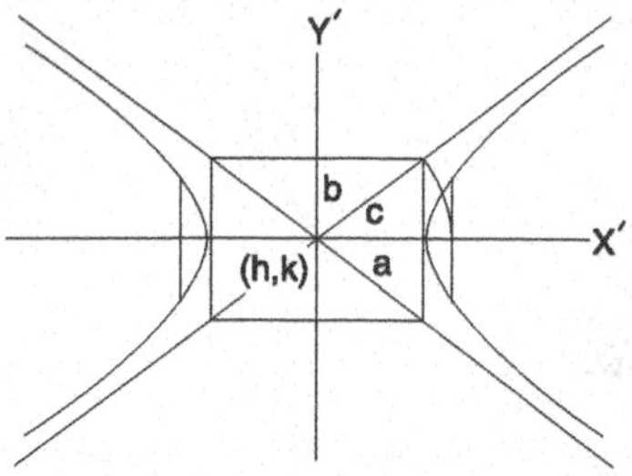

$$\frac{(x-h)^2}{a^2} - \frac{(y-k)^2}{b^2} = 1; \text{ Center at } (h, k)$$

is the standard form of the equation. When $h = k = 0$,

Eccentricity: $e = \sqrt{1 + (b^2/a^2)} = c/a$

$b = a\sqrt{e^2 - 1}$;

Focus: $(\pm ae, 0)$; Directrix: $x = \pm a/e$

Case 4. Circle $e = 0$:

$(x - h)^2 + (y - k)^2 = r^2$; Center at (h, k) is the standard form of the equation with radius

$$r = \sqrt{(x-h)^2 + (y-k)^2}$$

• 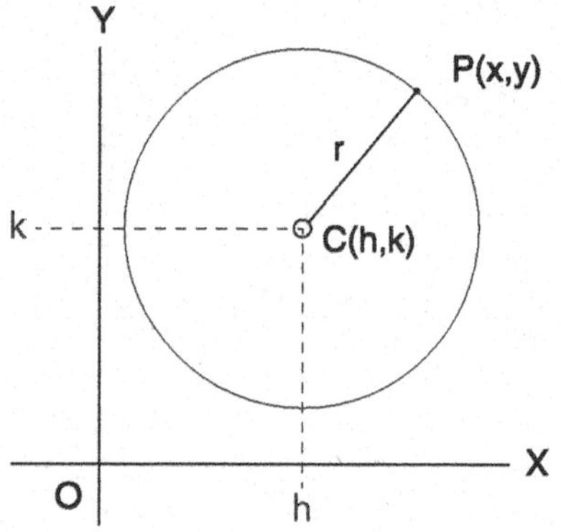

Length of the tangent line from a point on a circle to a point (x',y'):

$$t^2 = (x' - h)^2 + (y' - k)^2 - r^2$$

• 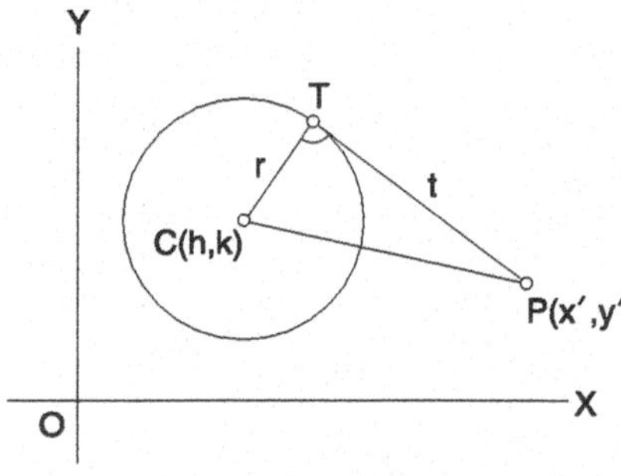

• Brink, R.W., *A First Year of College Mathematics*, D. Appleton-Century Co., Inc., 1937.

Conic Section Equation

The general form of the conic section equation is

$$Ax^2 + Bxy + Cy^2 + Dx + Ey + F = 0$$

where not both A and C are zero.

If $B^2 - 4AC < 0$, an *ellipse* is defined.

If $B^2 - 4AC > 0$, a *hyperbola* is defined.

If $B^2 - 4AC = 0$, the conic is a *parabola*.

If $A = C$ and $B = 0$, a *circle* is defined.

If $A = B = C = 0$, a *straight line* is defined.

$$x^2 + y^2 + 2ax + 2by + c = 0$$

is the normal form of the conic section equation, if that conic section has a principal axis parallel to a coordinate axis.

$h = -a$; $k = -b$

$r = \sqrt{a^2 + b^2 - c}$

If $a^2 + b^2 - c$ is positive, a *circle*, center $(-a, -b)$.

If $a^2 + b^2 - c$ equals zero, a *point* at $(-a, -b)$.

If $a^2 + b^2 - c$ is negative, locus is *imaginary*.

DIFFERENTIAL CALCULUS

The Derivative

For any function $y = f(x)$,

the derivative $= D_x y = dy/dx = y'$

$$y' = \lim_{\Delta x \to 0} [(\Delta y)/(\Delta x)]$$

$$= \lim_{\Delta x \to 0} \{[f(x + \Delta x) - f(x)]/(\Delta x)\}$$

y' = the slope of the curve $f(x)$.

Test for a Maximum

$y = f(x)$ is a maximum for $x = a$, if $f'(a) = 0$ and $f''(a) < 0$.

Test for a Minimum

$y = f(x)$ is a minimum for $x = a$, if $f'(a) = 0$ and $f''(a) > 0$.

Test for a Point of Inflection

$y = f(x)$ has a point of inflection at $x = a$, if $f''(a) = 0$, and if $f''(x)$ changes sign as x increases through $x = a$.

The Partial Derivative

In a function of two independent variables x and y, a derivative with respect to one of the variables may be found if the other variable is *assumed* to remain constant. If *y is kept fixed*, the function

$$z = f(x, y)$$

becomes a function of the *single variable x*, and its derivative (if it exists) can be found. This derivative is called the *partial derivative of z with respect to x*. The partial derivative with respect to x is denoted as follows:

$$\frac{\partial z}{\partial x} = \frac{\partial f(x,y)}{\partial x}$$

The Curvature of Any Curve

♦

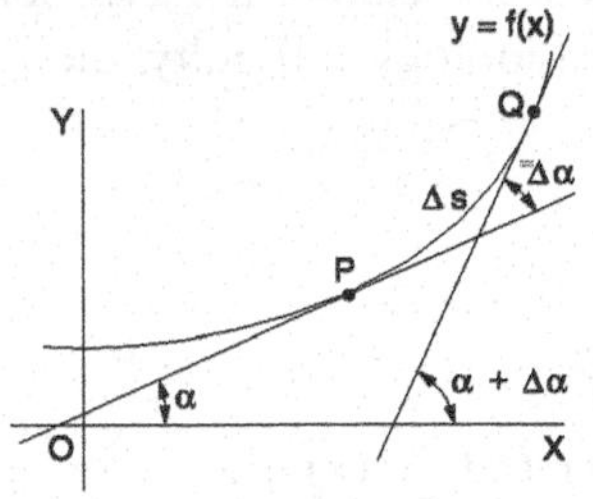

The curvature K of a curve at P is the limit of its average curvature for the arc PQ as Q approaches P. This is also expressed as: the curvature of a curve at a given point is the rate-of-change of its inclination with respect to its arc length.

$$K = \lim_{\Delta s \to 0} \frac{\Delta\alpha}{\Delta s} = \frac{d\alpha}{ds}$$

Curvature in Rectangular Coordinates

$$K = \frac{y''}{\left[1 + (y')^2\right]^{3/2}}$$

When it may be easier to differentiate the function with respect to y rather than x, the notation x' will be used for the derivative.

$$x' = dx/dy$$

$$K = \frac{-x''}{\left[1 + (x')^2\right]^{3/2}}$$

The Radius of Curvature

The *radius of curvature* R at any point on a curve is defined as the absolute value of the reciprocal of the curvature K at that point.

$$R = \frac{1}{|K|} \qquad (K \neq 0)$$

$$R = \left|\frac{\left[1 + (y')^2\right]^{3/2}}{|y''|}\right| \qquad (y'' \neq 0)$$

L'Hospital's Rule (L'Hôpital's Rule)

If the fractional function $f(x)/g(x)$ assumes one of the indeterminate forms 0/0 or ∞/∞ (where α is finite or infinite), then

$$\lim_{x \to \alpha} f(x)/g(x)$$

is equal to the first of the expressions

$$\lim_{x \to \alpha} \frac{f'(x)}{g'(x)}, \lim_{x \to \alpha} \frac{f''(x)}{g''(x)}, \lim_{x \to \alpha} \frac{f'''(x)}{g'''(x)}$$

which is not indeterminate, provided such first indicated limit exists.

INTEGRAL CALCULUS

The definite integral is defined as:

$$\lim_{n \to \infty} \sum_{i=1}^{n} f(x_i)\Delta x_i = \int_a^b f(x)\,dx$$

Also, $\Delta x_i \to 0$ for all i.

A table of derivatives and integrals is available in the Derivatives and Indefinite Integrals section. The integral equations can be used along with the following methods of integration:

A. Integration by Parts (integral equation #6),
B. Integration by Substitution, and
C. Separation of Rational Fractions into Partial Fractions.

♦ Wade, Thomas L., *Calculus*, Ginn & Company/Simon & Schuster Publishers, 1953.

DERIVATIVES AND INDEFINITE INTEGRALS

In these formulas, u, v, and w represent functions of x. Also, a, c, and n represent constants. All arguments of the trigonometric functions are in radians. A constant of integration should be added to the integrals. To avoid terminology difficulty, the following definitions are followed: $\arcsin u = \sin^{-1} u$, $(\sin u)^{-1} = 1/\sin u$.

1. $dc/dx = 0$
2. $dx/dx = 1$
3. $d(cu)/dx = c\, du/dx$
4. $d(u + v - w)/dx = du/dx + dv/dx - dw/dx$
5. $d(uv)/dx = u\, dv/dx + v\, du/dx$
6. $d(uvw)/dx = uv\, dw/dx + uw\, dv/dx + vw\, du/dx$
7. $\dfrac{d(u/v)}{dx} = \dfrac{v\, du/dx - u\, dv/dx}{v^2}$
8. $d(u^n)/dx = nu^{n-1}\, du/dx$
9. $d[f(u)]/dx = \{d[f(u)]/du\}\, du/dx$
10. $du/dx = 1/(dx/du)$
11. $\dfrac{d(\log_a u)}{dx} = (\log_a e)\dfrac{1}{u}\dfrac{du}{dx}$
12. $\dfrac{d(\ln u)}{dx} = \dfrac{1}{u}\dfrac{du}{dx}$
13. $\dfrac{d(a^u)}{dx} = (\ln a)a^u\dfrac{du}{dx}$
14. $d(e^u)/dx = e^u\, du/dx$
15. $d(u^v)/dx = vu^{v-1}\, du/dx + (\ln u)\, u^v\, dv/dx$
16. $d(\sin u)/dx = \cos u\, du/dx$
17. $d(\cos u)/dx = -\sin u\, du/dx$
18. $d(\tan u)/dx = \sec^2 u\, du/dx$
19. $d(\cot u)/dx = -\csc^2 u\, du/dx$
20. $d(\sec u)/dx = \sec u \tan u\, du/dx$
21. $d(\csc u)/dx = -\csc u \cot u\, du/dx$
22. $\dfrac{d(\sin^{-1}u)}{dx} = \dfrac{1}{\sqrt{1-u^2}}\dfrac{du}{dx}$ $\quad(-\pi/2 \le \sin^{-1}u \le \pi/2)$
23. $\dfrac{d(\cos^{-1}u)}{dx} = -\dfrac{1}{\sqrt{1-u^2}}\dfrac{du}{dx}$ $\quad(0 \le \cos^{-1}u \le \pi)$
24. $\dfrac{d(\tan^{-1}u)}{dx} = \dfrac{1}{1+u^2}\dfrac{du}{dx}$ $\quad(-\pi/2 < \tan^{-1}u < \pi/2)$
25. $\dfrac{d(\cot^{-1}u)}{dx} = -\dfrac{1}{1+u^2}\dfrac{du}{dx}$ $\quad(0 < \cot^{-1}u < \pi)$
26. $\dfrac{d(\sec^{-1}u)}{dx} = \dfrac{1}{u\sqrt{u^2-1}}\dfrac{du}{dx}$
    $(0 < \sec^{-1}u < \pi/2)(-\pi \le \sec^{-1}u < -\pi/2)$
27. $\dfrac{d(\csc^{-1}u)}{dx} = -\dfrac{1}{u\sqrt{u^2-1}}\dfrac{du}{dx}$
    $(0 < \csc^{-1}u \le \pi/2)(-\pi < \csc^{-1}u \le -\pi/2)$

1. $\int d f(x) = f(x)$
2. $\int dx = x$
3. $\int a f(x)\, dx = a \int f(x)\, dx$
4. $\int [u(x) \pm v(x)]\, dx = \int u(x)\, dx \pm \int v(x)\, dx$
5. $\int x^m dx = \dfrac{x^{m+1}}{m+1}$ $\quad(m \ne -1)$
6. $\int u(x)\, dv(x) = u(x)\, v(x) - \int v(x)\, du(x)$
7. $\int \dfrac{dx}{ax+b} = \dfrac{1}{a}\ln|ax+b|$
8. $\int \dfrac{dx}{\sqrt{x}} = 2\sqrt{x}$
9. $\int a^x dx = \dfrac{a^x}{\ln a}$
10. $\int \sin x\, dx = -\cos x$
11. $\int \cos x\, dx = \sin x$
12. $\int \sin^2 x dx = \dfrac{x}{2} - \dfrac{\sin 2x}{4}$
13. $\int \cos^2 x dx = \dfrac{x}{2} + \dfrac{\sin 2x}{4}$
14. $\int x \sin x\, dx = \sin x - x \cos x$
15. $\int x \cos x\, dx = \cos x + x \sin x$
16. $\int \sin x \cos x\, dx = (\sin^2 x)/2$
17. $\int \sin ax \cos bx\, dx = -\dfrac{\cos(a-b)x}{2(a-b)} - \dfrac{\cos(a+b)x}{2(a+b)}$ $\quad(a^2 \ne b^2)$
18. $\int \tan x\, dx = -\ln|\cos x| = \ln|\sec x|$
19. $\int \cot x\, dx = -\ln|\csc x| = \ln|\sin x|$
20. $\int \tan^2 x\, dx = \tan x - x$
21. $\int \cot^2 x\, dx = -\cot x - x$
22. $\int e^{ax} dx = (1/a)\, e^{ax}$
23. $\int xe^{ax} dx = (e^{ax}/a^2)(ax - 1)$
24. $\int \ln x\, dx = x[\ln(x) - 1]$ $\quad(x > 0)$
25. $\int \dfrac{dx}{a^2+x^2} = \dfrac{1}{a}\tan^{-1}\dfrac{x}{a}$ $\quad(a \ne 0)$
26. $\int \dfrac{dx}{ax^2+c} = \dfrac{1}{\sqrt{ac}}\tan^{-1}\left(x\sqrt{\dfrac{a}{c}}\right),$ $\quad(a > 0, c > 0)$

27a. $\int \dfrac{dx}{ax^2+bx+c} = \dfrac{2}{\sqrt{4ac-b^2}}\tan^{-1}\dfrac{2ax+b}{\sqrt{4ac-b^2}}$ $\quad(4ac - b^2 > 0)$

27b. $\int \dfrac{dx}{ax^2+bx+c} = \dfrac{1}{\sqrt{b^2-4ac}}\ln\left|\dfrac{2ax+b-\sqrt{b^2-4ac}}{2ax+b+\sqrt{b^2-4ac}}\right|$ $\quad(b^2 - 4ac > 0)$

27c. $\int \dfrac{dx}{ax^2+bx+c} = -\dfrac{2}{2ax+b},$ $\quad(b^2 - 4ac = 0)$

CENTROIDS AND MOMENTS OF INERTIA

The *location of the centroid of an area*, bounded by the axes and the function $y = f(x)$, can be found by integration.

$$x_c = \frac{\int x dA}{A}$$

$$y_c = \frac{\int y dA}{A}$$

$$A = \int f(x)\,dx$$

$$dA = f(x)\,dx = g(y)\,dy$$

The *first moment of area* with respect to the y-axis and the x-axis, respectively, are:

$$M_y = \int x\,dA = x_c A$$

$$M_x = \int y\,dA = y_c A$$

The *moment of inertia* (*second moment of area*) with respect to the y-axis and the x-axis, respectively, are:

$$I_y = \int x^2\,dA$$

$$I_x = \int y^2\,dA$$

The moment of inertia taken with respect to an axis passing through the area's centroid is the *centroidal moment of inertia*. The *parallel axis theorem* for the moment of inertia with respect to another axis parallel with and located d units from the centroidal axis is expressed by

$$I_{\text{parallel axis}} = I_c + Ad^2$$

In a plane, $J = \int r^2 dA = I_x + I_y$

PROGRESSIONS AND SERIES

Arithmetic Progression

To determine whether a given finite sequence of numbers is an arithmetic progression, subtract each number from the following number. If the differences are equal, the series is arithmetic.

1. The first term is a.
2. The common difference is d.
3. The number of terms is n.
4. The last or nth term is l.
5. The sum of n terms is S.

$$l = a + (n - 1)d$$

$$S = n(a + l)/2 = n\,[2a + (n - 1)\,d]/2$$

Geometric Progression

To determine whether a given finite sequence is a geometric progression (G.P.), divide each number after the first by the preceding number. If the quotients are equal, the series is geometric:

1. The first term is a.
2. The common ratio is r.
3. The number of terms is n.
4. The last or nth term is l.
5. The sum of n terms is S.

$$l = ar^{n-1}$$

$$S = a\,(1 - r^n)/(1 - r);\ r \neq 1$$

$$S = (a - rl)/(1 - r);\ r \neq 1$$

$$\lim_{n\to\infty} S_n = a/(1 - r);\ r < 1$$

A G.P. converges if $|r| < 1$ and it diverges if $|r| > 1$.

Properties of Series

$$\sum_{i=1}^{n} c = nc; \qquad c = \text{constant}$$

$$\sum_{i=1}^{n} cx_i = c\sum_{i=1}^{n} x_i$$

$$\sum_{i=1}^{n} \left(x_i + y_i - z_i\right) = \sum_{i=1}^{n} x_i + \sum_{i=1}^{n} y_i - \sum_{i=1}^{n} z_i$$

$$\sum_{x=1}^{n} x = \left(n + n^2\right)/2$$

$$\prod_{i=1}^{n} x_i = x_1 x_2 x_3 \ldots x_n$$

Power Series

$$\Sigma_{i=0}^{\infty} a_i (x - a)^i$$

1. A power series, which is convergent in the interval $-R < x < R$, defines a function of x that is continuous for all values of x within the interval and is said to represent the function in that interval.
2. A power series may be differentiated term by term within its interval of convergence. The resulting series has the same interval of convergence as the original series (except possibly at the end points of the series).
3. A power series may be integrated term by term provided the limits of integration are within the interval of convergence of the series.
4. Two power series may be added, subtracted, or multiplied, and the resulting series in each case is convergent, at least, in the interval common to the two series.
5. Using the process of long division (as for polynomials), two power series may be divided one by the other within their common interval of convergence.

Taylor's Series

$$f(x) = f(a) + \frac{f'(a)}{1!}(x - a) + \frac{f''(a)}{2!}(x - a)^2 + \ldots + \frac{f^{(n)}(a)}{n!}(x - a)^n + \ldots$$

is called *Taylor's series*, and the function $f(x)$ is said to be expanded about the point a in a Taylor's series.

If $a = 0$, the Taylor's series equation becomes a *Maclaurin's series*.

DIFFERENTIAL EQUATIONS

A common class of ordinary linear differential equations is

$$b_n \frac{d^n y(x)}{dx^n} + \ldots + b_1 \frac{dy(x)}{dx} + b_0 y(x) = f(x)$$

where $b_n, \ldots, b_i, \ldots, b_1, b_0$ are constants.

When the equation is a homogeneous differential equation, $f(x) = 0$, the solution is

$$y_h(x) = C_1 e^{r_1 x} + C_2 e^{r_2 x} + \ldots + C_i e^{r_i x} + \ldots + C_n e^{r_n x}$$

where r_n is the nth distinct root of the characteristic polynomial $P(x)$ with

$$P(r) = b_n r^n + b_{n-1} r^{n-1} + \ldots + b_1 r + b_0$$

If the root $r_1 = r_2$, then $C_2 e^{r_2 x}$ is replaced with $C_2 x e^{r_1 x}$.

Higher orders of multiplicity imply higher powers of x. The complete solution for the differential equation is

$$y(x) = y_h(x) + y_p(x),$$

where $y_p(x)$ is any particular solution with $f(x)$ present. If $f(x)$ has $e^{r_n x}$ terms, then resonance is manifested. Furthermore, specific $f(x)$ forms result in specific $y_p(x)$ forms, some of which are:

$f(x)$	$y_p(x)$
A	B
$Ae^{\alpha x}$	$Be^{\alpha x}$, $\alpha \neq r_n$
$A_1 \sin \omega x + A_2 \cos \omega x$	$B_1 \sin \omega x + B_2 \cos \omega x$

If the independent variable is time t, then transient dynamic solutions are implied.

First-Order Linear Homogeneous Differential Equations with Constant Coefficients

$y' + ay = 0$, where a is a real constant:

Solution, $y = Ce^{-at}$

where C = a constant that satisfies the initial conditions.

First-Order Linear Nonhomogeneous Differential Equations

$$\tau \frac{dy}{dt} + y = Kx(t) \qquad x(t) = \begin{Bmatrix} A & t < 0 \\ B & t > 0 \end{Bmatrix}$$

$$y(0) = KA$$

τ is the time constant

K is the gain

The solution is

$$y(t) = KA + (KB - KA)\left(1 - \exp\left(\frac{-t}{\tau}\right)\right) \text{ or}$$

$$\frac{t}{\tau} = \ln\left[\frac{KB - KA}{KB - y}\right]$$

Second-Order Linear Homogeneous Differential Equations with Constant Coefficients

An equation of the form

$$y'' + ay' + by = 0$$

can be solved by the method of undetermined coefficients where a solution of the form $y = Ce^{rx}$ is sought. Substitution of this solution gives

$$(r^2 + ar + b)\, Ce^{rx} = 0$$

and since Ce^{rx} cannot be zero, the characteristic equation must vanish or

$$r^2 + ar + b = 0$$

The roots of the characteristic equation are

$$r_{1,2} = \frac{-a \pm \sqrt{a^2 - 4b}}{2}$$

and can be real and distinct for $a^2 > 4b$, real and equal for $a^2 = 4b$, and complex for $a^2 < 4b$.

If $a^2 > 4b$, the solution is of the form (overdamped)

$$y = C_1 e^{r_1 x} + C_2 e^{r_2 x}$$

If $a^2 = 4b$, the solution is of the form (critically damped)

$$y = (C_1 + C_2 x) e^{r_1 x}$$

If $a^2 < 4b$, the solution is of the form (underdamped)

$y = e^{\alpha x}(C_1 \cos \beta x + C_2 \sin \beta x)$, where

$$\alpha = -a/2$$

$$\beta = \frac{\sqrt{4b - a^2}}{2}$$

FOURIER TRANSFORM

The Fourier transform pair, one form of which is

$$F(\omega) = \int_{-\infty}^{\infty} f(t) e^{-j\omega t} dt$$

$$f(t) = [1/(2\pi)] \int_{-\infty}^{\infty} F(\omega) e^{j\omega t} d\omega$$

can be used to characterize a broad class of signal models in terms of their frequency or spectral content. Some useful transform pairs are:

$f(t)$	$F(\omega)$
$\delta(t)$	1
$u(t)$	$\pi\delta(\omega) + 1/j\omega$
$u\left(t + \frac{\tau}{2}\right) - u\left(t - \frac{\tau}{2}\right) = r_{rect} \frac{t}{\tau}$	$\tau \frac{\sin(\omega\tau/2)}{\omega\tau/2}$
$e^{j\omega_o t}$	$2\pi\delta(\omega - \omega_o)$

Some mathematical liberties are required to obtain the second and fourth form. Other Fourier transforms are derivable from the Laplace transform by replacing s with $j\omega$ provided

$$f(t) = 0, t < 0$$

$$\int_0^{\infty} |f(t)| dt < \infty$$

FOURIER SERIES

Every periodic function $f(t)$ which has the period $T = 2\pi/\omega_0$ and has certain continuity conditions can be represented by a series plus a constant

$$f(t) = a_0 + \sum_{n=1}^{\infty}\left[a_n\cos(n\omega_0 t) + b_n\sin(n\omega_0 t)\right]$$

The above holds if $f(t)$ has a continuous derivative $f'(t)$ for all t. It should be noted that the various sinusoids present in the series are orthogonal on the interval 0 to T and as a result the coefficients are given by

$$a_0 = (1/T)\int_0^T f(t)\,dt$$

$$a_n = (2/T)\int_0^T f(t)\cos(n\omega_0 t)\,dt \qquad n = 1,2,\ldots$$

$$b_n = (2/T)\int_0^T f(t)\sin(n\omega_0 t)\,dt \qquad n = 1,2,\ldots$$

The constants a_n and b_n are the *Fourier coefficients* of $f(t)$ for the interval 0 to T and the corresponding series is called the *Fourier series of* $f(t)$ over the same interval.

The integrals have the same value when evaluated over any interval of length T.

If a Fourier series representing a periodic function is truncated after term $n = N$ the mean square value F_N^2 of the truncated series is given by Parseval's relation. This relation says that the mean-square value is the sum of the mean-square values of the Fourier components, or

$$F_N^2 = a_0^2 + (1/2)\sum_{n=1}^{N}\left(a_n^2 + b_n^2\right)$$

and the RMS value is then defined to be the square root of this quantity or F_N.

Three useful and common Fourier series forms are defined in terms of the following graphs (with $\omega_0 = 2\pi/T$).

Given:

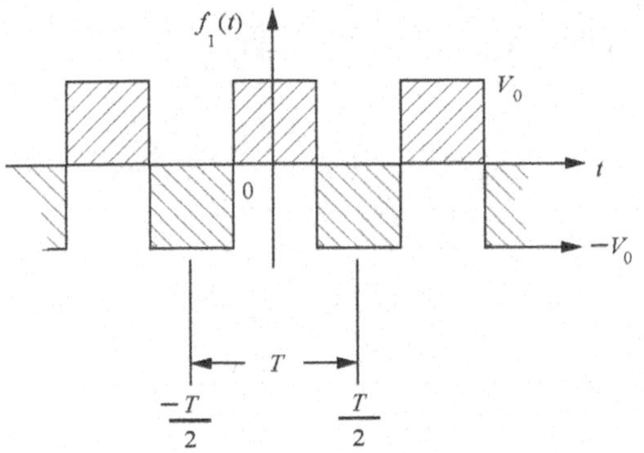

then

$$f_1(t) = \sum_{\substack{n=1 \\ (n\text{ odd})}}^{\infty}(-1)^{(n-1)/2}(4V_0/n\pi)\cos(n\omega_0 t)$$

Given:

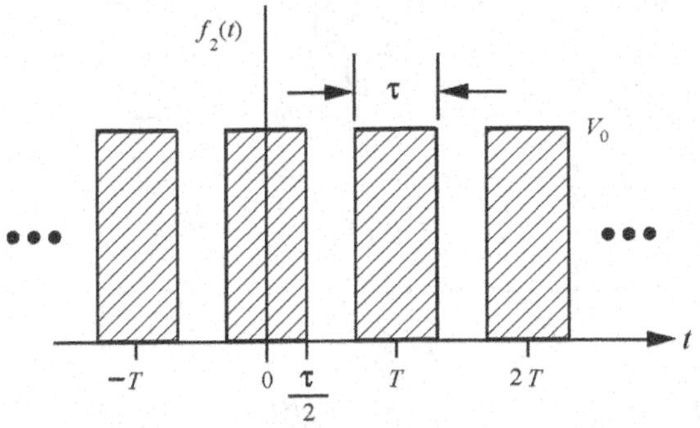

then

$$f_2(t) = \frac{V_0\tau}{T} + \frac{2V_0\tau}{T}\sum_{n=1}^{\infty}\frac{\sin(n\pi\tau/T)}{(n\pi\tau/T)}\cos(n\omega_0 t)$$

$$f_2(t) = \frac{V_0\tau}{T}\sum_{n=-\infty}^{\infty}\frac{\sin(n\pi\tau/T)}{(n\pi\tau/T)}e^{jn\omega_0 t}$$

Given:

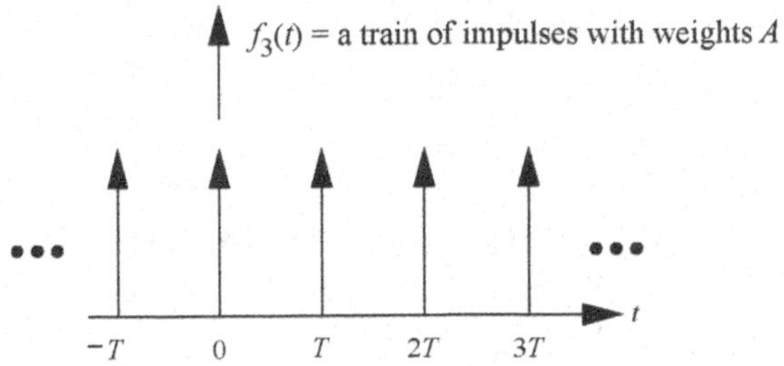

then

$$f_3(t) = \sum_{n=-\infty}^{\infty} A\delta(t - nT)$$

$$f_3(t) = (A/T) + (2A/T)\sum_{n=1}^{\infty}\cos(n\omega_0 t)$$

$$f_3(t) = (A/T)\sum_{n=-\infty}^{\infty} e^{jn\omega_0 t}$$

The Fourier Transform and its Inverse

$$X(f) = \int_{-\infty}^{+\infty} x(t)e^{-j2\pi ft}\,dt$$

$$x(t) = \int_{-\infty}^{+\infty} X(f)e^{j2\pi ft}\,df$$

We say that $x(t)$ and $X(f)$ form a *Fourier transform pair*:

$$x(t) \leftrightarrow X(f)$$

Fourier Transform Pairs

$x(t)$	$X(f)$
1	$\delta(f)$
$\delta(t)$	1
$u(t)$	$\frac{1}{2}\delta(f)+\frac{1}{j2\pi f}$
$\Pi(t/\tau)$	$\tau \operatorname{sinc}(\tau f)$
$\operatorname{sinc}(Bt)$	$\frac{1}{B}\Pi(f/B)$
$\Lambda(t/\tau)$	$\tau \operatorname{sinc}^2(\tau f)$
$e^{-at}u(t)$	$\frac{1}{a+j2\pi f} \quad a>0$
$te^{-at}u(t)$	$\frac{1}{(a+j2\pi f)^2} \quad a>0$
$e^{-a\|t\|}$	$\frac{2a}{a^2+(2\pi f)^2} \quad a>0$
$e^{-(at)^2}$	$\frac{\sqrt{\pi}}{a}e^{-(\pi f/a)^2}$
$\cos(2\pi f_0 t+\theta)$	$\frac{1}{2}[e^{j\theta}\delta(f-f_0)+e^{-j\theta}\delta(f+f_0)]$
$\sin(2\pi f_0 t+\theta)$	$\frac{1}{2j}[e^{j\theta}\delta(f-f_0)-e^{-j\theta}\delta(f+f_0)]$
$\sum_{n=-\infty}^{n=+\infty}\delta(t-nT_s)$	$f_s\sum_{k=-\infty}^{k=+\infty}\delta(f-kf_s) \quad f_s=\frac{1}{T_s}$

Fourier Transform Theorems

Linearity	$ax(t)+by(t)$	$aX(f)+bY(f)$
Scale change	$x(at)$	$\frac{1}{\|a\|}X\left(\frac{f}{a}\right)$
Time reversal	$x(-t)$	$X(-f)$
Duality	$X(t)$	$x(-f)$
Time shift	$x(t-t_0)$	$X(f)e^{-j2\pi f t_0}$
Frequency shift	$x(t)e^{j2\pi f_0 t}$	$X(f-f_0)$
Modulation	$x(t)\cos 2\pi f_0 t$	$\frac{1}{2}X(f-f_0)$ $+\frac{1}{2}X(f+f_0)$
Multiplication	$x(t)y(t)$	$X(f)*Y(f)$
Convolution	$x(t)*y(t)$	$X(f)Y(f)$
Differentiation	$\frac{d^n x(t)}{dt^n}$	$(j2\pi f)^n X(f)$
Integration	$\int_{-\infty}^{t} x(\lambda)d\lambda$	$\frac{1}{j2\pi f}X(f)$ $+\frac{1}{2}X(0)\delta(f)$

where:

$$\operatorname{sinc}(t)=\frac{\sin(\pi t)}{\pi t}$$

$$\Pi(t)=\begin{cases}1, |t|\le \frac{1}{2}\\ 0, \text{otherwise}\end{cases}$$

$$\Lambda(t)=\begin{cases}1-|t|, |t|\le 1\\ 0, \text{otherwise}\end{cases}$$

LAPLACE TRANSFORMS

The unilateral Laplace transform pair

$$F(s)=\int_{0^-}^{\infty} f(t)e^{-st}dt$$

$$f(t)=\frac{1}{2\pi j}\int_{\sigma-j\infty}^{\sigma+j\infty} F(s)e^{st}dt$$

where $s=\sigma+j\omega$

represents a powerful tool for the transient and frequency response of linear time invariant systems. Some useful Laplace transform pairs are:

$f(t)$	$F(s)$
$\delta(t)$, Impulse at $t = 0$	1
$u(t)$, Step at $t = 0$	$1/s$
$t[u(t)]$, Ramp at $t = 0$	$1/s^2$
$e^{-\alpha t}$	$1/(s+\alpha)$
$te^{-\alpha t}$	$1/(s+\alpha)^2$
$e^{-\alpha t}\sin\beta t$	$\beta/[(s+\alpha)^2+\beta^2]$
$e^{-\alpha t}\cos\beta t$	$(s+\alpha)/[(s+\alpha)^2+\beta^2]$
$\frac{d^n f(t)}{dt^n}$	$s^nF(s)-\sum_{m=0}^{n-1}s^{n-m-1}\frac{d^m f(0)}{dt^m}$
$\int_0^t f(\tau)d\tau$	$(1/s)F(s)$
$\int_0^\infty x(t-\tau)h(\tau)d\tau$	$H(s)X(s)$
$f(t-\tau)\,u(t-\tau)$	$e^{-\tau s}F(s)$
$\lim_{t\to\infty} f(t)$	$\lim_{s\to 0} sF(s)$
$\lim_{t\to 0} f(t)$	$\lim_{s\to\infty} sF(s)$

The last two transforms represent the Final Value Theorem (F.V.T.) and Initial Value Theorem (I.V.T.), respectively. It is assumed that the limits exist.

MATRICES

A matrix is an ordered rectangular array of numbers with m rows and n columns. The element a_{ij} refers to row i and column j.

The rank of a matrix is equal to the number of rows that are linearly independent.

Multiplication of Two Matrices

$$A=\begin{bmatrix}A & B\\ C & D\\ E & F\end{bmatrix}\quad A_{3,2} \text{ is a 3-row, 2-column matrix}$$

$$B=\begin{bmatrix}H & I\\ J & K\end{bmatrix}\quad B_{2,2} \text{ is a 2-row, 2-column matrix}$$

In order for multiplication to be possible, the number of columns in A must equal the number of rows in B.

Multiplying matrix B by matrix A occurs as follows:

$$C=\begin{bmatrix}A & B\\ C & D\\ E & F\end{bmatrix}\cdot\begin{bmatrix}H & I\\ J & K\end{bmatrix}$$

$$C=\begin{bmatrix}(A\cdot H+B\cdot J) & (A\cdot I+B\cdot K)\\ (C\cdot H+D\cdot J) & (C\cdot I+D\cdot K)\\ (E\cdot H+F\cdot J) & (E\cdot I+F\cdot K)\end{bmatrix}$$

Matrix multiplication is not commutative.

Addition

$$\begin{bmatrix}A & B & C\\ D & E & F\end{bmatrix}+\begin{bmatrix}G & H & I\\ J & K & L\end{bmatrix}=\begin{bmatrix}A+G & B+H & C+I\\ D+J & E+K & F+L\end{bmatrix}$$

Identity Matrix

The matrix $\mathbf{I}=(a_{ij})$ is a square $n\times n$ matrix with 1's on the diagonal and 0's everywhere else.

Matrix Transpose

Rows become columns. Columns become rows.

$$A=\begin{bmatrix}A & B & C\\ D & E & F\end{bmatrix}\quad A^T=\begin{bmatrix}A & D\\ B & E\\ C & F\end{bmatrix}$$

Inverse $[\;]^{-1}$

The inverse B of a square $n\times n$ matrix A is

$$B=A^{-1}=\frac{\text{adj}(A)}{|A|}, \text{ where}$$

adj(A) = adjoint of A (obtained by replacing A^T elements with their cofactors) and $|A|$= determinant of A.

$\mathbf{[A][A]^{-1}=[A]^{-1}[A]=[I]}$ where $\mathbf{I}$ is the identity matrix.

DETERMINANTS

A *determinant of order n* consists of n^2 numbers, called the *elements* of the determinant, arranged in n rows and n columns and enclosed by two vertical lines.

In any determinant, the *minor* of a given element is the determinant that remains after all of the elements are struck out that lie in the same row and in the same column as the given element. Consider an element which lies in the jth column and the ith row. The *cofactor* of this element is the value of the minor of the element (if $i + j$ is *even*), and it is the negative of the value of the minor of the element (if $i + j$ is *odd*).

If n is greater than 1, the *value* of a determinant of order n is the sum of the n products formed by multiplying each element of some specified row (or column) by its cofactor. This sum is called the *expansion of the determinant* [according to the elements of the specified row (or column)]. For a second-order determinant:

$$\begin{vmatrix} a_1 & a_2 \\ b_1 & b_2 \end{vmatrix} = a_1 b_2 - a_2 b_1$$

For a third-order determinant:

$$\begin{vmatrix} a_1 & a_2 & a_3 \\ b_1 & b_2 & b_3 \\ c_1 & c_2 & c_3 \end{vmatrix} = a_1 b_2 c_3 + a_2 b_3 c_1 + a_3 b_1 c_2 - a_3 b_2 c_1 - a_2 b_1 c_3 - a_1 b_3 c_2$$

VECTORS

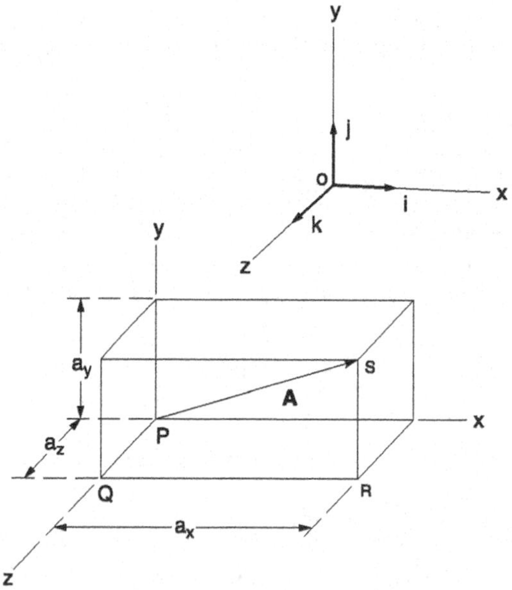

$$\mathbf{A} = a_x\mathbf{i} + a_y\mathbf{j} + a_z\mathbf{k}$$

Addition and *subtraction:*

$$\mathbf{A} + \mathbf{B} = (a_x + b_x)\mathbf{i} + (a_y + b_y)\mathbf{j} + (a_z + b_z)\mathbf{k}$$
$$\mathbf{A} - \mathbf{B} = (a_x - b_x)\mathbf{i} + (a_y - b_y)\mathbf{j} + (a_z - b_z)\mathbf{k}$$

The *dot product* is a *scalar product* and represents the projection of **B** onto **A** times |**A**|. It is given by

$$\mathbf{A}\bullet\mathbf{B} = a_x b_x + a_y b_y + a_z b_z$$
$$= |\mathbf{A}||\mathbf{B}|\cos\theta = \mathbf{B}\bullet\mathbf{A}$$

The *cross product* is a *vector product* of magnitude |**B**| |**A**| sin θ which is perpendicular to the plane containing **A** and **B**. The product is

$$\mathbf{A} \times \mathbf{B} = \begin{vmatrix} \mathbf{i} & \mathbf{j} & \mathbf{k} \\ a_x & a_y & a_z \\ b_x & b_y & b_z \end{vmatrix} = -\mathbf{B} \times \mathbf{A}$$

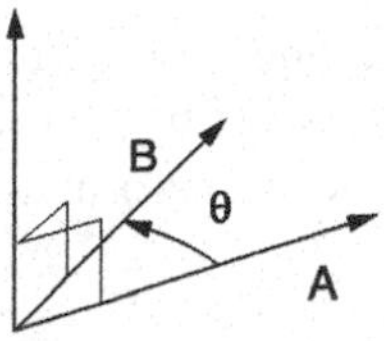

The sense of **A** × **B** is determined by the right-hand rule.

$$\mathbf{A} \times \mathbf{B} = |\mathbf{A}||\mathbf{B}|\mathbf{n}\sin\theta, \text{ where}$$

n = unit vector perpendicular to the plane of **A** and **B**.

Gradient, Divergence, and Curl

$$\nabla\phi = \left(\frac{\partial}{\partial x}\mathbf{i} + \frac{\partial}{\partial y}\mathbf{j} + \frac{\partial}{\partial z}\mathbf{k}\right)\phi$$

$$\nabla \bullet \mathbf{V} = \left(\frac{\partial}{\partial x}\mathbf{i} + \frac{\partial}{\partial y}\mathbf{j} + \frac{\partial}{\partial z}\mathbf{k}\right) \bullet (V_1\mathbf{i} + V_2\mathbf{j} + V_3\mathbf{k})$$

$$\nabla \times \mathbf{V} = \left(\frac{\partial}{\partial x}\mathbf{i} + \frac{\partial}{\partial y}\mathbf{j} + \frac{\partial}{\partial z}\mathbf{k}\right) \times (V_1\mathbf{i} + V_2\mathbf{j} + V_3\mathbf{k})$$

The Laplacian of a scalar function ϕ is

$$\nabla^2\phi = \frac{\partial^2\phi}{\partial x^2} + \frac{\partial^2\phi}{\partial y^2} + \frac{\partial^2\phi}{\partial z^2}$$

Identities

$\mathbf{A} \bullet \mathbf{B} = \mathbf{B} \bullet \mathbf{A}$; $\mathbf{A} \bullet (\mathbf{B} + \mathbf{C}) = \mathbf{A} \bullet \mathbf{B} + \mathbf{A} \bullet \mathbf{C}$

$\mathbf{A} \bullet \mathbf{A} = |\mathbf{A}|^2$

$\mathbf{i} \bullet \mathbf{i} = \mathbf{j} \bullet \mathbf{j} = \mathbf{k} \bullet \mathbf{k} = 1$

$\mathbf{i} \bullet \mathbf{j} = \mathbf{j} \bullet \mathbf{k} = \mathbf{k} \bullet \mathbf{i} = 0$

If $\mathbf{A} \bullet \mathbf{B} = 0$, then either $\mathbf{A} = 0$, $\mathbf{B} = 0$, or **A** is perpendicular to **B**.

$\mathbf{A} \times \mathbf{B} = -\mathbf{B} \times \mathbf{A}$

$\mathbf{A} \times (\mathbf{B} + \mathbf{C}) = (\mathbf{A} \times \mathbf{B}) + (\mathbf{A} \times \mathbf{C})$

$(\mathbf{B} + \mathbf{C}) \times \mathbf{A} = (\mathbf{B} \times \mathbf{A}) + (\mathbf{C} \times \mathbf{A})$

$\mathbf{i} \times \mathbf{i} = \mathbf{j} \times \mathbf{j} = \mathbf{k} \times \mathbf{k} = \mathbf{0}$

$\mathbf{i} \times \mathbf{j} = \mathbf{k} = -\mathbf{j} \times \mathbf{i}$; $\mathbf{j} \times \mathbf{k} = \mathbf{i} = -\mathbf{k} \times \mathbf{j}$

$\mathbf{k} \times \mathbf{i} = \mathbf{j} = -\mathbf{i} \times \mathbf{k}$

If $\mathbf{A} \times \mathbf{B} = \mathbf{0}$, then either $\mathbf{A} = \mathbf{0}$, $\mathbf{B} = \mathbf{0}$, or **A** is parallel to **B**.

$\nabla^2\phi = \nabla \bullet (\nabla\phi) = (\nabla \bullet \nabla)\phi$

$\nabla \times \nabla\phi = \mathbf{0}$

$\nabla \bullet (\nabla \times \mathbf{A}) = \mathbf{0}$

$\nabla \times (\nabla \times \mathbf{A}) = \nabla(\nabla \bullet \mathbf{A}) - \nabla^2\mathbf{A}$

DIFFERENCE EQUATIONS

Any system whose input $v(t)$ and output $y(t)$ are defined only at the equally spaced intervals

$$f(t) = y' = \frac{y_{i+1} - y_i}{t_{i+1} - t_i}$$

can be described by a difference equation.

First-Order Linear Difference Equation

$\Delta t = t_{i+1} - t_i$

$y_{i+1} = y_i + y'(\Delta t)$

NUMERICAL METHODS

Newton's Method for Root Extraction

Given a function $f(x)$ which has a simple root of $f(x) = 0$ at $x = a$ an important computational task would be to find that root. If $f(x)$ has a continuous first derivative then the $(j+1)$st estimate of the root is

$$a^{j+1} = a^j - \left.\frac{f(x)}{\frac{df(x)}{dx}}\right|_{x = a^j}$$

The initial estimate of the root a^0 must be near enough to the actual root to cause the algorithm to converge to the root.

Newton's Method of Minimization

Given a scalar value function

$$h(\mathbf{x}) = h(x_1, x_2, \ldots, x_n)$$

find a vector $\mathbf{x}^* \in R_n$ such that

$$h(\mathbf{x}^*) \le h(\mathbf{x}) \text{ for all } \mathbf{x}$$

Newton's algorithm is

$$\mathbf{x}_{k+1} = \mathbf{x}_k - \left(\left.\frac{\partial^2 h}{\partial \mathbf{x}^2}\right|_{\mathbf{x} = \mathbf{x}_k}\right)^{-1} \left.\frac{\partial h}{\partial \mathbf{x}}\right|_{\mathbf{x} = \mathbf{x}_k}, \text{ where}$$

$$\frac{\partial h}{\partial \mathbf{x}} = \begin{bmatrix} \frac{\partial h}{\partial x_1} \\ \frac{\partial h}{\partial x_2} \\ \ldots \\ \ldots \\ \frac{\partial h}{\partial x_n} \end{bmatrix}$$

and

$$\frac{\partial^2 h}{\partial \mathbf{x}^2} = \begin{bmatrix} \frac{\partial^2 h}{\partial x_1^2} & \frac{\partial^2 h}{\partial x_1 \partial x_2} & \ldots\ \ldots & \frac{\partial^2 h}{\partial x_1 \partial x_n} \\ \frac{\partial^2 h}{\partial x_1 \partial x_2} & \frac{\partial^2 h}{\partial x_2^2} & \ldots\ \ldots & \frac{\partial^2 h}{\partial x_2 \partial x_n} \\ \ldots & \ldots & \ldots\ \ldots & \ldots \\ \ldots & \ldots & \ldots\ \ldots & \ldots \\ \frac{\partial^2 h}{\partial x_1 \partial x_n} & \frac{\partial^2 h}{\partial x_2 \partial x_n} & \ldots\ \ldots & \frac{\partial^2 h}{\partial x_n^2} \end{bmatrix}$$

Numerical Integration

Three of the more common numerical integration algorithms used to evaluate the integral

$$\int_a^b f(x)\,dx$$

are:

Euler's or Forward Rectangular Rule

$$\int_a^b f(x)\,dx \approx \Delta x \sum_{k=0}^{n-1} f(a + k\Delta x)$$

Trapezoidal Rule

for $n = 1$

$$\int_a^b f(x)\,dx \approx \Delta x\left[\frac{f(a) + f(b)}{2}\right]$$

for $n > 1$

$$\int_a^b f(x)\,dx \approx \frac{\Delta x}{2}\left[f(a) + 2\sum_{k=1}^{n-1} f(a + k\Delta x) + f(b)\right]$$

Simpson's Rule/Parabolic Rule (n must be an even integer)

for $n = 2$

$$\int_a^b f(x)\,dx \approx \left(\frac{b-a}{6}\right)\left[f(a) + 4f\left(\frac{a+b}{2}\right) + f(b)\right]$$

for $n \ge 4$

$$\int_a^b f(x)\,dx \approx \frac{\Delta x}{3}\begin{bmatrix} f(a) + 2\sum\limits_{k=2,4,6,\ldots}^{n-2} f(a + k\Delta x) \\ +4\sum\limits_{k=1,3,5,\ldots}^{n-1} f(a + k\Delta x) + f(b) \end{bmatrix}$$

with $\Delta x = (b - a)/n$

n = number of intervals between data points

Numerical Solution of Ordinary Differential Equations

Euler's Approximation

Given a differential equation

$$dx/dt = f(x, t) \text{ with } x(0) = x_o$$

At some general time $k\Delta t$

$$x[(k+1)\Delta t] \cong x(k\Delta t) + \Delta t f[x(k\Delta t), k\Delta t]$$

which can be used with starting condition x_o to solve recursively for $x(\Delta t), x(2\Delta t), \ldots, x(n\Delta t)$.

The method can be extended to nth order differential equations by recasting them as n first-order equations.

In particular, when $dx/dt = f(x)$

$$x[(k+1)\Delta t] \cong x(k\Delta t) + \Delta t f[x(k\Delta t)]$$

which can be expressed as the recursive equation

$$x_{k+1} = x_k + \Delta t\,(dx_k/dt)$$

$$x_{k+1} = x + \Delta t\,[f(x(k), t(k))]$$

ENGINEERING PROBABILITY AND STATISTICS

DISPERSION, MEAN, MEDIAN, AND MODE VALUES

If $X_1, X_2, \ldots, X_n$ represent the values of a random sample of n items or observations, the *arithmetic mean* of these items or observations, denoted $\overline{X}$, is defined as

$$\overline{X} = (1/n)(X_1 + X_2 + \ldots + X_n) = (1/n)\sum_{i=1}^{n} X_i$$

$\overline{X} \rightarrow \mu$ for sufficiently large values of n.

The *weighted arithmetic mean* is

$$\overline{X}_w = \frac{\sum w_i X_i}{\sum w_i}, \text{ where}$$

X_i = the value of the *i*th observation, and
w_i = the weight applied to X_i.

The *variance* of the population is the *arithmetic mean* of the *squared deviations from the population mean*. If μ is the arithmetic mean of a discrete population of size N, the *population variance* is defined by

$$\sigma^2 = (1/N)\left[(X_1 - \mu)^2 + (X_2 - \mu)^2 + \ldots + (X_N - \mu)^2\right]$$

$$= (1/N)\sum_{i=1}^{N}(X_i - \mu)^2$$

Standard deviation formulas are

$$\sigma_{\text{population}} = \sqrt{(1/N)\Sigma(X_i - \mu)^2}$$

$$\sigma_{\text{sum}} = \sqrt{\sigma_1^2 + \sigma_2^2 + \ldots + \sigma_n^2}$$

$$\sigma_{\text{series}} = \sigma\sqrt{n}$$

$$\sigma_{\text{mean}} = \frac{\sigma}{\sqrt{n}}$$

$$\sigma_{\text{product}} = \sqrt{A^2\sigma_b^2 + B^2\sigma_a^2}$$

The *sample variance* is

$$s^2 = [1/(n-1)]\sum_{i=1}^{n}(X_i - \overline{X})^2$$

The *sample standard deviation* is

$$s = \sqrt{[1/(n-1)]\sum_{i=1}^{n}(X_i - \overline{X})^2}$$

The *sample coefficient of variation* = $CV = s/\overline{X}$

The *sample geometric mean* = $\sqrt[n]{X_1 X_2 X_3 \ldots X_n}$

The *sample root-mean-square value* = $\sqrt{(1/n)\Sigma X_i^2}$

When the discrete data are rearranged in increasing order and n is odd, the median is the value of the $\left(\frac{n+1}{2}\right)^{\text{th}}$ item

When n is even, the median is the average of the $\left(\frac{n}{2}\right)^{\text{th}}$ and $\left(\frac{n}{2} + 1\right)^{\text{th}}$ items.

The *mode* of a set of data is the value that occurs with greatest frequency.

The *sample range R* is the largest sample value minus the smallest sample value.

PERMUTATIONS AND COMBINATIONS

A *permutation* is a particular sequence of a given set of objects. A *combination* is the set itself without reference to order.

1. The number of different *permutations* of *n* distinct objects *taken r at a time* is

$$P(n,r) = \frac{n!}{(n-r)!}$$

   nPr is an alternative notation for $P(n,r)$

2. The number of different *combinations* of *n* distinct objects *taken r at a time* is

$$C(n,r) = \frac{P(n,r)}{r!} = \frac{n!}{[r!(n-r)!]}$$

   nCr and $\binom{n}{r}$ are alternative notations for $C(n,r)$

3. The number of different *permutations* of *n* objects *taken n at a time*, given that n_i are of type i, where $i = 1, 2, \ldots, k$ and $\Sigma n_i = n$, is

$$P(n; n_1, n_2, \ldots, n_k) = \frac{n!}{n_1!n_2!\ldots n_k!}$$

SETS

De Morgan's Law

$$\overline{A \cup B} = \overline{A} \cap \overline{B}$$

$$\overline{A \cap B} = \overline{A} \cup \overline{B}$$

Associative Law

$$A \cup (B \cup C) = (A \cup B) \cup C$$

$$A \cap (B \cap C) = (A \cap B) \cap C$$

Distributive Law

$$A \cup (B \cap C) = (A \cup B) \cap (A \cup C)$$

$$A \cap (B \cup C) = (A \cap B) \cup (A \cap C)$$

LAWS OF PROBABILITY

Property 1. General Character of Probability

The probability $P(E)$ of an event E is a real number in the range of 0 to 1. The probability of an impossible event is 0 and that of an event certain to occur is 1.

Property 2. Law of Total Probability

$$P(A + B) = P(A) + P(B) - P(A, B), \text{ where}$$

$P(A + B)$ = the probability that either A or B occur alone or that both occur together

$P(A)$ = the probability that A occurs

$P(B)$ = the probability that B occurs

$P(A, B)$ = the probability that both A and B occur simultaneously

Property 3. Law of Compound or Joint Probability

If neither $P(A)$ nor $P(B)$ is zero,

$$P(A, B) = P(A)P(B \mid A) = P(B)P(A \mid B), \text{ where}$$

$P(B \mid A) =$ the probability that B occurs given the fact that A has occurred

$P(A \mid B) =$ the probability that A occurs given the fact that B has occurred

If either $P(A)$ or $P(B)$ is zero, then $P(A, B) = 0$.

Bayes' Theorem

$$P(B_j|A) = \frac{P(B_j)P(A|B_j)}{\sum_{i=1}^{n} P(A|B_i)P(B_i)}$$

where $P(A_j)$ is the probability of event A_j within the population of A

$P(B_j)$ is the probability of event B_j within the population of B

PROBABILITY FUNCTIONS, DISTRIBUTIONS, AND EXPECTED VALUES

A random variable X has a probability associated with each of its possible values. The probability is termed a discrete probability if X can assume only discrete values, or

$$X = x_1, x_2, x_3, \ldots, x_n$$

The *discrete probability* of any single event, $X = x_i$, occurring is defined as $P(x_i)$ while the *probability mass function* of the random variable X is defined by

$$f(x_k) = P(X = x_k), k = 1, 2, \ldots, n$$

Probability Density Function

If X is continuous, the *probability density function, f,* is defined such that

$$P(a \le X \le b) = \int_a^b f(x)dx$$

Cumulative Distribution Functions

The *cumulative distribution function, F,* of a discrete random variable X that has a probability distribution described by $P(x_i)$ is defined as

$$F(x_m) = \sum_{k=1}^{m} P(x_k) = P(X \le x_m), m = 1, 2, \ldots, n$$

If X is continuous, the *cumulative distribution function, F,* is defined by

$$F(x) = \int_{-\infty}^{x} f(t)dt$$

which implies that $F(a)$ is the probability that $X \le a$.

Expected Values

Let X be a discrete random variable having a probability mass function

$$f(x_k), k = 1, 2, \ldots, n$$

The expected value of X is defined as

$$\mu = E[X] = \sum_{k=1}^{n} x_k f(x_k)$$

The variance of X is defined as

$$\sigma^2 = V[X] = \sum_{k=1}^{n} (x_k - \mu)^2 f(x_k)$$

Let X be a continuous random variable having a density function $f(X)$ and let $Y = g(X)$ be some general function. The expected value of Y is:

$$E[Y] = E[g(X)] = \int_{-\infty}^{\infty} g(x)f(x)dx$$

The mean or expected value of the random variable X is now defined as

$$\mu = E[X] = \int_{-\infty}^{\infty} xf(x)dx$$

while the variance is given by

$$\sigma^2 = V[X] = E[(X - \mu)^2] = \int_{-\infty}^{\infty} (x - \mu)^2 f(x)dx$$

The standard deviation is given by

$$\sigma = \sqrt{V[X]}$$

The coefficient of variation is defined as σ/μ.

Combinations of Random Variables

$$Y = a_1X_1 + a_2X_2 + \ldots + a_nX_n$$

The expected value of Y is:

$$\mu_y = E(Y) = a_1E(X_1) + a_2E(X_2) + \ldots + a_nE(X_n)$$

If the random variables are statistically *independent*, then the variance of Y is:

$$\sigma_y^2 = V(Y) = a_1^2V(X_1) + a_2^2V(X_2) + \ldots + a_n^2V(X_n)$$
$$= a_1^2\sigma_1^2 + a_2^2\sigma_2^2 + \ldots + a_n^2\sigma_n^2$$

Also, the standard deviation of Y is:

$$\sigma_y = \sqrt{\sigma_y^2}$$

When $Y = f(X_1, X_2, \ldots, X_n)$ and X_i are independent, the standard deviation of Y is expressed as:

$$\sigma_y = \sqrt{\left(\frac{\partial f}{\partial X_1}\sigma_{X_1}\right)^2 + \left(\frac{\partial f}{\partial X_2}\sigma_{X_2}\right)^2 + \ldots + \left(\frac{\partial f}{\partial X_n}\sigma_{X_n}\right)^2}$$

Binomial Distribution

$P(x)$ is the probability that x successes will occur in n trials. If p = probability of success and q = probability of failure = $1 - p$, then

$$P_n(x) = C(n,x)p^x q^{n-x} = \frac{n!}{x!(n-x)!}p^x q^{n-x},$$

where

x = 0, 1, 2, ..., n

$C(n, x)$ = the number of combinations

n, p = parameters

The variance is given by the form:

$$\sigma^2 = npq$$

Normal Distribution (Gaussian Distribution)

This is a unimodal distribution, the mode being $x = \mu$, with two points of inflection (each located at a distance σ to either side of the mode). The averages of n observations tend to become normally distributed as n increases. The variate x is said to be normally distributed if its density function $f(x)$ is given by an expression of the form

$$f(x) = \frac{1}{\sigma\sqrt{2\pi}}e^{-\frac{1}{2}\left(\frac{x-\mu}{\sigma}\right)^2}, \text{ where}$$

μ = the population mean

σ = the standard deviation of the population

$-\infty \le x \le \infty$

When $\mu = 0$ and $\sigma^2 = \sigma = 1$, the distribution is called a *standardized* or *unit normal* distribution. Then

$$f(x) = \frac{1}{\sqrt{2\pi}}e^{-x^2/2}, \text{ where } -\infty \le x \le \infty.$$

It is noted that $Z = \frac{x-\mu}{\sigma}$ follows a standardized normal distribution function.

A unit normal distribution table is included at the end of this section. In the table, the following notations are utilized:

$F(x)$ = the area under the curve from $-\infty$ to x

$R(x)$ = the area under the curve from x to ∞

$W(x)$ = the area under the curve between $-x$ and x

$F(-x) = 1 - F(x)$

The Central Limit Theorem

Let $X_1, X_2, ..., X_n$ be a sequence of independent and identically distributed random variables each having mean μ and variance σ^2. Then for large n, the Central Limit Theorem asserts that the sum

$Y = X_1 + X_2 + ... X_n$ is approximately normal.

$$\mu_{\bar{y}} = \mu$$

and the standard deviation

$$\sigma_{\bar{y}} = \frac{\sigma}{\sqrt{n}}$$

***t*-Distribution**

Student's t-distribution has the probability density function given by:

$$f(t) = \frac{\Gamma\left(\frac{v+1}{2}\right)}{\sqrt{v\pi}\,\Gamma\left(\frac{v}{2}\right)}\left(1 + \frac{t^2}{v}\right)^{-\frac{v+1}{2}}$$

where

v = number of *degrees of freedom*

n = sample size

$v = n - 1$

Γ = gamma function

$$t = \frac{\bar{x} - \mu}{s/\sqrt{n}}$$

$-\infty \le t \le \infty$

A table later in this section gives the values of $t_{\alpha,v}$ for values of α and v. Note that, in view of the symmetry of the t-distribution, $t_{1-\alpha,v} = -t_{\alpha,v}$

The function for α follows:

$$\alpha = \int_{t_{\alpha,v}}^{\infty} f(t)\,dt$$

χ^2 - Distribution

If $Z_1, Z_2, ..., Z_n$ are independent unit normal random variables, then

$$\chi^2 = Z_1^2 + Z_2^2 + \ldots + Z_n^2$$

is said to have a chi-square distribution with n degrees of freedom.

A table at the end of this section gives values of $\chi^2_{\alpha,n}$ for selected values of α and n.

Gamma Function

$$\Gamma(n) = \int_0^{\infty} t^{n-1}e^{-t}dt,\ n > 0$$

PROPAGATION OF ERROR

Measurement Error

$$x = x_{true} + x_{bias} + x_{re}, \text{ where}$$

x = measured value for dimension

x_{true} = true value for dimension

x_{bias} = bias or systematic error in measuring dimension

x_{re} = random error in measuring dimension

$$\mu = \left(\frac{1}{N}\right)\sum_{i=1}^{n} x_i = \left(\frac{1}{N}\right)\sum_{i=1}^{n}(x_{true} + x_{bias})_i, \text{ where}$$

μ = population or arithmetic mean

N = number of observations of measured values for population

$$\sigma_{population} = \sigma_{re} = \sqrt{\left(\frac{1}{N}\Sigma(x_i - \mu)^2\right)}, \text{ where}$$

σ = standard deviation of uncertainty

Linear Combinations

$$\sigma_{cx} = |c|\sigma_x, \text{ where}$$

c = constant

$$\sigma_{c_1x_1+\ldots c_nx_n} = \sqrt{c_1^2\sigma_1^2 + \ldots c_n^2\sigma_{x_n}^2}, \text{ where}$$

n = number of observations of measured values for sample

Measurement Uncertainty

$$\sigma_u = \sqrt{\left(\frac{\partial u}{\partial x_1}\right)^2\sigma_{x_1}^2 + \left(\frac{\partial u}{\partial x_2}\right)^2\sigma_{x_2}^2 + \ldots + \left(\frac{\partial u}{\partial x_n}\right)^2\sigma_{x_n}^2}$$

$$u(x_1, x_2, \ldots x_n) = f(x_1, x_2, \ldots x_n)$$

where

$f(x_1, x_2, \ldots x_n)$ = the functional relationship between the desired state u and the measured or available states x_i

σ_{x_i} = standard deviation of the state x_i

σ_u = computed standard deviation of u with multiple states x_i

When the state variable x_1 is to be transformed to $u = f(x_1)$ the following relation holds:

$$\sigma_u \approx \left|\frac{du}{dx_1}\right|\sigma_{x_1}$$

LINEAR REGRESSION AND GOODNESS OF FIT

Least Squares

$$\hat{y} = \hat{a} + \hat{b}x, \text{ where}$$

$$\hat{b} = S_{xy}/S_{xx}$$

$$\hat{a} = \bar{y} - \hat{b}\bar{x}$$

$$S_{xy} = \sum_{i=1}^{n} x_i y_i - (1/n)\left(\sum_{i=1}^{n} x_i\right)\left(\sum_{i=1}^{n} y_i\right)$$

$$S_{xx} = \sum_{i=1}^{n} x_i^2 - (1/n)\left(\sum_{i=1}^{n} x_i\right)^2$$

$$\bar{y} = (1/n)\left(\sum_{i=1}^{n} y_i\right)$$

$$\bar{x} = (1/n)\left(\sum_{i=1}^{n} x_i\right)$$

where

n = sample size

S_{xx} = sum of squares of x

S_{yy} = sum of squares of y

S_{xy} = sum of x-y products

Residual

$$e_i = y_i - \hat{y} = y_i - (\hat{a} + \hat{b}x_i)$$

Standard Error of Estimate (S_e^2):

$$S_e^2 = \frac{S_{xx}S_{yy} - S_{xy}^2}{S_{xx}(n-2)} = MSE, \text{ where}$$

$$S_{yy} = \sum_{i=1}^{n} y_i^2 - (1/n)\left(\sum_{i=1}^{n} y_i\right)^2$$

Confidence Interval for Intercept $(\hat{a})$:

$$\hat{a} \pm t_{\alpha/2,n-2}\sqrt{\left(\frac{1}{n} + \frac{\bar{x}^2}{S_{xx}}\right)MSE}$$

Confidence Interval for Slope $(\hat{b})$:

$$\hat{b} \pm t_{\alpha/2,n-2}\sqrt{\frac{MSE}{S_{xx}}}$$

Sample Correlation Coefficient (R) and Coefficient of Determination (R^2):

$$R = \frac{S_{xy}}{\sqrt{S_{xx}S_{yy}}}$$

$$R^2 = \frac{S_{xy}^2}{S_{xx}S_{yy}}$$

HYPOTHESIS TESTING

Let a "dot" subscript indicate summation over the subscript. Thus:

$$y_{i\bullet} = \sum_{j=1}^{n} y_{ij} \quad \text{and} \quad y_{\bullet\bullet} = \sum_{i=1}^{a}\sum_{j=1}^{n} y_{ij}$$

♦ One-Way Analysis of Variance (ANOVA)

Given independent random samples of size n_i from k populations, then:

$$\sum_{i=1}^{k}\sum_{j=1}^{n_i}\left(y_{ij} - \bar{y}_{\bullet\bullet}\right)^2 = \sum_{i=1}^{k} n_i\left(\bar{y}_{i\bullet} - \bar{y}_{\bullet\bullet}\right)^2 + \sum_{i=1}^{k}\sum_{i=1}^{n_i}\left(y_{ij} - \bar{y}_{i\bullet}\right)^2$$

$$SS_{\text{total}} = SS_{\text{treatments}} + SS_{\text{error}}$$

If N = total number observations

$$N = \sum_{i=1}^{k} n_i, \textit{ then}$$

$$SS_{\text{total}} = \sum_{i=1}^{k}\sum_{j=1}^{n_i} y_{ij}^2 - \frac{y_{\bullet\bullet}^2}{N}$$

$$SS_{\text{treatments}} = \sum_{i=1}^{k}\frac{y_{i\bullet}^2}{n_i} - \frac{y_{\bullet\bullet}^2}{N}$$

$$SS_{\text{error}} = SS_{\text{total}} - SS_{\text{treatments}}$$

♦ Randomized Complete Block Design

For k treatments and b blocks

$$\sum_{i=1}^{k}\sum_{j=1}^{b}\left(y_{ij} - \bar{y}_{\bullet\bullet}\right)^2 = b\sum_{i=1}^{k}\left(\bar{y}_{i\bullet} - \bar{y}_{\bullet\bullet}\right)^2 + k\sum_{j=1}^{b}\left(\bar{y}_{\bullet j} - \bar{y}_{\bullet\bullet}\right)^2 + \sum_{i=1}^{k}\sum_{j=1}^{b}\left(\bar{y}_{ij} - \bar{y}_{\bullet j} - \bar{y}_{i\bullet} + \bar{y}_{\bullet\bullet}\right)^2$$

$$SS_{\text{total}} = SS_{\text{treatments}} + SS_{\text{blocks}} + SS_{\text{error}}$$

$$SS_{\text{total}} = \sum_{i=1}^{k}\sum_{j=1}^{b} y_{ij}^2 - \frac{y_{\bullet\bullet}^2}{kb}$$

$$SS_{\text{treatments}} = \frac{1}{b}\sum_{i=1}^{k} y_{i\bullet}^2 - \frac{y_{\bullet\bullet}^2}{bk}$$

$$SS_{\text{blocks}} = \frac{1}{k}\sum_{j=1}^{b} y_{\bullet j}^2 - \frac{y_{\bullet\bullet}^2}{bk}$$

$$SS_{\text{error}} = SS_{\text{total}} - SS_{\text{treatments}} - SS_{\text{blocks}}$$

♦ From Montgomery, Douglas C., and George C. Runger, *Applied Statistics and Probability for Engineers*, 4th ed., Wiley, 2007.

Two-factor Factorial Designs

For a levels of Factor A, b levels of Factor B, and n repetitions per cell:

$$\sum_{i=1}^{a}\sum_{j=1}^{b}\sum_{k=1}^{n}\left(y_{ijk}-\bar{y}_{\ldots}\right)^2 = bn\sum_{i=1}^{a}\left(\bar{y}_{i\cdot\cdot}-\bar{y}_{\ldots}\right)^2 + an\sum_{j=1}^{b}\left(\bar{y}_{\cdot j\cdot}-\bar{y}_{\ldots}\right)^2$$

$$+ n\sum_{i=1}^{a}\sum_{j=1}^{b}\left(\bar{y}_{ij\cdot}-\bar{y}_{i\cdot\cdot}-\bar{y}_{\cdot j\cdot}+\bar{y}_{\ldots}\right)^2 + \sum_{i=1}^{a}\sum_{j=1}^{b}\sum_{k=1}^{n}\left(y_{ijk}-\bar{y}_{ij\cdot}\right)^2$$

$$SS_{\text{total}} = SS_{\text{A}} + SS_{\text{B}} + SS_{\text{AB}} + SS_{\text{error}}$$

$$SS_{\text{total}} = \sum_{i=1}^{a}\sum_{j=1}^{b}\sum_{k=1}^{n} y_{ijk}^2 - \frac{y_{\ldots}^2}{abn}$$

$$SS_{\text{A}} = \sum_{i=1}^{a}\frac{y_{i\cdot\cdot}^2}{bn} - \frac{y_{\ldots}^2}{abn}$$

$$SS_{\text{B}} = \sum_{j=1}^{b}\frac{y_{\cdot j\cdot}^2}{an} - \frac{y_{\ldots}^2}{abn}$$

$$SS_{\text{AB}} = \sum_{i=1}^{a}\sum_{j=1}^{b}\frac{y_{ij\cdot}^2}{n} - \frac{y_{\ldots}^2}{abn} - SS_{\text{A}} - SS_{\text{B}}$$

$$SS_{\text{error}} = SS_{\text{T}} - SS_{\text{A}} - SS_{\text{B}} - SS_{\text{AB}}$$

From Montgomery, Douglas C., and George C. Runger, *Applied Statistics and Probability for Engineers*, 4th ed., Wiley, 2007.

One-Way ANOVA Table

Source of Variation	Degrees of Freedom	Sum of Squares	Mean Square	F
Between Treatments	$k-1$	$SS_{\text{treatments}}$	$MST = \dfrac{SS_{\text{treatments}}}{k-1}$	$\dfrac{MST}{MSE}$
Error	$N-k$	SS_{error}	$MSE = \dfrac{SS_{\text{error}}}{N-k}$	
Total	$N-1$	SS_{total}		

Randomized Complete Block ANOVA Table

Source of Variation	Degrees of Freedom	Sum of Squares	Mean Square	F
Between Treatments	$k-1$	$SS_{\text{treatments}}$	$MST = \dfrac{SS_{\text{treatments}}}{k-1}$	$\dfrac{MST}{MSE}$
Between Blocks	$n-1$	SS_{blocks}	$MSB = \dfrac{SS_{\text{blocks}}}{n-1}$	$\dfrac{MSB}{MSE}$
Error	$(k-1)(n-1)$	SS_{error}	$MSE = \dfrac{SS_{\text{error}}}{(k-1)(n-1)}$	
Total	$N-1$	SS_{total}		

Two-Way Factorial ANOVA Table

Source of Variation	Degrees of Freedom	Sum of Squares	Mean Square	F
A Treatments	$a-1$	SS_A	$MSA = \frac{SS_A}{a-1}$	$\frac{MSA}{MSE}$
B Treatments	$b-1$	SS_B	$MSB = \frac{SS_B}{b-1}$	$\frac{MSB}{MSE}$
AB Interaction	$(a-1)(b-1)$	SS_{AB}	$MSAB = \frac{SS_{AB}}{(a-1)(b-1)}$	$\frac{MSAB}{MSE}$
Error	$ab(n-1)$	SS_{error}	$MSE = \frac{SS_E}{ab(n-1)}$	
Total	$abn-1$	SS_{total}		

Consider an unknown parameter θ of a statistical distribution. Let the null hypothesis be

H_0: $\mu = \mu_0$

and let the alternative hypothesis be

H_1: $\mu \neq \mu_0$

Rejecting H_0 when it is true is known as a type I error, while accepting H_0 when it is wrong is known as a type II error. Furthermore, the probabilities of type I and type II errors are usually represented by the symbols α and β, respectively:

α = probability (type I error)

β = probability (type II error)

The probability of a type I error is known as the level of significance of the test.

Table A. Tests on Means of Normal Distribution—Variance Known

Hypothesis	*Test Statistic*	*Criteria for Rejection*
H_0: $\mu = \mu_0$ H_1: $\mu \neq \mu_0$		$\lvert Z_0 \rvert > Z_{\alpha/2}$
H_0: $\mu = \mu_0$ H_1: $\mu < \mu_0$	$Z_0 \equiv \frac{\bar{X} - \mu_0}{\sigma/\sqrt{n}}$	$Z_0 < -Z_\alpha$
H_0: $\mu = \mu_0$ H_1: $\mu > \mu_0$		$Z_0 > Z_\alpha$
H_0: $\mu_1 - \mu_2 = \gamma$ H_1: $\mu_1 - \mu_2 \neq \gamma$		$\lvert Z_0 \rvert > Z_{\alpha/2}$
H_0: $\mu_1 - \mu_2 = \gamma$ H_1: $\mu_1 - \mu_2 < \gamma$	$Z_0 \equiv \frac{\bar{X}_1 - \bar{X}_2 - \gamma}{\sqrt{\frac{\sigma_1^2}{n_1} + \frac{\sigma_2^2}{n_2}}}$	$Z_0 < -Z_\alpha$
H_0: $\mu_1 - \mu_2 = \gamma$ H_1: $\mu_1 - \mu_2 > \gamma$		$Z_0 > Z_\alpha$

Table B. Tests on Means of Normal Distribution—Variance Unknown

Hypothesis	*Test Statistic*	*Criteria for Rejection*
H_0: $\mu = \mu_0$ H_1: $\mu \neq \mu_0$		$\lvert t_0 \rvert > t_{\alpha/2,\, n-1}$
H_0: $\mu = \mu_0$ H_1: $\mu < \mu_0$	$t_0 = \dfrac{\overline{X} - \mu_0}{s/\sqrt{n}}$	$t_0 < -t_{\alpha,\, n-1}$
H_0: $\mu = \mu_0$ H_1: $\mu > \mu_0$		$t_0 > t_{\alpha,\, n-1}$
H_0: $\mu_1 - \mu_2 = \gamma$ H_1: $\mu_1 - \mu_2 \neq \gamma$	Variances equal: $t_0 = \dfrac{\overline{X_1} - \overline{X_2} - \gamma}{s_p \sqrt{\dfrac{1}{n_1} + \dfrac{1}{n_2}}}$ $v = n_1 + n_2 - 2$	$\lvert t_0 \rvert > t_{\alpha/2,\, v}$
H_0: $\mu_1 - \mu_2 = \gamma$ H_1: $\mu_1 - \mu_2 < \gamma$	Variances unequal: $t_0 = \dfrac{\overline{X_1} - \overline{X_2} - \gamma}{\sqrt{\dfrac{s_1^2}{n_1} + \dfrac{s_2^2}{n_2}}}$	$t_0 < -t_{\alpha,\, v}$
H_0: $\mu_1 - \mu_2 = \gamma$ H_1: $\mu_1 - \mu_2 > \gamma$	$v = \dfrac{\left(\dfrac{s_1^2}{n_1} + \dfrac{s_2^2}{n_2}\right)^2}{\dfrac{\left(s_1^2/n_1\right)^2}{n_1 - 1} + \dfrac{\left(s_2^2/n_2\right)^2}{n_2 - 1}}$	$t_0 > t_{\alpha,\, v}$

In Table B, $s_p^2 = [(n_1 - 1)s_1^2 + (n_2 - 1)s_2^2]/v$

Table C. Tests on Variances of Normal Distribution with Unknown Mean

Hypothesis	*Test Statistic*	*Criteria for Rejection*
H_0: $\sigma^2 = \sigma_0^2$ H_1: $\sigma^2 \neq \sigma_0^2$		$\chi_0^2 > \chi^2_{\alpha/2,\, n-1}$ or $\chi_0^2 < \chi^2_{1-\alpha/2,\, n-1}$
H_0: $\sigma^2 = \sigma_0^2$ H_1: $\sigma^2 < \sigma_0^2$	$\chi_0^2 = \dfrac{(n-1)s^2}{\sigma_0^2}$	$\chi_0^2 < \chi^2_{1-\alpha,\, n-1}$
H_0: $\sigma^2 = \sigma_0^2$ H_1: $\sigma^2 > \sigma_0^2$		$\chi_0^2 > \chi^2_{\alpha,\, n-1}$
H_0: $\sigma_1^2 = \sigma_2^2$ H_1: $\sigma_1^2 \neq \sigma_2^2$	$F_0 = \dfrac{s_1^2}{s_2^2}$	$F_0 > F_{\alpha/2,\, n_1-1,\, n_2-1}$ $F_0 < F_{1-\alpha/2,\, n_1-1,\, n_2-1}$
H_0: $\sigma_1^2 = \sigma_2^2$ H_1: $\sigma_1^2 < \sigma_2^2$	$F_0 = \dfrac{s_2^2}{s_1^2}$	$F_0 > F_{\alpha,\, n_2-1,\, n_1-1}$
H_0: $\sigma_1^2 = \sigma_2^2$ H_1: $\sigma_1^2 > \sigma_2^2$	$F_0 = \dfrac{s_1^2}{s_2^2}$	$F_0 > F_{\alpha,\, n_1-1,\, n_2-1}$

Assume that the values of α and β are given. The sample size can be obtained from the following relationships. In (A) and (B), μ_1 is the value assumed to be the true mean.

(A) $H_0: \mu = \mu_0;\ H_1: \mu \neq \mu_0$

$$\beta = \Phi\left(\frac{\mu_0 - \mu}{\sigma/\sqrt{n}} + Z_{\alpha/2}\right) - \Phi\left(\frac{\mu_0 - \mu}{\sigma/\sqrt{n}} - Z_{\alpha/2}\right)$$

An approximate result is

$$n \simeq \frac{\left(Z_{\alpha/2} + Z_\beta\right)^2 \sigma^2}{\left(\mu_1 - \mu_0\right)^2}$$

(B) $H_0: \mu = \mu_0;\ H_1: \mu > \mu_0$

$$\beta = \Phi\left(\frac{\mu_0 - \mu}{\sigma/\sqrt{n}} + Z_\alpha\right)$$

$$n = \frac{\left(Z_\alpha + Z_\beta\right)^2 \sigma^2}{\left(\mu_1 - \mu_0\right)^2}$$

CONFIDENCE INTERVALS, SAMPLE DISTRIBUTIONS AND SAMPLE SIZE

Confidence Interval for the Mean μ of a Normal Distribution

(A) Standard deviation σ is known

$$\overline{X} - Z_{\alpha/2}\frac{\sigma}{\sqrt{n}} \le \mu \le \overline{X} + Z_{\alpha/2}\frac{\sigma}{\sqrt{n}}$$

(B) Standard deviation σ is not known

$$\overline{X} - t_{\alpha/2}\frac{s}{\sqrt{n}} \le \mu \le \overline{X} + t_{\alpha/2}\frac{s}{\sqrt{n}}$$

where $t_{\alpha/2}$ corresponds to $n - 1$ degrees of freedom.

Confidence Interval for the Difference Between Two Means μ_1 and μ_2

(A) Standard deviations σ_1 and σ_2 known

$$\overline{X}_1 - \overline{X}_2 - Z_{\alpha/2}\sqrt{\frac{\sigma_1^2}{n_1} + \frac{\sigma_2^2}{n_2}} \le \mu_1 - \mu_2 \le \overline{X}_1 - \overline{X}_2 + Z_{\alpha/2}\sqrt{\frac{\sigma_1^2}{n_1} + \frac{\sigma_2^2}{n_2}}$$

(B) Standard deviations σ_1 and σ_2 are not known

$$\overline{X}_1 - \overline{X}_2 - t_{\alpha/2}\sqrt{\frac{\left(\frac{1}{n_1} + \frac{1}{n_2}\right)\left[\left(n_1 - 1\right)s_1^2 + \left(n_2 - 1\right)s_2^2\right]}{n_1 + n_2 - 2}} \le \mu_1 - \mu_2 \le \overline{X}_1 - \overline{X}_2 + t_{\alpha/2}\sqrt{\frac{\left(\frac{1}{n_1} + \frac{1}{n_2}\right)\left[\left(n_1 - 1\right)s_1^2 + \left(n_2 - 1\right)s_2^2\right]}{n_1 + n_2 - 2}}$$

where $t_{\alpha/2}$ corresponds to $n_1 + n_2 - 2$ degrees of freedom.

Confidence Intervals for the Variance σ^2 of a Normal Distribution

$$\frac{(n - 1)s^2}{x^2_{\alpha/2,\, n-1}} \le \sigma^2 \le \frac{(n - 1)s^2}{x^2_{1-\alpha/2,\, n-1}}$$

Sample Size

$$z = \frac{\overline{X} - \mu}{\sigma/\sqrt{n}} \qquad n = \left[\frac{z_{\alpha/2}\sigma}{\overline{x} - \mu}\right]^2$$

TEST STATISTICS

The following definitions apply.

$$Z_{var} = \frac{\overline{X} - \mu_o}{\frac{\sigma}{\sqrt{n}}}$$

$$t_{var} = \frac{\overline{X} - \mu_o}{\frac{s}{\sqrt{n}}}$$

where

Z_{var} is the standard normal Z score

t_{var} is the sample distribution test statistic

σ is known standard deviation

μ_o is population mean

$\overline{X}$ is hypothesized mean or sample mean

n is sample size

s is computed sample standard deviation

The Z score is applicable when the standard deviation(s) is known. The test statistic is applicable when the standard deviation(s) is computed at time of sampling.

Z_α corresponds to the appropriate probability under the normal probability curve for a given Z_{var}.

$t_{\alpha, n-1}$ corresponds to the appropriate probability under the t distribution with $n - 1$ degrees of freedom for a given t_{var}.

Values of $Z_{\alpha/2}$

Confidence Interval	$Z_{\alpha/2}$
80%	1.2816
90%	1.6449
95%	1.9600
96%	2.0537
98%	2.3263
99%	2.5758

Unit Normal Distribution

	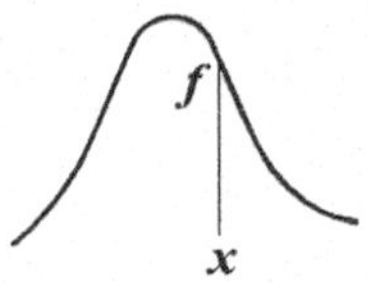	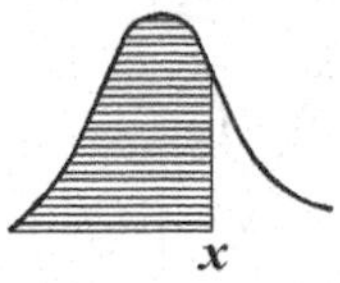	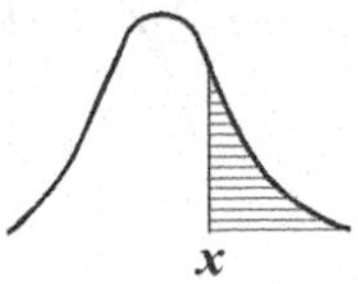	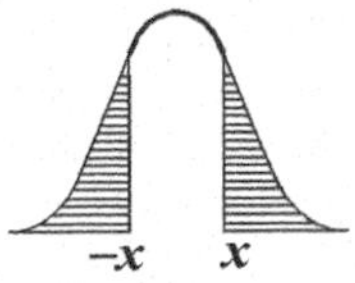	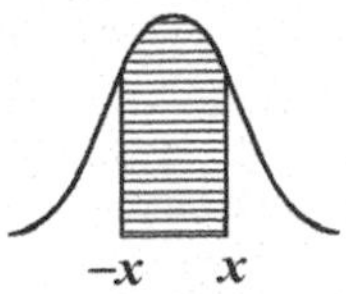
x	$f(x)$	$F(x)$	$R(x)$	$2R(x)$	$W(x)$
0.0	0.3989	0.5000	0.5000	1.0000	0.0000
0.1	0.3970	0.5398	0.4602	0.9203	0.0797
0.2	0.3910	0.5793	0.4207	0.8415	0.1585
0.3	0.3814	0.6179	0.3821	0.7642	0.2358
0.4	0.3683	0.6554	0.3446	0.6892	0.3108
0.5	0.3521	0.6915	0.3085	0.6171	0.3829
0.6	0.3332	0.7257	0.2743	0.5485	0.4515
0.7	0.3123	0.7580	0.2420	0.4839	0.5161
0.8	0.2897	0.7881	0.2119	0.4237	0.5763
0.9	0.2661	0.8159	0.1841	0.3681	0.6319
1.0	0.2420	0.8413	0.1587	0.3173	0.6827
1.1	0.2179	0.8643	0.1357	0.2713	0.7287
1.2	0.1942	0.8849	0.1151	0.2301	0.7699
1.3	0.1714	0.9032	0.0968	0.1936	0.8064
1.4	0.1497	0.9192	0.0808	0.1615	0.8385
1.5	0.1295	0.9332	0.0668	0.1336	0.8664
1.6	0.1109	0.9452	0.0548	0.1096	0.8904
1.7	0.0940	0.9554	0.0446	0.0891	0.9109
1.8	0.0790	0.9641	0.0359	0.0719	0.9281
1.9	0.0656	0.9713	0.0287	0.0574	0.9426
2.0	0.0540	0.9772	0.0228	0.0455	0.9545
2.1	0.0440	0.9821	0.0179	0.0357	0.9643
2.2	0.0355	0.9861	0.0139	0.0278	0.9722
2.3	0.0283	0.9893	0.0107	0.0214	0.9786
2.4	0.0224	0.9918	0.0082	0.0164	0.9836
2.5	0.0175	0.9938	0.0062	0.0124	0.9876
2.6	0.0136	0.9953	0.0047	0.0093	0.9907
2.7	0.0104	0.9965	0.0035	0.0069	0.9931
2.8	0.0079	0.9974	0.0026	0.0051	0.9949
2.9	0.0060	0.9981	0.0019	0.0037	0.9963
3.0	0.0044	0.9987	0.0013	0.0027	0.9973
Fractiles					
1.2816	0.1755	0.9000	0.1000	0.2000	0.8000
1.6449	0.1031	0.9500	0.0500	0.1000	0.9000
1.9600	0.0584	0.9750	0.0250	0.0500	0.9500
2.0537	0.0484	0.9800	0.0200	0.0400	0.9600
2.3263	0.0267	0.9900	0.0100	0.0200	0.9800
2.5758	0.0145	0.9950	0.0050	0.0100	0.9900

Student's *t*-Distribution

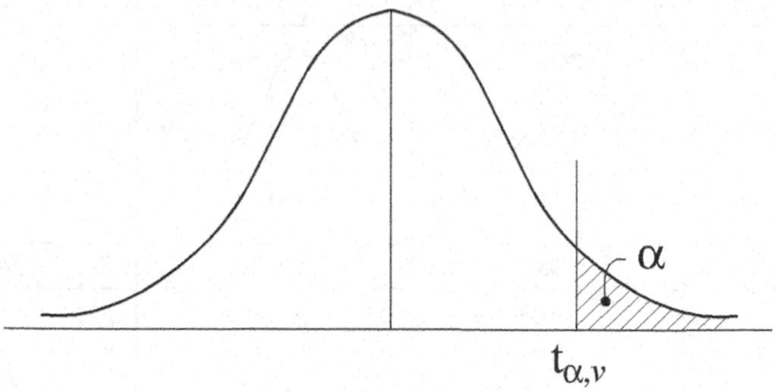

VALUES OF $t_{\alpha,\nu}$

ν	α 0.25	0.20	0.15	0.10	0.05	0.025	0.01	0.005	ν
1	1.000	1.376	1.963	3.078	6.314	12.706	31.821	63.657	1
2	0.816	1.061	1.386	1.886	2.920	4.303	6.965	9.925	2
3	0.765	0.978	1.350	1.638	2.353	3.182	4.541	5.841	3
4	0.741	0.941	1.190	1.533	2.132	2.776	3.747	4.604	4
5	**0.727**	**0.920**	**1.156**	**1.476**	**2.015**	**2.571**	**3.365**	**4.032**	**5**
6	0.718	0.906	1.134	1.440	1.943	2.447	3.143	3.707	6
7	0.711	0.896	1.119	1.415	1.895	2.365	2.998	3.499	7
8	0.706	0.889	1.108	1.397	1.860	2.306	2.896	3.355	8
9	0.703	0.883	1.100	1.383	1.833	2.262	2.821	3.250	9
10	**0.700**	**0.879**	**1.093**	**1.372**	**1.812**	**2.228**	**2.764**	**3.169**	**10**
11	0.697	0.876	1.088	1.363	1.796	2.201	2.718	3.106	11
12	0.695	0.873	1.083	1.356	1.782	2.179	2.681	3.055	12
13	0.694	0.870	1.079	1.350	1.771	2.160	2.650	3.012	13
14	0.692	0.868	1.076	1.345	1.761	2.145	2.624	2.977	14
15	**0.691**	**0.866**	**1.074**	**1.341**	**1.753**	**2.131**	**2.602**	**2.947**	**15**
16	0.690	0.865	1.071	1.337	1.746	2.120	2.583	2.921	16
17	0.689	0.863	1.069	1.333	1.740	2.110	2.567	2.898	17
18	0.688	0.862	1.067	1.330	1.734	2.101	2.552	2.878	18
19	0.688	0.861	1.066	1.328	1.729	2.093	2.539	2.861	19
20	**0.687**	**0.860**	**1.064**	**1.325**	**1.725**	**2.086**	**2.528**	**2.845**	**20**
21	0.686	0.859	1.063	1.323	1.721	2.080	2.518	2.831	21
22	0.686	0.858	1.061	1.321	1.717	2.074	2.508	2.819	22
23	0.685	0.858	1.060	1.319	1.714	2.069	2.500	2.807	23
24	0.685	0.857	1.059	1.318	1.711	2.064	2.492	2.797	24
25	**0.684**	**0.856**	**1.058**	**1.316**	**1.708**	**2.060**	**2.485**	**2.787**	**25**
26	0.684	0.856	1.058	1.315	1.706	2.056	2.479	2.779	26
27	0.684	0.855	1.057	1.314	1.703	2.052	2.473	2.771	27
28	0.683	0.855	1.056	1.313	1.701	2.048	2.467	2.763	28
29	0.683	0.854	1.055	1.311	1.699	2.045	2.462	2.756	29
30	**0.683**	**0.854**	**1.055**	**1.310**	**1.697**	**2.042**	**2.457**	**2.750**	**30**
∞	0.674	0.842	1.036	1.282	1.645	1.960	2.326	2.576	∞

CRITICAL VALUES OF THE *F* DISTRIBUTION – TABLE

For a particular combination of numerator and denominator degrees of freedom, entry represents the critical values of *F* corresponding to a specified upper tail area (α).

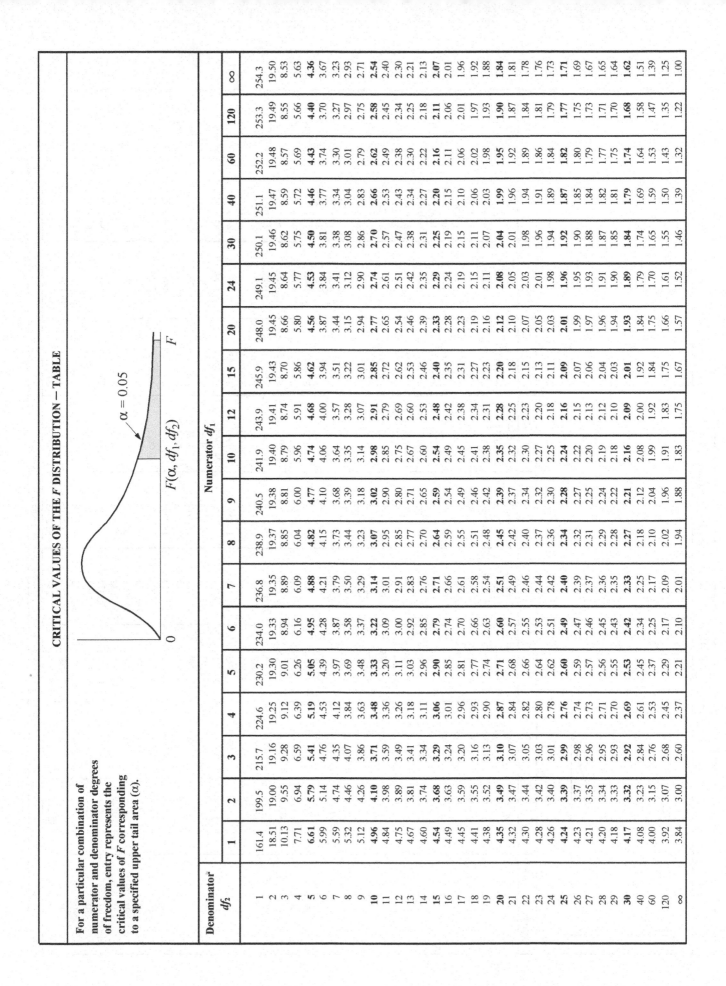

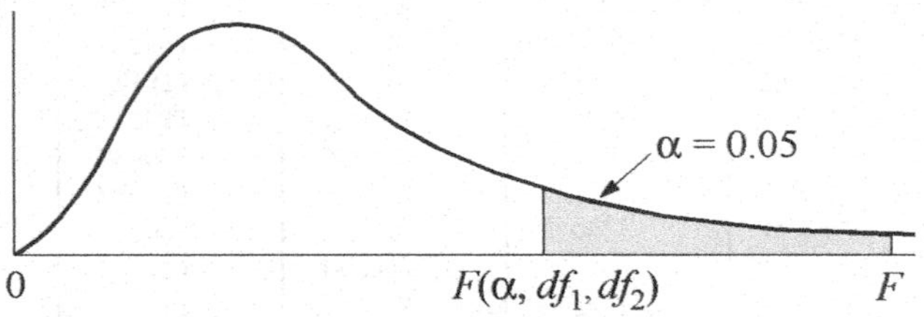

Denominator df_2	Numerator df_1: 1	2	3	4	5	6	7	8	9	10	12	15	20	24	30	40	60	120	∞
1	161.4	199.5	215.7	224.6	230.2	234.0	236.8	238.9	240.5	241.9	243.9	245.9	248.0	249.1	250.1	251.1	252.2	253.3	254.3
2	18.51	19.00	19.16	19.25	19.30	19.33	19.35	19.37	19.38	19.40	19.41	19.43	19.45	19.45	19.46	19.47	19.48	19.49	19.50
3	10.13	9.55	9.28	9.12	9.01	8.94	8.89	8.85	8.81	8.79	8.74	8.70	8.66	8.64	8.62	8.59	8.57	8.55	8.53
4	7.71	6.94	6.59	6.39	6.26	6.16	6.09	6.04	6.00	5.96	5.91	5.86	5.80	5.77	5.75	5.72	5.69	5.66	5.63
5	**6.61**	**5.79**	**5.41**	**5.19**	**5.05**	**4.95**	**4.88**	**4.82**	**4.77**	**4.74**	**4.68**	**4.62**	**4.56**	**4.53**	**4.50**	**4.46**	**4.43**	**4.40**	**4.36**
6	5.99	5.14	4.76	4.53	4.39	4.28	4.21	4.15	4.10	4.06	4.00	3.94	3.87	3.84	3.81	3.77	3.74	3.70	3.67
7	5.59	4.74	4.35	4.12	3.97	3.87	3.79	3.73	3.68	3.64	3.57	3.51	3.44	3.41	3.38	3.34	3.30	3.27	3.23
8	5.32	4.46	4.07	3.84	3.69	3.58	3.50	3.44	3.39	3.35	3.28	3.22	3.15	3.12	3.08	3.04	3.01	2.97	2.93
9	5.12	4.26	3.86	3.63	3.48	3.37	3.29	3.23	3.18	3.14	3.07	3.01	2.94	2.90	2.86	2.83	2.79	2.75	2.71
10	**4.96**	**4.10**	**3.71**	**3.48**	**3.33**	**3.22**	**3.14**	**3.07**	**3.02**	**2.98**	**2.91**	**2.85**	**2.77**	**2.74**	**2.70**	**2.66**	**2.62**	**2.58**	**2.54**
11	4.84	3.98	3.59	3.36	3.20	3.09	3.01	2.95	2.90	2.85	2.79	2.72	2.65	2.61	2.57	2.53	2.49	2.45	2.40
12	4.75	3.89	3.49	3.26	3.11	3.00	2.91	2.85	2.80	2.75	2.69	2.62	2.54	2.51	2.47	2.43	2.38	2.34	2.30
13	4.67	3.81	3.41	3.18	3.03	2.92	2.83	2.77	2.71	2.67	2.60	2.53	2.46	2.42	2.38	2.34	2.30	2.25	2.21
14	4.60	3.74	3.34	3.11	2.96	2.85	2.76	2.70	2.65	2.60	2.53	2.46	2.39	2.35	2.31	2.27	2.22	2.18	2.13
15	**4.54**	**3.68**	**3.29**	**3.06**	**2.90**	**2.79**	**2.71**	**2.64**	**2.59**	**2.54**	**2.48**	**2.40**	**2.33**	**2.29**	**2.25**	**2.20**	**2.16**	**2.11**	**2.07**
16	4.49	3.63	3.24	3.01	2.85	2.74	2.66	2.59	2.54	2.49	2.42	2.35	2.28	2.24	2.19	2.15	2.11	2.06	2.01
17	4.45	3.59	3.20	2.96	2.81	2.70	2.61	2.55	2.49	2.45	2.38	2.31	2.23	2.19	2.15	2.10	2.06	2.01	1.96
18	4.41	3.55	3.16	2.93	2.77	2.66	2.58	2.51	2.46	2.41	2.34	2.27	2.19	2.15	2.11	2.06	2.02	1.97	1.92
19	4.38	3.52	3.13	2.90	2.74	2.63	2.54	2.48	2.42	2.38	2.31	2.23	2.16	2.11	2.07	2.03	1.98	1.93	1.88
20	**4.35**	**3.49**	**3.10**	**2.87**	**2.71**	**2.60**	**2.51**	**2.45**	**2.39**	**2.35**	**2.28**	**2.20**	**2.12**	**2.08**	**2.04**	**1.99**	**1.95**	**1.90**	**1.84**
21	4.32	3.47	3.07	2.84	2.68	2.57	2.49	2.42	2.37	2.32	2.25	2.18	2.10	2.05	2.01	1.96	1.92	1.87	1.81
22	4.30	3.44	3.05	2.82	2.66	2.55	2.46	2.40	2.34	2.30	2.23	2.15	2.07	2.03	1.98	1.94	1.89	1.84	1.78
23	4.28	3.42	3.03	2.80	2.64	2.53	2.44	2.37	2.32	2.27	2.20	2.13	2.05	2.01	1.96	1.91	1.86	1.81	1.76
24	4.26	3.40	3.01	2.78	2.62	2.51	2.42	2.36	2.30	2.25	2.18	2.11	2.03	1.98	1.94	1.89	1.84	1.79	1.73
25	**4.24**	**3.39**	**2.99**	**2.76**	**2.60**	**2.49**	**2.40**	**2.34**	**2.28**	**2.24**	**2.16**	**2.09**	**2.01**	**1.96**	**1.92**	**1.87**	**1.82**	**1.77**	**1.71**
26	4.23	3.37	2.98	2.74	2.59	2.47	2.39	2.32	2.27	2.22	2.15	2.07	1.99	1.95	1.90	1.85	1.80	1.75	1.69
27	4.21	3.35	2.96	2.73	2.57	2.46	2.37	2.31	2.25	2.20	2.13	2.06	1.97	1.93	1.88	1.84	1.79	1.73	1.67
28	4.20	3.34	2.95	2.71	2.56	2.45	2.36	2.29	2.24	2.19	2.12	2.04	1.96	1.91	1.87	1.82	1.77	1.71	1.65
29	4.18	3.33	2.93	2.70	2.55	2.43	2.35	2.28	2.22	2.18	2.10	2.03	1.94	1.90	1.85	1.81	1.75	1.70	1.64
30	**4.17**	**3.32**	**2.92**	**2.69**	**2.53**	**2.42**	**2.33**	**2.27**	**2.21**	**2.16**	**2.09**	**2.01**	**1.93**	**1.89**	**1.84**	**1.79**	**1.74**	**1.68**	**1.62**
40	4.08	3.23	2.84	2.61	2.45	2.34	2.25	2.18	2.12	2.08	2.00	1.92	1.84	1.79	1.74	1.69	1.64	1.58	1.51
60	4.00	3.15	2.76	2.53	2.37	2.25	2.17	2.10	2.04	1.99	1.92	1.84	1.75	1.70	1.65	1.59	1.53	1.47	1.39
120	3.92	3.07	2.68	2.45	2.29	2.17	2.09	2.02	1.96	1.91	1.83	1.75	1.66	1.61	1.55	1.50	1.43	1.35	1.25
∞	3.84	3.00	2.60	2.37	2.21	2.10	2.01	1.94	1.88	1.83	1.75	1.67	1.57	1.52	1.46	1.39	1.32	1.22	1.00

CRITICAL VALUES OF X^2 DISTRIBUTION

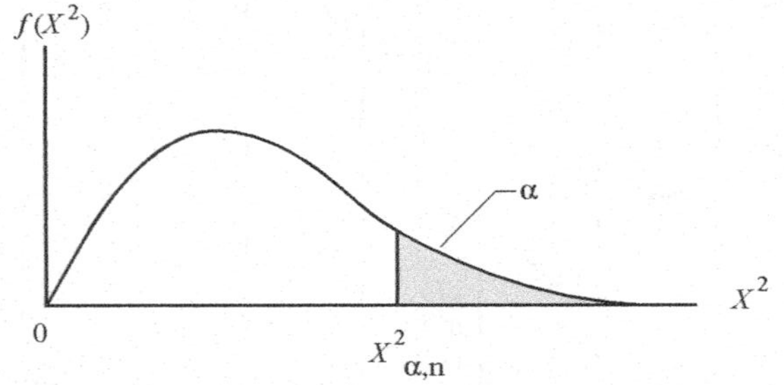

Degrees of Freedom	$X^2_{.995}$	$X^2_{.990}$	$X^2_{.975}$	$X^2_{.950}$	$X^2_{.900}$	$X^2_{.100}$	$X^2_{.050}$	$X^2_{.025}$	$X^2_{.010}$	$X^2_{.005}$
1	0.0000393	0.0001571	0.0009821	0.0039321	0.0157908	2.70554	3.84146	5.02389	6.63490	7.87944
2	0.0100251	0.0201007	0.0506356	0.102587	0.210720	4.60517	5.99147	7.37776	9.21034	10.5966
3	0.0717212	0.114832	0.215795	0.351846	0.584375	6.25139	7.81473	9.34840	11.3449	12.8381
4	0.206990	0.297110	0.484419	0.710721	1.063623	7.77944	9.48773	11.1433	13.2767	14.8602
5	0.411740	0.554300	0.831211	1.145476	1.61031	9.23635	11.0705	12.8325	15.0863	16.7496
6	0.675727	0.872085	1.237347	1.63539	2.20413	10.6446	12.5916	14.4494	16.8119	18.5476
7	0.989265	1.239043	1.68987	2.16735	2.83311	12.0170	14.0671	16.0128	18.4753	20.2777
8	1.344419	1.646482	2.17973	2.73264	3.48954	13.3616	15.5073	17.5346	20.0902	21.9550
9	1.734926	2.087912	2.70039	3.32511	4.16816	14.6837	16.9190	19.0228	21.6660	23.5893
10	2.15585	2.55821	3.24697	3.94030	4.86518	15.9871	18.3070	20.4831	23.2093	25.1882
11	2.60321	3.05347	3.81575	4.57481	5.57779	17.2750	19.6751	21.9200	24.7250	26.7569
12	3.07382	3.57056	4.40379	5.22603	6.30380	18.5494	21.0261	23.3367	26.2170	28.2995
13	3.56503	4.10691	5.00874	5.89186	7.04150	19.8119	22.3621	24.7356	27.6883	29.8194
14	4.07468	4.66043	5.62872	6.57063	7.78953	21.0642	23.6848	26.1190	29.1413	31.3193
15	4.60094	5.22935	6.26214	7.26094	8.54675	22.3072	24.9958	27.4884	30.5779	32.8013
16	5.14224	5.81221	6.90766	7.96164	9.31223	23.5418	26.2962	28.8454	31.9999	34.2672
17	5.69724	6.40776	7.56418	8.67176	10.0852	24.7690	27.5871	30.1910	33.4087	35.7185
18	6.26481	7.01491	8.23075	9.39046	10.8649	25.9894	28.8693	31.5264	34.8053	37.1564
19	6.84398	7.63273	8.90655	10.1170	11.6509	27.2036	30.1435	32.8523	36.1908	38.5822
20	7.43386	8.26040	9.59083	10.8508	12.4426	28.4120	31.4104	34.1696	37.5662	39.9968
21	8.03366	8.89720	10.28293	11.5913	13.2396	29.6151	32.6705	35.4789	38.9321	41.4010
22	8.64272	9.54249	10.9823	12.3380	14.0415	30.8133	33.9244	36.7807	40.2894	42.7956
23	9.26042	10.19567	11.6885	13.0905	14.8479	32.0069	35.1725	38.0757	41.6384	44.1813
24	9.88623	10.8564	12.4011	13.8484	15.6587	33.1963	36.4151	39.3641	42.9798	45.5585
25	10.5197	11.5240	13.1197	14.6114	16.4734	34.3816	37.6525	40.6465	44.3141	46.9278
26	11.1603	12.1981	13.8439	15.3791	17.2919	35.5631	38.8852	41.9232	45.6417	48.2899
27	11.8076	12.8786	14.5733	16.1513	18.1138	36.7412	40.1133	43.1944	46.9630	49.6449
28	12.4613	13.5648	15.3079	16.9279	18.9392	37.9159	41.3372	44.4607	48.2782	50.9933
29	13.1211	14.2565	16.0471	17.7083	19.7677	39.0875	42.5569	45.7222	49.5879	52.3356
30	13.7867	14.9535	16.7908	18.4926	20.5992	40.2560	43.7729	46.9792	50.8922	53.6720
40	20.7065	22.1643	24.4331	26.5093	29.0505	51.8050	55.7585	59.3417	63.6907	66.7659
50	27.9907	29.7067	32.3574	34.7642	37.6886	63.1671	67.5048	71.4202	76.1539	79.4900
60	35.5346	37.4848	40.4817	43.1879	46.4589	74.3970	79.0819	83.2976	88.3794	91.9517
70	43.2752	45.4418	48.7576	51.7393	55.3290	85.5271	90.5312	95.0231	100.425	104.215
80	51.1720	53.5400	57.1532	60.3915	64.2778	96.5782	101.879	106.629	112.329	116.321
90	59.1963	61.7541	65.6466	69.1260	73.2912	107.565	113.145	118.136	124.116	128.299
100	67.3276	70.0648	74.2219	77.9295	82.3581	118.498	124.342	129.561	135.807	140.169

Source: Thompson, C. M., "Tables of the Percentage Points of the X^2–Distribution," *Biometrika*, ©1941, 32, 188-189. Reproduced by permission of Oxford University Press.

Cumulative Binomial Probabilities $P(X \le x)$

		P										
n	*x*	0.1	0.2	0.3	0.4	0.5	0.6	0.7	0.8	0.9	0.95	0.99
1	0	0.9000	0.8000	0.7000	0.6000	0.5000	0.4000	0.3000	0.2000	0.1000	0.0500	0.0100
2	0	0.8100	0.6400	0.4900	0.3600	0.2500	0.1600	0.0900	0.0400	0.0100	0.0025	0.0001
	1	0.9900	0.9600	0.9100	0.8400	0.7500	0.6400	0.5100	0.3600	0.1900	0.0975	0.0199
3	0	0.7290	0.5120	0.3430	0.2160	0.1250	0.0640	0.0270	0.0080	0.0010	0.0001	0.0000
	1	0.9720	0.8960	0.7840	0.6480	0.5000	0.3520	0.2160	0.1040	0.0280	0.0073	0.0003
	2	0.9990	0.9920	0.9730	0.9360	0.8750	0.7840	0.6570	0.4880	0.2710	0.1426	0.0297
4	0	0.6561	0.4096	0.2401	0.1296	0.0625	0.0256	0.0081	0.0016	0.0001	0.0000	0.0000
	1	0.9477	0.8192	0.6517	0.4752	0.3125	0.1792	0.0837	0.0272	0.0037	0.0005	0.0000
	2	0.9963	0.9728	0.9163	0.8208	0.6875	0.5248	0.3483	0.1808	0.0523	0.0140	0.0006
	3	0.9999	0.9984	0.9919	0.9744	0.9375	0.8704	0.7599	0.5904	0.3439	0.1855	0.0394
5	0	0.5905	0.3277	0.1681	0.0778	0.0313	0.0102	0.0024	0.0003	0.0000	0.0000	0.0000
	1	0.9185	0.7373	0.5282	0.3370	0.1875	0.0870	0.0308	0.0067	0.0005	0.0000	0.0000
	2	0.9914	0.9421	0.8369	0.6826	0.5000	0.3174	0.1631	0.0579	0.0086	0.0012	0.0000
	3	0.9995	0.9933	0.9692	0.9130	0.8125	0.6630	0.4718	0.2627	0.0815	0.0226	0.0010
	4	1.0000	0.9997	0.9976	0.9898	0.9688	0.9222	0.8319	0.6723	0.4095	0.2262	0.0490
6	0	0.5314	0.2621	0.1176	0.0467	0.0156	0.0041	0.0007	0.0001	0.0000	0.0000	0.0000
	1	0.8857	0.6554	0.4202	0.2333	0.1094	0.0410	0.0109	0.0016	0.0001	0.0000	0.0000
	2	0.9842	0.9011	0.7443	0.5443	0.3438	0.1792	0.0705	0.0170	0.0013	0.0001	0.0000
	3	0.9987	0.9830	0.9295	0.8208	0.6563	0.4557	0.2557	0.0989	0.0159	0.0022	0.0000
	4	0.9999	0.9984	0.9891	0.9590	0.8906	0.7667	0.5798	0.3446	0.1143	0.0328	0.0015
	5	1.0000	0.9999	0.9993	0.9959	0.9844	0.9533	0.8824	0.7379	0.4686	0.2649	0.0585
7	0	0.4783	0.2097	0.0824	0.0280	0.0078	0.0016	0.0002	0.0000	0.0000	0.0000	0.0000
	1	0.8503	0.5767	0.3294	0.1586	0.0625	0.0188	0.0038	0.0004	0.0000	0.0000	0.0000
	2	0.9743	0.8520	0.6471	0.4199	0.2266	0.0963	0.0288	0.0047	0.0002	0.0000	0.0000
	3	0.9973	0.9667	0.8740	0.7102	0.5000	0.2898	0.1260	0.0333	0.0027	0.0002	0.0000
	4	0.9998	0.9953	0.9712	0.9037	0.7734	0.5801	0.3529	0.1480	0.0257	0.0038	0.0000
	5	1.0000	0.9996	0.9962	0.9812	0.9375	0.8414	0.6706	0.4233	0.1497	0.0444	0.0020
	6	1.0000	1.0000	0.9998	0.9984	0.9922	0.9720	0.9176	0.7903	0.5217	0.3017	0.0679
8	0	0.4305	0.1678	0.0576	0.0168	0.0039	0.0007	0.0001	0.0000	0.0000	0.0000	0.0000
	1	0.8131	0.5033	0.2553	0.1064	0.0352	0.0085	0.0013	0.0001	0.0000	0.0000	0.0000
	2	0.9619	0.7969	0.5518	0.3154	0.1445	0.0498	0.0113	0.0012	0.0000	0.0000	0.0000
	3	0.9950	0.9437	0.8059	0.5941	0.3633	0.1737	0.0580	0.0104	0.0004	0.0000	0.0000
	4	0.9996	0.9896	0.9420	0.8263	0.6367	0.4059	0.1941	0.0563	0.0050	0.0004	0.0000
	5	1.0000	0.9988	0.9887	0.9502	0.8555	0.6846	0.4482	0.2031	0.0381	0.0058	0.0001
	6	1.0000	0.9999	0.9987	0.9915	0.9648	0.8936	0.7447	0.4967	0.1869	0.0572	0.0027
	7	1.0000	1.0000	0.9999	0.9993	0.9961	0.9832	0.9424	0.8322	0.5695	0.3366	0.0773
9	0	0.3874	0.1342	0.0404	0.0101	0.0020	0.0003	0.0000	0.0000	0.0000	0.0000	0.0000
	1	0.7748	0.4362	0.1960	0.0705	0.0195	0.0038	0.0004	0.0000	0.0000	0.0000	0.0000
	2	0.9470	0.7382	0.4628	0.2318	0.0898	0.0250	0.0043	0.0003	0.0000	0.0000	0.0000
	3	0.9917	0.9144	0.7297	0.4826	0.2539	0.0994	0.0253	0.0031	0.0001	0.0000	0.0000
	4	0.9991	0.9804	0.9012	0.7334	0.5000	0.2666	0.0988	0.0196	0.0009	0.0000	0.0000
	5	0.9999	0.9969	0.9747	0.9006	0.7461	0.5174	0.2703	0.0856	0.0083	0.0006	0.0000
	6	1.0000	0.9997	0.9957	0.9750	0.9102	0.7682	0.5372	0.2618	0.0530	0.0084	0.0001
	7	1.0000	1.0000	0.9996	0.9962	0.9805	0.9295	0.8040	0.5638	0.2252	0.0712	0.0034
	8	1.0000	1.0000	1.0000	0.9997	0.9980	0.9899	0.9596	0.8658	0.6126	0.3698	0.0865

Montgomery, Douglas C., and George C. Runger, *Applied Statistics and Probability for Engineers*, 4th ed. Reproduced by permission of John Wiley & Sons, 2007.

Cumulative Binomial Probabilities $P(X \le x)$ (*continued*)

		P										
n	x	0.1	0.2	0.3	0.4	0.5	0.6	0.7	0.8	0.9	0.95	0.99
10	0	0.3487	0.1074	0.0282	0.0060	0.0010	0.0001	0.0000	0.0000	0.0000	0.0000	0.0000
	1	0.7361	0.3758	0.1493	0.0464	0.0107	0.0017	0.0001	0.0000	0.0000	0.0000	0.0000
	2	0.9298	0.6778	0.3828	0.1673	0.0547	0.0123	0.0016	0.0001	0.0000	0.0000	0.0000
	3	0.9872	0.8791	0.6496	0.3823	0.1719	0.0548	0.0106	0.0009	0.0000	0.0000	0.0000
	4	0.9984	0.9672	0.8497	0.6331	0.3770	0.1662	0.0473	0.0064	0.0001	0.0000	0.0000
	5	0.9999	0.9936	0.9527	0.8338	0.6230	0.3669	0.1503	0.0328	0.0016	0.0001	0.0000
	6	1.0000	0.9991	0.9894	0.9452	0.8281	0.6177	0.3504	0.1209	0.0128	0.0010	0.0000
	7	1.0000	0.9999	0.9984	0.9877	0.9453	0.8327	0.6172	0.3222	0.0702	0.0115	0.0001
	8	1.0000	1.0000	0.9999	0.9983	0.9893	0.9536	0.8507	0.6242	0.2639	0.0861	0.0043
	9	1.0000	1.0000	1.0000	0.9999	0.9990	0.9940	0.9718	0.8926	0.6513	0.4013	0.0956
15	0	0.2059	0.0352	0.0047	0.0005	0.0000	0.0000	0.0000	0.0000	0.0000	0.0000	0.0000
	1	0.5490	0.1671	0.0353	0.0052	0.0005	0.0000	0.0000	0.0000	0.0000	0.0000	0.0000
	2	0.8159	0.3980	0.1268	0.0271	0.0037	0.0003	0.0000	0.0000	0.0000	0.0000	0.0000
	3	0.9444	0.6482	0.2969	0.0905	0.0176	0.0019	0.0001	0.0000	0.0000	0.0000	0.0000
	4	0.9873	0.8358	0.5155	0.2173	0.0592	0.0093	0.0007	0.0000	0.0000	0.0000	0.0000
	5	0.9978	0.9389	0.7216	0.4032	0.1509	0.0338	0.0037	0.0001	0.0000	0.0000	0.0000
	6	0.9997	0.9819	0.8689	0.6098	0.3036	0.0950	0.0152	0.0008	0.0000	0.0000	0.0000
	7	1.0000	0.9958	0.9500	0.7869	0.5000	0.2131	0.0500	0.0042	0.0000	0.0000	0.0000
	8	1.0000	0.9992	0.9848	0.9050	0.6964	0.3902	0.1311	0.0181	0.0003	0.0000	0.0000
	9	1.0000	0.9999	0.9963	0.9662	0.8491	0.5968	0.2784	0.0611	0.0022	0.0001	0.0000
	10	1.0000	1.0000	0.9993	0.9907	0.9408	0.7827	0.4845	0.1642	0.0127	0.0006	0.0000
	11	1.0000	1.0000	0.9999	0.9981	0.9824	0.9095	0.7031	0.3518	0.0556	0.0055	0.0000
	12	1.0000	1.0000	1.0000	0.9997	0.9963	0.9729	0.8732	0.6020	0.1841	0.0362	0.0004
	13	1.0000	1.0000	1.0000	1.0000	0.9995	0.9948	0.9647	0.8329	0.4510	0.1710	0.0096
	14	1.0000	1.0000	1.0000	1.0000	1.0000	0.9995	0.9953	0.9648	0.7941	0.5367	0.1399
20	0	0.1216	0.0115	0.0008	0.0000	0.0000	0.0000	0.0000	0.0000	0.0000	0.0000	0.0000
	1	0.3917	0.0692	0.0076	0.0005	0.0000	0.0000	0.0000	0.0000	0.0000	0.0000	0.0000
	2	0.6769	0.2061	0.0355	0.0036	0.0002	0.0000	0.0000	0.0000	0.0000	0.0000	0.0000
	3	0.8670	0.4114	0.1071	0.0160	0.0013	0.0000	0.0000	0.0000	0.0000	0.0000	0.0000
	4	0.9568	0.6296	0.2375	0.0510	0.0059	0.0003	0.0000	0.0000	0.0000	0.0000	0.0000
	5	0.9887	0.8042	0.4164	0.1256	0.0207	0.0016	0.0000	0.0000	0.0000	0.0000	0.0000
	6	0.9976	0.9133	0.6080	0.2500	0.0577	0.0065	0.0003	0.0000	0.0000	0.0000	0.0000
	7	0.9996	0.9679	0.7723	0.4159	0.1316	0.0210	0.0013	0.0000	0.0000	0.0000	0.0000
	8	0.9999	0.9900	0.8867	0.5956	0.2517	0.0565	0.0051	0.0001	0.0000	0.0000	0.0000
	9	1.0000	0.9974	0.9520	0.7553	0.4119	0.1275	0.0171	0.0006	0.0000	0.0000	0.0000
	10	1.0000	0.9994	0.9829	0.8725	0.5881	0.2447	0.0480	0.0026	0.0000	0.0000	0.0000
	11	1.0000	0.9999	0.9949	0.9435	0.7483	0.4044	0.1133	0.0100	0.0001	0.0000	0.0000
	12	1.0000	1.0000	0.9987	0.9790	0.8684	0.5841	0.2277	0.0321	0.0004	0.0000	0.0000
	13	1.0000	1.0000	0.9997	0.9935	0.9423	0.7500	0.3920	0.0867	0.0024	0.0000	0.0000
	14	1.0000	1.0000	1.0000	0.9984	0.9793	0.8744	0.5836	0.1958	0.0113	0.0003	0.0000
	15	1.0000	1.0000	1.0000	0.9997	0.9941	0.9490	0.7625	0.3704	0.0432	0.0026	0.0000
	16	1.0000	1.0000	1.0000	1.0000	0.9987	0.9840	0.8929	0.5886	0.1330	0.0159	0.0000
	17	1.0000	1.0000	1.0000	1.0000	0.9998	0.9964	0.9645	0.7939	0.3231	0.0755	0.0010
	18	1.0000	1.0000	1.0000	1.0000	1.0000	0.9995	0.9924	0.9308	0.6083	0.2642	0.0169
	19	1.0000	1.0000	1.0000	1.0000	1.0000	1.0000	0.9992	0.9885	0.8784	0.6415	0.1821

Montgomery, Douglas C., and George C. Runger, *Applied Statistics and Probability for Engineers*, 4th ed. Reproduced by permission of John Wiley & Sons, 2007.

STATISTICAL QUALITY CONTROL

Average and Range Charts

n	A_2	D_3	D_4
2	1.880	0	3.268
3	1.023	0	2.574
4	0.729	0	2.282
5	0.577	0	2.114
6	0.483	0	2.004
7	0.419	0.076	1.924
8	0.373	0.136	1.864
9	0.337	0.184	1.816
10	0.308	0.223	1.777

X_i = an individual observation

n = the sample size of a group

k = the number of groups

R = (range) the difference between the largest and smallest observations in a sample of size n.

$$\bar{X} = \frac{X_1 + X_2 + \ldots + X_n}{n}$$

$$\bar{\bar{X}} = \frac{\bar{X}_1 + \bar{X}_2 + \ldots + \bar{X}_k}{k}$$

$$\bar{R} = \frac{R_1 + R_2 + \ldots + R_k}{k}$$

The R Chart formulas are:

$$CL_R = \bar{R}$$

$$UCL_R = D_4\bar{R}$$

$$LCL_R = D_3\bar{R}$$

The $\bar{X}$ Chart formulas are:

$$CL_X = \bar{\bar{X}}$$

$$UCL_X = \bar{\bar{X}} + A_2\bar{R}$$

$$LCL_X = \bar{\bar{X}} - A_2\bar{R}$$

Standard Deviation Charts

n	A_3	B_3	B_4
2	2.659	0	3.267
3	1.954	0	2.568
4	1.628	0	2.266
5	1.427	0	2.089
6	1.287	0.030	1.970
7	1.182	0.119	1.882
8	1.099	0.185	1.815
9	1.032	0.239	1.761
10	0.975	0.284	1.716

$$UCL_X = \bar{\bar{X}} + A_3\bar{S}$$

$$CL_X = \bar{\bar{X}}$$

$$LCL_X = \bar{\bar{X}} - A_3\bar{S}$$

$$UCL_S = B_4\bar{S}$$

$$CL_S = \bar{S}$$

$$LCL_S = B_3\bar{S}$$

Approximations

The following table and equations may be used to generate initial approximations of the items indicated.

n	c_4	d_2	d_3
2	0.7979	1.128	0.853
3	0.8862	1.693	0.888
4	0.9213	2.059	0.880
5	0.9400	2.326	0.864
6	0.9515	2.534	0.848
7	0.9594	2.704	0.833
8	0.9650	2.847	0.820
9	0.9693	2.970	0.808
10	0.9727	3.078	0.797

$$\hat{\sigma} = \bar{R}/d_2$$

$$\hat{\sigma} = \bar{S}/c_4$$

$$\sigma_R = d_3\hat{\sigma}$$

$$\sigma_S = \hat{\sigma}\sqrt{1 - c_4^2}, \text{ where}$$

$\hat{\sigma}$ = an estimate of σ

σ_R = an estimate of the standard deviation of the ranges of the samples

σ_S = an estimate of the standard deviation of the standard deviations of the samples

Tests for Out of Control

1. A single point falls outside the (three sigma) control limits.
2. Two out of three successive points fall on the same side of and more than two sigma units from the center line.
3. Four out of five successive points fall on the same side of and more than one sigma unit from the center line.
4. Eight successive points fall on the same side of the center line.

Probability and Density Functions: Means and Variances

Variable	Equation	Mean	Variance
Binomial Coefficient	$\binom{n}{x}=\dfrac{n!}{x!(n-x)!}$		
Binomial	$b(x;n,p)=\binom{n}{x}p^x(1-p)^{n-x}$	np	$np(1-p)$
Hyper Geometric	$h(x;\, n,\, r,\, N)=\binom{r}{x}\dfrac{\binom{N-r}{n-x}}{\binom{N}{n}}$	$\dfrac{nr}{N}$	$\dfrac{r(N-r)n(N-n)}{N^2(N-1)}$
Poisson	$f(x;\lambda)=\dfrac{\lambda^x e^{-\lambda}}{x!}$	λ	λ
Geometric	$g(x;\, p)=p\,(1-p)^{x-1}$	$1/p$	$(1-p)/p^2$
Negative Binomial	$f(y;r,\, p)=\binom{y+r-1}{r-1}p^r(1-p)^y$	r/p	$r\,(1-p)/p^2$
Multinomial	$f(x_1,\ldots,x_k)=\dfrac{n!}{x_1!,\ldots,x_k!}p_1^{x_1}\ldots p_k^{x_k}$	np_i	$np_i\,(1-p_i)$
Uniform	$f(x)=1/(b-a)$	$(a+b)/2$	$(b-a)^2/12$
Gamma	$f(x)=\dfrac{x^{\alpha-1}e^{-x/\beta}}{\beta^\alpha\Gamma(\alpha)};\quad \alpha>0,\beta>0$	$\alpha\beta$	$\alpha\beta^2$
Exponential	$f(x)=\dfrac{1}{\beta}e^{-x/\beta}$	β	β^2
Weibull	$f(x)=\dfrac{\alpha}{\beta}x^{\alpha-1}e^{-x^\alpha/\beta}$	$\beta^{1/\alpha}\Gamma[(\alpha+1)/\alpha]$	$\beta^{2/\alpha}\left[\Gamma\left(\dfrac{\alpha+1}{\alpha}\right)-\Gamma^2\left(\dfrac{\alpha+1}{\alpha}\right)\right]$
Normal	$f(x)=\dfrac{1}{\sigma\sqrt{2\pi}}e^{-\frac{1}{2}\left(\frac{x-\mu}{\sigma}\right)^2}$	μ	σ^2
Triangular	$f(x)=\begin{cases}\dfrac{2(x-a)}{(b-a)(m-a)} & \text{if } a\le x\le m\\ \dfrac{2(b-x)}{(b-a)(b-m)} & \text{if } m<x\le b\end{cases}$	$\dfrac{a+b+m}{3}$	$\dfrac{a^2+b^2+m^2-ab-am-bm}{18}$

CHEMISTRY

DEFINITIONS

Avogadro's Number – The number of elementary particles in a mol of a substance.

1 mol = 1 gram mole

1 mol = 6.02×10^{23} particles

Mol – The amount of a substance that contains as many particles as 12 grams of ^{12}C (carbon 12).

Molarity of Solutions – The number of gram moles of a substance dissolved in a liter of solution.

Molality of Solutions – The number of gram moles of a substance per 1,000 grams of solvent.

Normality of Solutions – The product of the molarity of a solution and the number of valence changes taking place in a reaction.

Molar Volume of an Ideal Gas [at 0°C (32°F) and 1 atm (14.7 psia)]; 22.4 L/(g mole) [359 ft^3/(lb mole)].

Mole Fraction of a Substance – The ratio of the number of moles of a substance to the total moles present in a mixture of substances.

Equilibrium Constant of a Chemical Reaction

$$aA + bB \rightleftharpoons cC + dD$$

$$K_{eq} = \frac{[C]^c[D]^d}{[A]^a[B]^b}$$

Le Chatelier's Principle for Chemical Equilibrium – When a stress (such as a change in concentration, pressure, or temperature) is applied to a system in equilibrium, the equilibrium shifts in such a way that tends to relieve the stress.

Heats of Reaction, Solution, Formation, and Combustion – Chemical processes generally involve the absorption or evolution of heat. In an endothermic process, heat is absorbed (enthalpy change is positive). In an exothermic process, heat is evolved (enthalpy change is negative).

Solubility Product of a slightly soluble substance *AB:*

$$A_mB_n \rightarrow mA^{n+} + nB^{m-}$$

Solubility Product Constant = $K_{SP} = [A^+]^m [B^-]^n$

Faraday's Law – One gram equivalent weight of matter is chemically altered at each electrode for 96,485 coulombs, or 1 Faraday, of electricity passed through the electrolyte.

Faraday's Equation

$$m = \left(\frac{Q}{F}\right)\left(\frac{M}{z}\right), \text{ where}$$

m = mass (grams) of substance liberated at electrode

Q = total electric charge passed through electrolyte (coulomb or ampere•second)

F = 96,485 coulombs/mol

M = molar mass of the substance (g/mol)

z = valence number

A *catalyst* is a substance that alters the rate of a chemical reaction. The catalyst does not affect the position of equilibrium of a reversible reaction.

The *atomic number* is the number of protons in the atomic nucleus.

Boiling Point Elevation – The presence of a nonvolatile solute in a solvent raises the boiling point of the resulting solution.

Freezing Point Depression – The presence of a solute lowers the freezing point of the resulting solution.

Nernst Equation

$$\Delta E = \left(E_2^0 - E_1^0\right) - \frac{RT}{nF}\ln\left[\frac{M_1^{n+}}{M_2^{n+}}\right]$$

E_1^0 = half-cell potential (volts)

R = ideal gas constant

n = number of electrons participating in either half-cell reaction

T = absolute temperature (K)

M_1^{n+} and M_2^{n+} = molar ion concentration

ACIDS, BASES, and pH (aqueous solutions)

$$\text{pH} = \log_{10}\left(\frac{1}{[H^+]}\right), \text{ where}$$

$[H^+]$ = molar concentration of hydrogen ion, in gram moles per liter. *Acids* have pH < 7. *Bases* have pH > 7.

$$HA \longleftrightarrow A^- + H^+$$

$$K_a = \frac{[A^-][H^+]}{[HA]}$$

$$pK_a = -\log(K_a)$$

For water $[H^+][OH^-] = 10^{-14}$

[ ] denotes molarity

Periodic Table of Elements

Atomic Number Symbol Atomic Weight

I	II											III	IV	V	VI	VII	VIII
1 **H** 1.0079																	2 **He** 4.0026
3 **Li** 6.941	4 **Be** 9.0122											5 **B** 10.811	6 **C** 12.011	7 **N** 14.007	8 **O** 15.999	9 **F** 18.998	10 **Ne** 20.179
11 **Na** 22.990	12 **Mg** 24.305											13 **Al** 26.981	14 **Si** 28.086	15 **P** 30.974	16 **S** 32.066	17 **Cl** 35.453	18 **Ar** 39.948
19 **K** 39.098	20 **Ca** 40.078	21 **Sc** 44.956	22 **Ti** 47.88	23 **V** 50.941	24 **Cr** 51.996	25 **Mn** 54.938	26 **Fe** 55.847	27 **Co** 58.933	28 **Ni** 58.69	29 **Cu** 63.546	30 **Zn** 65.39	31 **Ga** 69.723	32 **Ge** 72.61	33 **As** 74.921	34 **Se** 78.96	35 **Br** 79.904	36 **Kr** 83.80
37 **Rb** 85.468	38 **Sr** 87.62	39 **Y** 88.906	**Ac** 40 **Zr** 91.224	41 **Nb** 92.906	42 **Mo** 95.94	43 **Tc** (98)	44 **Ru** 101.07	45 **Rh** 102.91	46 **Pd** 106.42	47 **Ag** 107.87	48 **Cd** 112.41	49 **In** 114.82	50 **Sn** 118.71	51 **Sb** 121.75	52 **Te** 127.60	53 **I** 126.90	54 **Xe** 131.29
55 **Cs** 132.91	56 **Ba** 137.33	57–71	72 **Hf** 178.49	73 **Ta** 180.95	74 **W** 183.85	75 **Re** 186.21	76 **Os** 190.2	77 **Ir** 192.22	78 **Pt** 195.08	79 **Au** 196.97	80 **Hg** 200.59	81 **Tl** 204.38	82 **Pb** 207.2	83 **Bi** 208.98	84 **Po** (209)	85 **At** (210)	86 **Rn** (222)
87 **Fr** (223)	88 **Ra** 226.02	89–103	104 **Rf** (261)	105 **Db** (262)	106 **Sg** (266)	107 **Bh** (264)	108 **Hs** (269)	109 **Mt** (268)	110 **Ds** (269)	111 **Rg** (272)	112 **Cn** (277)	113 **Uut** unknown	114 **Fl** (289)	115 **Uup** unknown	116 **Lv** (298)	117 **Uus** unknown	118 **Uuo** unknown

Lanthanide Series	57 **La** 138.91	58 **Ce** 140.12	59 **Pr** 140.91	60 **Nd** 144.24	61 **Pm** (145)	62 **Sm** 150.36	63 **Eu** 151.96	64 **Gd** 157.25	65 **Tb** 158.92	66 **Dy** 162.50	67 **Ho** 164.93	68 **Er** 167.26	69 **Tm** 168.93	70 **Yb** 173.04	71 **Lu** 174.97
Actinide Series	89 **Ac** 227.03	90 **Th** 232.04	91 **Pa** 231.04	92 **U** 238.03	93 **Np** 237.05	94 **Pu** (244)	95 **Am** (243)	96 **Cm** (247)	97 **Bk** (247)	98 **Cf** (251)	99 **Es** (252)	100 **Fm** (257)	101 **Md** (258)	102 **No** (259)	103 **Lr** (260)

SELECTED RULES OF NOMENCLATURE IN ORGANIC CHEMISTRY

Alcohols

Three systems of nomenclature are in general use. In the first, the alkyl group attached to the hydroxyl group is named and the separate word *alcohol* is added. In the second system, the higher alcohols are considered as derivatives of the first member of the series, which is called *carbinol*. The third method is the modified Geneva system in which (1) the longest carbon chain containing the hydroxyl group determines the surname, (2) the ending *e* of the corresponding saturated hydrocarbon is replaced by *ol*, (3) the carbon chain is numbered from the end that gives the hydroxyl group the smaller number, and (4) the side chains are named and their positions indicated by the proper number. Alcohols in general are divided into three classes. In *primary* alcohols the hydroxyl group is united to a primary carbon atom, that is, a carbon atom united directly to only one other carbon atom. *Secondary* alcohols have the hydroxyl group united to a secondary carbon atom, that is, one united to two other carbon atoms. *Tertiary* alcohols have the hydroxyl group united to a tertiary carbon atom, that is, one united to three other carbon atoms.

Ethers

Ethers are generally designated by naming the alkyl groups and adding the word *ether*. The group RO is known as an *alkoxyl group*. Ethers may also be named as alkoxy derivatives of hydrocarbons.

Carboxylic Acids

The name of each linear carboxylic acid is unique to the number of carbon atoms it contains. 1: (one carbon atom) Formic. 2: Acetic. 3: Propionic. 4: Butyric. 5: Valeric. 6: Caproic. 7: Enanthic. 8: Caprylic. 9: Pelargonic. 10: Capric.

Aldehydes

The common names of aldehydes are derived from the acids that would be formed on oxidation, that is, the acids having the same number of carbon atoms. In general the *ic acid* is dropped and *aldehyde* added.

Ketones

The common names of ketones are derived from the acid which on pyrolysis would yield the ketone. A second method, especially useful for naming mixed ketones, simply names the alkyl groups and adds the word *ketone*. The name is written as three separate words.

Unsaturated Acyclic Hydrocarbons

The simplest compounds in this class of hydrocarbon chemicals are olefins or alkenes with a single carbon-carbon double bond, having the general formula of C_nH_{2n}. The simplest example in this category is ethylene, C_2H_4.

Dienes are acyclic hydrocarbons with two carbon-carbon double bonds, having the general formula of C_nH_{2n-2}; butadiene (C_4H_6) is an example of such.

Similarly, trienes have three carbon-carbon double bonds with the general formula of C_nH_{2n-4}; hexatriene (C_6H_8) is such an example.

The simplest alkynes have a single carbon-carbon triple bond with the general formula of C_nH_{2n-2}. This series of compounds begins with acetylene, or C_2H_2.

Important Families of Organic Compounds

	FAMILY											
	Alkane	**Alkene**	**Alkyne**	**Arene**	**Haloalkane**	**Alcohol**	**Ether**	**Amine**	**Aldehyde**	**Ketone**	**Carboxylic Acid**	**Ester**
Specific Example	CH_3CH_3	$H_2C = CH_2$	$HC \equiv CH$		CH_3CH_2Cl	CH_3CH_2OH	CH_3OCH_3	CH_3NH_2	$CH_3\overset{\overset{O}{\parallel}}{C}H$	$CH_3\overset{\overset{O}{\parallel}}{C}CH_3$	$CH_3\overset{\overset{O}{\parallel}}{C}OH$	$CH_3\overset{\overset{O}{\parallel}}{C}OCH_3$
IUPAC Name	Ethane	Ethene or Ethylene	Ethyne or Acetylene	Benzene	Chloroethane	Ethanol	Methoxy-methane	Methan-amine	Ethanal	Acetone	Ethanoic Acid	Methyl ethanoate
Common Name	Ethane	Ethylene	Acetylene	Benzene	Ethyl chloride	Ethyl alcohol	Dimethyl ether	Methyl-amine	Acetal-dehyde	Dimethyl ketone	Acetic Acid	Methyl acetate
General Formula	RH	$RCH = CH_2$ $RCH = CHR$ $R_2C = CHR$ $R_2C = CR_2$	$RC \equiv CH$ $RC \equiv CR$	ArH	RX	ROH	ROR	RNH_2 R_2NH R_3N	$R\overset{\overset{O}{\parallel}}{C}H$	$R_1\overset{\overset{O}{\parallel}}{C}R_2$	$R\overset{\overset{O}{\parallel}}{C}OH$	$R\overset{\overset{O}{\parallel}}{C}OR$
Functional Group	C–H and C–C bonds	$\diagdown C = C \diagup$ (with $\diagup$, $\diagdown$ below)	$-C \equiv C-$	Aromatic Ring	$-\overset{\mid}{\underset{\mid}{C}}-X$	$-\overset{\mid}{\underset{\mid}{C}}-OH$	$-\overset{\mid}{\underset{\mid}{C}}-O-\overset{\mid}{\underset{\mid}{C}}-$	$-\overset{\mid}{\underset{\mid}{C}}-\underset{\mid}{N}-$	$-\overset{\overset{O}{\parallel}}{C}-H$	$-\overset{\overset{O}{\parallel}}{C}-$	$-\overset{\overset{O}{\parallel}}{C}-OH$	$-\overset{\overset{O}{\parallel}}{C}-O-\overset{\mid}{\underset{\mid}{C}}-$

Common Names and Molecular Formulas of Some Industrial (Inorganic and Organic) Chemicals

Common Name	Chemical Name	Molecular Formula
Muriatic acid	Hydrochloric acid	HCl
Cumene	Isopropyl benzene	$C_6H_5CH(CH_3)_2$
Styrene	Vinyl benzene	$C_6H_5CH{=}CH_2$
—	Hypochlorite ion	OCl^{-1}
—	Chlorite ion	ClO_2^{-1}
—	Chlorate ion	ClO_3^{-1}
—	Perchlorate ion	ClO_4^{-1}
Gypsum	Calcium sulfate	$CaSO_4$
Limestone	Calcium carbonate	$CaCO_3$
Dolomite	Magnesium carbonate	$MgCO_3$
Bauxite	Aluminum oxide	Al_2O_3
Anatase	Titanium dioxide	TiO_2
Rutile	Titanium dioxide	TiO_2
—	Vinyl chloride	$CH_2{=}CHCl$
—	Ethylene oxide	C_2H_4O
Pyrite	Ferrous sulfide	FeS
Epsom salt	Magnesium sulfate	$MgSO_4$
Hydroquinone	p-Dihydroxy benzene	$C_6H_4(OH)_2$
Soda ash	Sodium carbonate	Na_2CO_3
Salt	Sodium chloride	$NaCl$
Potash	Potassium carbonate	K_2CO_3
Baking soda	Sodium bicarbonate	$NaHCO_3$
Lye	Sodium hydroxide	$NaOH$
Caustic soda	Sodium hydroxide	$NaOH$
—	Vinyl alcohol	$CH_2{=}CHOH$
Carbolic acid	Phenol	C_6H_5OH
Aniline	Aminobenzene	$C_6H_5NH_2$
—	Urea	$(NH_2)_2CO$
Toluene	Methyl benzene	$C_6H_5CH_3$
Xylene	Dimethyl benzene	$C_6H_4(CH_3)_2$
—	Silane	SiH_4
—	Ozone	O_3
Neopentane	2,2-Dimethylpropane	$CH_3C(CH_3)_2CH_3$
Magnetite	Ferrous/ferric oxide	Fe_3O_4
Quicksilver	Mercury	Hg
Heavy water	Deuterium oxide	$(H^2)_2O$
—	Borane	BH_3
Eyewash	Boric acid (solution)	H_3BO_3
—	Deuterium	H^2
—	Tritium	H^3
Laughing gas	Nitrous oxide	N_2O
—	Phosgene	$COCl_2$
Wolfram	Tungsten	W
—	Permanganate ion	MnO_4^{-1}
—	Dichromate ion	$Cr_2O_7^{-2}$
—	Hydronium ion	H_3O^{+1}
Brine	Sodium chloride (solution)	$NaCl$
Battery acid	Sulfuric acid	H_2SO_4

ELECTROCHEMISTRY

Cathode – The electrode at which reduction occurs.

Anode – The electrode at which oxidation occurs.

Oxidation – The loss of electrons.

Reduction – The gaining of electrons.

Cation – Positive ion

Anion – Negative ion

Standard Oxidation Potentials for Corrosion Reactions*	
Corrosion Reaction	**Potential, E_o, Volts vs. Normal Hydrogen Electrode**
$Au \rightarrow Au^{3+} + 3e^-$	–1.498
$2H_2O \rightarrow O_2 + 4H^+ + 4e^-$	–1.229
$Pt \rightarrow Pt^{2+} + 2e^-$	–1.200
$Pd \rightarrow Pd^{2+} + 2e^-$	–0.987
$Ag \rightarrow Ag^+ + e^-$	–0.799
$2Hg \rightarrow Hg_2^{2+} + 2e^-$	–0.788
$Fe^{2+} \rightarrow Fe^{3+} + e^-$	–0.771
$4(OH)^- \rightarrow O_2 + 2H_2O + 4e^-$	–0.401
$Cu \rightarrow Cu^{2+} + 2e^-$	–0.337
$Sn^{2+} \rightarrow Sn^{4+} + 2e^-$	–0.150
$H_2 \rightarrow 2H^+ + 2e^-$	0.000
$Pb \rightarrow Pb^{2+} + 2e^-$	+0.126
$Sn \rightarrow Sn^{2+} + 2e^-$	+0.136
$Ni \rightarrow Ni^{2+} + 2e^-$	+0.250
$Co \rightarrow Co^{2+} + 2e^-$	+0.277
$Cd \rightarrow Cd^{2+} + 2e^-$	+0.403
$Fe \rightarrow Fe^{2+} + 2e^-$	+0.440
$Cr \rightarrow Cr^{3+} + 3e^-$	+0.744
$Zn \rightarrow Zn^{2+} + 2e^-$	+0.763
$Al \rightarrow Al^{3+} + 3e^-$	+1.662
$Mg \rightarrow Mg^{2+} + 2e^-$	+2.363
$Na \rightarrow Na^+ + e^-$	+2.714
$K \rightarrow K^+ + e^-$	+2.925

* Measured at 25°C. Reactions are written as anode half-cells. Arrows are reversed for cathode half-cells.

Flinn, Richard A., and Paul K. Trojan, *Engineering Materials and Their Applications*, 4th ed., Houghton Mifflin Company, 1990.

NOTE: In some chemistry texts, the reactions and the signs of the values (in this table) are reversed; for example, the half-cell potential of zinc is given as –0.763 volt for the reaction $Zn^{2+} + 2e^- \rightarrow Zn$. When the potential E_o is positive, the reaction proceeds spontaneously as written.

MATERIALS SCIENCE/STRUCTURE OF MATTER

ATOMIC BONDING

Primary Bonds

Ionic (e.g., salts, metal oxides)

Covalent (e.g., within polymer molecules)

Metallic (e.g., metals)

CORROSION

A table listing the standard electromotive potentials of metals is shown on the previous page.

For corrosion to occur, there must be an anode and a cathode in electrical contact in the presence of an electrolyte.

Anode Reaction (Oxidation) of a Typical Metal, M

$M^o \rightarrow M^{n+} + ne^-$

Possible Cathode Reactions (Reduction)

$$½\, O_2 + 2\, e^- + H_2O \rightarrow 2\, OH^-$$

$$½\, O_2 + 2\, e^- + 2\, H_3O^+ \rightarrow 3\, H_2O$$

$$2\, e^- + 2\, H_3O^+ \rightarrow 2\, H_2O + H_2$$

When dissimilar metals are in contact, the more electropositive one becomes the anode in a corrosion cell. Different regions of carbon steel can also result in a corrosion reaction: e.g., cold-worked regions are anodic to noncold-worked; different oxygen concentrations can cause oxygen-deficient regions to become cathodic to oxygen-rich regions; grain boundary regions are anodic to bulk grain; in multiphase alloys, various phases may not have the same galvanic potential.

DIFFUSION

Diffusion Coefficient

$D = D_o e^{-Q/(RT)}$, where

D = diffusion coefficient

D_o = proportionality constant

Q = activation energy

R = gas constant [8.314 J/(mol•K)]

T = absolute temperature

THERMAL AND MECHANICAL PROCESSING

Cold working (plastically deforming) a metal increases strength and lowers ductility.

Raising temperature causes (1) recovery (stress relief), (2) recrystallization, and (3) grain growth. *Hot working* allows these processes to occur simultaneously with deformation.

Quenching is rapid cooling from elevated temperature, preventing the formation of equilibrium phases.

In steels, quenching austenite [FCC (γ) iron] can result in martensite instead of equilibrium phases—ferrite [BCC (α) iron] and cementite (iron carbide).

PROPERTIES OF MATERIALS

Electrical

Capacitance: The charge-carrying capacity of an insulating material

Charge held by a capacitor

$$q = CV$$

q = charge
C = capacitance
V = voltage

Capacitance of a parallel plate capacitor

$$C = \frac{\varepsilon A}{d}$$

C = capacitance
ε = permittivity of material
A = cross-sectional area of the plates
d = distance between the plates
ε is also expressed as the product of the dielectric constant (κ) and the permittivity of free space ($\varepsilon_0 = 8.85 \times 10^{-12}$ F/m)

Resistivity: The material property that determines the resistance of a resistor

Resistivity of a material within a resistor

$$\rho = \frac{RA}{L}$$

ρ = resistivity of the material
R = resistance of the resistor
A = cross-sectional area of the resistor
L = length of the resistor

Conductivity is the reciprocal of the resistivity

Photoelectric effect–electrons are emitted from matter (metals and nonmetallic solids, liquids or gases) as a consequence of their absorption of energy from electromagnetic radiation of very short wavelength and high frequency.

Piezoelectric effect–the electromechanical and the electrical state in crystalline materials.

Mechanical

Strain is defined as change in length per unit length; for pure tension the following apply:

Engineering strain

$$\varepsilon = \frac{\Delta L}{L_0}$$

ε = engineering strain
ΔL = change in length
L_0 = initial length

True strain

$$\varepsilon_T = \frac{dL}{L}$$

ε_T = true strain
dL = differential change in length
L = initial length
$\varepsilon_T = \ln(1 + \varepsilon)$

Properties of Metals

Metal	Symbol	Atomic Weight	Density ρ (kg/m^3) Water = 1000	Melting Point (°C)	Melting Point (°F)	Specific Heat [J/(kg·K)]	Electrical Resistivity (10^{-8} Ω·m) at 0°C (273.2 K)	Heat Conductivity λ[W/(m·K)] at 0°C (273.2 K)
Aluminum	Al	26.98	2,698	660	1,220	895.9	2.5	236
Antimony	Sb	121.75	6,692	630	1,166	209.3	39	25.5
Arsenic	As	74.92	5,776	subl. 613	subl. 1,135	347.5	26	–
Barium	Ba	137.33	3,594	710	1,310	284.7	36	–
Beryllium	Be	9.012	1,846	1,285	2,345	2,051.5	2.8	218
Bismuth	Bi	208.98	9,803	271	519	125.6	107	8.2
Cadmium	Cd	112.41	8,647	321	609	234.5	6.8	97
Caesium	Cs	132.91	1,900	29	84	217.7	18.8	36
Calcium	Ca	40.08	1,530	840	1,544	636.4	3.2	–
Cerium	Ce	140.12	6,711	800	1,472	188.4	7.3	11
Chromium	Cr	52	7,194	1,860	3,380	406.5	12.7	96.5
Cobalt	Co	58.93	8,800	1,494	2,721	431.2	5.6	105
Copper	Cu	63.54	8,933	1,084	1,983	389.4	1.55	403
Gallium	Ga	69.72	5,905	30	86	330.7	13.6	41
Gold	Au	196.97	19,281	1,064	1,947	129.8	2.05	319
Indium	In	114.82	7,290	156	312	238.6	8	84
Iridium	Ir	192.22	22,550	2,447	4,436	138.2	4.7	147
Iron	Fe	55.85	7,873	1,540	2,804	456.4	8.9	83.5
Lead	Pb	207.2	11,343	327	620	129.8	19.2	36
Lithium	Li	6.94	533	180	356	4,576.2	8.55	86
Magnesium	Mg	24.31	1,738	650	1,202	1,046.7	3.94	157
Manganese	Mn	54.94	7,473	1,250	2,282	502.4	138	8
Mercury	Hg	200.59	13,547	−39	−38	142.3	94.1	7.8
Molybendum	Mo	95.94	10,222	2,620	4,748	272.1	5	139
Nickel	Ni	58.69	8,907	1,455	2,651	439.6	6.2	94
Niobium	Nb	92.91	8,578	2,425	4,397	267.9	15.2	53
Osmium	Os	190.2	22,580	3,030	5,486	129.8	8.1	88
Palladium	Pd	106.4	11,995	1,554	2,829	230.3	10	72
Platinum	Pt	195.08	21,450	1,772	3,221	134	9.81	72
Potassium	K	39.09	862	63	145	753.6	6.1	104
Rhodium	Rh	102.91	12,420	1,963	3,565	242.8	4.3	151
Rubidium	Rb	85.47	1,533	38.8	102	330.7	11	58
Ruthenium	Ru	101.07	12,360	2,310	4,190	255.4	7.1	117
Silver	Ag	107.87	10,500	961	1,760	234.5	1.47	428
Sodium	Na	22.989	966	97.8	208	1,235.1	4.2	142
Strontium	Sr	87.62	2,583	770	1,418	–	20	–
Tantalum	Ta	180.95	16,670	3,000	5,432	150.7	12.3	57
Thallium	Tl	204.38	11,871	304	579	138.2	10	10
Thorium	Th	232.04	11,725	1,700	3,092	117.2	14.7	54
Tin	Sn	118.69	7,285	232	449	230.3	11.5	68
Titanium	Ti	47.88	4,508	1,670	3,038	527.5	39	22
Tungsten	W	183.85	19,254	3,387	6,128	142.8	4.9	177
Uranium	U	238.03	19,050	1,135	2,075	117.2	28	27
Vanadium	V	50.94	6,090	1,920	3,488	481.5	18.2	31
Zinc	Zn	65.38	7,135	419	786	393.5	5.5	117
Zirconium	Zr	91.22	6,507	1,850	3,362	284.7	40	23

♦

Some Extrinsic, Elemental Semiconductors

Element	Dopant	Periodic table group of dopant	Maximum solid solubility of dopant (atoms/m³)
Si	B	III A	600×10^{24}
	Al	III A	20×10^{24}
	Ga	III A	40×10^{24}
	P	V A	$1{,}000 \times 10^{24}$
	As	V A	$2{,}000 \times 10^{24}$
	Sb	V A	70×10^{24}
Ge	Al	III A	400×10^{24}
	Ga	III A	500×10^{24}
	In	III A	4×10^{24}
	As	V A	80×10^{24}
	Sb	V A	10×10^{24}

♦

Impurity Energy Levels for Extrinsic Semiconductors

Semiconductor	Dopant	$E_g - E_d$ (eV)	E_a (eV)
Si	P	0.044	–
	As	0.049	–
	Sb	0.039	–
	Bi	0.069	–
	B	–	0.045
	Al	–	0.057
	Ga	–	0.065
	In	–	0.160
	Tl	–	0.260
Ge	P	0.012	–
	As	0.013	–
	Sb	0.096	–
	B	–	0.010
	Al	–	0.010
	Ga	–	0.010
	In	–	0.011
	Tl	–	0.01
GaAs	Se	0.005	–
	Te	0.003	–
	Zn	–	0.024
	Cd	–	0.021

Stress is defined as force per unit area; for pure tension the following apply:

Engineering stress

$$\sigma = \frac{F}{A_0}$$

σ = engineering stress
F = applied force
A_0 = initial cross-sectional area

True stress

$$\sigma_T = \frac{F}{A}$$

σ_T = true stress
F = applied force
A = actual cross-sectional area

The elastic modulus (also called modulus, modulus of elasticity, Young's modulus) describes the relationship between engineering stress and engineering strain during elastic loading. Hooke's Law applies in such a case.

$\sigma = E\varepsilon$ where E is the elastic modulus.

Key mechanical properties obtained from a tensile test curve:

- Elastic modulus
- Ductility (also called percent elongation): Permanent engineering strain after failure
- Ultimate tensile strength (also called tensile strength): Maximum engineering stress
- Yield strength: Engineering stress at which permanent deformation is first observed, calculated by 0.2% offset method.

Other mechanical properties:

- Creep: Time-dependent deformation under load. Usually measured by strain rate. For steady-state creep this is:

$$\frac{d\varepsilon}{dt} = A\sigma^n e^{-\frac{Q}{RT}}$$

A = pre-exponential constant
n = stress sensitivity
Q = activation energy for creep
R = ideal gas law constant
T = absolute temperature

- Fatigue: Time-dependent failure under cyclic load. Fatigue life is the number of cycles to failure. The endurance limit is the stress below which fatigue failure is unlikely.

- Fracture toughness: The combination of applied stress and the crack length in a brittle material. It is the stress intensity when the material will fail.

$$K_{IC} = Y\sigma\sqrt{\pi a}$$

K_{IC} = fracture toughness
σ = applied engineering stress
a = crack length
Y = geometrical factor

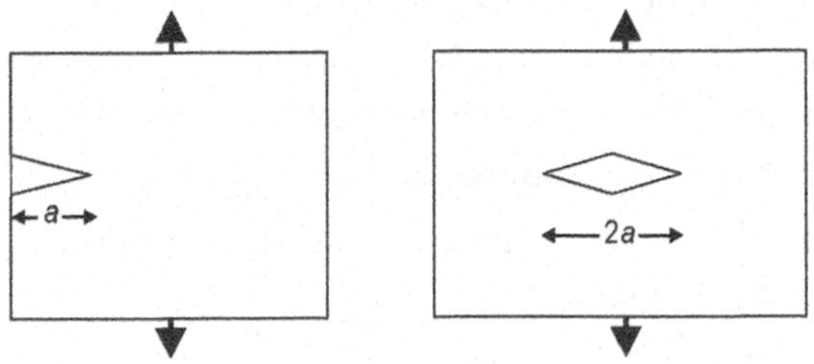

EXTERIOR CRACK (Y = 1.1)   INTERIOR CRACK (Y = 1)

The critical value of stress intensity at which catastrophic crack propagation occurs, K_{Ic}, is a material property.

♦ W.R. Runyan and S.B. Watelski, in *Handbook of Materials and Processes for Electronics*, C.A. Harper, ed., McGraw-Hill, 1970.

Representative Values of Fracture Toughness

Material	K_{Ic} (MPa•m$^{1/2}$)	K_{Ic} (ksi•in$^{1/2}$)
A1 2014-T651	24.2	22
A1 2024-T3	44	40
52100 Steel	14.3	13
4340 Steel	46	42
Alumina	4.5	4.1
Silicon Carbide	3.5	3.2

RELATIONSHIP BETWEEN HARDNESS AND TENSILE STRENGTH

For plain carbon steels, there is a general relationship between Brinell hardness and tensile strength as follows:

$$TS(psi) \simeq 500\ BHN$$

$$TS(MPa) \simeq 3.5\ BHN$$

ASTM GRAIN SIZE

$$S_V = 2P_L$$

$$N_{(0.0645\ mm^2)} = 2^{(n-1)}$$

$$\frac{N_{actual}}{\text{Actual Area}} = \frac{N}{(0.0645\ mm^2)}, \text{ where}$$

S_V = grain-boundary surface per unit volume

P_L = number of points of intersection per unit length between the line and the boundaries

N = number of grains observed in an area of 0.0645 mm^2

n = grain size (nearest integer > 1)

COMPOSITE MATERIALS

$$\rho_c = \Sigma f_i \rho_i$$

$$C_c = \Sigma f_i c_i$$

$$\left[\Sigma \frac{f_i}{E_i}\right]^{-1} \le E_c \le \Sigma f_i E_i$$

$$\sigma_c = \Sigma f_i \sigma_i$$

ρ_c = density of composite

C_c = heat capacity of composite per unit volume

E_c = Young's modulus of composite

f_i = volume fraction of individual material

c_i = heat capacity of individual material per unit volume

E_i = Young's modulus of individual material

σ_c = strength parallel to fiber direction

Also, for axially oriented, long, fiber-reinforced composites, the strains of the two components are equal.

$$(\Delta L/L)_1 = (\Delta L/L)_2$$

ΔL = change in length of the composite

L = original length of the composite

Hardness: Resistance to penetration. Measured by denting a material under known load and measuring the size of the dent.

Hardenability: The "ease" with which hardness can be obtained.

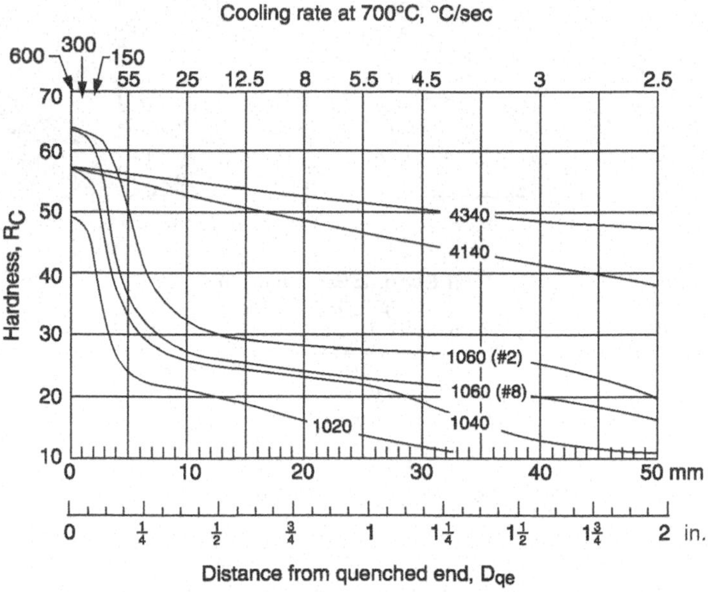

(#2) and (#8) indicate grain size

JOMINY HARDENABILITY CURVES FOR SIX STEELS

Van Vlack, L., *Elements of Materials Science & Engineering*, Addison-Wesley, 1989.

The following two graphs show cooling curves for four different positions in the bar.
C = Center
M-R = Halfway between center and surface

3/4-R = 75% of the distance between the center and the surface

S = Surface

These positions are shown in the following figure.

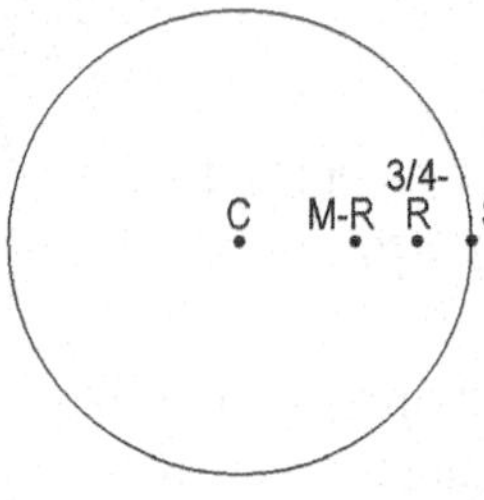

♦

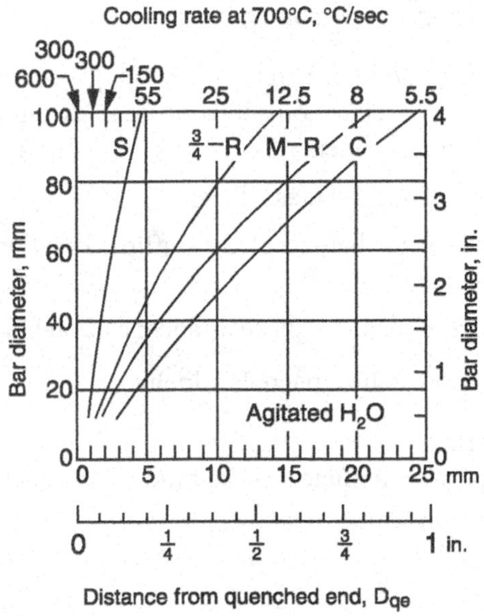

COOLING RATES FOR BARS QUENCHED IN AGITATED WATER

♦

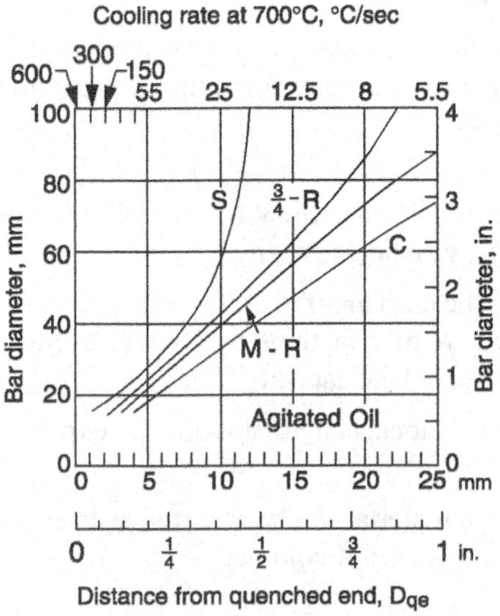

COOLING RATES FOR BARS QUENCHED IN AGITATED OIL

Impact Test

The *Charpy Impact Test* is used to find energy required to fracture and to identify ductile to brittle transition.

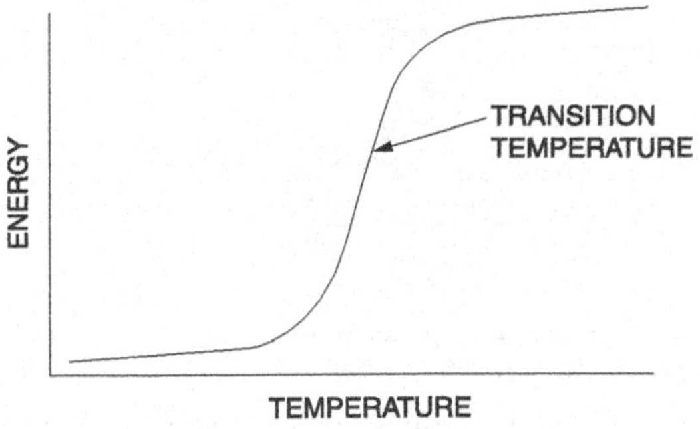

Impact tests determine the amount of energy required to cause failure in standardized test samples. The tests are repeated over a range of temperatures to determine the *ductile to brittle transition temperature.*

♦ Van Vlack, L., *Elements of Materials Science & Engineering*, Addison-Wesley, 1989.

Concrete

♦

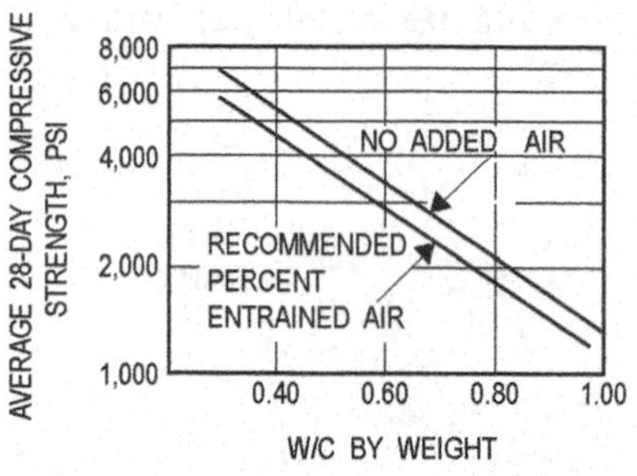

Concrete strength decreases with increases in water-cement ratio for concrete with and without entrained air.

Water-cement (W/C) ratio is the primary factor affecting the strength of concrete. The figure above shows how W/C expressed as a ratio of weight of water and cement by weight of concrete mix affects the compressive strength of both air-entrained and non-air-entrained concrete.

•

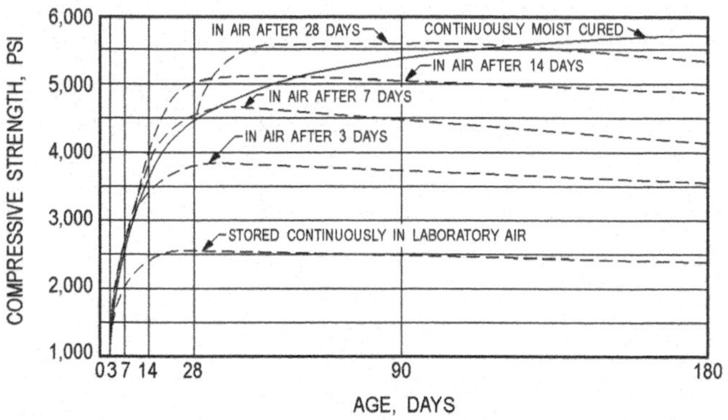

Concrete compressive strength varies with moist-curing conditions. Mixes tested had a water-cement ratio of 0.50, a slump of 3.5 in., cement content of 556 lb/yd^3, sand content of 36%, and air content of 4%.

Water content affects workability. However, an increase in water without a corresponding increase in cement reduces the concrete strength. Superplasticizers are the most typical way to increase workability. Air entrainment is used to improve durability.

Amorphous Materials

Amorphous materials such as glass are non-crystalline solids. Thermoplastic polymers are either semicrystalline or amorphous.

Below the glass transition temperature (T_g) the amorphous material will be a brittle solid.

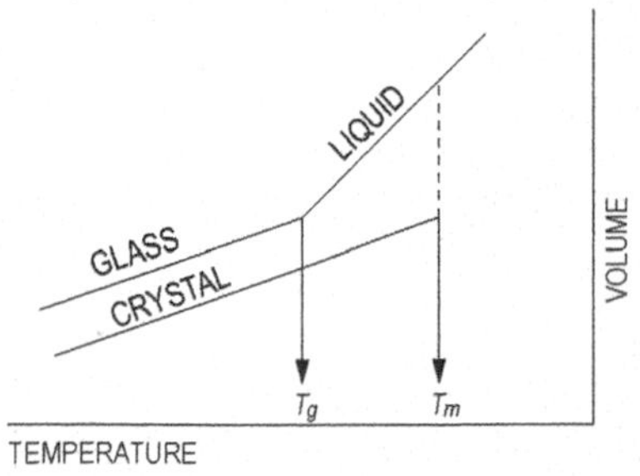

The volume temperature curve as shown above is often used to show the difference between amorphous and crystalline solids.

Polymers

Polymers are classified as thermoplastics that can be melted and reformed. Thermosets cannot be melted and reformed.

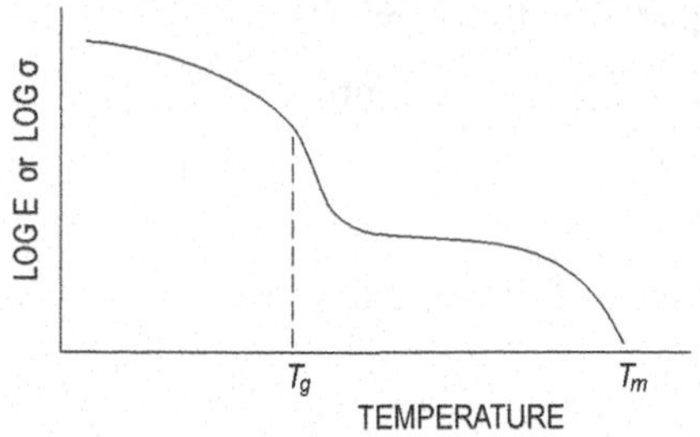

The above curve shows the temperature dependent strength (σ) or modulus (E) for a thermoplastic polymer.

Polymer Additives

Chemicals and compounds are added to polymers to improve properties for commercial use. These substances, such as plasticizers, improve formability during processing, while others increase strength or durability.

Examples of common additives are:

- Plasticizers: vegetable oils, low molecular weight polymers or monomers
- Fillers: talc, chopped glass fibers
- Flame retardants: halogenated paraffins, zinc borate, chlorinated phosphates
- Ultraviolet or visible light resistance: carbon black
- Oxidation resistance: phenols, aldehydes

Thermal Properties

The thermal expansion coefficient is the ratio of engineering strain to the change in temperature.

$$\alpha = \frac{\varepsilon}{\Delta T}$$

α = thermal expansion coefficient
ε = engineering strain
ΔT = change in temperature

Specific heat (also called heat capacity) is the amount of heat required to raise the temperature of something or an amount of something by 1 degree.

At constant pressure the amount of heat (Q) required to increase the temperature of something by ΔT is $C_p \Delta T$, where C_p is the constant pressure heat capacity.

At constant volume the amount of heat (Q) required to increase the temperature of something by ΔT is $C_v \Delta T$, where C_v is the constant volume heat capacity.

An object can have a heat capacity that would be expressed as energy/degree.

The heat capacity of a material can be reported as energy/degree per unit mass or per unit volume.

♦ *Concrete Manual*, 8th ed., U.S. Bureau of Reclamation, 1975.
• Merritt, Frederick S., *Standard Handbook for Civil Engineers*, 3rd ed., McGraw-Hill, 1983.

BINARY PHASE DIAGRAMS

Allows determination of (1) what phases are present at equilibrium at any temperature and average composition, (2) the compositions of those phases, and (3) the fractions of those phases.

Eutectic reaction (liquid → two solid phases)

Eutectoid reaction (solid → two solid phases)

Peritectic reaction (liquid + solid → solid)

Peritectoid reaction (two solid phases → solid)

Lever Rule

The following phase diagram and equations illustrate how the weight of each phase in a two-phase system can be determined:

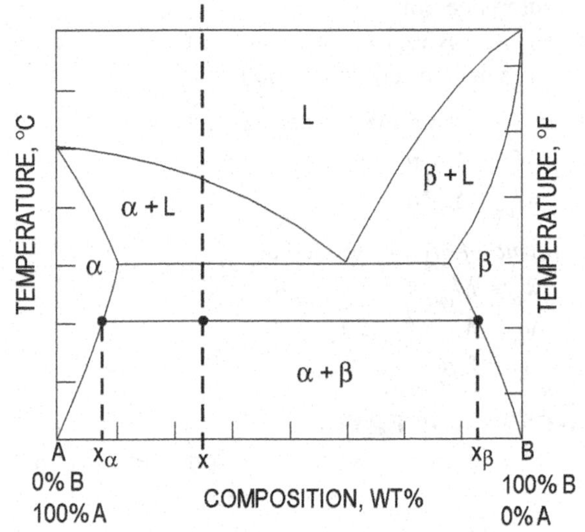

(In diagram, L = liquid.) If x = the average composition at temperature *T*, then

$$\text{wt\%}\ \alpha = \frac{x_\beta - x}{x_\beta - x_\alpha} \times 100$$

$$\text{wt\%}\ \beta = \frac{x - x_\alpha}{x_\beta - x_\alpha} \times 100$$

Iron-Iron Carbide Phase Diagram

♦

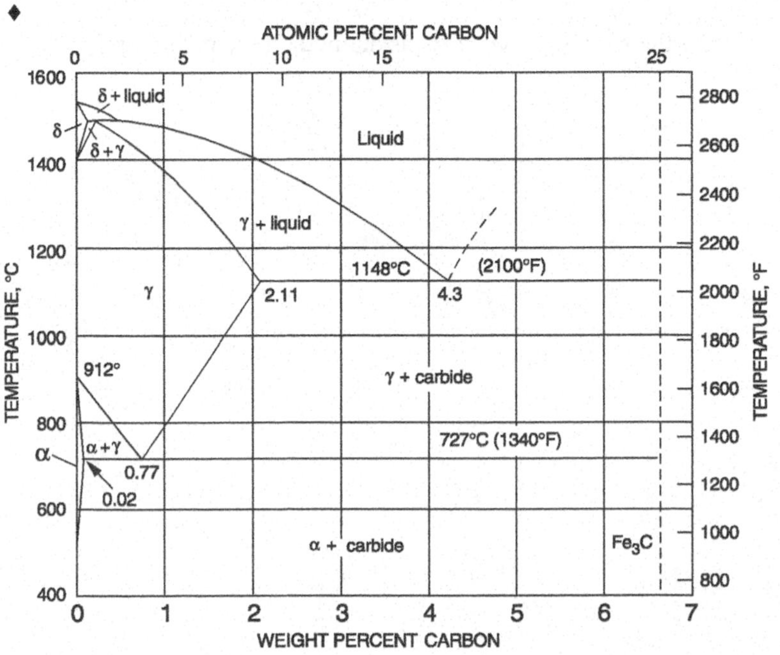

♦ Van Vlack, L., *Elements of Materials Science & Engineering*, Addison-Wesley, Boston, 1989.

STATICS

FORCE

A *force* is a *vector* quantity. It is defined when its **(1)** magnitude, **(2)** point of application, and **(3)** direction are known.

The vector form of a force is

$$\boldsymbol{F} = F_x\mathbf{i} + F_y\mathbf{j}$$

RESULTANT (TWO DIMENSIONS)

The *resultant*, F, of n forces with components $F_{x,i}$ and $F_{y,i}$ has the magnitude of

$$F = \left[\left(\sum_{i=1}^{n} F_{x,i}\right)^2 + \left(\sum_{i=1}^{n} F_{y,i}\right)^2\right]^{1/2}$$

The resultant direction with respect to the x-axis is

$$\theta = \arctan\left(\sum_{i=1}^{n} F_{y,i} \Big/ \sum_{i=1}^{n} F_{x,i}\right)$$

RESOLUTION OF A FORCE

$F_x = F\cos\theta_x$ $\quad F_y = F\cos\theta_y$ $\quad F_z = F\cos\theta_z$

$\cos\theta_x = F_x/F$ $\quad \cos\theta_y = F_y/F$ $\quad \cos\theta_z = F_z/F$

Separating a force into components when the geometry of force is known and $R = \sqrt{x^2 + y^2 + z^2}$

$F_x = (x/R)F$ $\quad F_y = (y/R)\mathrm{F}$ $\quad F_z = (z/R)F$

MOMENTS (COUPLES)

A system of two forces that are equal in magnitude, opposite in direction, and parallel to each other is called a *couple*. A *moment* $\boldsymbol{M}$ is defined as the cross product of the *radius vector* $\boldsymbol{r}$ and the *force* $\boldsymbol{F}$ from a point to the line of action of the force.

$$\boldsymbol{M} = \boldsymbol{r} \times \boldsymbol{F} \qquad M_x = yF_z - zF_y$$
$$M_y = zF_x - xF_z$$
$$M_z = xF_y - yF_x$$

SYSTEMS OF FORCES

$$\boldsymbol{F} = \Sigma\,\boldsymbol{F}_n$$
$$\boldsymbol{M} = \Sigma\,(\boldsymbol{r}_n \times \boldsymbol{F}_n)$$

Equilibrium Requirements

$$\Sigma\,\boldsymbol{F}_n = 0$$
$$\Sigma\,\boldsymbol{M}_n = 0$$

CENTROIDS OF MASSES, AREAS, LENGTHS, AND VOLUMES

The following formulas are for discrete masses, areas, lengths, and volumes:

$$\boldsymbol{r}_c = \Sigma\, m_n\boldsymbol{r}_n / \Sigma\, m_n, \text{ where}$$

m_n = the *mass of each particle* making up the system,

$\boldsymbol{r}_n$ = the *radius vector* to each particle from a selected reference point, and

$\boldsymbol{r}_c$ = the *radius vector* to the *centroid of the total mass* from the selected reference point.

The *moment of area* (M_a) is defined as

$$M_{ay} = \Sigma\, x_n a_n$$
$$M_{ax} = \Sigma\, y_n a_n$$

The *centroid of area* is defined as

$$x_{ac} = M_{ay}/A = \Sigma\, x_n a_n/A$$
$$y_{ac} = M_{ax}/A = \Sigma\, y_n a_n/A$$

where $A = \Sigma\, a_n$

MOMENT OF INERTIA

The *moment of inertia*, or the second moment of area, is defined as

$$I_y = \int x^2\, dA$$
$$I_x = \int y^2\, dA$$

The *polar moment of inertia* J of an area about a point is equal to the sum of the moments of inertia of the area about any two perpendicular axes in the area and passing through the same point.

$$I_z = J = I_y + I_x = \int (x^2 + y^2)\, dA$$
$$= r_p^{\,2} A, \text{ where}$$

r_p = the *radius of gyration* (as defined on the next page)

Moment of Inertia Parallel Axis Theorem

The moment of inertia of an area about any axis is defined as the moment of inertia of the area about a parallel centroidal axis plus a term equal to the area multiplied by the square of the perpendicular distance d from the centroidal axis to the axis in question.

$$I'_x = I_{x_c} + d_y^2 A$$

$$I'_y = I_{y_c} + d_x^2 A, \text{ where}$$

d_x, d_y = distance between the two axes in question

I_{x_c}, I_{y_c} = the moment of inertia about the centroidal axis

I'_x, I'_y = the moment of inertia about the new axis

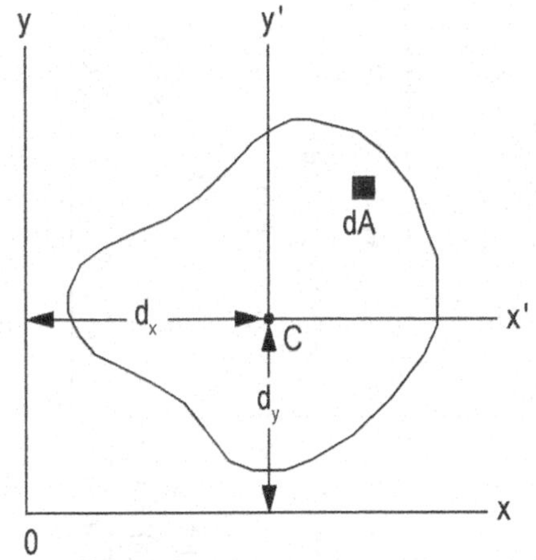

Hibbeler, R.C., *Engineering Mechanics: Statics and Dynamics*, 10 ed., Pearson Prentice Hall, 2004.

Radius of Gyration

The *radius of gyration* r_p, r_x, r_y is the distance from a reference axis at which all of the area can be considered to be concentrated to produce the moment of inertia.

$$r_x = \sqrt{I_x/A} \quad r_y = \sqrt{I_y/A} \quad r_p = \sqrt{J/A}$$

Product of Inertia

The *product of inertia* (I_{xy}, etc.) is defined as:

$I_{xy} = \int xydA$, with respect to the xy-coordinate system,

The *parallel-axis theorem* also applies:

$I'_{xy} = I_{x_c y_c} + d_x d_y A$ for the xy-coordinate system, etc.

where

d_x = x-axis distance between the two axes in question, and

d_y = y-axis distance between the two axes in question.

FRICTION

The largest frictional force is called the *limiting friction*. Any further increase in applied forces will cause motion.

$$F \le \mu_s N, \text{ where}$$

F = friction force,

μ_s = *coefficient of static friction*, and

N = normal force between surfaces in contact.

SCREW THREAD

For a *screw-jack, square thread*,

$$M = Pr \tan(\alpha \pm \phi), \text{ where}$$

+ is for screw tightening,

– is for screw loosening,

M = external moment applied to axis of screw,

P = load on jack applied along and on the line of the axis,

r = the mean thread radius,

α = the *pitch angle* of the thread, and

$\mu = \tan \phi$ = the appropriate coefficient of friction.

BELT FRICTION

$$F_1 = F_2 e^{\mu\theta}, \text{ where}$$

F_1 = force being applied in the direction of impending motion,

F_2 = force applied to resist impending motion,

μ = coefficient of static friction, and

θ = the total *angle of contact* between the surfaces expressed in radians.

STATICALLY DETERMINATE TRUSS

Plane Truss: Method of Joints

The method consists of solving for the forces in the members by writing the two equilibrium equations for each joint of the truss.

$$\Sigma F = 0 \text{ and } \Sigma F = 0, \text{ where}$$

F = horizontal forces and member components and

F = vertical forces and member components.

Plane Truss: Method of Sections

The method consists of drawing a free-body diagram of a portion of the truss in such a way that the unknown truss member force is exposed as an external force.

CONCURRENT FORCES

A concurrent-force system is one in which the lines of action of the applied forces all meet at one point.

A *two-force* body in static equilibrium has two applied forces that are equal in magnitude, opposite in direction, and collinear.

Figure	Area & Centroid	Area Moment of Inertia	(Radius of Gyration)^2	Product of Inertia
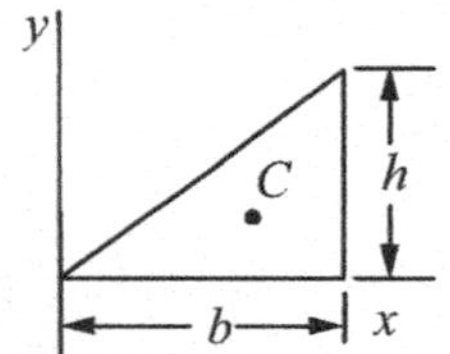	$A = bh/2$ $x_c = 2b/3$ $y_c = h/3$	$I_{x_c} = bh^3/36$ $I_{y_c} = b^3h/36$ $I_x = bh^3/12$ $I_y = b^3h/4$	$r_{x_c}^2 = h^2/18$ $r_{y_c}^2 = b^2/18$ $r_x^2 = h^2/6$ $r_y^2 = b^2/2$	$I_{x_c y_c} = Abh/36 = b^2h^2/72$ $I_{xy} = Abh/4 = b^2h^2/8$
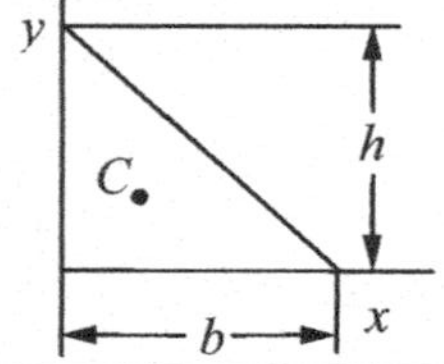	$A = bh/2$ $x_c = b/3$ $y_c = h/3$	$I_{x_c} = bh^3/36$ $I_{y_c} = b^3h/36$ $I_x = bh^3/12$ $I_y = b^3h/12$	$r_{x_c}^2 = h^2/18$ $r_{y_c}^2 = b^2/18$ $r_x^2 = h^2/6$ $r_y^2 = b^2/6$	$I_{x_c y_c} = -Abh/36 = -b^2h^2/72$ $I_{xy} = Abh/12 = b^2h^2/24$
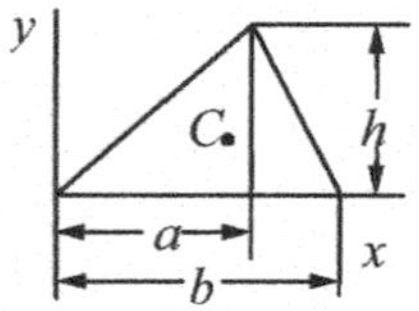	$A = bh/2$ $x_c = (a+b)/3$ $y_c = h/3$	$I_{x_c} = bh^3/36$ $I_{y_c} = \left[bh\left(b^2 - ab + a^2\right)\right]/36$ $I_x = bh^3/12$ $I_y = \left[bh\left(b^2 + ab + a^2\right)\right]/12$	$r_{x_c}^2 = h^2/18$ $r_{y_c}^2 = \left(b^2 - ab + a^2\right)/18$ $r_x^2 = h^2/6$ $r_y^2 = \left(b^2 + ab + a^2\right)/6$	$I_{x_c y_c} = \left[Ah(2a-b)\right]/36$ $= \left[bh^2(2a-b)\right]/72$ $I_{xy} = \left[Ah(2a+b)\right]/12$ $= \left[bh^2(2a+b)\right]/24$
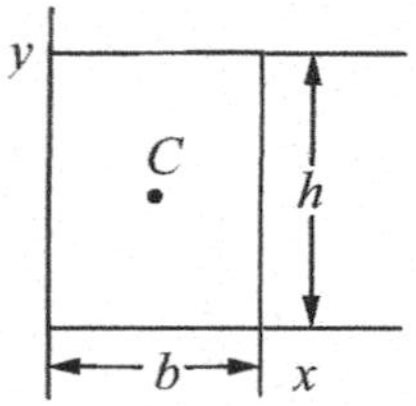	$A = bh$ $x_c = b/2$ $y_c = h/2$	$I_{x_c} = bh^3/12$ $I_{y_c} = b^3h/12$ $I_x = bh^3/3$ $I_y = b^3h/3$ $J = \left[bh\left(b^2 + h^2\right)\right]/12$	$r_{x_c}^2 = h^2/12$ $r_{y_c}^2 = b^2/12$ $r_x^2 = h^2/3$ $r_y^2 = b^2/3$ $r_p^2 = \left(b^2 + h^2\right)/12$	$I_{x_c y_c} = 0$ $I_{xy} = Abh/4 = b^2h^2/4$
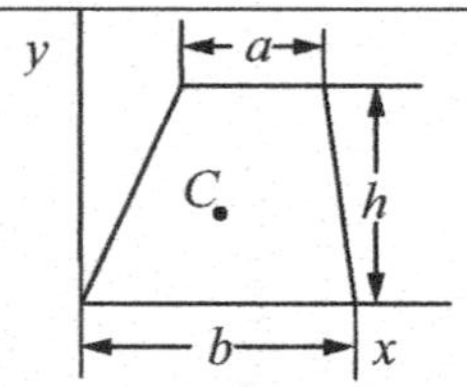	$A = h(a+b)/2$ $y_c = \frac{h(2a+b)}{3(a+b)}$	$I_{x_c} = \frac{h^3\left(a^2 + 4ab + b^2\right)}{36(a+b)}$ $I_x = \frac{h^3(3a+b)}{12}$	$r_{x_c}^2 = \frac{h^2\left(a^2 + 4ab + b^2\right)}{18(a+b)}$ $r_x^2 = \frac{h^2(3a+b)}{6(a+b)}$	
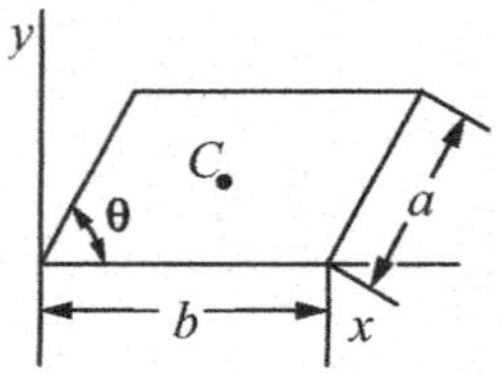	$A = ab \sin\theta$ $x_c = (b + a\cos\theta)/2$ $y_c = (a\sin\theta)/2$	$I_{x_c} = \left(a^3 b \sin^3\theta\right)/12$ $I_{y_c} = \left[ab\sin\theta\left(b^2 + a^2\cos^2\theta\right)\right]/12$ $I_x = \left(a^3 b \sin^3\theta\right)/3$ $I_y = \left[ab\sin\theta(b + a\cos\theta)^2\right]/3$ $-\left(a^2b^2\sin\theta\cos\theta\right)/6$	$r_{x_c}^2 = (a\sin\theta)^2/12$ $r_{y_c}^2 = \left(b^2 + a^2\cos^2\theta\right)/12$ $r_x^2 = (a\sin\theta)^2/3$ $r_y^2 = (b + a\cos\theta)^2/3$ $-(ab\cos\theta)/6$	$I_{x_c y_c} = \left(a^3 b \sin^2\theta\cos\theta\right)/12$

Housner, George W., and Donald E. Hudson, *Applied Mechanics Dynamics*, D. Van Nostrand Company, Inc., Princeton, NJ, 1959. Table reprinted by permission of G.W. Housner & D.E. Hudson.

Figure	Area & Centroid	Area Moment of Inertia	(Radius of Gyration)^2	Product of Inertia
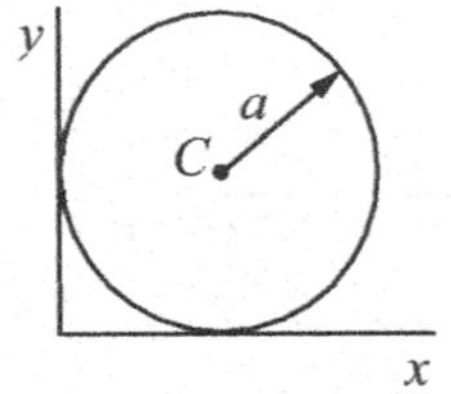	$A = \pi a^2$ $x_c = a$ $y_c = a$	$I_{x_c} = I_{y_c} = \pi a^4/4$ $I_x = I_y = 5\pi a^4/4$ $J = \pi a^4/2$	$r_{x_c}^2 = r_{y_c}^2 = a^2/4$ $r_x^2 = r_y^2 = 5a^2/4$ $r_p^2 = a^2/2$	$I_{x_c y_c} = 0$ $I_{xy} = Aa^2$
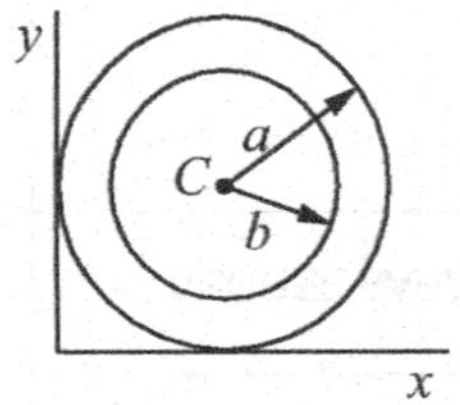	$A = \pi(a^2 - b^2)$ $x_c = a$ $y_c = a$	$I_{x_c} = I_{y_c} = \pi(a^4 - b^4)/4$ $I_x = I_y = \frac{5\pi a^4}{4} - \pi a^2 b^2 - \frac{\pi b^4}{4}$ $J = \pi(a^4 - b^4)/2$	$r_{x_c}^2 = r_{y_c}^2 = (a^2 + b^2)/4$ $r_x^2 = r_y^2 = (5a^2 + b^2)/4$ $r_p^2 = (a^2 + b^2)/2$	$I_{x_c y_c} = 0$ $I_{xy} = Aa^2$ $= \pi a^2(a^2 - b^2)$
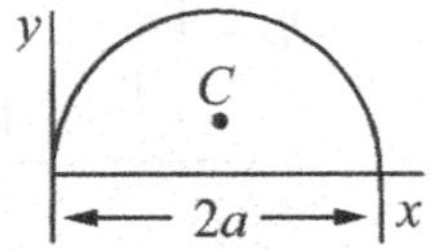	$A = \pi a^2/2$ $x_c = a$ $y_c = 4a/(3\pi)$	$I_{x_c} = \frac{a^4(9\pi^2 - 64)}{72\pi}$ $I_{y_c} = \pi a^4/8$ $I_x = \pi a^4/8$ $I_y = 5\pi a^4/8$	$r_{x_c}^2 = \frac{a^2(9\pi^2 - 64)}{36\pi^2}$ $r_{y_c}^2 = a^2/4$ $r_x^2 = a^2/4$ $r_y^2 = 5a^2/4$	$I_{x_c y_c} = 0$ $I_{xy} = 2a^4/3$
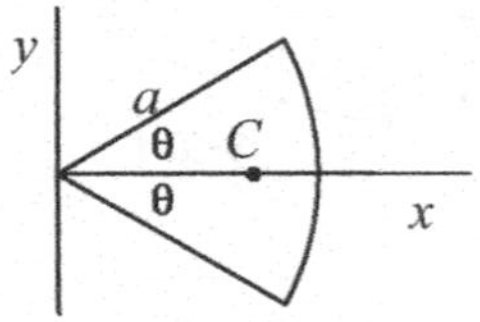 CIRCULAR SECTOR	$A = a^2\theta$ $x_c = \frac{2a}{3}\frac{\sin\theta}{\theta}$ $y_c = 0$	$I_x = a^4(\theta - \sin\theta\cos\theta)/4$ $I_y = a^4(\theta + \sin\theta\cos\theta)/4$	$r_x^2 = \frac{a^2}{4}\frac{(\theta - \sin\theta\cos\theta)}{\theta}$ $r_y^2 = \frac{a^2}{4}\frac{(\theta + \sin\theta\cos\theta)}{\theta}$	$I_{x_c y_c} = 0$ $I_{xy} = 0$
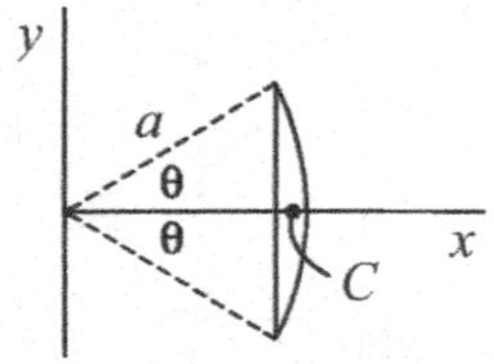 CIRCULAR SEGMENT	$A = a^2\left[\theta - \frac{\sin 2\theta}{2}\right]$ $x_c = \frac{2a}{3}\frac{\sin^3\theta}{\theta - \sin\theta\cos\theta}$ $y_c = 0$	$I_x = \frac{Aa^2}{4}\left[1 - \frac{2\sin^3\theta\cos\theta}{3\theta - 3\sin\theta\cos\theta}\right]$ $I_y = \frac{Aa^2}{4}\left[1 + \frac{2\sin^3\theta\cos\theta}{\theta - \sin\theta\cos\theta}\right]$	$r_x^2 = \frac{a^2}{4}\left[1 - \frac{2\sin^3\theta\cos\theta}{3\theta - 3\sin\theta\cos\theta}\right]$ $r_y^2 = \frac{a^2}{4}\left[1 + \frac{2\sin^3\theta\cos\theta}{\theta - \sin\theta\cos\theta}\right]$	$I_{x_c y_c} = 0$ $I_{xy} = 0$

Housner, George W., and Donald E. Hudson, *Applied Mechanics Dynamics*, D. Van Nostrand Company, Inc., Princeton, NJ, 1959. Table reprinted by permission of G.W. Housner & D.E. Hudson.

Figure	Area & Centroid	Area Moment of Inertia	(Radius of Gyration)^2	Product of Inertia
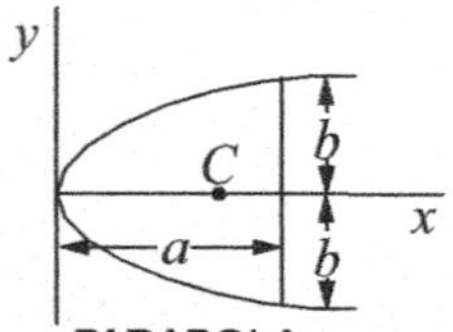 PARABOLA	$A = 4ab/3$ $x_c = 3a/5$ $y_c = 0$	$I_{x_c} = I_x = 4ab^3/15$ $I_{y_c} = 16a^3b/175$ $I_y = 4a^3b/7$	$r_{x_c}^2 = r_x^2 = b^2/5$ $r_{y_c}^2 = 12a^2/175$ $r_y^2 = 3a^2/7$	$I_{x_c y_c} = 0$ $I_{xy} = 0$
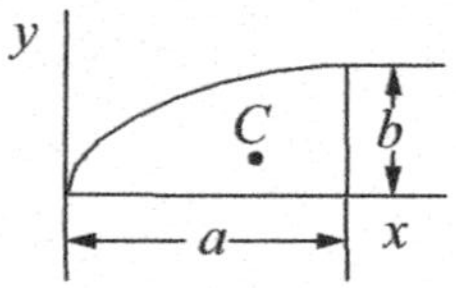 HALF A PARABOLA	$A = 2ab/3$ $x_c = 3a/5$ $y_c = 3b/8$	$I_x = 2ab^3/15$ $I_y = 2ba^3/7$	$r_x^2 = b^2/5$ $r_y^2 = 3a^2/7$	$I_{xy} = Aab/4 = a^2b^2$
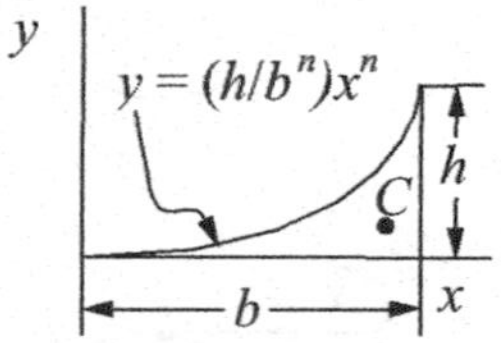 n[th] DEGREE PARABOLA	$A = bh/(n+1)$ $x_c = \frac{n+1}{n+2}b$ $y_c = \frac{h}{2}\frac{n+1}{2n+1}$	$I_x = \frac{bh^3}{3(3n+1)}$ $I_y = \frac{hb^3}{n+3}$	$r_x^2 = \frac{h^2(n+1)}{3(3n+1)}$ $r_y^2 = \frac{n+1}{n+3}b^2$	
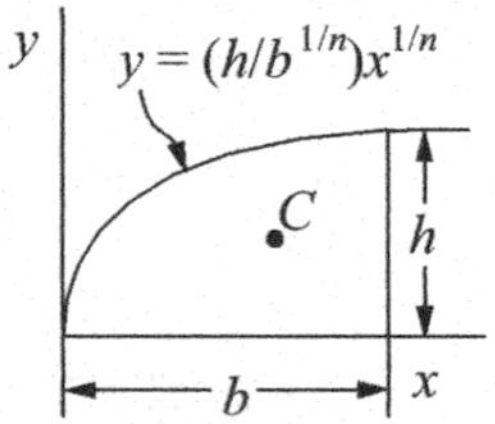 n[th] DEGREE PARABOLA	$A = \frac{n}{n+1}bh$ $x_c = \frac{n+1}{2n+1}b$ $y_c = \frac{n+1}{2(n+2)}h$	$I_x = \frac{n}{3(n+3)}bh^3$ $I_y = \frac{n}{3n+1}b^3h$	$r_x^2 = \frac{n+1}{3(n+1)}h^2$ $r_y^2 = \frac{n+1}{3n+1}b^2$	

Housner, George W., and Donald E. Hudson, *Applied Mechanics Dynamics*, D. Van Nostrand Company, Inc., Princeton, NJ, 1959. Table reprinted by permission of G.W. Housner & D.E. Hudson.

DYNAMICS

COMMON NOMENCLATURE

t = time

s = position coordinate, measured along a curve from an origin

v = velocity

a = acceleration

a_n = normal acceleration

a_t = tangential acceleration

θ = angular position coordinate

ω = angular velocity

α = angular acceleration

Ω = angular velocity of x,y,z reference axis measured from the X,Y,Z reference

$\dot{\Omega}$ = angular acceleration of x,y,z reference axis measured from the X,Y,Z reference

$\mathbf{r}_{A/B}$ = relative position of "A" with respect to "B"

$\mathbf{v}_{A/B}$ = relative velocity of "A" with respect to "B"

$\mathbf{a}_{A/B}$ = relative acceleration of "A" with respect to "B"

PARTICLE KINEMATICS

Kinematics is the study of motion without consideration of the mass of, or the forces acting on, a system. For particle motion, let $\mathbf{r}(t)$ be the position vector of the particle in an inertial reference frame. The velocity and acceleration of the particle are defined, respectively, as

$\mathbf{v} = d\mathbf{r}/dt$

$\mathbf{a} = d\mathbf{v}/dt$, where

$\mathbf{v}$ = the instantaneous velocity

$\mathbf{a}$ = the instantaneous acceleration

t = time

Cartesian Coordinates

$$\mathbf{r} = x\mathbf{i} + y\mathbf{j} + z\mathbf{k}$$

$$\mathbf{v} = \dot{x}\mathbf{i} + \dot{y}\mathbf{j} + \dot{z}\mathbf{k}$$

$$\mathbf{a} = \ddot{x}\mathbf{i} + \ddot{y}\mathbf{j} + \ddot{z}\mathbf{k}, \text{ where}$$

$\dot{x} = dx/dt = v_x$, etc.

$\ddot{x} = d^2x/dt^2 = a_x$, etc.

Radial and Transverse Components for Planar Motion

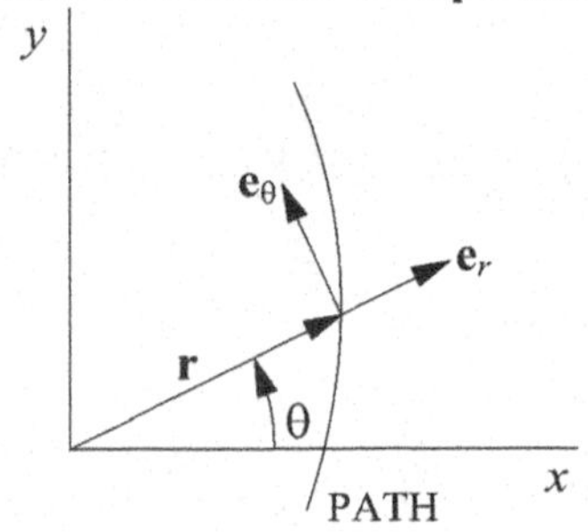

Unit vectors $\mathbf{e}_r$ and $\mathbf{e}_\theta$ are, respectively, collinear with and normal to the position vector $\mathbf{r}$. Thus:

$$\mathbf{r} = r\mathbf{e}_r$$

$$\mathbf{v} = \dot{r}\mathbf{e}_r + r\dot{\theta}\mathbf{e}_\theta$$

$$\mathbf{a} = \left(\ddot{r} - r\dot{\theta}^2\right)\mathbf{e}_r + \left(r\ddot{\theta} + 2\dot{r}\dot{\theta}\right)\mathbf{e}_\theta, \text{ where}$$

r = radial position coordinate

θ = angle from the x axis to $\mathbf{r}$

$\dot{r} = dr/dt$, etc., $\ddot{r} = d^2r/dt^2$, etc.

♦ Particle Rectilinear Motion

Variable a	*Constant $a = a_0$*
$a = \dfrac{dv}{dt}$	$v = v_0 + a_0 t$
$v = \dfrac{ds}{dt}$	$s = s_0 + v_0 t + \frac{1}{2} a_0 t^2$
$a\,ds = v\,dv$	$v^2 = v_0^2 + 2a_0\left(s - s_0\right)$

♦ Particle Curvilinear Motion

x, y, z Coordinates		*r, θ, z Coordinates*	
$v_x = \dot{x}$	$a_x = \ddot{x}$	$v_r = \dot{r}$	$a_r = \ddot{r} - r\dot{\theta}^2$
$v_y = \dot{y}$	$a_y = \ddot{y}$	$v_\theta = r\dot{\theta}$	$a_\theta = r\ddot{\theta} + 2\dot{r}\dot{\theta}$
$v_z = \dot{z}$	$a_z = \ddot{z}$	$v_z = \dot{z}$	$a_z = \ddot{z}$

n, t, b Coordinates

$v = \dot{s}$	$a_t = \dot{v} = \dfrac{dv}{dt} = v\dfrac{dv}{ds}$
	$a_n = \dfrac{v^2}{\rho}$ $\quad \rho = \dfrac{\left[1 + (dy/dx)^2\right]^{3/2}}{\left\lvert \dfrac{d^2y}{dx^2} \right\rvert}$

♦ Relative Motion

$\mathbf{r}_A = \mathbf{r}_B + \mathbf{r}_{A/B} \qquad \mathbf{v}_A = \mathbf{v}_B + \mathbf{v}_{A/B} \qquad \mathbf{a}_A = \mathbf{a}_B + \mathbf{a}_{A/B}$

Translating Axes x-y

The equations that relate the absolute and relative position, velocity, and acceleration vectors of two particles A and B, in plane motion, and separated at a constant distance, may be written as

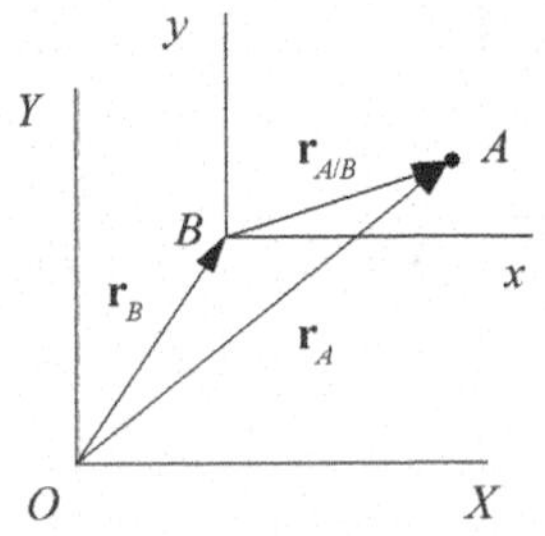

$$\mathbf{r}_A = \mathbf{r}_B + \mathbf{r}_{A/B}$$

$$\mathbf{v}_A = \mathbf{v}_B + \boldsymbol{\omega} \times \mathbf{r}_{A/B} = \mathbf{v}_B + \mathbf{v}_{A/B}$$

$$\mathbf{a}_A = \mathbf{a}_B + \boldsymbol{\alpha} \times \mathbf{r}_{A/B} + \boldsymbol{\omega} \times (\boldsymbol{\omega} \times \mathbf{r}_{A/B}) = \mathbf{a}_B + \mathbf{a}_{A/B}$$

♦ Adapted from Hibbeler, R.C., *Engineering Mechanics*, 10th ed., Prentice Hall, 2003.

where ω and α are the absolute angular velocity and absolute angular acceleration of the relative position vector $\mathbf{r}_{A/B}$ of constant length, respectively.

Rotating Axis

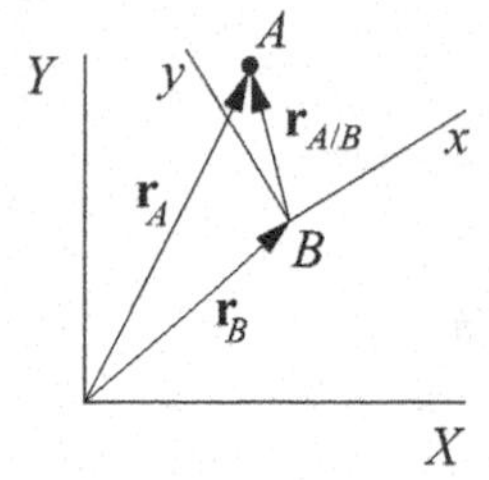

$$\mathbf{r}_A = \mathbf{r}_B + \mathbf{r}_{A/B}$$
$$\mathbf{v}_A = \mathbf{v}_B + \mathbf{\Omega} \times \mathbf{r}_{A/B} + (\mathbf{v}_{A/B})_{xyz}$$
$$\mathbf{a}_A = \mathbf{a}_B + \dot{\mathbf{\Omega}} \times \mathbf{r}_{A/B} + \mathbf{\Omega} \times (\mathbf{\Omega} \times \mathbf{r}_{A/B}) + 2\mathbf{\Omega} \times (\mathbf{v}_{A/B})_{xyz} + (\mathbf{a}_{A/B})_{xyz}$$

where $\mathbf{\Omega}$ and $\dot{\mathbf{\Omega}}$ are, respectively, the total angular velocity and acceleration of the relative position vector $\mathbf{r}_{A/B}$.

Plane Circular Motion

A special case of radial and transverse components is for constant radius rotation about the origin, or plane circular motion.

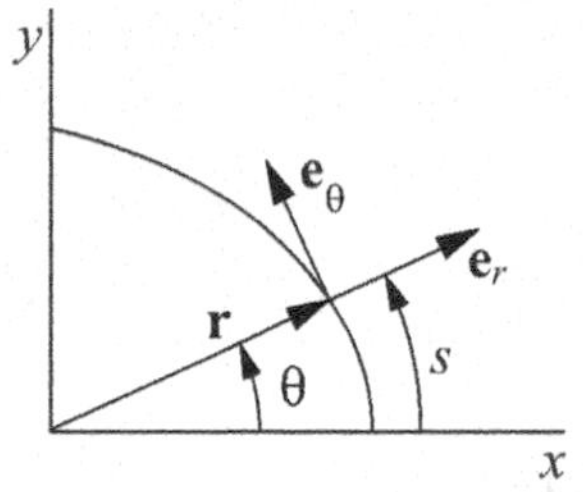

Here the vector quantities are defined as

$\mathbf{r} = r\mathbf{e}_r$

$\mathbf{v} = r\omega\mathbf{e}_\theta$

$\mathbf{a} = \left(-r\omega^2\right)\mathbf{e}_r + r\alpha\mathbf{e}_\theta$, where

r = the radius of the circle

θ = the angle from the x axis to $\mathbf{r}$

The values of the angular velocity and acceleration, respectively, are defined as

$$\omega = \dot{\theta}$$
$$\alpha = \dot{\omega} = \ddot{\theta}$$

Arc length, transverse velocity, and transverse acceleration, respectively, are

$$s = r\theta$$
$$v_\theta = r\omega$$
$$a_\theta = r\alpha$$

The radial acceleration is given by

$$a_r = -r\omega^2 \text{ (towards the center of the circle)}$$

Normal and Tangential Components

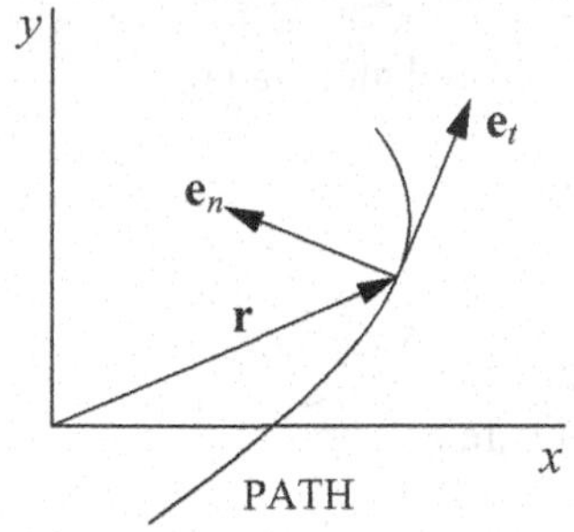

Unit vectors $\mathbf{e}_t$ and $\mathbf{e}_n$ are, respectively, tangent and normal to the path with $\mathbf{e}_n$ pointing to the center of curvature. Thus

$\mathbf{v} = v(t)\mathbf{e}_t$

$\mathbf{a} = a(t)\mathbf{e}_t + \left(v_t^2/\rho\right)\mathbf{e}_n$, where

ρ = instantaneous radius of curvature

Constant Acceleration

The equations for the velocity and displacement when acceleration is a constant are given as

$a(t) = a_0$
$v(t) = a_0(t - t_0) + v_0$
$s(t) = a_0(t - t_0)^2/2 + v_0(t - t_0) + s_0$, where
s = displacement at time t, along the line of travel
s_0 = displacement at time t_0
v = velocity along the direction of travel
v_0 = velocity at time t_0
a_0 = constant acceleration
t = time
t_0 = some initial time

For a free-falling body, $a_0 = g$ (downward towards earth). An additional equation for velocity as a function of position may be written as

$$v^2 = v_0^2 + 2a_0(s - s_0)$$

For constant angular acceleration, the equations for angular velocity and displacement are

$$\alpha(t) = \alpha_0$$
$$\omega(t) = \alpha_0(t - t_0) + \omega_0$$
$$\theta(t) = \alpha_0(t - t_0)^2/2 + \omega_0(t - t_0) + \theta_0, \quad \text{where}$$

θ = angular displacement
θ_0 = angular displacement at time t_0
ω = angular velocity
ω_0 = angular velocity at time t_0
α_0 = constant angular acceleration
t = time
t_0 = some initial time

An additional equation for angular velocity as a function of angular position may be written as

$$\omega^2 = \omega_0^2 + 2\alpha_0(\theta - \theta_0)$$

Projectile Motion

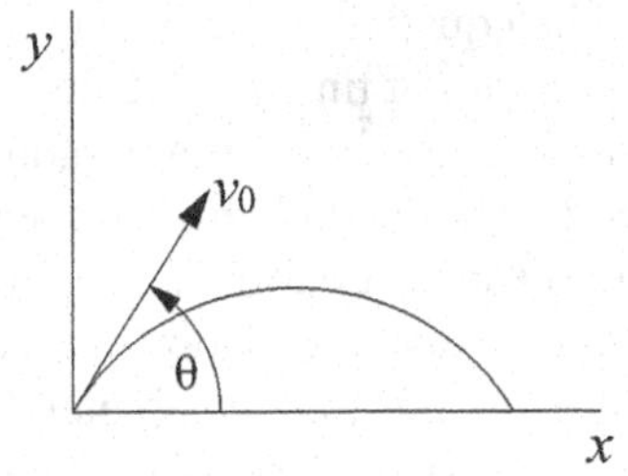

The equations for common projectile motion may be obtained from the constant acceleration equations as

$a_x = 0$

$v_x = v_0 \cos(\theta)$

$x = v_0 \cos(\theta)t + x_0$

$a_y = -g$

$v_y = -gt + v_0 \sin(\theta)$

$y = -gt^2/2 + v_0 \sin(\theta)t + y_0$

Non-constant Acceleration

When non-constant acceleration, $a(t)$, is considered, the equations for the velocity and displacement may be obtained from

$$v(t) = \int_{t_0}^{t} a(\tau)d\tau + v_{t_0}$$

$$s(t) = \int_{t_0}^{t} v(\tau)d\tau + s_{t_0}$$

For variable angular acceleration

$$\omega(t) = \int_{t_0}^{t} \alpha(\tau)d\tau + \omega_{t_0}$$

$$\theta(t) = \int_{t_0}^{t} \omega(\tau)d\tau + \theta_{t_0}$$

where τ is the variable of integration

CONCEPT OF WEIGHT

$W = mg$, where

W = weight, N (lbf)

m = mass, kg (lbf-sec^2/ft)

g = local acceleration of gravity, m/s^2 (ft/sec^2)

PARTICLE KINETICS

Newton's second law for a particle is

$\Sigma\mathbf{F} = d(m\mathbf{v})/dt$, where

$\Sigma\mathbf{F}$ = the sum of the applied forces acting on the particle

m = the mass of the particle

$\mathbf{v}$ = the velocity of the particle

For constant mass,

$$\Sigma\mathbf{F} = m\, d\mathbf{v}/dt = m\mathbf{a}$$

One-Dimensional Motion of a Particle (Constant Mass)

When motion exists only in a single dimension then, without loss of generality, it may be assumed to be in the x direction, and

$a_x = F_x/m$, where

F_x = the resultant of the applied forces, which in general can depend on t, x, and v_x.

If F_x only depends on t, then

$$a_x(t) = F_x(t)/m$$

$$v_x(t) = \int_{t_0}^{t} a_x(\tau)d\tau + v_{xt_0}$$

$$x(t) = \int_{t_0}^{t} v_x(\tau)d\tau + x_{t_0}$$

where τ is the variable of integration

If the force is constant (i.e., independent of time, displacement, and velocity) then

$$a_x = F_x/m$$

$$v_x = a_x(t - t_0) + v_{xt_0}$$

$$x = a_x(t - t_0)^2/2 + v_{xt_0}(t - t_0) + x_{t_0}$$

Normal and Tangential Kinetics for Planar Problems

When working with normal and tangential directions, the scalar equations may be written as

$$\Sigma F_t = ma_t = mdv_t/dt$$

$$\Sigma F_n = ma_n = m(v_t^2/\rho)$$

PRINCIPLE OF WORK AND ENERGY

If T_i and V_i are, respectively, the kinetic and potential energy of a particle at state i, then for conservative systems (no energy dissipation or gain), the law of conservation of energy is

$$T_2 + V_2 = T_1 + V_1$$

If nonconservative forces are present, then the work done by these forces must be accounted for. Hence

$T_2 + V_2 = T_1 + V_1 + U_{1\to2}$, where

$U_{1\to2}$ = the work done by the nonconservative forces in moving between state 1 and state 2. Care must be exercised during computations to correctly compute the algebraic sign of the work term. If the forces serve to increase the energy of the system, $U_{1\to2}$ is positive. If the forces, such as friction, serve to dissipate energy, $U_{1\to2}$ is negative.

♦ **Kinetic Energy**

Particle	$T = \frac{1}{2}mv^2$
Rigid Body (*Plane Motion*)	$T = \frac{1}{2}mv_c^2 + \frac{1}{2}I_c\,\omega^2$

subscript c represents the center of mass

♦ **Potential Energy**

$V = V_g + V_e$, where $V_g = Wy$, $V_e = 1/2\ ks^2$

The work done by an external agent in the presence of a conservative field is termed the change in potential energy.

Potential Energy in Gravity Field

$$V_g = mgh, \text{ where}$$

h = the elevation above some specified datum.

Elastic Potential Energy

For a linear elastic spring with modulus, stiffness, or spring constant, k, the force in the spring is

$$F_s = k\,s, \text{ where}$$

s = the change in length of the spring from the undeformed length of the spring.

In changing the deformation in the spring from position s_1 to s_2, the change in the potential energy stored in the spring is

$$V_2 - V_1 = k\left(s_2^2 - s_1^2\right)/2$$

♦ **Work**

Work U is defined as

$$U = \int \mathbf{F} \bullet d\mathbf{r}$$

Variable force	$U_F = \int F \cos\theta\, ds$
Constant force	$U_F = \left(F_c \cos\theta\right)\Delta s$
Weight	$U_W = -W\Delta y$
Spring	$U_s = -\left(\frac{1}{2}k s_2^2 - \frac{1}{2}k s_1^2\right)$
Couple moment	$U_M = M\Delta\theta$

♦ **Power and Efficiency**

$$P = \frac{dU}{dt} = \mathbf{F} \bullet \mathbf{v} \qquad \varepsilon = \frac{P_{out}}{P_{in}} = \frac{U_{out}}{U_{in}}$$

IMPULSE AND MOMENTUM

Linear

Assuming constant mass, the equation of motion of a particle may be written as

$$m d\mathbf{v}/dt = \mathbf{F}$$
$$m d\mathbf{v} = \mathbf{F}dt$$

For a system of particles, by integrating and summing over the number of particles, this may be expanded to

$$\Sigma m_i \left(\mathbf{v}_i\right)_{t_2} = \Sigma m_i \left(\mathbf{v}_i\right)_{t_1} + \Sigma \int_{t_1}^{t_2} \mathbf{F}_i\, dt$$

The term on the left side of the equation is the linear momentum of a system of particles at time t_2. The first term on the right side of the equation is the linear momentum of a system of particles at time t_1. The second term on the right side of the equation is the impulse of the force $\boldsymbol{F}$ from time t_1 to t_2. It should be noted that the above equation is a vector equation. Component scalar equations may be obtained by considering the momentum and force in a set of orthogonal directions.

Angular Momentum or Moment of Momentum

The angular momentum or the moment of momentum about point 0 for a particle is defined as

$$\mathbf{H}_0 = \mathbf{r} \times m\mathbf{v}, \text{ or}$$
$$\mathbf{H}_0 = I_0\,\omega$$

Taking the time derivative of the above, the equation of motion may be written as

$$\dot{\mathbf{H}}_0 = d\left(I_0\,\omega\right)/dt = \mathbf{M}_0, \text{ where}$$

$\mathbf{M}_0$ is the moment applied to the particle. Now by integrating and summing over a system of any number of particles, this may be expanded to

$$\Sigma\left(\mathbf{H}_{0i}\right)_{t_2} = \Sigma\left(\mathbf{H}_{0i}\right)_{t_1} + \Sigma\int_{t_1}^{t_2} \mathbf{M}_{0i}\,dt$$

The term on the left side of the equation is the angular momentum of a system of particles at time t_2. The first term on the right side of the equation is the angular momentum of a system of particles at time t_1. The second term on the right side of the equation is the angular impulse of the moment $\mathbf{M}_0$ from time t_1 to t_2.

Impact

During an impact, momentum is conserved while energy may or may not be conserved. For direct central impact with no external forces

$$m_1 v_1 + m_2 v_2 = m_1 v'_1 + m_2 v'_2, \text{ where}$$

m_1, m_2 = the masses of the two bodies

v_1, v_2 = the velocities of the bodies just before impact

v'_1, v'_2 = the velocities of the bodies just after impact

♦ Adapted from Hibbeler, R.C., *Engineering Mechanics*, 10th ed., Prentice Hall, 2003.

For impacts, the relative velocity expression is

$$e = \frac{(v'_2)_n - (v'_1)_n}{(v_1)_n - (v_2)_n}, \text{ where}$$

e = coefficient of restitution

$(v_i)_n$ = the velocity normal to the plane of impact just **before** impact

$(v'_i)_n$ = the velocity normal to the plane of impact just **after** impact

The value of e is such that

$0 \le e \le 1$, with limiting values

$e = 1$, perfectly elastic (energy conserved)

$e = 0$, perfectly plastic (no rebound)

Knowing the value of e, the velocities after the impact are given as

$$(v'_1)_n = \frac{m_2 (v_2)_n (1 + e) + (m_1 - e m_2)(v_1)_n}{m_1 + m_2}$$

$$(v'_2)_n = \frac{m_1 (v_1)_n (1 + e) - (e m_1 - m_2)(v_2)_n}{m_1 + m_2}$$

Friction

The Laws of Friction are

1. The total friction force F that can be developed is independent of the magnitude of the area of contact.
2. The total friction force F that can be developed is proportional to the normal force N.
3. For low velocities of sliding, the total frictional force that can be developed is practically independent of the sliding velocity, although experiments show that the force F necessary to initiate slip is greater than that necessary to maintain the motion.

The formula expressing the Laws of Friction is

$F \le \mu N$, where

μ = the coefficient of friction.

In general

$F < \mu_s N$, no slip occurring

$F = \mu_s N$, at the point of impending slip

$F = \mu_k N$, when slip is occurring

Here,

μ_s = the coefficient of static friction

μ_k = the coefficient of kinetic friction

PLANE MOTION OF A RIGID BODY

Kinematics of a Rigid Body

Rigid Body Rotation

For rigid body rotation θ

$\omega = d\theta/dt$

$\alpha = d\omega/dt$

$\alpha d\theta = \omega d\omega$

Instantaneous Center of Rotation (Instant Centers)

An instantaneous center of rotation (instant center) is a point, common to two bodies, at which each has the same velocity (magnitude and direction) at a given instant. It is also a point in space about which a body rotates, instantaneously.

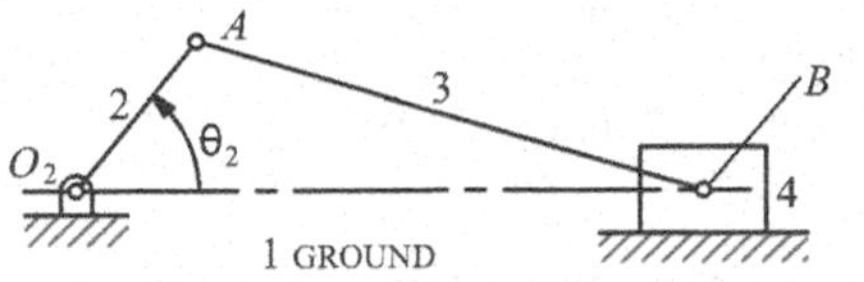

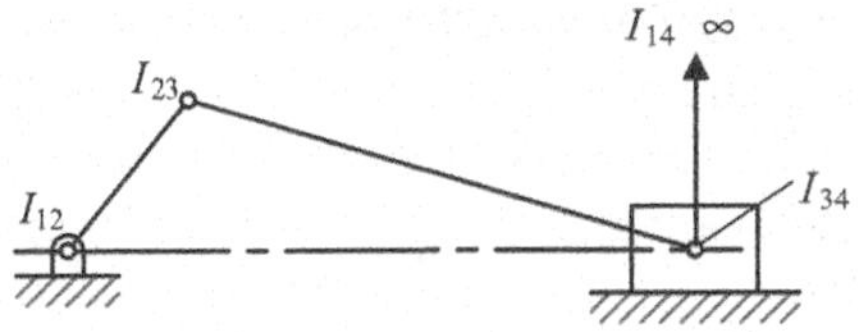

The figure shows a fourbar slider-crank. Link 2 (the crank) rotates about the fixed center, O_2. Link 3 couples the crank to the slider (link 4), which slides against ground (link 1). Using the definition of an instant center (IC), we see that the pins at O_2, A, and B are ICs that are designated I_{12}, I_{23}, and I_{34}. The easily observable IC is I_{14}, which is located at infinity with its direction perpendicular to the interface between links 1 and 4 (the direction of sliding). To locate the remaining two ICs (for a fourbar) we must make use of Kennedy's rule.

Kennedy's Rule: When three bodies move relative to one another they have three instantaneous centers, all of which lie on the same straight line.

To apply this rule to the slider-crank mechanism, consider links 1, 2, and 3 whose ICs are I_{12}, I_{23}, and I_{13}, all of which lie on a straight line. Consider also links 1, 3, and 4 whose ICs are I_{13}, I_{34}, and I_{14}, all of which lie on a straight line. Extending the line through I_{12} and I_{23} and the line through I_{34} and I_{14} to their intersection locates I_{13}, which is common to the two groups of links that were considered.

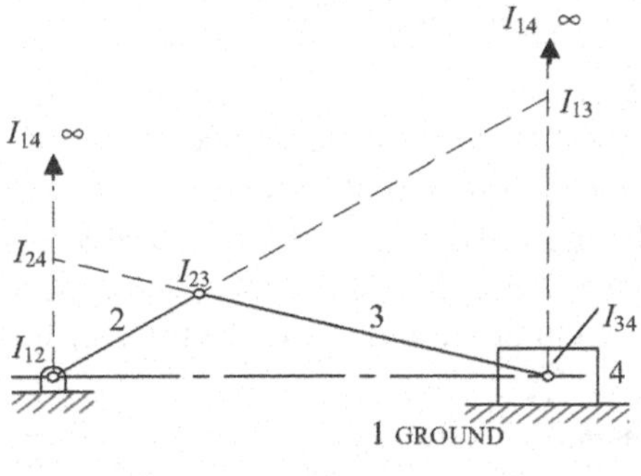

Similarly, if body groups 1, 2, 4 and 2, 3, 4 are considered, a line drawn through known ICs I_{12} and I_{14} to the intersection of a line drawn through known ICs I_{23} and I_{34} locates I_{24}.

The number of ICs, c, for a given mechanism is related to the number of links, n, by

$$c = \frac{n(n-1)}{2}$$

Kinetics of a Rigid Body

In general, Newton's second law for a rigid body, with constant mass and mass moment of inertia, in plane motion may be written in vector form as

$$\Sigma \mathbf{F} = m\mathbf{a}_c$$
$$\Sigma \mathbf{M}_c = I_c \boldsymbol{\alpha}$$
$$\Sigma \mathbf{M}_p = I_c \boldsymbol{\alpha} + \boldsymbol{\rho}_{pc} \times m\mathbf{a}_c, \text{ where}$$

$\mathbf{F}$ are forces and $\mathbf{a}_c$ is the acceleration of the body's mass center both in the plane of motion, $\mathbf{M}_c$ are moments and α is the angular acceleration both about an axis normal to the plane of motion, I_c is the mass moment of inertia about the normal axis through the mass center, and $\boldsymbol{\rho}_{pc}$ is a vector from point p to point c.

♦ **Mass Moment of Inertia** $I = \int r^2\, dm$

Parallel-Axis Theorem $I = I_c + md^2$

Radius of Gyration $r_m = \sqrt{\dfrac{I}{m}}$

♦ **Equations of Motion**

Rigid Body	$\Sigma F_x = m(a_c)_x$
(Plane Motion)	$\Sigma F_y = m(a_c)_y$
	$\Sigma M_c = I_c\alpha$ *or* $\Sigma M_P = \Sigma (M_k)_P$

Subscript c indicates center of mass.

Mass Moment of Inertia

The definitions for the mass moments of inertia are

$$I_x = \int \left(y^2 + z^2\right) dm$$
$$I_y = \int \left(x^2 + z^2\right) dm$$
$$I_z = \int \left(x^2 + y^2\right) dm$$

A table listing moment of inertia formulas for some standard shapes is at the end of this section.

Parallel-Axis Theorem

The mass moments of inertia may be calculated about any axis through the application of the above definitions. However, once the moments of inertia have been determined about an axis passing through a body's mass center, it may be transformed to another parallel axis. The transformation equation is

$$I_{new} = I_c + md^2, \text{ where}$$

I_{new} = the mass moment of inertia about any specified axis

I_c = the mass moment of inertia about an axis that is parallel to the above specified axis but passes through the body's mass center

m = the mass of the body

d = the normal distance from the body's mass center to the above-specified axis

Mass Radius of Gyration

The mass radius of gyration is defined as

$$r_m = \sqrt{I/m}$$

Without loss of generality, the body may be assumed to be in the x-y plane. The scalar equations of motion may then be written as

$$\Sigma F_x = ma_{xc}$$
$$\Sigma F_y = ma_{yc}$$
$$\Sigma M_{zc} = I_{zc}\alpha, \text{ where}$$

zc indicates the z axis passing through the body's mass center, a_{xc} and a_{yc} are the acceleration of the body's mass center in the x and y directions, respectively, and α is the angular acceleration of the body about the z axis.

Rotation about an Arbitrary Fixed Axis

♦ Rigid Body Motion About a Fixed Axis

Variable α	*Constant* $\alpha = \alpha_0$
$\alpha = \dfrac{d\omega}{dt}$	$\omega = \omega_0 + \alpha_0 t$
$\omega = \dfrac{d\theta}{dt}$	$\theta = \theta_0 + \omega_0 t + \dfrac{1}{2}\alpha_0 t^2$
$\omega\, d\omega = \alpha\, d\theta$	$\omega^2 = \omega_0^2 + 2\alpha_0(\theta - \theta_0)$

For rotation about some arbitrary fixed axis q

$$\Sigma M_q = I_q \alpha$$

If the applied moment acting about the fixed axis is constant then integrating with respect to time, from $t = 0$ yields

$$\alpha = M_q / I_q$$
$$\omega = \omega_0 + \alpha t$$
$$\theta = \theta_0 + \omega_0 t + \alpha t^2/2$$

where ω_0 and θ_0 are the values of angular velocity and angular displacement at time $t = 0$, respectively.

The change in kinetic energy is the work done in accelerating the rigid body from ω_0 to ω

$$I_q \omega^2/2 = I_q \omega_0^2/2 + \int_{\theta_0}^{\theta} M_q d\theta$$

Kinetic Energy

In general the kinetic energy for a rigid body may be written as

$$T = mv^2/2 + I_c \omega^2/2$$

For motion in the xy plane this reduces to

$$T = m\left(v_{cx}^2 + v_{cy}^2\right)/2 + I_c \omega_z^2/2$$

For motion about an instant center,

$$T = I_{IC}\omega^2/2$$

♦ Adapted from Hibbeler, R.C., *Engineering Mechanics*, 10th ed., Prentice Hall, 2003.

♦ Principle of Angular Impulse and Momentum

Rigid Body	$(\mathbf{H}_c)_1 + \Sigma\int \mathbf{M}_c dt = (\mathbf{H}_c)_2$
(*Plane Motion*)	where $\mathbf{H}_c = I_c\omega$
	$(\mathbf{H}_0)_1 + \Sigma\int \mathbf{M}_0 dt = (\mathbf{H}_0)_2$
	where $\mathbf{H}_0 = I_0\omega$

Subscript c indicates center of mass.

♦ Conservation of Angular Momentum

$\Sigma(\text{syst. }\mathbf{H})_1 = \Sigma(\text{syst. }\mathbf{H})_2$

Free Vibration

The figure illustrates a single degree-of-freedom system.

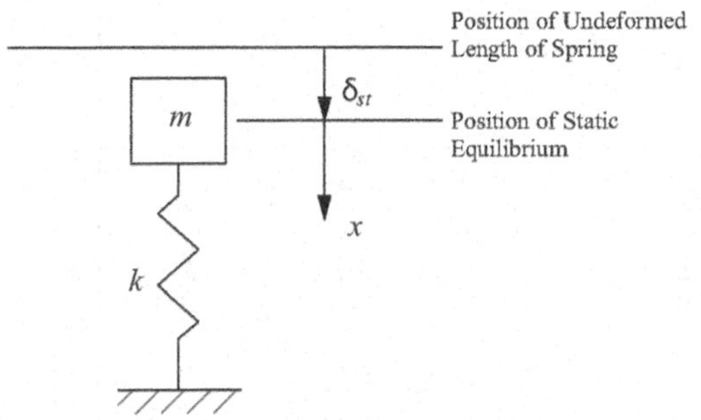

The equation of motion may be expressed as

$$m\ddot{x} = mg - k(x + \delta_{st})$$

where m is mass of the system, k is the spring constant of the system, δ_{st} is the static deflection of the system, and x is the displacement of the system from static equilibrium.

From statics it may be shown that

$$mg = k\delta_{st}$$

thus the equation of motion may be written as

$$m\ddot{x} + kx = 0, \text{ or}$$

$$\ddot{x} + (k/m)x = 0$$

The solution of this differential equation is

$$x(t) = C_1 \cos(\omega_n t) + C_2 \sin(\omega_n t)$$

where $\omega_n = \sqrt{k/m}$ is the undamped natural circular frequency and C_1 and C_2 are constants of integration whose values are determined from the initial conditions.

If the initial conditions are denoted as $x(0) = x_0$ and $\dot{x}(0) = v_0$, then

$$x(t) = x_0 \cos(\omega_n t) + (v_0/\omega_n)\sin(\omega_n t)$$

It may also be shown that the undamped natural frequency may be expressed in terms of the static deflection of the system as

$$\omega_n = \sqrt{g/\delta_{st}}$$

The undamped natural period of vibration may now be written as

$$\tau_n = 2\pi/\omega_n = \frac{2\pi}{\sqrt{\frac{k}{m}}} = \frac{2\pi}{\sqrt{\frac{g}{\delta_{st}}}}$$

Torsional Vibration

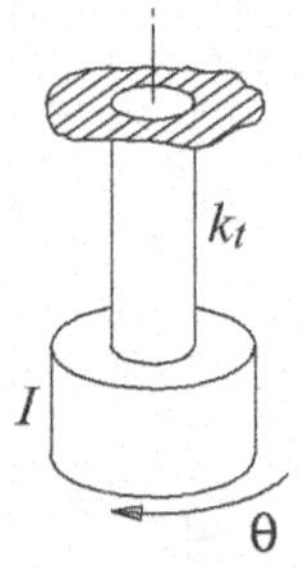

For torsional free vibrations it may be shown that the differential equation of motion is

$$\ddot{\theta} + (k_t/I)\theta = 0, \text{ where}$$

θ = the angular displacement of the system

k_t = the torsional stiffness of the massless rod

I = the mass moment of inertia of the end mass

The solution may now be written in terms of the initial conditions $\theta(0) = \theta_0$ and $\dot{\theta}(0) = \dot{\theta}_0$ as

$$\theta(t) = \theta_0 \cos(\omega_n t) + (\dot{\theta}_0/\omega_n)\sin(\omega_n t)$$

where the undamped natural circular frequency is given by

$$\omega_n = \sqrt{k_t/I}$$

The torsional stiffness of a solid round rod with associated polar moment-of-inertia J, length L, and shear modulus of elasticity G is given by

$$k_t = GJ/L$$

Thus the undamped circular natural frequency for a system with a solid round supporting rod may be written as

$$\omega_n = \sqrt{GJ/IL}$$

Similar to the linear vibration problem, the undamped natural period may be written as

$$\tau_n = 2\pi/\omega_n = \frac{2\pi}{\sqrt{\frac{k_t}{I}}} = \frac{2\pi}{\sqrt{\frac{GJ}{IL}}}$$

♦ Adapted from Hibbeler, R.C., *Engineering Mechanics*, 10th ed., Prentice Hall, 2003.

Figure	Mass & Centroid	Mass Moment of Inertia	(Radius of Gyration)^2	Product of Inertia
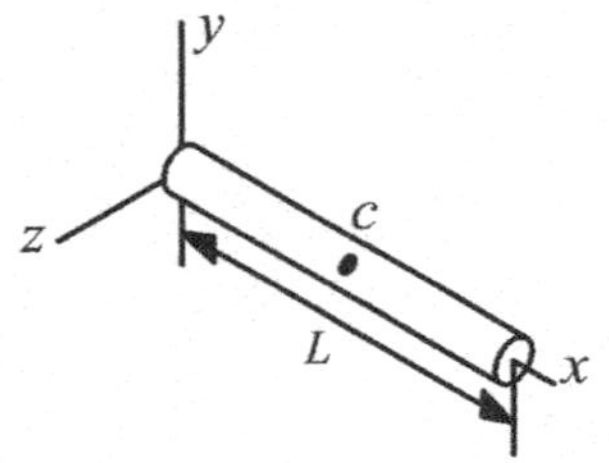	$M = \rho LA$ $x_c = L/2$ $y_c = 0$ $z_c = 0$ A = cross-sectional area of rod ρ = mass/vol.	$I_x = I_{x_c} = 0$ $I_{y_c} = I_{z_c} = ML^2/12$ $I_y = I_z = ML^2/3$	$r_x^2 = r_{x_c}^2 = 0$ $r_{y_c}^2 = r_{z_c}^2 = L^2/12$ $r_y^2 = r_z^2 = L^2/3$	$I_{x_c y_c}$, etc. = 0 I_{xy}, etc. = 0
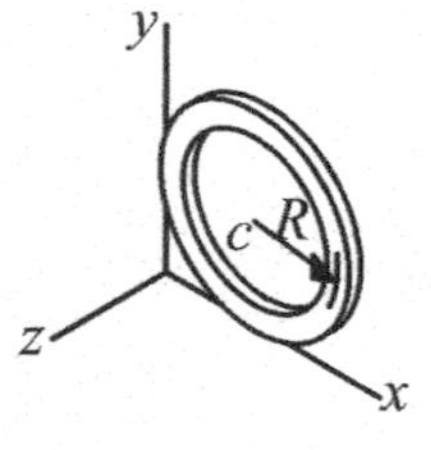	$M = 2\pi R\rho A$ x_c = R = mean radius y_c = R = mean radius $z_c = 0$ A = cross-sectional area of ring ρ = mass/vol.	$I_{x_c} = I_{y_c} = MR^2/2$ $I_{z_c} = MR^2$ $I_x = I_y = 3MR^2/2$ $I_z = 3MR^2$	$r_{x_c}^2 = r_{y_c}^2 = R^2/2$ $r_{z_c}^2 = R^2$ $r_x^2 = r_y^2 = 3R^2/2$ $r_z^2 = 3R^2$	$I_{x_c y_c}$, etc. = 0 $I_{z_c z_c} = MR^2$ $I_{xz} = I_{yz} = 0$
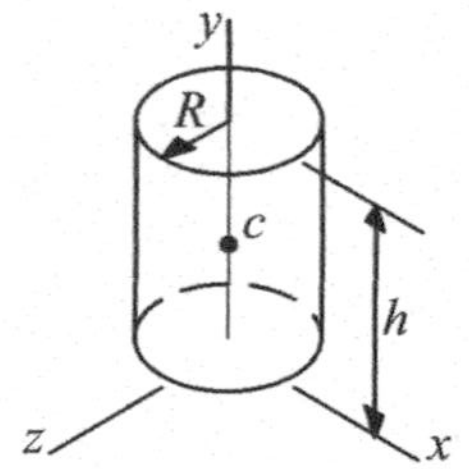	$M = \pi R^2 \rho h$ $x_c = 0$ $y_c = h/2$ $z_c = 0$ ρ = mass/vol.	$I_{x_c} = I_{z_c} = M(3R^2 + h^2)/12$ $I_{y_c} = I_y = MR^2/2$ $I_x = I_z = M(3R^2 + 4h^2)/12$	$r_{x_c}^2 = r_{z_c}^2 = (3R^2 + h^2)/12$ $r_{y_c}^2 = r_y^2 = R^2/2$ $r_x^2 = r_z^2 = (3R^2 + 4h^2)/12$	$I_{x_c y_c}$, etc. = 0 I_{xy}, etc. = 0
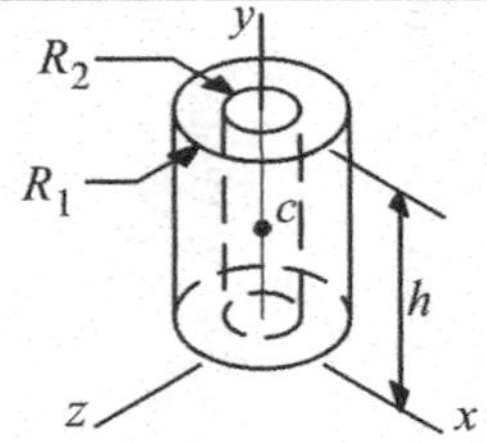	$M = \pi(R_1^2 - R_2^2)\rho h$ $x_c = 0$ $y_c = h/2$ $z_c = 0$ ρ = mass/vol.	$I_{x_c} = I_{z_c}$ $= M(3R_1^2 + 3R_2^2 + h^2)/12$ $I_{y_c} = I_y = M(R_1^2 + R_2^2)/2$ $I_x = I_z$ $= M(3R_1^2 + 3R_2^2 + 4h^2)/12$	$r_{x_c}^2 = r_{z_c}^2 = (3R_1^2 + 3R_2^2 + h^2)/12$ $r_{y_c}^2 = r_y^2 = (R_1^2 + R_2^2)/2$ $r_x^2 = r_z^2$ $= (3R_1^2 + 3R_2^2 + 4h^2)/12$	$I_{x_c y_c}$, etc. = 0 I_{xy}, etc. = 0
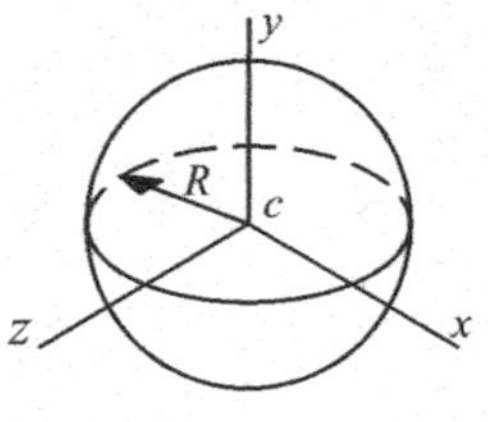	$M = \frac{4}{3}\pi R^3 \rho$ $x_c = 0$ $y_c = 0$ $z_c = 0$ ρ = mass/vol.	$I_{x_c} = I_x = 2MR^2/5$ $I_{y_c} = I_y = 2MR^2/5$ $I_{z_c} = I_z = 2MR^2/5$	$r_{x_c}^2 = r_x^2 = 2R^2/5$ $r_{y_c}^2 = r_y^2 = 2R^2/5$ $r_{z_c}^2 = r_z^2 = 2R^2/5$	$I_{x_c y_c}$, etc. = 0

Housner, George W., and Donald E. Hudson, *Applied Mechanics Dynamics*, D. Van Nostrand Company, Inc., Princeton, NJ, 1959. Table reprinted by permission of G.W. Housner & D.E. Hudson.

MECHANICS OF MATERIALS

UNIAXIAL STRESS-STRAIN

Stress-Strain Curve for Mild Steel

♦

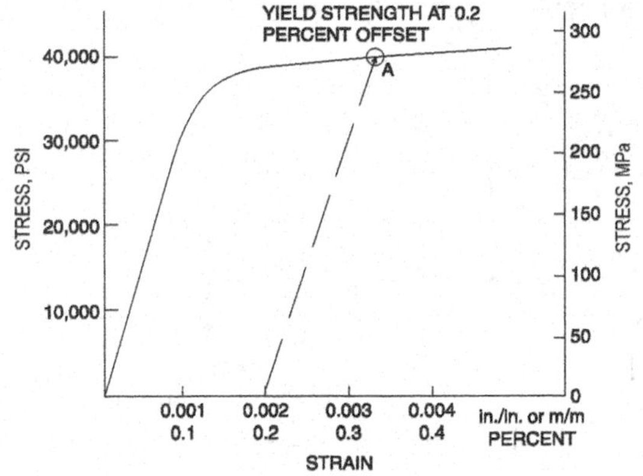

The slope of the linear portion of the curve equals the modulus of elasticity.

DEFINITIONS

Engineering Strain

$\varepsilon = \Delta L/L_o$, where

ε = engineering strain (units per unit)

ΔL = change in length (units) of member

L_o = original length (units) of member

Percent Elongation

$$\%\text{ Elongation} = \left(\frac{\Delta L}{L_o}\right) \times 100$$

Percent Reduction in Area (RA)

The % reduction in area from initial area, A_i, to final area, A_f, is:

$$\%RA = \left(\frac{A_i - A_f}{A_i}\right) \times 100$$

Shear Stress-Strain

$\gamma = \tau/G$, where

γ = shear strain

τ = shear stress

G = *shear modulus* (constant in linear torsion-rotation relationship)

$$G = \frac{E}{2(1 + v)}, \text{ where}$$

E = modulus of elasticity (Young's modulus)

v = *Poisson's ratio*
= – (lateral strain)/(longitudinal strain)

Uniaxial Loading and Deformation

$\sigma = P/A$, where

σ = stress on the cross section

P = loading

A = cross-sectional area

$\varepsilon = \delta/L$, where

δ = elastic longitudinal deformation

L = length of member

$$E = \sigma/\varepsilon = \frac{P/A}{\delta/L}$$

$$\delta = \frac{PL}{AE}$$

True stress is load divided by actual cross-sectional area whereas engineering stress is load divided by the initial area.

THERMAL DEFORMATIONS

$\delta_t = \alpha L(T - T_o)$, where

δ_t = deformation caused by a change in temperature

α = temperature coefficient of expansion

L = length of member

T = final temperature

T_o = initial temperature

CYLINDRICAL PRESSURE VESSEL

For internal pressure only, the stresses at the inside wall are:

$$\sigma_t = P_i \frac{r_o^2 + r_i^2}{r_o^2 - r_i^2} \quad \text{and} \quad \sigma_r = -P_i$$

For external pressure only, the stresses at the outside wall are:

$$\sigma_t = -P_o \frac{r_o^2 + r_i^2}{r_o^2 - r_i^2} \quad \text{and} \quad \sigma_r = -P_o, \text{ where}$$

σ_t = tangential (hoop) stress σ_t – hoop stress

σ_r = radial stress

P_i = internal pressure

P_o = external pressure

r_i = inside radius

r_o = outside radius

For vessels with end caps, the axial stress is:

$$\sigma_a = P_i \frac{r_i^2}{r_o^2 - r_i^2}$$

σ_t, σ_r, and σ_a are principal stresses.

♦ Flinn, Richard A., and Paul K. Trojan, *Engineering Materials & Their Applications*, 4th ed., Houghton Mifflin Co., Boston, 1990.

When the thickness of the cylinder wall is about one-tenth or less of inside radius, the cylinder can be considered as thin-walled. In which case, the internal pressure is resisted by the hoop stress and the axial stress.

$$\sigma_t = \frac{P_i r}{t} \quad \text{and} \quad \sigma_a = \frac{P_i r}{2t}$$

where t = wall thickness and $r = \frac{r_i + r_o}{2}$.

STRESS AND STRAIN

Principal Stresses

For the special case of a *two-dimensional* stress state, the equations for principal stress reduce to

$$\sigma_a, \sigma_b = \frac{\sigma_x + \sigma_y}{2} \pm \sqrt{\left(\frac{\sigma_x - \sigma_y}{2}\right)^2 + \tau_{xy}^2}$$

$$\theta = \tan^{-1}\left[\frac{2\tau_{xy}}{\sigma_x - \sigma_y}\right] \times \frac{1}{2}$$

$$\sigma_c = 0$$

The two nonzero values calculated from this equation are temporarily labeled σ_a and σ_b and the third value σ_c is always zero in this case. Depending on their values, the three roots are then labeled according to the convention:
algebraically largest = σ_1, *algebraically smallest* = σ_3, *other* = σ_2. A typical 2D stress element is shown below with all indicated components shown in their positive sense.

♦

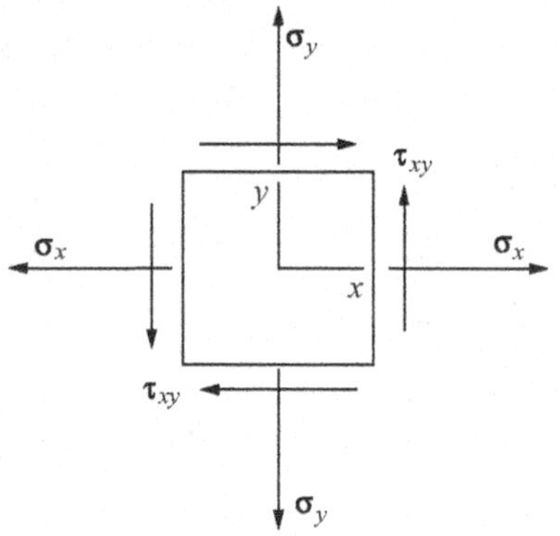

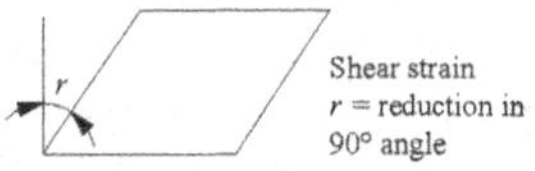

Mohr's Circle – Stress, 2D

To construct a Mohr's circle, the following sign conventions are used.

1. Tensile normal stress components are plotted on the horizontal axis and are considered positive. Compressive normal stress components are negative.
2. For constructing Mohr's circle only, shearing stresses are plotted above the normal stress axis when the pair of shearing stresses, acting on opposite and parallel faces of an element, forms a clockwise couple. Shearing stresses are plotted below the normal axis when the shear stresses form a counterclockwise couple.

The circle drawn with the center on the normal stress (horizontal) axis with center, C, and radius, R, where

$$C = \frac{\sigma_x + \sigma_y}{2}, \quad R = \sqrt{\left(\frac{\sigma_x - \sigma_y}{2}\right)^2 + \tau_{xy}^2}$$

The two nonzero principal stresses are then:

♦

$$\sigma_a = C + R$$
$$\sigma_b = C - R$$

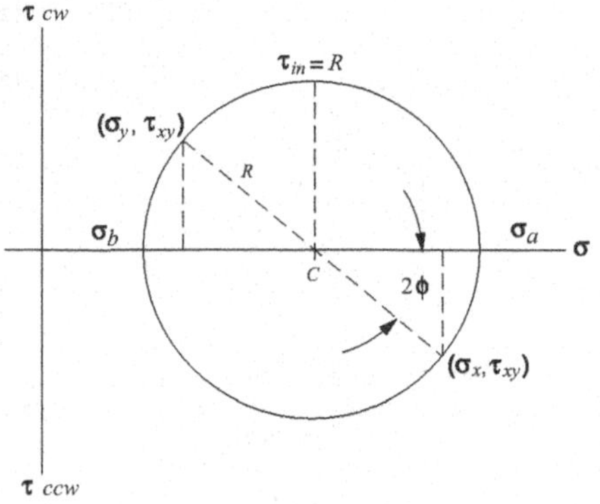

The maximum *inplane* shear stress is $\tau_{in} = R$. However, the maximum shear stress considering three dimensions is always

$$\tau_{max} = \frac{\sigma_1 - \sigma_3}{2}. \qquad \tau_{max} = \pm\sqrt{\left(\frac{\sigma_x - \sigma_y}{2}\right)^2 + (\tau_{xy})^2}$$

Hooke's Law

Three-dimensional case:

$\varepsilon_x = (1/E)[\sigma_x - v(\sigma_y + \sigma_z)]$ $\qquad \gamma_{xy} = \tau_{xy}/G$

$\varepsilon_y = (1/E)[\sigma_y - v(\sigma_z + \sigma_x)]$ $\qquad \gamma_{yz} = \tau_{yz}/G$

$\varepsilon_z = (1/E)[\sigma_z - v(\sigma_x + \sigma_y)]$ $\qquad \gamma_{zx} = \tau_{zx}/G$

Plane stress case ($\sigma_z = 0$):

$\varepsilon_x = (1/E)(\sigma_x - v\sigma_y)$

$\varepsilon_y = (1/E)(\sigma_y - v\sigma_x)$

$\varepsilon_z = -(1/E)(v\sigma_x + v\sigma_y)$

$$\begin{Bmatrix}\sigma_x\\ \sigma_y\\ \tau_{xy}\end{Bmatrix} = \frac{E}{1 - v^2}\begin{bmatrix}1 & v & 0\\ v & 1 & 0\\ 0 & 0 & \frac{1-v}{2}\end{bmatrix}\begin{Bmatrix}\varepsilon_x\\ \varepsilon_y\\ \gamma_{xy}\end{Bmatrix}$$

Uniaxial case ($\sigma_y = \sigma_z = 0$): $\quad \sigma_x = E\varepsilon_x$ or $\sigma = E\varepsilon$, where

$\varepsilon_x, \varepsilon_y, \varepsilon_z$ = normal strain

$\sigma_x, \sigma_y, \sigma_z$ = normal stress

$\gamma_{xy}, \gamma_{yz}, \gamma_{zx}$ = shear strain

$\tau_{xy}, \tau_{yz}, \tau_{zx}$ = shear stress

E = modulus of elasticity

G = shear modulus

v = Poisson's ratio

When there is a temperature change from an initial temperature T_i to a final temperature T_f there are also thermally-induced normal strains. In this case, ε_x, ε_y, and ε_z require modification. Thus,

$$\varepsilon_x = \frac{1}{E}\left[\sigma_x - v(\sigma_y + \sigma_z)\right] + \alpha(T_f - T_i)$$

and similarly for ε_y and ε_z, where α = coefficient of thermal expansion (CTE).

♦ Crandall, S.H., and N.C. Dahl, *An Introduction to Mechanics of Solids*, McGraw-Hill, New York, 1959.

TORSION

Torsion stress in circular solid or thick-walled ($t > 0.1\ r$) shafts:

$$\tau = \frac{Tr}{J}$$

where J = polar moment of inertia

TORSIONAL STRAIN

$$\gamma_{\phi z} = \lim_{\Delta z \to 0} r(\Delta\phi/\Delta z) = r(d\phi/dz)$$

The shear strain varies in direct proportion to the radius, from zero strain at the center to the greatest strain at the outside of the shaft. $d\phi/dz$ is the twist per unit length or the rate of twist.

$$\tau_{\phi z} = G\gamma_{\phi z} = Gr(d\phi/dz)$$

$$T = G(d\phi/dz) \int_A r^2 dA = GJ(d\phi/dz)$$

$$\phi = \int_o^L \frac{T}{GJ} dz = \frac{TL}{GJ}, \text{ where}$$

ϕ = total angle (radians) of twist

T = torque

L = length of shaft

T/ϕ gives the *twisting moment per radian of twist*. This is called the *torsional stiffness* and is often denoted by the symbol k or c.

For Hollow, Thin-Walled Shafts

$$\tau = \frac{T}{2A_m t}, \text{ where}$$

t = thickness of shaft wall

A_m = the area of a solid shaft of radius equal to the mean radius of the hollow shaft

BEAMS

Shearing Force and Bending Moment Sign Conventions

1. The bending moment is *positive* if it produces bending of the beam *concave upward* (compression in top fibers and tension in bottom fibers).
2. The shearing force is *positive* if the *right portion of the beam tends to shear downward with respect to the left*.

♦

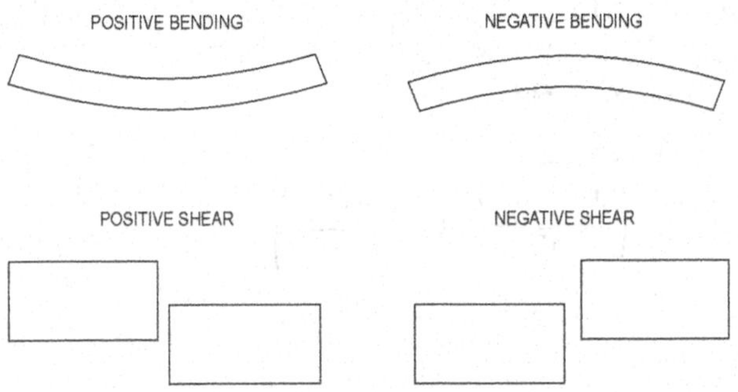

The relationship between the load (w), shear (V), and moment (M) equations are:

$$w(x) = -\frac{dV(x)}{dx}$$

$$V = \frac{dM(x)}{dx}$$

$$V_2 - V_1 = \int_{x_1}^{x_2} [-w(x)]dx$$

$$M_2 - M_1 = \int_{x_1}^{x_2} V(x)dx$$

Stresses in Beams

The normal stress in a beam due to bending:

$\sigma_x = -My/I$, where

M = the moment at the section

I = the *moment of inertia* of the cross section

y = the distance from the neutral axis to the fiber location above or below the neutral axis

The maximum normal stresses in a beam due to bending:

$\sigma_x = \pm Mc/I$, where

c = distance from the neutral axis to the outermost fiber of a symmetrical beam section.

$\sigma_x = -M/s$, where

s = I/c: the elastic section modulus of the beam.

Transverse shear stress:

$\tau_{xy} = VQ/(Ib)$, where $\rightarrow \tau_{max} = \frac{3V}{2A}$

V = shear force

Q = $A'\overline{y}'$, where

A' = area above the layer (or plane) upon which the desired transverse shear stress acts

$\overline{y}'$ = distance from neutral axis to area centroid

b = width or thickness or the cross-section

Transverse shear flow:

$q = VQ/I$

♦ Timoshenko, S., and Gleason H. MacCullough, *Elements of Strengths of Materials*, K. Van Nostrand Co./Wadsworth Publishing Co., 1949.

Deflection of Beams

Using $1/\rho = M/(EI)$,

$$EI\frac{d^2y}{dx^2} = M, \text{ differential equation of deflection curve}$$

$$EI\frac{d^3y}{dx^3} = dM(x)/dx = V$$

$$EI\frac{d^4y}{dx^4} = dV(x)/dx = -w$$

Determine the deflection curve equation by double integration (apply boundary conditions applicable to the deflection and/or slope).

$$EI\,(dy/dx) = \int M(x)\,dx$$

$$EIy = \int\left[\int M(x)\,dx\right]dx$$

The constants of integration can be determined from the physical geometry of the beam.

Composite Sections

The bending stresses in a beam composed of dissimilar materials (material 1 and material 2) where $E_1 > E_2$ are:

$$\sigma_1 = -nMy/I_T$$
$$\sigma_2 = -My/I_T, \text{ where}$$

I_T = the *moment of inertia* of the *transformed* section

n = the modular ratio E_1/E_2

E_1 = elastic modulus of material 1

E_2 = elastic modulus of material 2

y = distance from the neutral axis to the fiber location above or below the neutral axis

The composite section is transformed into a section composed of a single material. The centroid and then the moment of inertia are found on the *transformed section* for use in the bending stress equations.

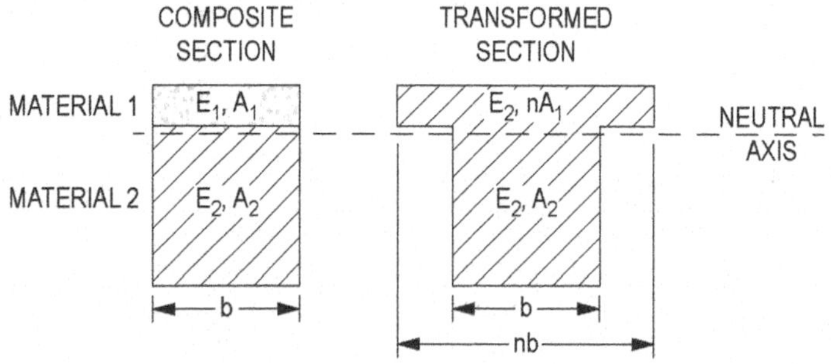

COLUMNS

Critical axial load for long column subject to buckling: Euler's Formula

$$P_{cr} = \frac{\pi^2 EI}{(K\ell)^2}, \text{ where}$$

ℓ = unbraced column length

K = effective-length factor to account for end supports

Theoretical effective-length factors for columns include:

Pinned-pinned, $K = 1.0$

Fixed-fixed, $K = 0.5$

Fixed-pinned, $K = 0.7$

Fixed-free, $K = 2.0$

Critical buckling stress for long columns:

$$\sigma_{cr} = \frac{P_{cr}}{A} = \frac{\pi^2 E}{(K\ell/r)^2}, \text{ where}$$

r = *radius of gyration* $\sqrt{I/A}$

$K\ell/r$ = effective *slenderness ratio* for the column

ELASTIC STRAIN ENERGY

If the strain remains within the elastic limit, the work done during deflection (extension) of a member will be transformed into potential energy and can be recovered.

If the final load is P and the corresponding elongation of a tension member is δ, then the total energy U stored is equal to the work W done during loading.

$$U = W = P\delta/2$$

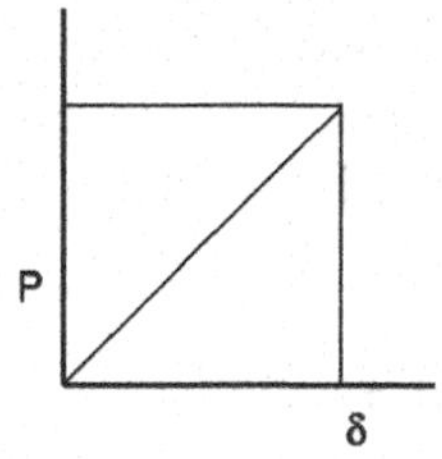

The strain energy per unit volume is

$$u = U/AL = \sigma^2/2E \qquad \text{(for tension)}$$

Eccentric load is applied →

$$\sigma = \frac{F}{A} \pm \frac{Fec}{I}$$

c → dist. of neutral axis from extreme fiber

Extreme Fibers →

$$\sigma = \frac{F}{A} \pm \left(\frac{F_x c_x e_x}{I} + \frac{F_y c_y e_y}{I}\right)$$

MATERIAL PROPERTIES

Table 1 - Typical Material Properties
(Use these values if the specific alloy and temper are not listed on Table 2 below)

Material	Modulus of Elasticity, E [Mpsi (GPa)]	Modulus of Rigidity, G [Mpsi (GPa)]	Poisson's Ratio, ν	Coefficient of Thermal Expansion, α [10^{-6}/°F (10^{-6}/°C)]	Density, ρ [lb/in^3 (Mg/m^3)]
Steel	29.0 (200.0)	11.5 (80.0)	0.30	6.5 (11.7)	0.282 (7.8)
Aluminum	10.0 (69.0)	3.8 (26.0)	0.33	13.1 (23.6)	0.098 (2.7)
Cast Iron	14.5 (100.0)	6.0 (41.4)	0.21	6.7 (12.1)	0.246–0.282 (6.8–7.8)
Wood (Fir)	1.6 (11.0)	0.6 (4.1)	0.33	1.7 (3.0)	–
Brass	14.8–18.1 (102–125)	5.8 (40)	0.33	10.4 (18.7)	0.303–0.313 (8.4–8.7)
Copper	17 (117)	6.5 (45)	0.36	9.3 (16.6)	0.322 (8.9)
Bronze	13.9–17.4 (96–120)	6.5 (45)	0.34	10.0 (18.0)	0.278–0.314 (7.7–8.7)
Magnesium	6.5 (45)	2.4 (16.5)	0.35	14 (25)	0.061 (1.7)
Glass	10.2 (70)	–	0.22	5.0 (9.0)	0.090 (2.5)
Polystyrene	0.3 (2)	–	0.34	38.9 (70.0)	0.038 (1.05)
Polyvinyl Chloride (PVC)	<0.6 (<4)	–	–	28.0 (50.4)	0.047 (1.3)
Alumina Fiber	58 (400)	–	–	–	0.141 (3.9)
Aramide Fiber	18.1 (125)	–	–	–	0.047 (1.3)
Boron Fiber	58 (400)	–	–	–	0.083 (2.3)
Beryllium Fiber	43.5 (300)	–	–	–	0.069 (1.9)
BeO Fiber	58 (400)	–	–	–	0.108 (3.0)
Carbon Fiber	101.5 (700)	–	–	–	0.083 (2.3)
Silicon Carbide Fiber	58 (400)	–	–	–	0.116 (3.2)

Table 2 - Average Mechanical Properties of Typical Engineering Materials
(U.S. Customary Units)
(Use these values for the specific alloys and temper listed. For all other materials refer to Table 1 above.)

Materials	Specific Weight γ (lb/in^3)	Modulus of Elasticity E (10^3 ksi)	Modulus of Rigidity G (10^3 ksi)	Yield Strength (ksi) σ_y			Ultimate Strength (ksi) σ_u			% Elongation in 2 in. specimen	Poisson's Ratio v	Coef. of Therm. Expansion α $(10^{-6})/°Φ$
				Tens.	Comp.	Shear	Tens.	Comp.	Shear			
Metallic												
Aluminum Wrought Alloys 2014-T6	0.101	10.6	3.9	60	60	25	68	68	42	10	0.35	12.8
Aluminum Wrought Alloys 6061-T6	0.098	10.0	3.7	37	37	19	42	42	27	12	0.35	13.1
Cast Iron Alloys Gray ASTM 20	0.260	10.0	3.9	—	—	—	26	97	—	0.6	0.28	6.70
Cast Iron Alloys Malleable ASTM A-197	0.263	25.0	9.8	—	—	—	40	83	—	5	0.28	6.60
Copper Alloys Red Brass C83400	0.316	14.6	5.4	11.4	11.4	—	35	35	—	35	0.35	9.80
Copper Alloys Bronze C86100	0.319	15.0	5.6	50	50	—	95	95	—	20	0.34	9.60
Magnesium Alloy [Am 1004-T611]	0.066	6.48	2.5	22	22	—	40	40	22	1	0.30	14.3
Steel Alloys Structural A36	0.284	29.0	11.0	36	36	—	58	58	—	30	0.32	6.60
Steel Alloys Stainless 304	0.284	28.0	11.0	30	30	—	75	75	—	40	0.27	9.60
Steel Alloys Tool L2	0.295	29.0	11.0	102	102	—	116	116	—	22	0.32	6.50
Titanium Alloy [Ti-6Al-4V]	0.160	17.4	6.4	134	134	—	145	145	—	16	0.36	5.20
Nonmetallic												
Concrete Low Strength	0.086	3.20	—	—	—	1.8	—	—	—	—	0.15	6.0
Concrete High Strength	0.086	4.20	—	—	—	5.5	—	—	—	—	0.15	6.0
Plastic Reinforced Kevlar 49	0.0524	19.0	—	—	—	—	104	70	10.2	2.8	0.34	—
Plastic Reinforced 30% Glass	0.0524	10.5	—	—	—	—	13	19	—	—	0.34	—
Wood Select Structural Grade Douglas Fir	0.017	1.90	—	—	—	—	0.30[c]	3.78[d]	0.90[d]	—	0.29[c]	—
Wood Select Structural Grade White Spruce	0.130	1.40	—	—	—	—	0.36[c]	5.18[d]	0.97[d]	—	0.31[c]	—

[a] SPECIFIC VALUES MAY VARY FOR A PARTICULAR MATERIAL DUE TO ALLOY OR MINERAL COMPOSITION, MECHANICAL WORKING OF THE SPECIMEN, OR HEAT TREATMENT. FOR A MORE EXACT VALUE REFERENCE BOOKS FOR THE MATERIAL SHOULD BE CONSULTED.

[b] THE YIELD AND ULTIMATE STRENGTHS FOR DUCTILE MATERIALS CAN BE ASSUMED EQUAL FOR BOTH TENSION AND COMPRESSION.

[c] MEASURED PERPENDICULAR TO THE GRAIN.

[d] MEASURED PARALLEL TO THE GRAIN.

[e] DEFORMATION MEASURED PERPENDICULAR TO THE GRAIN WHEN THE LOAD IS APPLIED ALONG THE GRAIN.

Hibbeler, R.C., *Mechanics of Materials*, 4th ed., Prentice Hall, 2000.

Simply Supported Beam Slopes and Deflections

BEAM	SLOPE	DEFLECTION	ELASTIC CURVE
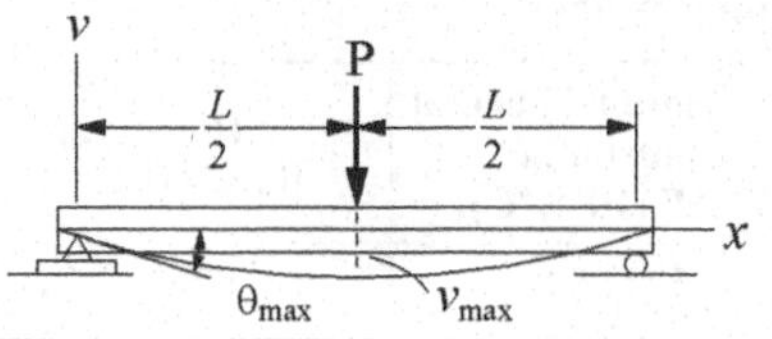	$\theta_{max} = \frac{-PL^2}{16EI}$	$v_{max} = \frac{-PL^3}{48EI}$	$v = \frac{-Px}{48EI}(3L^2 - 4x^2)$ $0 \le x \le L/2$
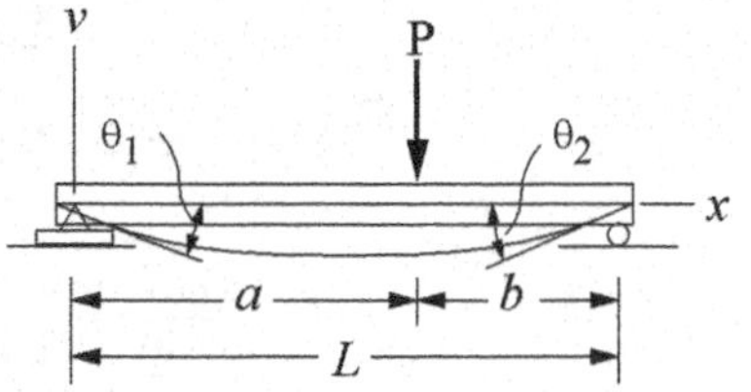	$\theta_1 = \frac{-Pab(L+b)}{6EIL}$ $\theta_2 = \frac{Pab(L+a)}{6EIL}$	$v\big\vert_{x=a} = \frac{-Pba}{6EIL}(L^2 - b^2 - a^2)$	$v = \frac{-Pbx}{6EIL}(L^2 - b^2 - x^2)$ $0 \le x \le a$
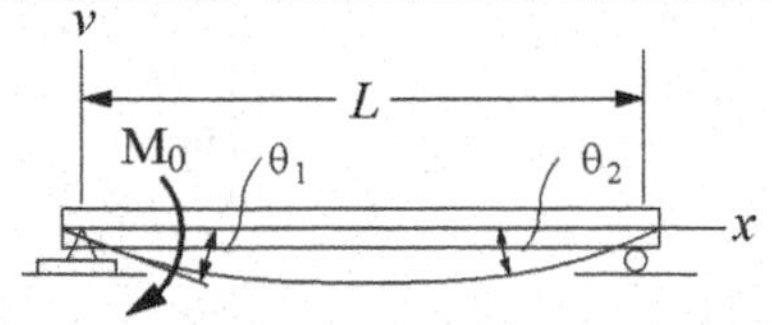	$\theta_1 = \frac{-M_0L}{3EI}$ $\theta_2 = \frac{M_0L}{6EI}$	$v_{max} = \frac{-M_0L^2}{\sqrt{243}EI}$	$v = \frac{-M_0x}{6EIL}(x^2 - 3Lx + 2L^2)$
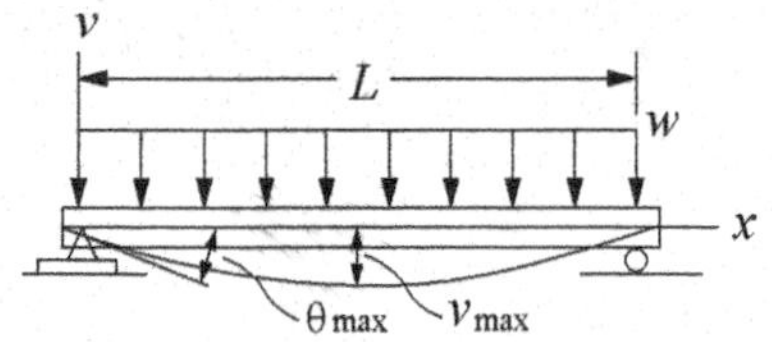	$\theta_{max} = \frac{-wL^3}{24EI}$	$v_{max} = \frac{-5wL^4}{384EI}$	$v = \frac{-wx}{24EI}(x^3 - 2Lx^2 + L^3)$
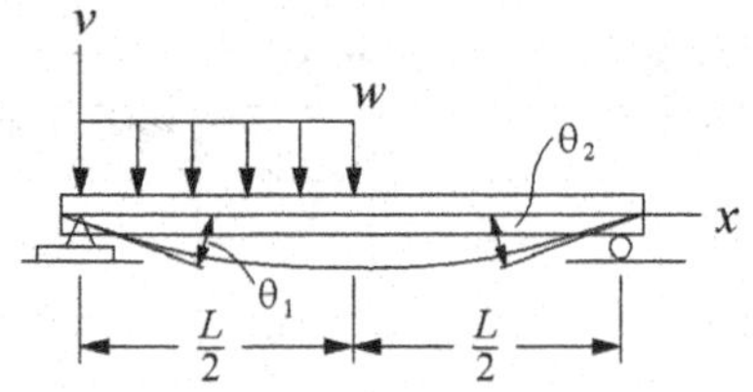	$\theta_1 = \frac{-3wL^3}{128EI}$ $\theta_2 = \frac{7wL^3}{384EI}$	$v\big\vert_{x=L/2} = \frac{-5wL^4}{768EI}$ $v_{max} = -0.006563\frac{wL^4}{EI}$ at $x = 0.4598L$	$v = \frac{-wx}{384EI}(16x^3 - 24Lx^2 + 9L^3)$ $0 \le x \le L/2$ $v = \frac{-wL}{384EI}(8x^3 - 24Lx^2 + 17L^2x - L^3)$ $L/2 \le x < L$
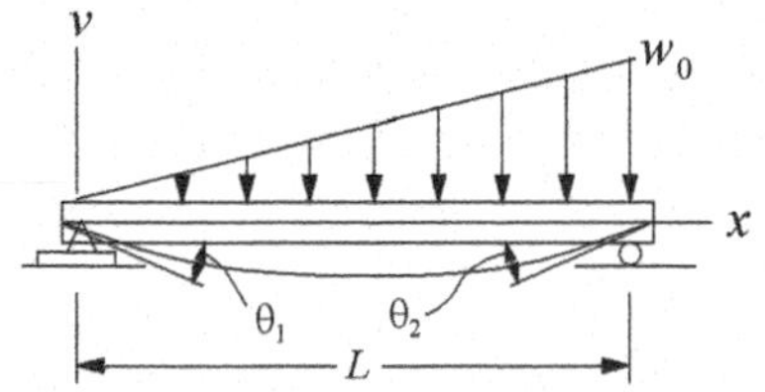	$\theta_1 = \frac{-7w_0L^3}{360EI}$ $\theta_2 = \frac{w_0L^3}{45EI}$	$v_{max} = -0.00652\frac{w_0L^4}{EI}$ at $x = 0.5193L$	$v = \frac{-w_0x}{360EIL}(3x^4 - 10L^2x^2 + 7L^4)$

Adapted from Hibbeler, R.C., *Mechanics of Materials*, 4th ed., Prentice Hall, 2000, p. 800.

Cantilevered Beam Slopes and Deflections

BEAM	SLOPE	DEFLECTION	ELASTIC CURVE
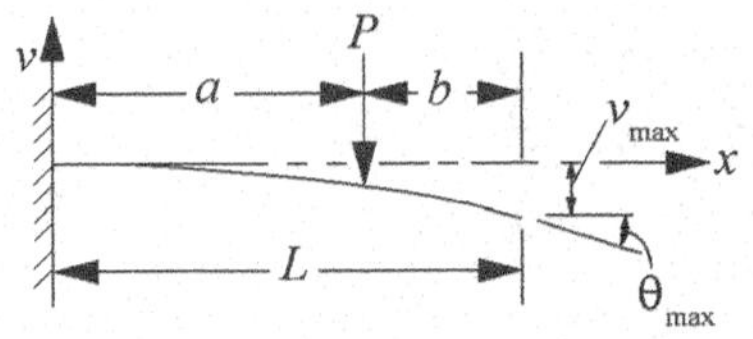	$\theta_{max} = \frac{-Pa^2}{2EI}$	$v_{max} = \frac{-Pa^2}{6EI}(3L - a)$	$v = \frac{-Pa^2}{6EI}(3x - a)$, for $x > a$ $v = \frac{-Px^2}{6EI}(-x + 3a)$, for $x \leq a$

Adapted from Crandall, S.H. and N.C. Dahl, *An Introduction to Mechanics of Solids*, McGraw-Hill, New York, 1959.

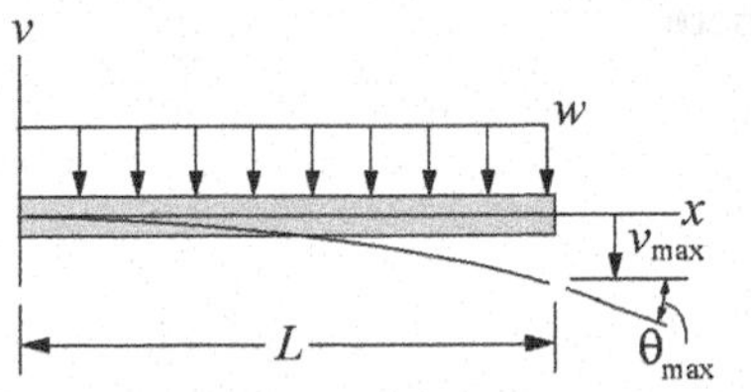	$\theta_{max} = \frac{-wL^3}{6EI}$	$v_{max} = \frac{-wL^4}{8EI}$	$v = \frac{-wx^2}{24EI}(x^2 - 4Lx + 6L^2)$
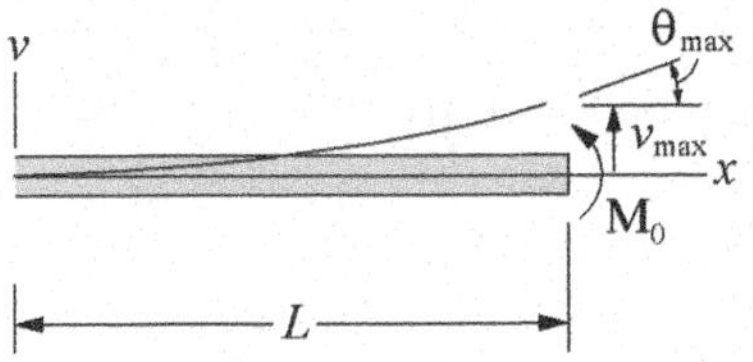	$\theta_{max} = \frac{M_0L}{EI}$	$v_{max} = \frac{M_0L^2}{2EI}$	$v = \frac{M_0x^2}{2EI}$
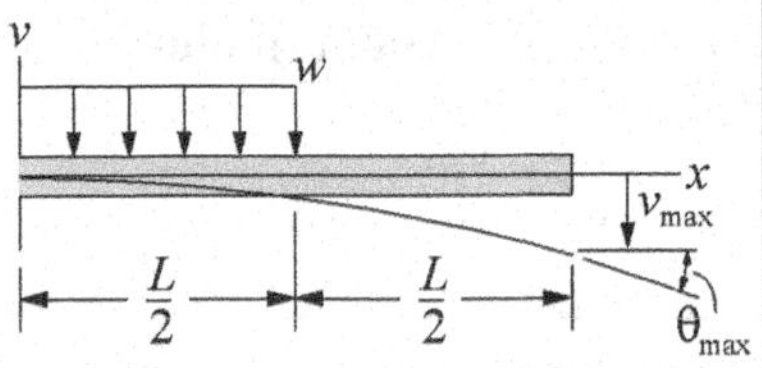	$\theta_{max} = \frac{-wL^3}{48EI}$	$v_{max} = \frac{-7wL^4}{384EI}$	$v = \frac{-wx^2}{24EI}\left(x^2 - 2Lx + \frac{3}{2}L^2\right)$ $0 \leq x \leq L/2$ $v = \frac{-wL^3}{192EI}(4x - L/2)$ $L/2 \leq x \leq L$
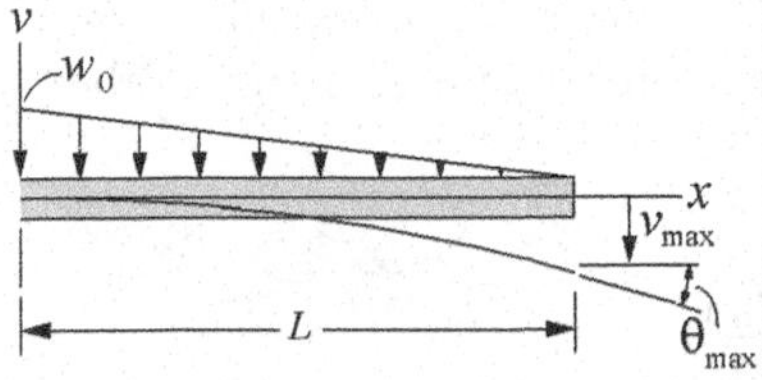	$\theta_{max} = \frac{-w_0L^3}{24EI}$	$v_{max} = \frac{-w_0L^4}{30EI}$	$v = \frac{-w_0x^2}{120EIL}(10L^3 - 10L^2x + 5Lx^2 - x^3)$

Adapted from Hibbeler, R.C., *Mechanics of Materials*, 4th ed., Prentice Hall, 2000, p. 801.

Piping Segment Slopes and Deflections

PIPE	SLOPE	DEFLECTION	ELASTIC CURVE
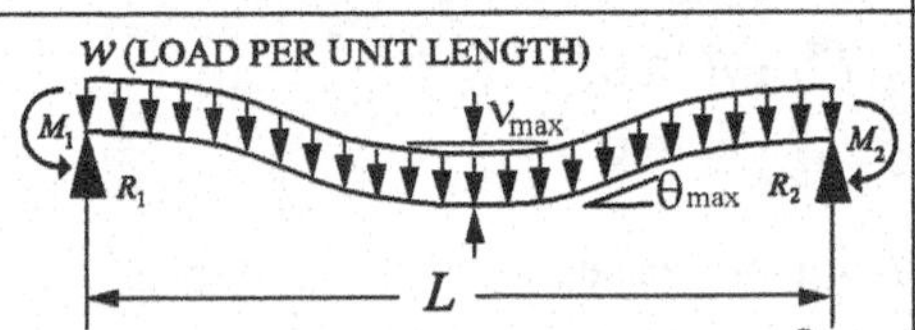 $R_1 = R_2 = \frac{wL}{2}$ and $M_1 = M_2 = \frac{wL^2}{12}$	$\lvert\theta_{max}\rvert = 0.008\frac{wL^3}{24EI}$ at $x = \frac{1}{2} \pm \frac{L}{\sqrt{12}}$	$\lvert v_{max}\rvert = \frac{wL^4}{384EI}$ at $x = \frac{L}{2}$	$v(x) = \frac{wx^2}{24EI}(L^2 - 2Lx + x^2)$

Adapted from Crandall, S.H. and N.C. Dahl, *An Introduction to Mechanics of Solids*, McGraw-Hill, New York, 1959.

THERMODYNAMICS

PROPERTIES OF SINGLE-COMPONENT SYSTEMS

Nomenclature

1. Intensive properties are independent of mass.
2. Extensive properties are proportional to mass.
3. Specific properties are lowercase (extensive/mass).

State Functions (properties)

Absolute Pressure, P	(lbf/in^2 or Pa)
Absolute Temperature, T	(°R or K)
Volume, V	(ft^3 or m^3)
Specific Volume, $v = V/m$	($\text{ft}^3\text{/lbm}$ or $\text{m}^3\text{/kg}$)
Internal Energy, U	(Btu or kJ)
Specific Internal Energy, $u = U/m$	(Btu/lbm or kJ/kg)
Enthalpy, H	(Btu or kJ)
Specific Enthalpy, $h = u + Pv = H/m$	(Btu/lbm or kJ/kg)
Entropy, S	(Btu/°R or kJ/K)
Specific Entropy, $s = S/m$	[Btu/(lbm-°R) or kJ/(kg•K)]
Gibbs Free Energy, $g = h - Ts$	(Btu/lbm or kJ/kg)
Helmholtz Free Energy, $a = u - Ts$	(Btu/lbm or kJ/kg)

For a single-phase pure component, specification of any two intensive, independent properties is sufficient to fix all the rest.

Heat Capacity at Constant Pressure,

$$c_p = \left(\frac{\partial h}{\partial T}\right)_P \quad \text{[Btu/(lbm-°R) or kJ/(kg•K)]}$$

Heat Capacity at Constant Volume,

$$c_v = \left(\frac{\partial u}{\partial T}\right)_v \quad \text{[Btu/(lbm-°R) or kJ/(kg•K)]}$$

The steam tables in this section provide T, P, v, u, h, and s data for saturated and superheated water.

A P-h diagram for refrigerant HFC-134a providing T, P, v, h, and s data in a graphical format is included in this section.

Thermal and physical property tables for selected gases, liquids, and solids are included in this section.

Properties for Two-Phase (vapor-liquid) Systems

Quality x (for liquid-vapor systems at saturation) is defined as the mass fraction of the vapor phase:

$$x = m_g/(m_g + m_f), \text{ where}$$

m_g = mass of vapor

m_f = mass of liquid

Specific volume of a two-phase system can be written:

$v = xv_g + (1 - x)v_f$ or $v = v_f + xv_{fg}$, where

v_f = specific volume of saturated liquid

v_g = specific volume of saturated vapor

v_{fg} = specific volume change upon vaporization $= v_g - v_f$

Similar expressions exist for u, h, and s:

$$u = xu_g + (1 - x)\,u_f \text{ or } u = u_f + xu_{fg}$$
$$h = xh_g + (1 - x)\,h_f \text{ or } h = h_f + xh_{fg}$$
$$s = xs_g + (1 - x)\,s_f \text{ or } s = s_f + xs_{fg}$$

PVT BEHAVIOR

Ideal Gas

For an ideal gas, $Pv = RT$ or $PV = mRT$, and

$$P_1v_1/T_1 = P_2v_2/T_2, \text{ where}$$

P = pressure

v = specific volume

m = mass of gas

R = gas constant

T = absolute temperature

V = volume

R is *specific to each gas* but can be found from

$$R_i = \frac{\bar{R}}{(mol.\ wt)_i}, \text{ where}$$

$\bar{R}$ = the universal gas constant

= 1,545 ft-lbf/(lbmol-°R) = 8,314 J/(kmol•K)

= 8.314 kPa•m³/(kmol•K) = 0.08206 L•atm/(mole•K)

For *ideal gases*, $c_p - c_v = R$

Ideal gas behavior is characterized by:

- no intermolecular interactions
- molecules occupy zero volume

The properties of an ideal gas reflect those of a single molecule and are attributable entirely to the structure of the molecule and the system T.

For *ideal gases*:

$$\left(\frac{\partial h}{\partial P}\right)_T = 0 \qquad \left(\frac{\partial u}{\partial v}\right)_T = 0$$

For cold air standard, *heat capacities are assumed to be constant* at their room temperature values. In that case, the following are true:

$$\Delta u = c_v\Delta T; \quad \Delta h = c_p\Delta T$$
$$\Delta s = c_p \ln (T_2/T_1) - R \ln (P_2/P_1)$$
$$\Delta s = c_v \ln (T_2/T_1) + R \ln (v_2/v_1)$$

Also, for *constant entropy* processes:

$$\frac{P_2}{P_1} = \left(\frac{v_1}{v_2}\right)^k; \qquad \frac{T_2}{T_1} = \left(\frac{P_2}{P_1}\right)^{\frac{k-1}{k}}$$

$$\frac{T_2}{T_1} = \left(\frac{v_1}{v_2}\right)^{k-1}, \text{ where } k = c_p/c_v$$

Ideal Gas Mixtures

$i = 1, 2, \ldots, n$ constituents. Each constituent is an ideal gas.

Mole Fraction:

$$x_i = N_i / N;\ N = \Sigma N_i;\ \Sigma x_i = 1$$

where N_i = number of moles of component i

N = total moles in the mixture

Mass Fraction: $y_i = m_i / m;\ m = \Sigma m_i;\ \Sigma y_i = 1$

Molecular Weight: $M = m/N = \Sigma x_i M_i$

To convert *mole fractions* x_i to *mass fractions* y_i:

$$y_i = \frac{x_i M_i}{\Sigma(x_i M_i)}$$

To convert *mass fractions* to *mole fractions*:

$$x_i = \frac{y_i / M_i}{\Sigma(y_i / M_i)}$$

Partial Pressures: $P_i = \dfrac{m_i R_i T}{V}$ and $P = \Sigma P_i$

Partial Volumes: $V_i = \dfrac{m_i R_i T}{P}$ and $V = \Sigma V_i$

where P, V, T = the pressure, volume, and temperature of the mixture and $R_i = \overline{R} / M_i$

Combining the above generates the following additional expressions for mole fraction.

$$x_i = P_i / P = V_i / V$$

Other Properties:

$c_p = \Sigma (y_i c_{p_i})$

$c_v = \Sigma (y_i c_{v_i})$

$u = \Sigma (y_i u_i);\ h = \Sigma (y_i h_i);\ s = \Sigma (y_i s_i)$

u_i and h_i are evaluated at T

s_i is evaluated at T and P_i

Real Gas

Most gases exhibit ideal gas behavior when the system pressure is less than 3 atm since the distance between molecules is large enough to produce negligible molecular interactions. The behavior of a real gas deviates from that of an ideal gas at higher pressures due to molecular interactions.

For a real gas, $Pv = ZRT$ where Z = compressibility factor.

$Z = 1$ for an ideal gas; $Z \neq 1$ for a real gas

Equations of State (EOS)

EOS are used to quantify PvT behavior

Ideal Gas EOS (applicable only to ideal gases)

$$P = \left(\frac{RT}{v}\right)$$

Generalized Compressibility EOS (applicable to all systems as gases, liquids, and/or solids)

$$P = \left(\frac{RT}{v}\right)Z$$

Virial EOS (applicable only to gases)

$$P = \left(\frac{RT}{v}\right)\left(1 + \frac{B}{v} + \frac{C}{v^2} + \ldots\right) \text{ where } B, C, \ldots$$

are virial coefficients obtained from PvT measurements or statistical mechanics.

Cubic EOS (theoretically motivated with intent to predict gas and liquid thermodynamic properties)

$$P = \frac{RT}{v - b} - \frac{a(T)}{(v + c_1 b)(v + c_2 b)}$$

where $a(T)$, b, and c_1 and c_2 are species specific.

An example of a cubic EOS is the Van der Waals equation with constants based on the critical point:

$$\left(P + \frac{a}{\overline{v}^2}\right)(\overline{v} - b) = \overline{R}T$$

$$\text{where } a = \left(\frac{27}{64}\right)\left(\frac{\overline{R}^2 T_c^2}{P_c}\right),\quad b = \frac{\overline{R}T_c}{8P_c}$$

where P_c and T_c are the pressure and temperature at the critical point, respectively, and $\overline{v}$ is the molar specific volume.

EOS are used to predict:

- P, v, or T when two of the three are specified
- other thermodynamic properties based on analytic manipulation of the EOS
- mixture properties using appropriate mixing rules to create a pseudo-component that mimics the mixture properties

The Theorem of Corresponding States asserts that all normal fluids have the same value of Z at the same reduced temperature T_r and pressure P_r.

$$T_r = \frac{T}{T_c} \quad P_r = \frac{P}{P_c}$$

where T_c and P_c are the critical temperature and pressure, respectively, expressed in absolute units. This is captured in the Generalized Compressibility Factor chart.

FIRST LAW OF THERMODYNAMICS

The *First Law of Thermodynamics* is a statement of conservation of energy in a thermodynamic system. The net energy crossing the system boundary is equal to the change in energy inside the system.

Heat Q ($q = Q/m$) is *energy transferred* due to temperature difference and is considered positive if it is inward or added to the system.

Work W ($w = W/m$) is considered *positive if it is outward* or *work done* by the system.

Closed Thermodynamic System

No mass crosses system boundary

$$Q - W = \Delta U + \Delta KE + \Delta PE$$

where ΔU = change in internal energy
ΔKE = change in kinetic energy
ΔPE = change in potential energy

Energy can cross the boundary only in the form of heat or work. Work can be boundary work, w_b, or other work forms (electrical work, etc.)

Reversible boundary work is given by $w_b = \int P\,dv$.

Special Cases of Closed Systems (with no change in kinetic or potential energy)

Constant System Pressure process (***Charles' Law***):

$w_b = P\Delta v$

(ideal gas) T/v = constant

Constant Volume process:

$w_b = 0$

(ideal gas) T/P = constant

Isentropic process (ideal gas):

Pv^k = constant

$w = (P_2v_2 - P_1v_1)/(1 - k)$

$= R(T_2 - T_1)/(1 - k)$

Constant Temperature process (***Boyle's Law***):

(ideal gas) Pv = constant

$w_b = RT\ln(v_2/v_1) = RT\ln(P_1/P_2)$

Polytropic process (ideal gas):

Pv^n = constant

$w = (P_2v_2 - P_1v_1)/(1 - n),\ n \neq 1$

Open Thermodynamic System

Mass crosses the system boundary.

There is flow work (Pv) done by mass entering the system.

The reversible flow work is given by:

$$w_{rev} = -\int v\,dP + \Delta KE + \Delta PE$$

First Law applies whether or not processes are reversible.

Open System First Law (energy balance)

$$\Sigma\dot{m}_i\left[h_i + V_i^2/2 + gZ_i\right] - \Sigma\dot{m}_e\left[h_e + V_e^2/2 + gZ_e\right] + \dot{Q}_{in} - \dot{W}_{net} = d(m_s u_s)/dt, \text{ where}$$

$\dot{W}_{net}$ = rate of net or shaft work

$\dot{m}$ = mass flowrate (subscripts *i* and *e* refer to inlet and exit states of system)

g = acceleration of gravity

Z = elevation

V = velocity

m_s = mass of fluid within the system

u_s = specific internal energy of system

$\dot{Q}_{in}$ = rate of heat transfer (neglecting kinetic and potential energy of the system)

Special Cases of Open Systems (with no change in kinetic or potential energy)

Constant Volume process:

$w_{rev} = -v(P_2 - P_1)$

Constant System Pressure process:

$w_{rev} = 0$

Constant Temperature process:

(ideal gas) Pv = constant

$w_{rev} = RT\ln(v_2/v_1) = RT\ln(P_1/P_2)$

Isentropic process (ideal gas):

Pv^k = constant

$w_{rev} = k(P_2v_2 - P_1v_1)/(1 - k)$

$= kR(T_2 - T_1)/(1 - k)$

$$w_{rev} = \frac{k}{k-1}RT_1\left[1 - \left(\frac{P_2}{P_1}\right)^{(k-1)/k}\right]$$

Polytropic process (ideal gas):

Pv^n = constant

Closed system

$w_{rev} = (P_2v_2 - P_1v_1)/(1 - n)$

One-inlet, one-exit control volume

$w_{rev} = n(P_2v_2 - P_1v_1)/(1 - n)$

Steady-Flow Systems

The system does not change state with time. This assumption is valid for steady operation of turbines, pumps, compressors, throttling valves, nozzles, and heat exchangers, including boilers and condensers.

$$\Sigma\dot{m}_i\left(h_i + V_i^2/2 + gZ_i\right) - \Sigma\dot{m}_e\left(h_e + V_e^2/2 + gZ_e\right) + \dot{Q}_{in} - \dot{W}_{out} = 0$$

and

$$\Sigma\dot{m}_i = \Sigma\dot{m}_e$$

where

$\dot{m}$ = mass flowrate (subscripts *i* and *e* refer to inlet and exit states of system)

g = acceleration of gravity

Z = elevation

V = velocity

$\dot{Q}_{in}$ = the net rate of heat transfer into the system

$\dot{W}_{out}$ = the net rate of work out of the system

Special Cases of Steady-Flow Energy Equation

Nozzles, Diffusers: Velocity terms are significant. No elevation change, no heat transfer, and no work. Single-mass stream.

$$h_i + V_i^2/2 = h_e + V_e^2/2$$

Isentropic Efficiency (nozzle) $= \dfrac{V_e^2 - V_i^2}{2(h_i - h_{es})}$, where

h_{es} = enthalpy at isentropic exit state.

Turbines, Pumps, Compressors: Often considered adiabatic (no heat transfer). Velocity terms usually can be ignored. There are significant work terms and a single-mass stream.

$$h_i = h_e + w$$

Isentropic Efficiency (turbine) $= \dfrac{h_i - h_e}{h_i - h_{es}}$

Isentropic Efficiency (compressor, pump) $= \dfrac{h_{es} - h_i}{h_e - h_i}$

For pump only, $h_{es} - h_i = v_i(P_e - P_i)$

Throttling Valves and Throttling Processes: No work, no heat transfer, and single-mass stream. Velocity terms are often insignificant.

$$h_i = h_e$$

Boilers, Condensers, Evaporators, One Side in a Heat Exchanger: Heat transfer terms are significant. For a single-mass stream, the following applies:

$$h_i + q = h_e$$

Heat Exchangers: No heat loss to the surroundings or work. Two separate flowrates $\dot{m}_1$ and $\dot{m}_2$:

$$\dot{m}_1(h_{1i} - h_{1e}) = \dot{m}_2(h_{2e} - h_{2i})$$

Mixers, Separators, Open or Closed Feedwater Heaters:

$$\Sigma \dot{m}_i h_i = \Sigma \dot{m}_e h_e \quad \text{and}$$
$$\Sigma \dot{m}_i = \Sigma \dot{m}_e$$

BASIC CYCLES

Heat engines take in heat Q_H at a high temperature T_H, produce a net amount of work W, and reject heat Q_L at a low temperature T_L. The efficiency η of a heat engine is given by:

$$\eta = W/Q_H = (Q_H - Q_L)/Q_H$$

The most efficient engine possible is the *Carnot Cycle*. Its efficiency is given by:

$$\eta_c = (T_H - T_L)/T_H, \text{ where}$$

T_H and T_L = absolute temperatures (Kelvin or Rankine).

The following heat-engine cycles are plotted on *P-v* and *T-s* diagrams in this section:

Carnot, Otto, Rankine

Refrigeration cycles are the reverse of heat-engine cycles. Heat is moved from low to high temperature requiring work, W. Cycles can be used either for refrigeration or as heat pumps.

Coefficient of Performance (COP) is defined as:

COP = Q_H/W for heat pumps, and as

COP = Q_L/W for refrigerators and air conditioners.

Upper limit of COP is based on reversed Carnot Cycle:

$COP_c = T_H/(T_H - T_L)$ for heat pumps and

$COP_c = T_L/(T_H - T_L)$ for refrigeration.

1 ton refrigeration = 12,000 Btu/hr = 3,516 W

The following refrigeration cycles are plotted on *T-s* diagrams in this section:

reversed rankine, two-stage refrigeration, air refrigeration

PSYCHROMETRICS

Properties of an air-water vapor mixture at a fixed pressure are given in graphical form on a psychrometric chart as provided in this section. When the system pressure is 1 atm, an ideal-gas mixture is assumed.

The definitions that follow use subscript *a* for dry air and *v* for water vapor.

P = pressure of the air-water mixture, normally 1 atm

T = dry-bulb temp (air/water mixture temperature)

P_a = partial pressure of dry air

P_v = partial pressure of water vapor

$$P = P_a + P_v$$

Specific Humidity (absolute humidity, humidity ratio) ω:

$\omega = m_v/m_a$, where

m_v = mass of water vapor

m_a = mass of dry air

$$\omega = 0.622P_v/P_a = 0.622P_v/(P - P_v)$$

Relative Humidity (rh) ϕ:

$\phi = P_v/P_g$, where

P_g = saturation pressure of water at T.

Enthalpy h: $h = h_a + \omega h_v$

Dew-Point Temperature T_{dp}:

$$T_{dp} = T_{sat} \text{ at } P_g = P_v$$

Wet-bulb temperature T_{wb} is the temperature indicated by a thermometer covered by a wick saturated with liquid water and in contact with moving air.

Humid Volume: Volume of moist air/mass of dry air.

SECOND LAW OF THERMODYNAMICS

Thermal Energy Reservoirs

$\Delta S_{\text{reservoir}} = Q/T_{\text{reservoir}}$, where

Q is measured with respect to the reservoir.

Kelvin-Planck Statement of Second Law

No heat engine can operate in a cycle while transferring heat with a single heat reservoir.

COROLLARY to Kelvin-Planck: No heat engine can have a higher efficiency than a Carnot Cycle operating between the same reservoirs.

Clausius' Statement of Second Law

No refrigeration or heat pump cycle can operate without a net work input.

COROLLARY: No refrigerator or heat pump can have a higher COP than a Carnot Cycle refrigerator or heat pump.

Entropy

$$ds = (1/T)\delta q_{\text{rev}}$$

$$s_2 - s_1 = \int_1^2 (1/T)\delta q_{\text{rev}}$$

Inequality of Clausius

$$\oint (1/T)\delta q_{\text{rev}} \le 0$$

$$\int_1^2 (1/T)\delta q \le s_2 - s_1$$

Isothermal, Reversible Process

$$\Delta s = s_2 - s_1 = q/T$$

Isentropic Process

$$\Delta s = 0;\ ds = 0$$

A reversible adiabatic process is isentropic.

Adiabatic Process

$$\delta q = 0;\ \Delta s \ge 0$$

Increase of Entropy Principle

$$\Delta s_{\text{total}} = \Delta s_{\text{system}} + \Delta s_{\text{surroundings}} \ge 0$$

$$\Delta \dot{s}_{\text{total}} = \Sigma \dot{m}_{\text{out}} s_{\text{out}} - \Sigma \dot{m}_{\text{in}} s_{\text{in}} - \Sigma\left(\dot{q}_{\text{external}}/T_{\text{external}}\right) \ge 0$$

Temperature-Entropy (T-s) Diagram

$q_{rev} = \int_1^2 T\,ds$

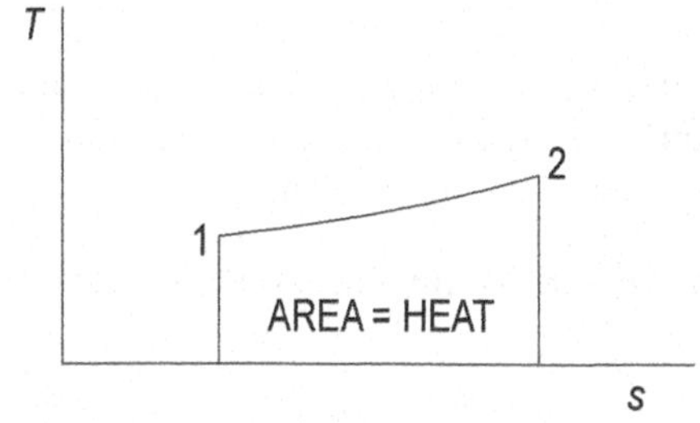

Entropy Change for Solids and Liquids

$$ds = c\,(dT/T)$$

$$s_2 - s_1 = \int c\,(dT/T) = c_{\text{mean}} \ln (T_2/T_1),$$

where c equals the heat capacity of the solid or liquid.

Exergy (Availability)

Exergy (also known as availability) is the maximum possible work that can be obtained from a cycle of a heat engine. The maximum possible work is obtained in a reversible process.

Closed-System Exergy (Availability)

(no chemical reactions)

$$\phi = (u - u_L) - T_L(s - s_L) + p_L(v - v_L)$$

where the subscript L designates environmental conditions and ϕ is availability function.

$$w_{max} = w_{rev} = \phi_i - \phi_2$$

Open-System Exergy (Availability)

$$\Psi = (h - h_L) - T_L(s - s_L) + V^2/2 + gZ$$

where V is velocity, g is acceleration of gravity, Z is elevation and Ψ is availability function.

$$w_{max} = w_{rev} = \Psi_i - \Psi_2$$

Gibbs Free Energy, ΔG

Energy released or absorbed in a reaction occurring reversibly at constant pressure and temperature.

Helmholtz Free Energy, ΔA

Energy released or absorbed in a reaction occurring reversibly at constant volume and temperature.

Irreversibility, I

$$I = w_{rev} - w_{actual} = T_L \Delta s_{total}$$

HEATS OF REACTION

For a chemical reaction the associated energy can be defined in terms of heats of formation of the individual species ΔH_f° at the standard state

$$\left(\Delta H_r^\circ\right) = \sum_{\text{products}} \nu_i \left(\Delta H_f^\circ\right)_i - \sum_{\text{reactants}} \nu_i \left(\Delta H_f^\circ\right)_i$$

ν_i = stoichiometric coefficient for species "i"

The standard state is 25°C and 1 bar.

The heat of formation is defined as the enthalpy change associated with the formation of a compound from its atomic species as they normally occur in nature [i.e., O_2(g), H_2(g), C(solid), etc.]

The heat of reaction varies with the temperature as follows:

$$\Delta H_r^\circ(T) = \Delta H_r^\circ(T_{\text{ref}}) + \int_{T_{\text{ref}}}^{T} \Delta c_p dT$$

where T_{ref} is some reference temperature (typically 25°C or 298 K), and:

$$\Delta c_p = \sum_{\text{products}} \nu_i c_{p,i} - \sum_{\text{reactants}} \nu_i c_{p,i}$$

and $c_{p,i}$ is the molar heat capacity of component i.

The heat of reaction for a combustion process using oxygen is also known as the heat of combustion. The principal products are CO_2(g) and $H_2O(l)$.

Combustion Processes

First, the combustion equation should be written and balanced. For example, for the stoichiometric combustion of methane in oxygen:

$$CH_4 + 2\,O_2 \rightarrow CO_2 + 2\,H_2O$$

Combustion in Air

For each mole of oxygen, there will be 3.76 moles of nitrogen. For stoichiometric combustion of methane in air:

$$CH_4 + 2\,O_2 + 2(3.76)\,N_2 \rightarrow CO_2 + 2\,H_2O + 7.52\,N_2$$

Combustion in Excess Air

The excess oxygen appears as oxygen on the right side of the combustion equation.

Incomplete Combustion

Some carbon is burned to create carbon monoxide (CO).

$$\textit{Molar Air-Fuel Ratio, } \overline{A/F} = \frac{\text{No. of moles of air}}{\text{No. of moles of fuel}}$$

$$\textit{Air-Fuel Ratio, } A/F = \frac{\text{Mass of air}}{\text{Mass of fuel}} = \left(\overline{A/F}\right)\left(\frac{M_{air}}{M_{fuel}}\right)$$

Stoichiometric (theoretical) air-fuel ratio is the air-fuel ratio calculated from the stoichiometric combustion equation.

$$\text{Percent Theoretical Air} = \frac{(A/F)_{\text{actual}}}{(A/F)_{\text{stoichiometric}}} \times 100$$

$$\text{Percent Excess Air} = \frac{(A/F)_{\text{actual}} - (A/F)_{\text{stoichiometric}}}{(A/F)_{\text{stoichiometric}}} \times 100$$

VAPOR-LIQUID EQUILIBRIUM (VLE)

Henry's Law at Constant Temperature

At equilibrium, the partial pressure of a gas is proportional to its concentration in a liquid. Henry's Law is valid for low concentrations; i.e., $x \approx 0$.

$$P_i = Py_i = hx_i, \text{ where}$$

h = Henry's Law constant

P_i = partial pressure of a gas in contact with a liquid

x_i = mol fraction of the gas in the liquid

y_i = mol fraction of the gas in the vapor

P = total pressure

Raoult's Law for Vapor-Liquid Equilibrium

Valid for concentrations near 1; i.e., $x_i \approx 1$ at low pressure (ideal gas behavior)

$$P_i = x_i P_i^*, \text{ where}$$

P_i = partial pressure of component i

x_i = mol fraction of component i in the liquid

P_i^* = vapor pressure of pure component i at the temperature of the mixture

Rigorous Vapor-Liquid Equilibrium

For a multicomponent mixture at equilibrium

$$\hat{f}_i^V = \hat{f}_i^L, \text{ where}$$

$\hat{f}_i^V$ = fugacity of component i in the vapor phase

$\hat{f}_i^L$ = fugacity of component i in the liquid phase

Fugacities of component i in a mixture are commonly calculated in the following ways:

For a liquid $\hat{f}_i^L = x_i \gamma_i f_i^L$, where

x_i = mole fraction of component i

γ_i = activity coefficient of component i

f_i^L = fugacity of pure liquid component i

For a vapor $\hat{f}_i^V = y_i \hat{\Phi}_i P$, where

y_i = mole fraction of component i in the vapor

$\hat{\Phi}_i$ = fugacity coefficient of component i in the vapor

P = system pressure

The activity coefficient γ_i is a correction for liquid phase non-ideality. Many models have been proposed for γ_i such as the Van Laar model:

$$\ln\gamma_1 = A_{12}\left(1 + \frac{A_{12}x_1}{A_{21}x_2}\right)^{-2}$$

$$\ln\gamma_2 = A_{21}\left(1 + \frac{A_{21}x_2}{A_{12}x_1}\right)^{-2}, \text{ where}$$

γ_1 = activity coefficient of component 1 in a two-component system

γ_2 = activity coefficient of component 2 in a two-component system

A_{12}, A_{21} = constants, typically fitted from experimental data

The pure component fugacity is calculated as:

$$f_i^L = \Phi_i^{sat} P_i^{sat} \exp\left\{v_i^L\left(P - P_i^{sat}\right)/(RT)\right\}, \text{ where}$$

Φ_i^{sat} = fugacity coefficient of pure saturated i

P_i^{sat} = saturation pressure of pure i

v_i^L = specific volume of pure liquid i

R = Ideal Gas Law Constant

T = absolute temperature

Often at system pressures close to atmospheric:

$$f_i^L \cong P_i^{sat}$$

The fugacity coefficient $\hat{\Phi}_i$ for component i in the vapor is calculated from an equation of state (e.g., Virial). Sometimes it is approximated by a pure component value from a correlation. Often at pressures close to atmospheric, $\hat{\Phi}_i = 1$. The fugacity coefficient is a correction for vapor phase non-ideality.

For sparingly soluble gases the liquid phase is sometimes represented as:

$$\hat{f}_i^L = x_i k_i$$

where k_i is a constant set by experiment (Henry's constant). Sometimes other concentration units are used besides mole fraction with a corresponding change in k_i.

PHASE RELATIONS

Clapeyron Equation for phase transitions:

$$\left(\frac{dP}{dT}\right)_{sat} = \frac{h_{fg}}{Tv_{fg}} = \frac{s_{fg}}{v_{fg}}, \text{ where}$$

h_{fg} = enthalpy change for phase transitions

v_{fg} = volume change

s_{fg} = entropy change

T = absolute temperature

$(dP/dT)_{sat}$ = slope of phase transition (e.g.,vapor-liquid) saturation line

Clausius-Clapeyron Equation
This equation results if it is assumed that (1) the volume change (v_{fg}) can be replaced with the vapor volume (v_g), (2) the latter can be replaced with $P/\bar{R}T$ from the ideal gas law, and (3) h_{fg} is independent of the temperature (T).

$$\ln_e\left(\frac{P_2}{P_1}\right) = \frac{h_{fg}}{R} \cdot \frac{T_2 - T_1}{T_1 T_2}$$

Gibbs Phase Rule (non-reacting systems)
P + F = C + 2, where
P = number of phases making up a system
F = degrees of freedom
C = number of components in a system

CHEMICAL REACTION EQUILIBRIA

Definitions

Conversion – moles reacted/moles fed

Extent – For each species in a reaction, the mole balance may be written:

$$\text{moles}_{i,out} = \text{moles}_{i,in} + v_i\xi \text{ where}$$

ξ is the extent in moles and v_i is the stoichiometric coefficient of the *i*th species, the sign of which is negative for reactants and positive for products.

Limiting reactant – Reactant that would be consumed first if the reaction proceeded to completion. Other reactants are excess reactants.

Selectivity – Moles of desired product formed/moles of undesired product formed.

Yield – Moles of desired product formed/moles that would have been formed if there were no side reactions and the limiting reactant had reacted completely.

Chemical Reaction Equilibrium
For the reaction

$$aA + bB \rightleftharpoons cC + dD$$

$$\Delta G^\circ = -RT \ln K_a$$

$$K_a = \frac{(\hat{a}_C^c)(\hat{a}_D^d)}{(\hat{a}_A^a)(\hat{a}_B^b)} = \prod_i (\hat{a}_i)^{v_i}, \text{ where}$$

$\hat{a}_i$ = activity of component $i = \dfrac{\hat{f}_i}{f_i^\circ}$

f_i° = fugacity of pure i in its standard state at the equilibrium reaction temperature, T

v_i = stoichiometric coefficient of component i

ΔG° = standard Gibbs energy change of reaction

K_a = chemical equilibrium constant

<u>For mixtures of ideal gases:</u>

f_i° = unit pressure, often 1 bar

$\hat{f}_i = y_i P = p_i$

where p_i = partial pressure of component i.

$$\text{Then } K_a = K_p = \frac{(p_C^c)(p_D^d)}{(p_A^a)(p_B^b)} = P^{c+d-a-b}\frac{(y_C^c)(y_D^d)}{(y_A^a)(y_B^b)}$$

<u>For solids</u> $\hat{a}_i = 1$

<u>For liquids</u> $\hat{a}_i = x_i \gamma_i$

The effect of temperature on the equilibrium constant is

$$\frac{d \ln K}{dT} = \frac{\Delta H^\circ}{RT^2}$$

where ΔH° = standard enthalpy change of reaction.

STEAM TABLES
Saturated Water - Temperature Table

Temp. °C T	Sat. Press. kPa p_{sat}	Specific Volume m³/kg Sat. liquid v_f	Specific Volume m³/kg Sat. vapor v_g	Internal Energy kJ/kg Sat. liquid u_f	Internal Energy kJ/kg Evap. u_{fg}	Internal Energy kJ/kg Sat. vapor u_g	Enthalpy kJ/kg Sat. liquid h_f	Enthalpy kJ/kg Evap. h_{fg}	Enthalpy kJ/kg Sat. vapor h_g	Entropy kJ/(kg·K) Sat. liquid s_f	Entropy kJ/(kg·K) Evap. s_{fg}	Entropy kJ/(kg·K) Sat. vapor s_g
0.01	0.6113	0.001 000	206.14	0.00	2375.3	2375.3	0.01	2501.3	2501.4	0.0000	9.1562	9.1562
5	0.8721	0.001 000	147.12	20.97	2361.3	2382.3	20.98	2489.6	2510.6	0.0761	8.9496	9.0257
10	1.2276	0.001 000	106.38	42.00	2347.2	2389.2	42.01	2477.7	2519.8	0.1510	8.7498	8.9008
15	1.7051	0.001 001	77.93	62.99	2333.1	2396.1	62.99	2465.9	2528.9	0.2245	8.5569	8.7814
20	**2.339**	**0.001 002**	**57.79**	**83.95**	**2319.0**	**2402.9**	**83.96**	**2454.1**	**2538.1**	**0.2966**	**8.3706**	**8.6672**
25	3.169	0.001 003	43.36	104.88	2304.9	2409.8	104.89	2442.3	2547.2	0.3674	8.1905	8.5580
30	4.246	0.001 004	32.89	125.78	2290.8	2416.6	125.79	2430.5	2556.3	0.4369	8.0164	8.4533
35	5.628	0.001 006	25.22	146.67	2276.7	2423.4	146.68	2418.6	2565.3	0.5053	7.8478	8.3531
40	7.384	0.001 008	19.52	167.56	2262.6	2430.1	167.57	2406.7	2574.3	0.5725	7.6845	8.2570
45	**9.593**	**0.001 010**	**15.26**	**188.44**	**2248.4**	**2436.8**	**188.45**	**2394.8**	**2583.2**	**0.6387**	**7.5261**	**8.1648**
50	12.349	0.001 012	12.03	209.32	2234.2	2443.5	209.33	2382.7	2592.1	0.7038	7.3725	8.0763
55	15.758	0.001 015	9.568	230.21	2219.9	2450.1	230.23	2370.7	2600.9	0.7679	7.2234	7.9913
60	19.940	0.001 017	7.671	251.11	2205.5	2456.6	251.13	2358.5	2609.6	0.8312	7.0784	7.9096
65	25.03	0.001 020	6.197	272.02	2191.1	2463.1	272.06	2346.2	2618.3	0.8935	6.9375	7.8310
70	**31.19**	**0.001 023**	**5.042**	**292.95**	**2176.6**	**2569.6**	**292.98**	**2333.8**	**2626.8**	**0.9549**	**6.8004**	**7.7553**
75	38.58	0.001 026	4.131	313.90	2162.0	2475.9	313.93	2321.4	2635.3	1.0155	6.6669	7.6824
80	47.39	0.001 029	3.407	334.86	2147.4	2482.2	334.91	2308.8	2643.7	1.0753	6.5369	7.6122
85	57.83	0.001 033	2.828	355.84	2132.6	2488.4	355.90	2296.0	2651.9	1.1343	6.4102	7.5445
90	70.14	0.001 036	2.361	376.85	2117.7	2494.5	376.92	2283.2	2660.1	1.1925	6.2866	7.4791
95	**84.55**	**0.001 040**	**1.982**	**397.88**	**2102.7**	**2500.6**	**397.96**	**2270.2**	**2668.1**	**1.2500**	**6.1659**	**7.4159**
	MPa											
100	0.101 35	0.001 044	1.6729	418.94	2087.6	2506.5	419.04	2257.0	2676.1	1.3069	6.0480	7.3549
105	0.120 82	0.001 048	1.4194	440.02	2072.3	2512.4	440.15	2243.7	2683.8	1.3630	5.9328	7.2958
110	0.143 27	0.001 052	1.2102	461.14	2057.0	2518.1	461.30	2230.2	2691.5	1.4185	5.8202	7.2387
115	0.169 06	0.001 056	1.0366	482.30	2041.4	2523.7	482.48	2216.5	2699.0	1.4734	5.7100	7.1833
120	**0.198 53**	**0.001 060**	**0.8919**	**503.50**	**2025.8**	**2529.3**	**503.71**	**2202.6**	**2706.3**	**1.5276**	**5.6020**	**7.1296**
125	0.2321	0.001 065	0.7706	524.74	2009.9	2534.6	524.99	2188.5	2713.5	1.5813	5.4962	7.0775
130	0.2701	0.001 070	0.6685	546.02	1993.9	2539.9	546.31	2174.2	2720.5	1.6344	5.3925	7.0269
135	0.3130	0.001 075	0.5822	567.35	1977.7	2545.0	567.69	2159.6	2727.3	1.6870	5.2907	6.9777
140	0.3613	0.001 080	0.5089	588.74	1961.3	2550.0	589.13	2144.7	2733.9	1.7391	5.1908	6.9299
145	**0.4154**	**0.001 085**	**0.4463**	**610.18**	**1944.7**	**2554.9**	**610.63**	**2129.6**	**2740.3**	**1.7907**	**5.0926**	**6.8833**
150	0.4758	0.001 091	0.3928	631.68	1927.9	2559.5	632.20	2114.3	2746.5	1.8418	4.9960	6.8379
155	0.5431	0.001 096	0.3468	653.24	1910.8	2564.1	653.84	2098.6	2752.4	1.8925	4.9010	6.7935
160	0.6178	0.001 102	0.3071	674.87	1893.5	2568.4	675.55	2082.6	2758.1	1.9427	4.8075	6.7502
165	0.7005	0.001 108	0.2727	696.56	1876.0	2572.5	697.34	2066.2	2763.5	1.9925	4.7153	6.7078
170	**0.7917**	**0.001 114**	**0.2428**	**718.33**	**1858.1**	**2576.5**	**719.21**	**2049.5**	**2768.7**	**2.0419**	**4.6244**	**6.6663**
175	0.8920	0.001 121	0.2168	740.17	1840.0	2580.2	741.17	2032.4	2773.6	2.0909	4.5347	6.6256
180	1.0021	0.001 127	0.194 05	762.09	1821.6	2583.7	763.22	2015.0	2778.2	2.1396	4.4461	6.5857
185	1.1227	0.001 134	0.174 09	784.10	1802.9	2587.0	785.37	1997.1	2782.4	2.1879	4.3586	6.5465
190	1.2544	0.001 141	0.156 54	806.19	1783.8	2590.0	807.62	1978.8	2786.4	2.2359	4.2720	6.5079
195	**1.3978**	**0.001 149**	**0.141 05**	**828.37**	**1764.4**	**2592.8**	**829.98**	**1960.0**	**2790.0**	**2.2835**	**4.1863**	**6.4698**
200	1.5538	0.001 157	0.127 36	850.65	1744.7	2595.3	852.45	1940.7	2793.2	2.3309	4.1014	6.4323
205	1.7230	0.001 164	0.115 21	873.04	1724.5	2597.5	875.04	1921.0	2796.0	2.3780	4.0172	6.3952
210	1.9062	0.001 173	0.104 41	895.53	1703.9	2599.5	897.76	1900.7	2798.5	2.4248	3.9337	6.3585
215	2.104	0.001 181	0.094 79	918.14	1682.9	2601.1	920.62	1879.9	2800.5	2.4714	3.8507	6.3221
220	**2.318**	**0.001 190**	**0.086 19**	**940.87**	**1661.5**	**2602.4**	**943.62**	**1858.5**	**2802.1**	**2.5178**	**3.7683**	**6.2861**
225	2.548	0.001 199	0.078 49	963.73	1639.6	2603.3	966.78	1836.5	2803.3	2.5639	3.6863	6.2503
230	2.795	0.001 209	0.071 58	986.74	1617.2	2603.9	990.12	1813.8	2804.0	2.6099	3.6047	6.2146
235	3.060	0.001 219	0.065 37	1009.89	1594.2	2604.1	1013.62	1790.5	2804.2	2.6558	3.5233	6.1791
240	3.344	0.001 229	0.059 76	1033.21	1570.8	2604.0	1037.32	1766.5	2803.8	2.7015	3.4422	6.1437
245	**3.648**	**0.001 240**	**0.054 71**	**1056.71**	**1546.7**	**2603.4**	**1061.23**	**1741.7**	**2803.0**	**2.7472**	**3.3612**	**6.1083**
250	3.973	0.001 251	0.050 13	1080.39	1522.0	2602.4	1085.36	1716.2	2801.5	2.7927	3.2802	6.0730
255	4.319	0.001 263	0.045 98	1104.28	1596.7	2600.9	1109.73	1689.8	2799.5	2.8383	3.1992	6.0375
260	4.688	0.001 276	0.042 21	1128.39	1470.6	2599.0	1134.37	1662.5	2796.9	2.8838	3.1181	6.0019
265	5.081	0.001 289	0.038 77	1152.74	1443.9	2596.6	1159.28	1634.4	2793.6	2.9294	3.0368	5.9662
270	**5.499**	**0.001 302**	**0.035 64**	**1177.36**	**1416.3**	**2593.7**	**1184.51**	**1605.2**	**2789.7**	**2.9751**	**2.9551**	**5.9301**
275	5.942	0.001 317	0.032 79	1202.25	1387.9	2590.2	1210.07	1574.9	2785.0	3.0208	2.8730	5.8938
280	6.412	0.001 332	0.030 17	1227.46	1358.7	2586.1	1235.99	1543.6	2779.6	3.0668	2.7903	5.8571
285	6.909	0.001 348	0.027 77	1253.00	1328.4	2581.4	1262.31	1511.0	2773.3	3.1130	2.7070	5.8199
290	7.436	0.001 366	0.025 57	1278.92	1297.1	2576.0	1289.07	1477.1	2766.2	3.1594	2.6227	5.7821
295	**7.993**	**0.001 384**	**0.023 54**	**1305.2**	**1264.7**	**2569.9**	**1316.3**	**1441.8**	**2758.1**	**3.2062**	**2.5375**	**5.7437**
300	8.581	0.001 404	0.021 67	1332.0	1231.0	2563.0	1344.0	1404.9	2749.0	3.2534	2.4511	5.7045
305	9.202	0.001 425	0.019 948	1359.3	1195.9	2555.2	1372.4	1366.4	2738.7	3.3010	2.3633	5.6643
310	9.856	0.001 447	0.018 350	1387.1	1159.4	2546.4	1401.3	1326.0	2727.3	3.3493	2.2737	5.6230
315	10.547	0.001 472	0.016 867	1415.5	1121.1	2536.6	1431.0	1283.5	2714.5	3.3982	2.1821	5.5804
320	**11.274**	**0.001 499**	**0.015 488**	**1444.6**	**1080.9**	**2525.5**	**1461.5**	**1238.6**	**2700.1**	**3.4480**	**2.0882**	**5.5362**
330	12.845	0.001 561	0.012 996	1505.3	993.7	2498.9	1525.3	1140.6	2665.9	3.5507	1.8909	5.4417
340	14.586	0.001 638	0.010 797	1570.3	894.3	2464.6	1594.2	1027.9	2622.0	3.6594	1.6763	5.3357
350	16.513	0.001 740	0.008 813	1641.9	776.6	2418.4	1670.6	893.4	2563.9	3.7777	1.4335	5.2112
360	18.651	0.001 893	0.006 945	1725.2	626.3	2351.5	1760.5	720.3	2481.0	3.9147	1.1379	5.0526
370	**21.03**	**0.002 213**	**0.004 925**	**1844.0**	**384.5**	**2228.5**	**1890.5**	**441.6**	**2332.1**	**4.1106**	**0.6865**	**4.7971**
374.14	22.09	0.003 155	0.003 155	2029.6	0	2029.6	2099.3	0	2099.3	4.4298	0	4.4298

Superheated Water Tables

T Temp. °C	v m³/kg	u kJ/kg	h kJ/kg	s kJ/(kg·K)	v m³/kg	u kJ/kg	h kJ/kg	s kJ/(kg·K)
	p = 0.01 MPa (45.81°C)				p = 0.05 MPa (81.33°C)			
Sat.	14.674	2437.9	2584.7	8.1502	3.240	2483.9	2645.9	7.5939
50	14.869	2443.9	2592.6	8.1749				
100	17.196	2515.5	2687.5	8.4479	3.418	2511.6	2682.5	7.6947
150	19.512	2587.9	2783.0	8.6882	3.889	2585.6	2780.1	7.9401
200	**21.825**	**2661.3**	**2879.5**	**8.9038**	**4.356**	**2659.9**	**2877.7**	**8.1580**
250	24.136	2736.0	2977.3	9.1002	4.820	2735.0	2976.0	8.3556
300	26.445	2812.1	3076.5	9.2813	5.284	2811.3	3075.5	8.5373
400	31.063	2968.9	3279.6	9.6077	6.209	2968.5	3278.9	8.8642
500	35.679	3132.3	3489.1	9.8978	7.134	3132.0	3488.7	9.1546
600	**40.295**	**3302.5**	**3705.4**	**10.1608**	**8.057**	**3302.2**	**3705.1**	**9.4178**
700	44.911	3479.6	3928.7	10.4028	8.981	3479.4	3928.5	9.6599
800	49.526	3663.8	4159.0	10.6281	9.904	3663.6	4158.9	9.8852
900	54.141	3855.0	4396.4	10.8396	10.828	3854.9	4396.3	10.0967
1000	58.757	4053.0	4640.6	11.0393	11.751	4052.9	4640.5	10.2964
1100	**63.372**	**4257.5**	**4891.2**	**11.2287**	**12.674**	**4257.4**	**4891.1**	**10.4859**
1200	67.987	4467.9	5147.8	11.4091	13.597	4467.8	5147.7	10.6662
1300	72.602	4683.7	5409.7	11.5811	14.521	4683.6	5409.6	10.8382
	p = 0.10 MPa (99.63°C)				p = 0.20 MPa (120.23°C)			
Sat.	1.6940	2506.1	2675.5	7.3594	0.8857	2529.5	2706.7	7.1272
100	1.6958	2506.7	2676.2	7.3614				
150	1.9364	2582.8	2776.4	7.6134	0.9596	2576.9	2768.8	7.2795
200	2.172	2658.1	2875.3	7.8343	1.0803	2654.4	2870.5	7.5066
250	**2.406**	**2733.7**	**2974.3**	**8.0333**	**1.1988**	**2731.2**	**2971.0**	**7.7086**
300	2.639	2810.4	3074.3	8.2158	1.3162	2808.6	3071.8	7.8926
400	3.103	2967.9	3278.2	8.5435	1.5493	2966.7	3276.6	8.2218
500	3.565	3131.6	3488.1	8.8342	1.7814	3130.8	3487.1	8.5133
600	4.028	3301.9	3704.4	9.0976	2.013	3301.4	3704.0	8.7770
700	**4.490**	**3479.2**	**3928.2**	**9.3398**	**2.244**	**3478.8**	**3927.6**	**9.0194**
800	4.952	3663.5	4158.6	9.5652	2.475	3663.1	4158.2	9.2449
900	5.414	3854.8	4396.1	9.7767	2.705	3854.5	4395.8	9.4566
1000	5.875	4052.8	4640.3	9.9764	2.937	4052.5	4640.0	9.6563
1100	6.337	4257.3	4891.0	10.1659	3.168	4257.0	4890.7	9.8458
1200	**6.799**	**4467.7**	**5147.6**	**10.3463**	**3.399**	**4467.5**	**5147.5**	**10.0262**
1300	7.260	4683.5	5409.5	10.5183	3.630	4683.2	5409.3	10.1982
	p = 0.40 MPa (143.63°C)				p = 0.60 MPa (158.85°C)			
Sat.	0.4625	2553.6	2738.6	6.8959	0.3157	2567.4	2756.8	6.7600
150	0.4708	2564.5	2752.8	6.9299				
200	0.5342	2646.8	2860.5	7.1706	0.3520	2638.9	2850.1	6.9665
250	0.5951	2726.1	2964.2	7.3789	0.3938	2720.9	2957.2	7.1816
300	**0.6548**	**2804.8**	**3066.8**	**7.5662**	**0.4344**	**2801.0**	**3061.6**	**7.3724**
350	0.7137	2884.6	3170.1	7.7324	0.4742	2881.2	3165.7	7.5464
400	0.7726	2964.4	3273.4	7.8985	0.5137	2962.1	3270.3	7.7079
500	0.8893	3129.2	3484.9	8.1913	0.5920	3127.6	3482.8	8.0021
600	1.0055	3300.2	3702.4	8.4558	0.6697	3299.1	3700.9	8.2674
700	**1.1215**	**3477.9**	**3926.5**	**8.6987**	**0.7472**	**3477.0**	**3925.3**	**8.5107**
800	1.2372	3662.4	4157.3	8.9244	0.8245	3661.8	4156.5	8.7367
900	1.3529	3853.9	4395.1	9.1362	0.9017	3853.4	4394.4	8.9486
1000	1.4685	4052.0	4639.4	9.3360	0.9788	4051.5	4638.8	9.1485
1100	1.5840	4256.5	4890.2	9.5256	1.0559	4256.1	4889.6	9.3381
1200	**1.6996**	**4467.0**	**5146.8**	**9.7060**	**1.1330**	**4466.5**	**5146.3**	**9.5185**
1300	1.8151	4682.8	5408.8	9.8780	1.2101	4682.3	5408.3	9.6906
	p = 0.80 MPa (170.43°C)				p = 1.00 MPa (179.91°C)			
Sat.	0.2404	2576.8	2769.1	6.6628	0.194 44	2583.6	2778.1	6.5865
200	0.2608	2630.6	2839.3	6.8158	0.2060	2621.9	2827.9	6.6940
250	0.2931	2715.5	2950.0	7.0384	0.2327	2709.9	2942.6	6.9247
300	0.3241	2797.2	3056.5	7.2328	0.2579	2793.2	3051.2	7.1229
350	**0.3544**	**2878.2**	**3161.7**	**7.4089**	**0.2825**	**2875.2**	**3157.7**	**7.3011**
400	0.3843	2959.7	3267.1	7.5716	0.3066	2957.3	3263.9	7.4651
500	0.4433	3126.0	3480.6	7.8673	0.3541	3124.4	3478.5	7.7622
600	0.5018	3297.9	3699.4	8.1333	0.4011	3296.8	3697.9	8.0290
700	0.5601	3476.2	3924.2	8.3770	0.4478	3475.3	3923.1	8.2731
800	**0.6181**	**3661.1**	**4155.6**	**8.6033**	**0.4943**	**3660.4**	**4154.7**	**8.4996**
900	0.6761	3852.8	4393.7	8.8153	0.5407	3852.2	4392.9	8.7118
1000	0.7340	4051.0	4638.2	9.0153	0.5871	4050.5	4637.6	8.9119
1100	0.7919	4255.6	4889.1	9.2050	0.6335	4255.1	4888.6	9.1017
1200	0.8497	4466.1	5145.9	9.3855	0.6798	4465.6	5145.4	9.2822
1300	**0.9076**	**4681.8**	**5407.9**	**9.5575**	**0.7261**	**4681.3**	**5407.4**	**9.4543**

P-*h* Diagram for Refrigerant HFC-134a
(metric units)

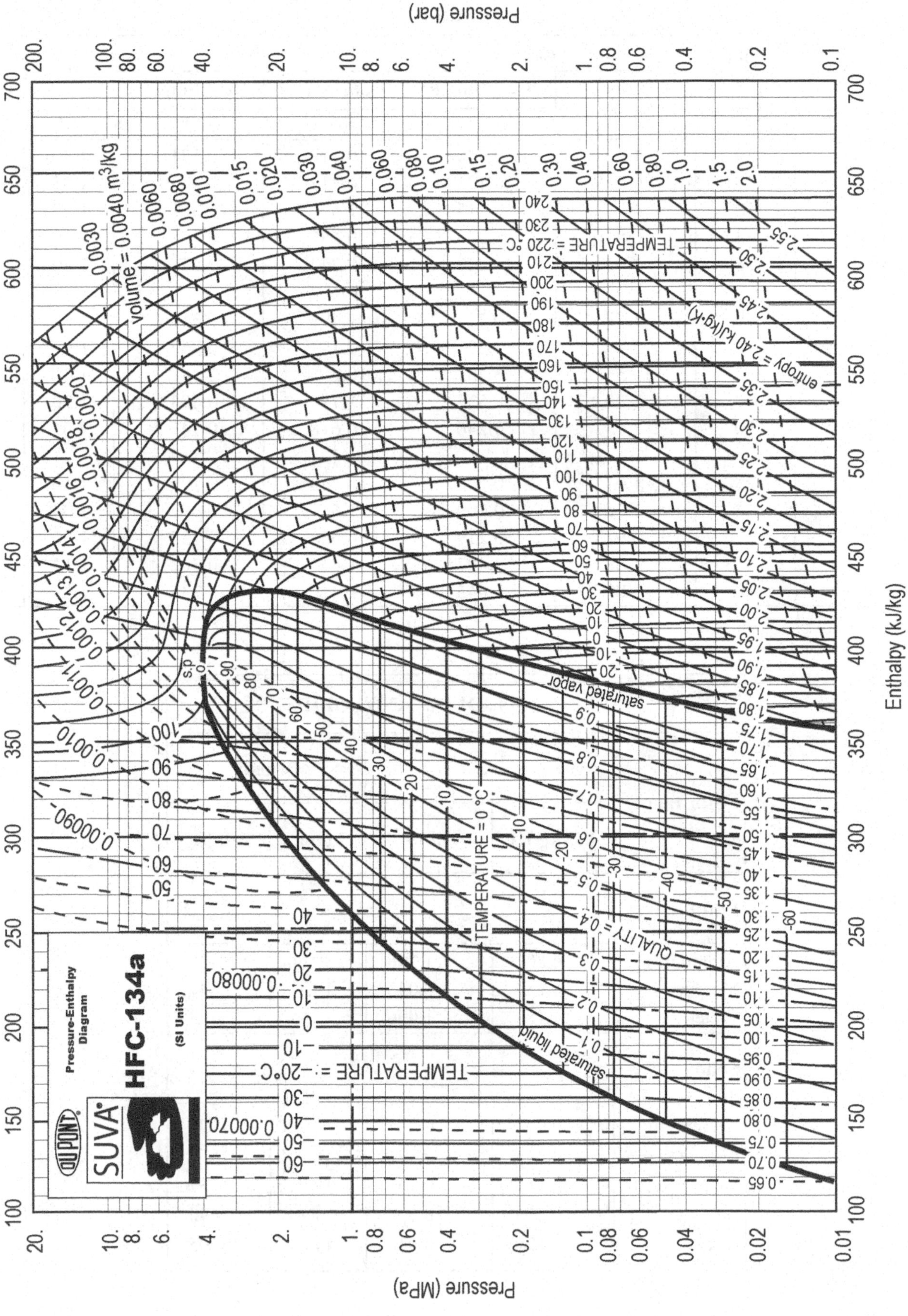

Data provided by DuPont Refrigerants, a division of E.I. duPont de Nemours and Co., Inc.

Thermal and Physical Property Tables
(at room temperature)

GASES							
Substance	Mol wt	c_p		c_v		k	R
		kJ/(kg·K)	Btu/(lbm-°R)	kJ/(kg·K)	Btu/(lbm-°R)		kJ/(kg·K)
Gases							
Air	29	1.00	0.240	0.718	0.171	1.40	0.2870
Argon	40	0.520	0.125	0.312	0.0756	1.67	0.2081
Butane	58	1.72	0.415	1.57	0.381	1.09	0.1430
Carbon dioxide	44	0.846	0.203	0.657	0.158	1.29	0.1889
Carbon monoxide	28	1.04	0.249	0.744	0.178	1.40	0.2968
Ethane	30	1.77	0.427	1.49	0.361	1.18	0.2765
Helium	4	5.19	1.25	3.12	0.753	1.67	2.0769
Hydrogen	2	14.3	3.43	10.2	2.44	1.40	4.1240
Methane	16	2.25	0.532	1.74	0.403	1.30	0.5182
Neon	20	1.03	0.246	0.618	0.148	1.67	0.4119
Nitrogen	28	1.04	0.248	0.743	0.177	1.40	0.2968
Octane vapor	114	1.71	0.409	1.64	0.392	1.04	0.0729
Oxygen	32	0.918	0.219	0.658	0.157	1.40	0.2598
Propane	44	1.68	0.407	1.49	0.362	1.12	0.1885
Steam	18	1.87	0.445	1.41	0.335	1.33	0.4615

SELECTED LIQUIDS AND SOLIDS				
Substance	c_p		Density	
	kJ/(kg·K)	Btu/(lbm-°R)	kg/m^3	lbm/ft^3
Liquids				
Ammonia	4.80	1.146	602	38
Mercury	0.139	0.033	13,560	847
Water	4.18	1.000	997	62.4
Solids				
Aluminum	0.900	0.215	2,700	170
Copper	0.386	0.092	8,900	555
Ice (0°C; 32°F)	2.11	0.502	917	57.2
Iron	0.450	0.107	7,840	490
Lead	0.128	0.030	11,310	705

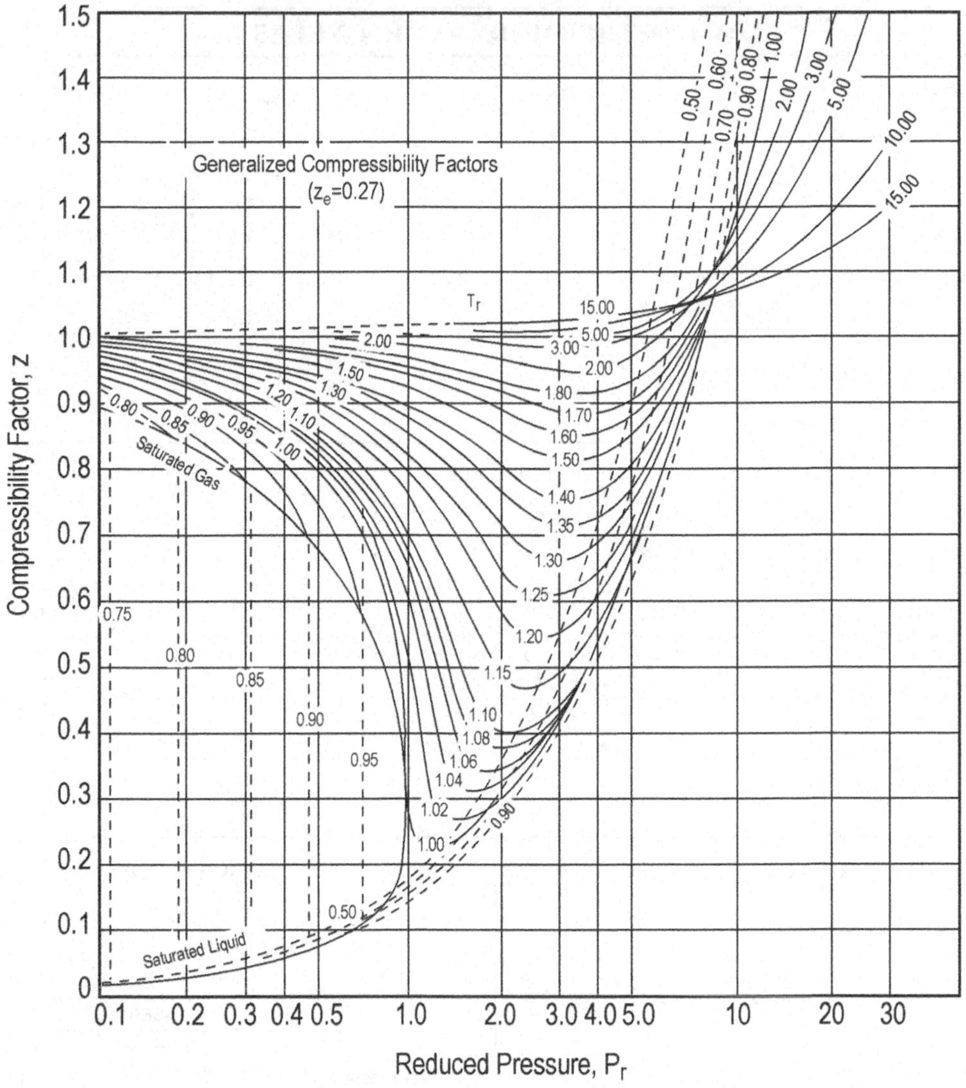

de Nevers, Noel, *Physical and Chemical Equilibrium for Chemical Engineers*, 2nd ed., Wiley, 2012.

COMMON THERMODYNAMIC CYCLES

Carnot Cycle

$$\eta = 1 - \frac{T_L}{T_H}$$

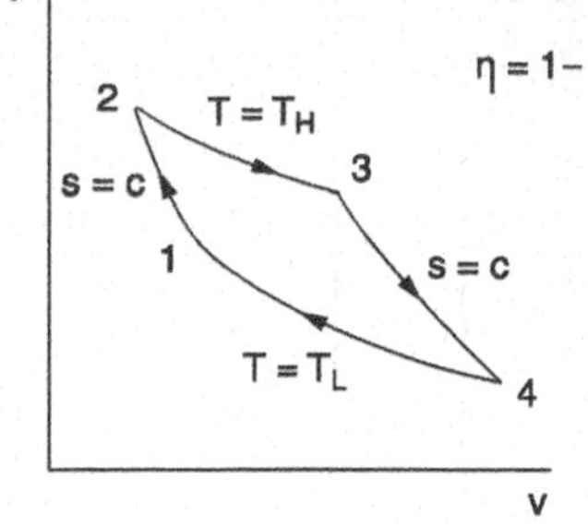

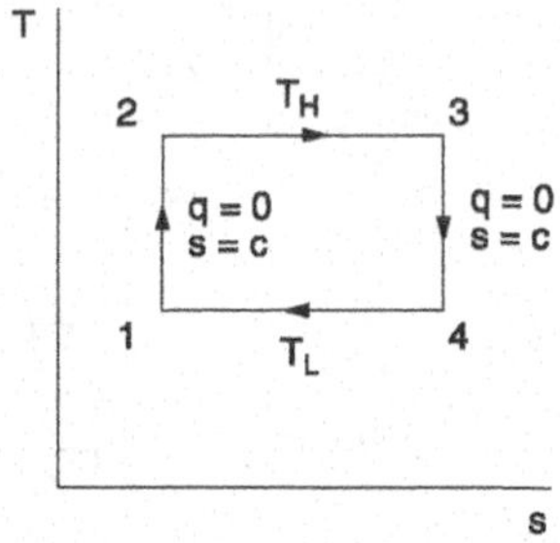

Reversed Carnot

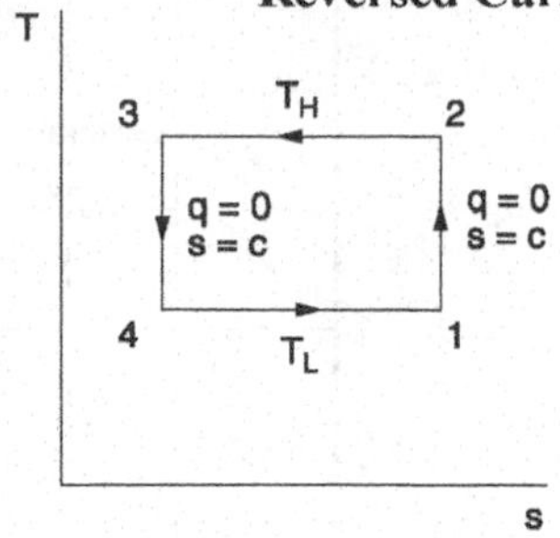

Otto Cycle
(Gasoline Engine)

$$\eta = 1 - r^{1-k}$$

$$r = v_1/v_2$$

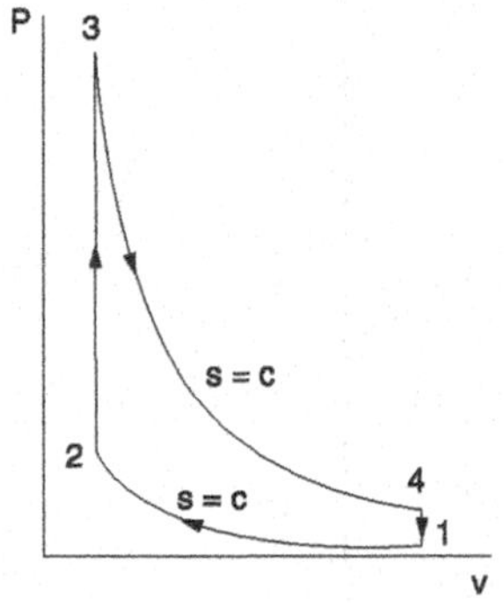

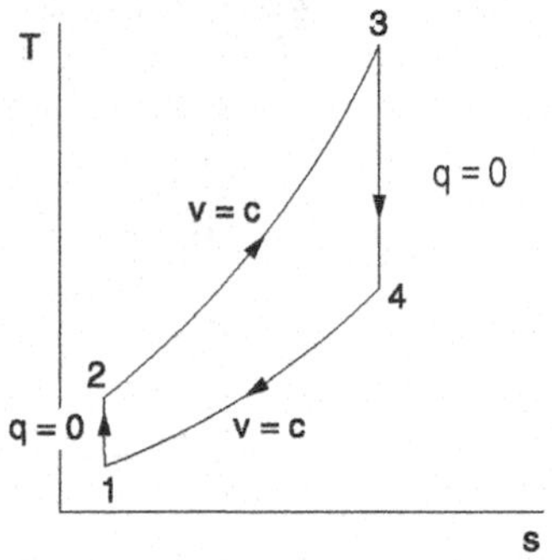

Rankine Cycle

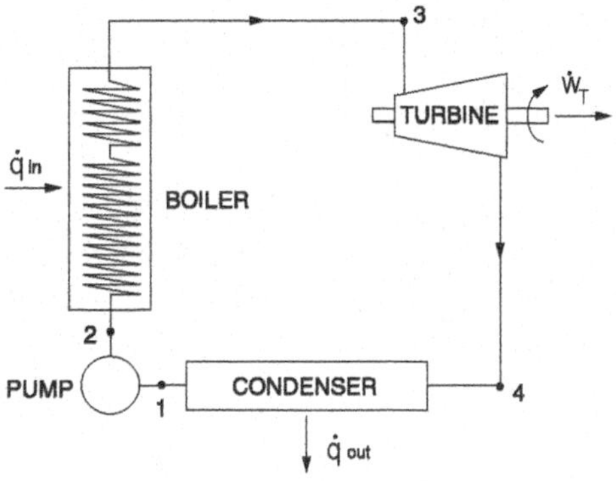

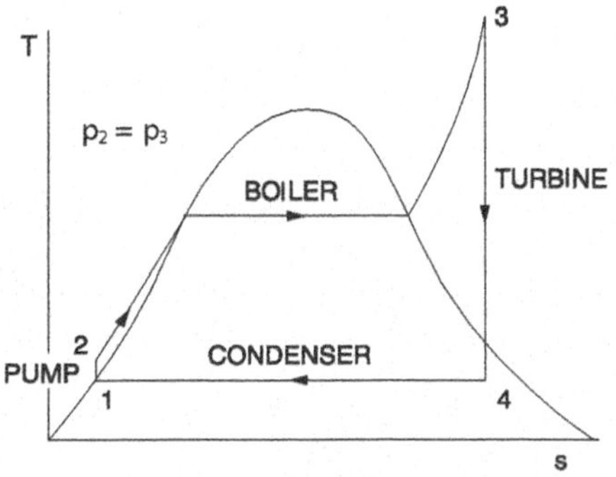

$$\eta = \frac{(h_3 - h_4) - (h_2 - h_1)}{h_3 - h_2}$$

Refrigeration

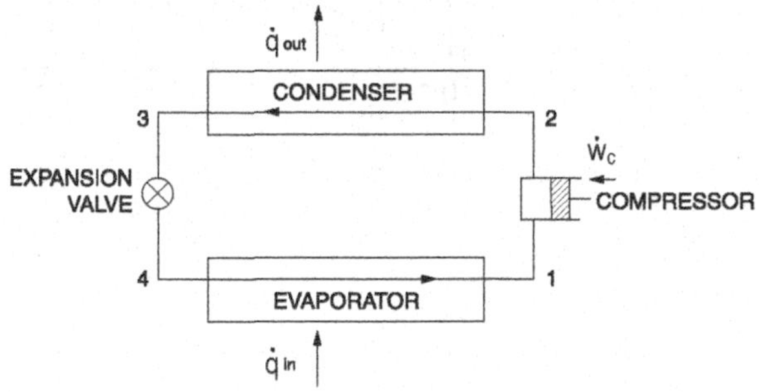

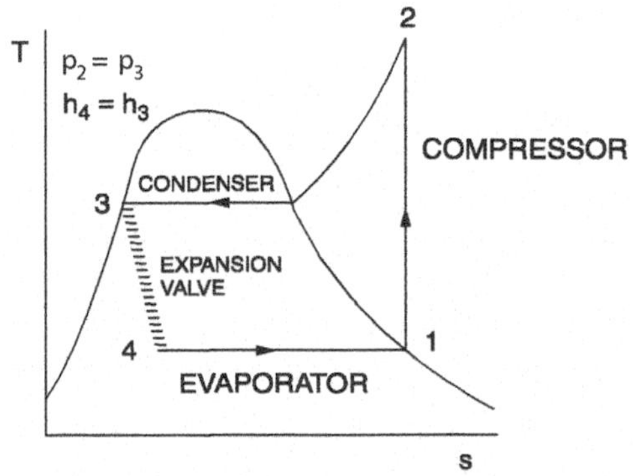

$$COP_{ref} = \frac{h_1 - h_4}{h_2 - h_1} \qquad COP_{HP} = \frac{h_2 - h_3}{h_2 - h_1}$$

REFRIGERATION AND HVAC

Cycles

Refrigeration and HVAC

Two-Stage Cycle

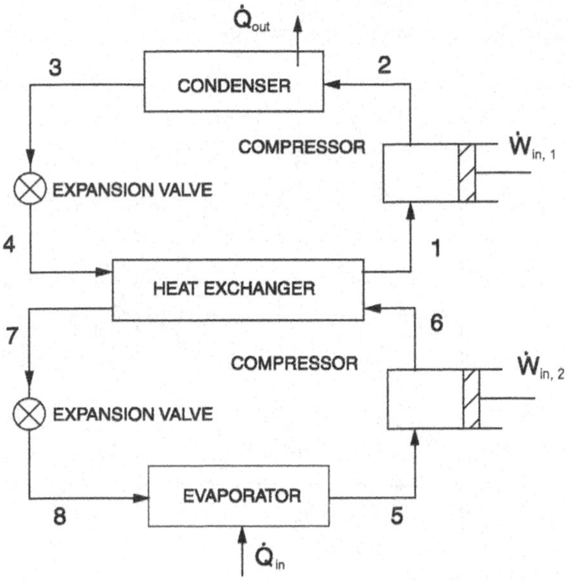

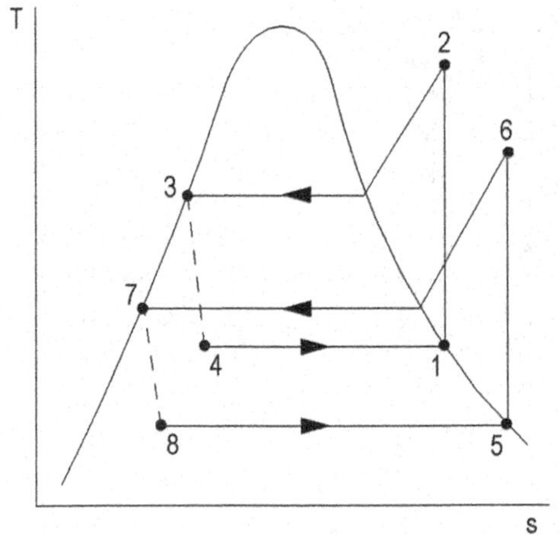

The following equations are valid if the mass flows are the same in each stage.

$$COP_{\text{ref}} = \frac{\dot{Q}_{\text{in}}}{\dot{W}_{\text{in},1} + \dot{W}_{\text{in},2}} = \frac{h_5 - h_8}{h_2 - h_1 + h_6 - h_5}$$

$$COP_{\text{HP}} = \frac{\dot{Q}_{\text{out}}}{\dot{W}_{\text{in},1} + \dot{W}_{\text{in},2}} = \frac{h_2 - h_3}{h_2 - h_1 + h_6 - h_5}$$

Air Refrigeration Cycle

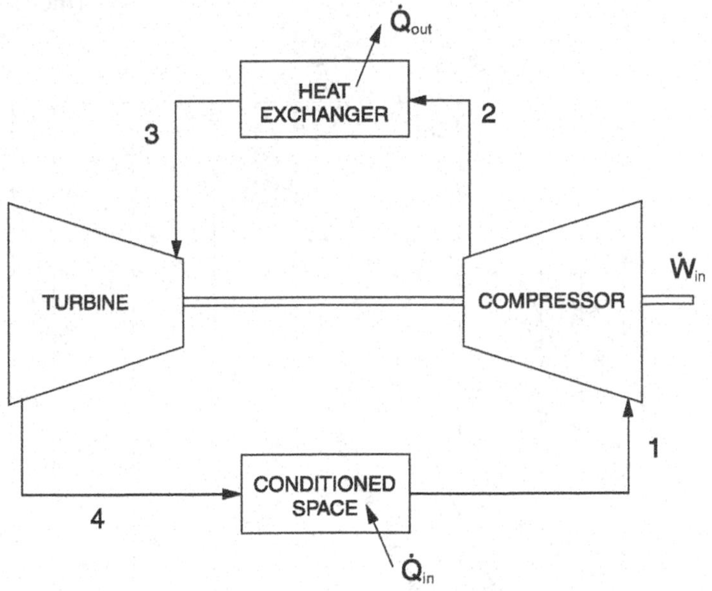

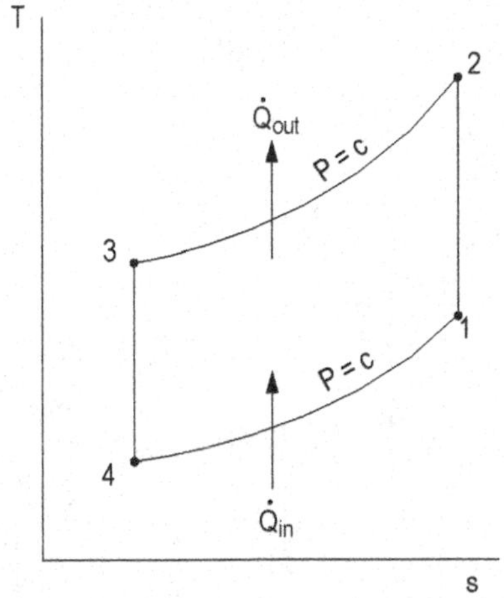

$$COP_{ref} = \frac{h_1 - h_4}{(h_2 - h_1) - (h_3 - h_4)}$$

$$COP_{HP} = \frac{h_2 - h_3}{(h_2 - h_1) - (h_3 - h_4)}$$

ASHRAE Psychrometric Chart No. 1
(metric units)

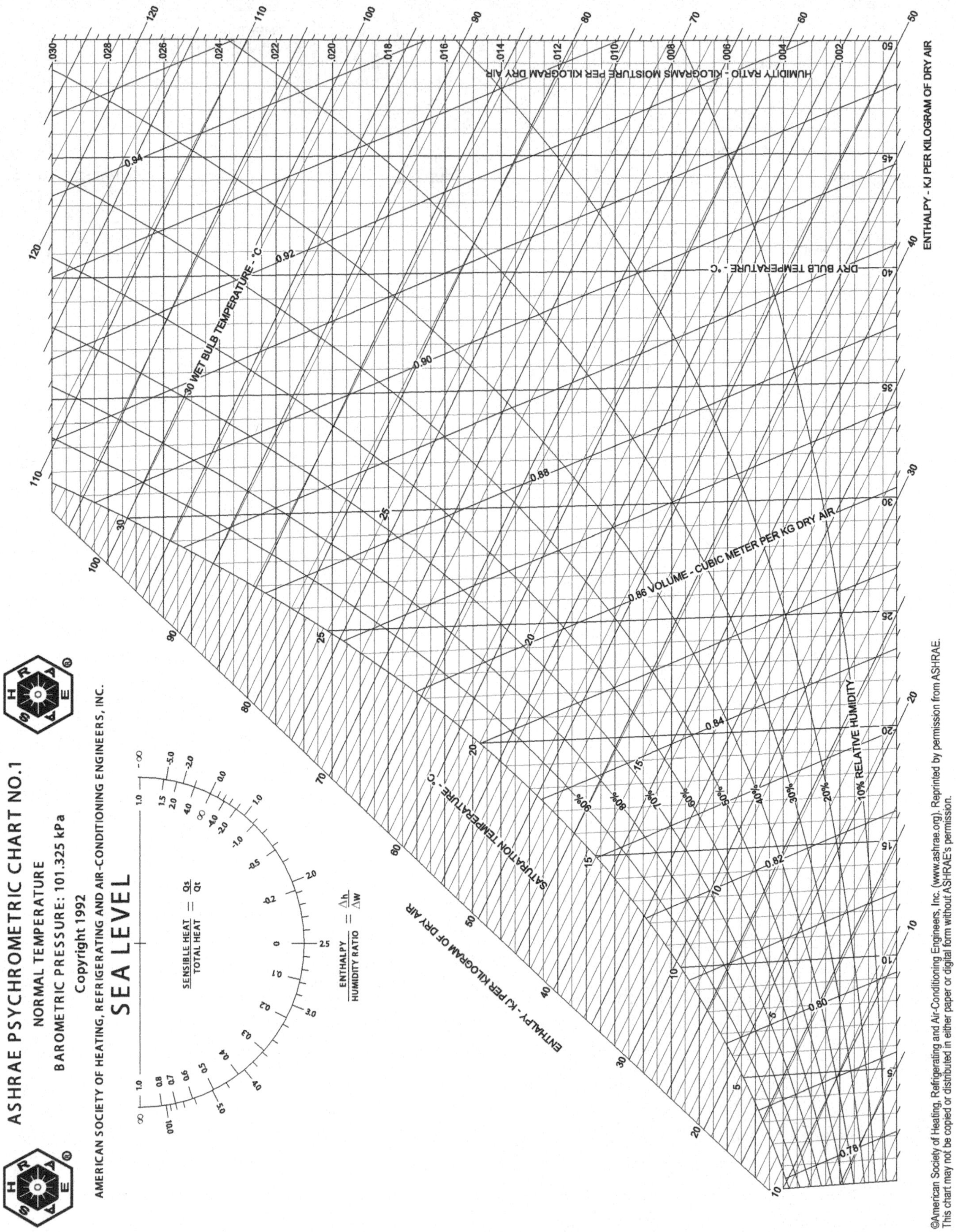

ASHRAE Psychrometric Chart No. 1
(English units)

ASHRAE PSYCHROMETRIC CHART NO.1

NORMAL TEMPERATURE

BAROMETRIC PRESSURE: 29.921 INCHES OF MERCURY

Copyright 1992

AMERICAN SOCIETY OF HEATING, REFRIGERATING AND AIR-CONDITIONING ENGINEERS, INC.

SEA LEVEL

$\dfrac{\text{SENSIBLE HEAT}}{\text{TOTAL HEAT}} = \dfrac{Q_s}{Q_t}$

$\dfrac{\text{ENTHALPY}}{\text{HUMIDITY RATIO}} = \dfrac{\Delta h}{\Delta W}$

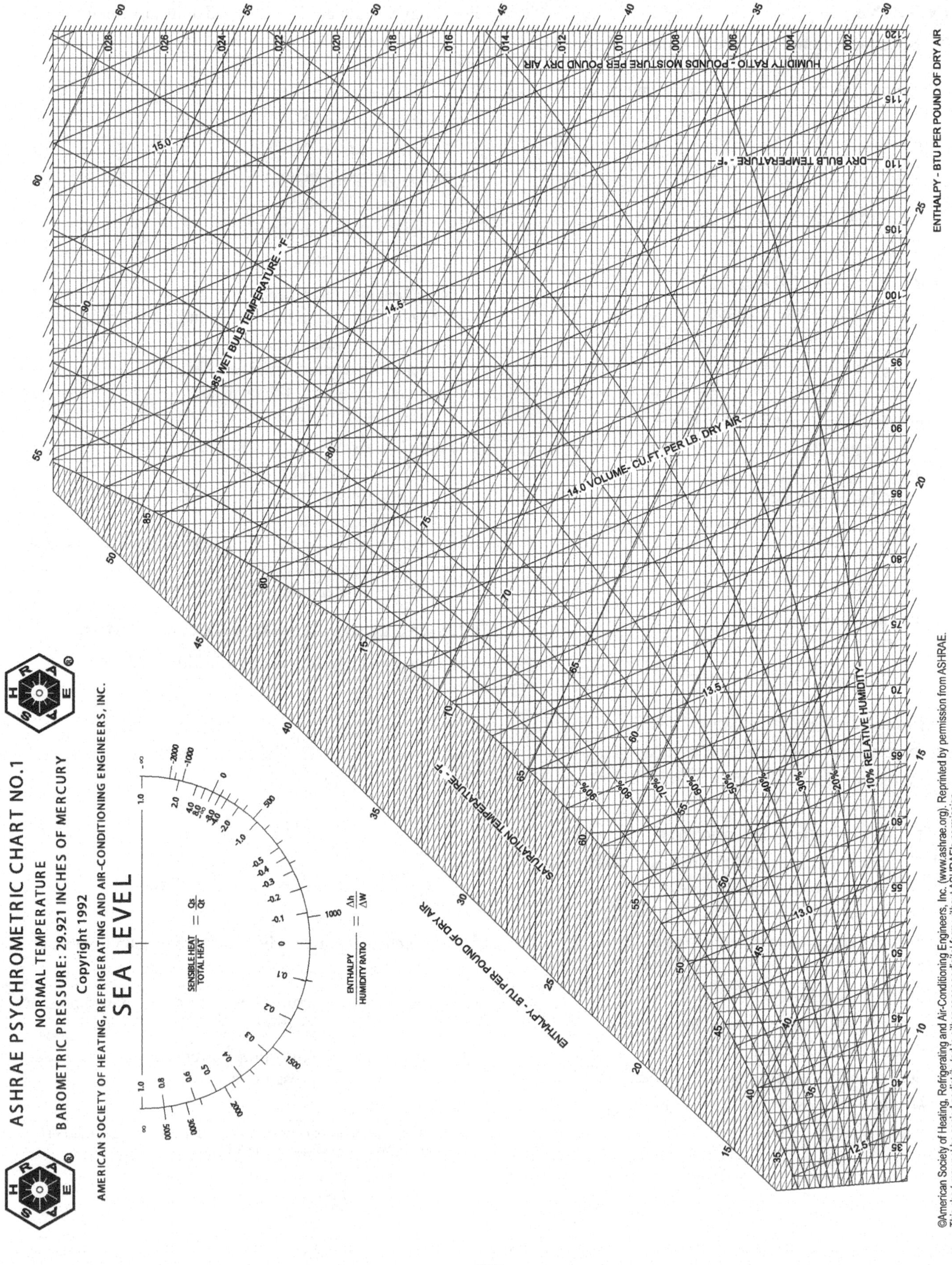

FLUID MECHANICS

DEFINITIONS

Density, Specific Volume, Specific Weight, and Specific Gravity

The definitions of density, specific weight, and specific gravity follow:

$$\rho = \lim_{\Delta V \to 0} \Delta m/\Delta V$$

$$\gamma = \lim_{\Delta V \to 0} \Delta W/\Delta V$$

$$\gamma = \lim_{\Delta V \to 0} g \cdot \Delta m/\Delta V = \rho g$$

also $SG = \gamma/\gamma_w = \rho/\rho_w$, where

ρ = *density* (also called *mass density*)

Δm = mass of infinitesimal volume

ΔV = volume of infinitesimal object considered

γ = *specific weight*
= ρg

ΔW = weight of an infinitesimal volume

SG = *specific gravity*

ρ_w = density of water at standard conditions
= 1,000 kg/m^3 (62.4 lbm/ft^3)

γ_ω = specific weight of water at standard conditions
= 9,810 N/m^3 (62.4 lbf/ft^3)
= 9,810 kg/(m^2•s^2)

Stress, Pressure, and Viscosity

Stress is defined as

$$\tau(1) = \lim_{\Delta A \to 0} \Delta F/\Delta A, \text{ where}$$

$\tau(1)$ = surface stress vector at point 1

ΔF = force acting on infinitesimal area ΔA

ΔA = infinitesimal area at point 1

$$\tau_n = -P$$

$$\tau_t = \mu(dv/dy) \text{ (one-dimensional; i.e., } y\text{), where}$$

τ_n and τ_t = normal and tangential stress components at point 1, respectively

P = pressure at point 1

μ = *absolute dynamic viscosity* of the fluid
N•s/m^2 [lbm/(ft-sec)]

dv = differential velocity

dy = differential distance, normal to boundary

v = velocity at boundary condition

y = normal distance, measured from boundary

ν = *kinematic viscosity*; m^2/s (ft^2/sec)
where $\nu = \mu/\rho$

For a thin Newtonian fluid film and a linear velocity profile,

$$v(y) = vy/\delta;\ dv/dy = v/\delta, \text{ where}$$

v = velocity of plate on film

δ = thickness of fluid film

For a power law (non-Newtonian) fluid

$$\tau_t = K(dv/dy)^n, \text{ where}$$

K = consistency index

n = power law index

$n < 1 \equiv$ pseudo plastic

$n > 1 \equiv$ dilatant

Surface Tension and Capillarity

Surface tension σ is the force per unit contact length

$$\sigma = F/L, \text{ where}$$

σ = surface tension, force/length

F = surface force at the interface

L = length of interface

The *capillary rise* h is approximated by

$$h = (4\sigma \cos \beta)/(\gamma d), \text{ where}$$

h = height of the liquid in the vertical tube

σ = surface tension

β = angle made by the liquid with the wetted tube wall

γ = specific weight of the liquid

d = the diameter of the capillary tube

CHARACTERISTICS OF A STATIC LIQUID

The Pressure Field in a Static Liquid

♦

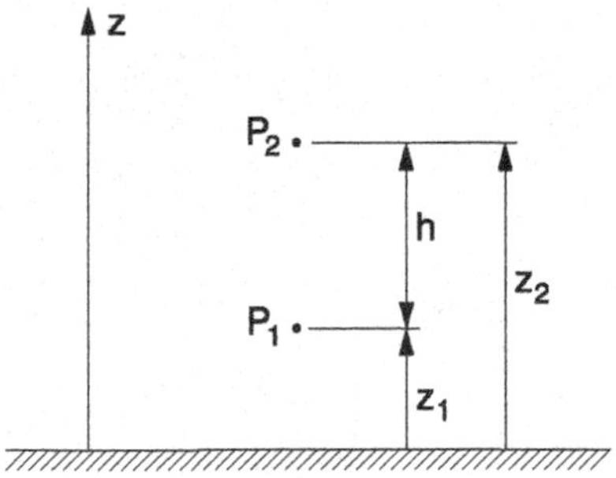

The difference in pressure between two different points is

$$P_2 - P_1 = -\gamma(z_2 - z_1) = -\gamma h = -\rho g h$$

Absolute pressure = atmospheric pressure + gauge pressure reading

Absolute pressure = atmospheric pressure – vacuum gauge pressure reading

♦ Bober, W., and R.A. Kenyon, *Fluid Mechanics*, Wiley, 1980. Diagrams reprinted by permission of William Bober and Richard A. Kenyon.

Manometers

♦

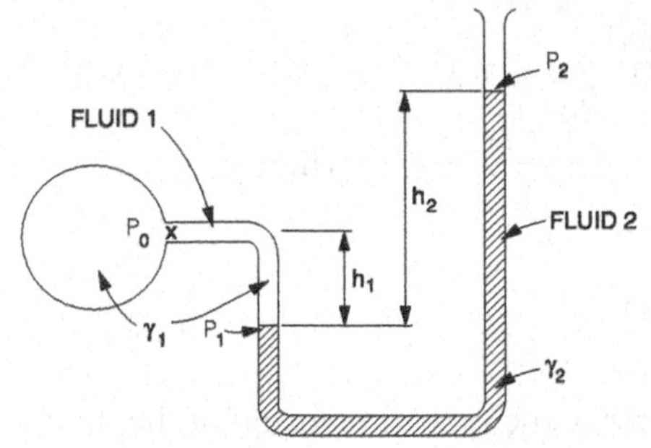

For a simple manometer,

$$P_0 = P_2 + \gamma_2 h_2 - \gamma_1 h_1 = P_2 + g(\rho_2 h_2 - \rho_1 h_1)$$

If $h_1 = h_2 = h$

$$P_0 = P_2 + (\gamma_2 - \gamma_1)h = P_2 + (\rho_2 - \rho_1)gh$$

Note that the difference between the two densities is used.

P = pressure
γ = specific weight of fluid
h = height
g = acceleration of gravity
ρ = fluid density

Another device that works on the same principle as the manometer is the simple barometer.

$$P_{atm} = P_A = P_v + \gamma h = P_B + \gamma h = P_B + \rho gh$$

♦

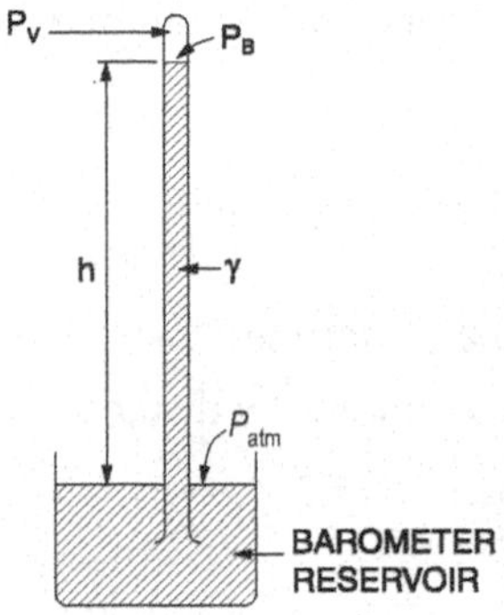

P_v = vapor pressure of the barometer fluid

Forces on Submerged Surfaces and the Center of Pressure

•

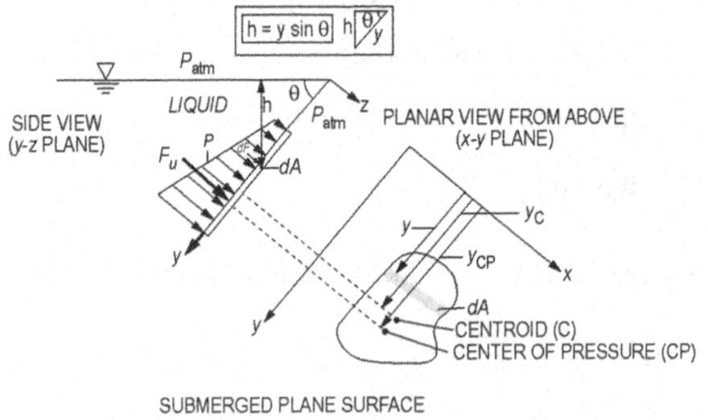

The pressure on a point at a vertical distance h below the surface is:

$$P = P_{atm} + \rho gh, \text{ for } h \geq 0$$

♦ Bober, W., and R.A. Kenyon, *Fluid Mechanics*, Wiley, 1980. Diagrams reprinted by permission of William Bober and Richard A. Kenyon.
• Elger, Donald F., et al, *Engineering Fluid Mechanics*, 10th ed., 2012. Reproduced with permission of John Wiley & Sons, Inc.

P = pressure
P_{atm} = atmospheric pressure
P_C = pressure at the centroid of area
P_{CP} = pressure at center of pressure
y_C = slant distance from liquid surface to the centroid of area
y_C = $h_C/\sin\theta$
h_C = vertical distance from liquid surface to centroid of area
y_{CP} = slant distance from liquid surface to center of pressure
h_{CP} = vertical distance from liquid surface to center of pressure
θ = angle between liquid surface and edge of submerged surface
I_{xC} = moment of inertia about the centroidal x-axis

If atmospheric pressure acts above the liquid surface and on the non-wetted side of the submerged surface:

$$y_{CP} = y_C + I_{xC}/y_C A$$

$$y_{CP} = y_C + \rho g \sin\theta \, I_{xC}/P_C A$$

Wetted side: $F_R = (P_{atm} + \rho g y_C \sin\theta)A$

P_{atm} acting both sides: $F_{R_{net}} = (\rho g y_C \sin\theta)A$

Archimedes Principle and Buoyancy

1. The buoyant force exerted on a submerged or floating body is equal to the weight of the fluid displaced by the body.
2. A floating body displaces a weight of fluid equal to its own weight; i.e., a floating body is in equilibrium.

The *center of buoyancy* is located at the centroid of the displaced fluid volume.

In the case of a body lying at the *interface of two immiscible fluids*, the buoyant force equals the sum of the weights of the fluids displaced by the body.

PRINCIPLES OF ONE-DIMENSIONAL FLUID FLOW

The Continuity Equation

So long as the flow Q is continuous, the *continuity equation*, as applied to one-dimensional flows, states that the flow passing two points (1 and 2) in a stream is equal at each point, $A_1 v_1 = A_2 v_2$.

$$Q = Av$$

$$\dot{m} = \rho Q = \rho Av, \text{ where}$$

Q = volumetric flowrate
$\dot{m}$ = mass flowrate
A = cross-sectional area of flow
v = average flow velocity
ρ = the fluid density

For steady, one-dimensional flow, $\dot{m}$ is a constant. If, in addition, the density is constant, then Q is constant.

The Energy Equation

The energy equation for steady incompressible flow with no shaft device is

$$\frac{P_1}{\gamma}+z_1+\frac{v_1^2}{2g}=\frac{P_2}{\gamma}+z_2+\frac{v_2^2}{2g}+h_f \text{ or}$$

$$\frac{P_1}{\rho g}+z_1+\frac{v_1^2}{2g}=\frac{P_2}{\rho g}+z_2+\frac{v_2^2}{2g}+h_f$$

h_f = the head loss, considered a friction effect, and all remaining terms are defined above.

If the cross-sectional area and the elevation of the pipe are the same at both sections (1 and 2), then $z_1 = z_2$ and $v_1 = v_2$.

The pressure drop $P_1 - P_2$ is given by the following:

$$P_1 - P_2 = \gamma h_f = \rho g h_f$$

The Field Equation

The field equation is derived when the energy equation is applied to one-dimensional flows. Assuming no friction losses and that no pump or turbine exists between sections 1 and 2 in the system,

$$\frac{P_2}{\gamma}+\frac{v_2^2}{2g}+z_2=\frac{P_1}{\gamma}+\frac{v_1^2}{2g}+z_1 \text{ or}$$

$$\frac{P_2}{\rho}+\frac{v_2^2}{2}+z_2 g=\frac{P_1}{\rho}+\frac{v_1^2}{2}+z_1 g, \text{ where}$$

P_1, P_2 = pressure at sections 1 and 2

v_1, v_2 = average velocity of the fluid at the sections

z_1, z_2 = the vertical distance from a datum to the sections (the potential energy)

γ = the specific weight of the fluid (ρg)

g = the acceleration of gravity

ρ = fluid density

Hydraulic Gradient (Grade Line)

Hydraulic grade line is the line connecting the sum of pressure and elevation heads at different points in conveyance systems. If a row of piezometers were placed at intervals along the pipe, the grade line would join the water levels in the piezometer water columns.

Energy Line (Bernoulli Equation)

The Bernoulli equation states that the sum of the pressure, velocity, and elevation heads is constant. The energy line is this sum or the "total head line" above a horizontal datum. The difference between the hydraulic grade line and the energy line is the $v^2/2g$ term.

FLUID FLOW CHARACTERIZATION

Reynolds Number

$$\text{Re} = vD\rho/\mu = vD/\nu$$

$$\text{Re}' = \frac{v^{(2-n)}D^n\rho}{K\left(\frac{3n+1}{4n}\right)^n 8^{(n-1)}}, \text{ where}$$

v = fluid velocity

ρ = the mass density

D = the diameter of the pipe, dimension of the fluid streamline, or characteristic length

μ = the dynamic viscosity

ν = the kinematic viscosity

Re = the Reynolds number (Newtonian fluid)

Re′ = the Reynolds number (Power law fluid)

K and n are defined in the Stress, Pressure, and Viscosity section.

The critical Reynolds number $(\text{Re})_c$ is defined to be the minimum Reynolds number at which a flow will turn turbulent.

Flow through a pipe is generally characterized as laminar for Re < 2,100 and fully turbulent for Re > 10,000, and transitional flow for 2,100 < Re < 10,000.

The velocity distribution for *laminar flow* in circular tubes or between planes is

$$v(r) = v_{\max}\left[1-\left(\frac{r}{R}\right)^2\right], \text{ where}$$

r = the distance (m) from the centerline

R = the radius (m) of the tube or half the distance between the parallel planes

v = the local velocity (m/s) at r

$v_{\max}$ = the velocity (m/s) at the centerline of the duct

$v_{\max}$ = $1.18\bar{v}$, for fully turbulent flow

$v_{\max}$ = $2\bar{v}$, for circular tubes in laminar flow and

$v_{\max}$ = $1.5\bar{v}$, for parallel planes in laminar flow, where

$\bar{v}$ = the average velocity (m/s) in the duct

The shear stress distribution is

$$\frac{\tau}{\tau_w} = \frac{r}{R}, \text{ where}$$

τ and τ_w are the shear stresses at radii r and R respectively.

CONSEQUENCES OF FLUID FLOW

Head Loss Due to Flow

The *Darcy-Weisbach equation* is

$$h_f = f\frac{L}{D}\frac{v^2}{2g}, \text{ where}$$

f = $f(\text{Re}, \varepsilon/D)$, the Moody, Darcy, or Stanton friction factor

D = diameter of the pipe

L = length over which the pressure drop occurs

ε = roughness factor for the pipe, and other symbols are defined as before

An alternative formulation employed by chemical engineers is

$$h_f = \left(4f_{\text{Fanning}}\right)\frac{Lv^2}{D2g} = \frac{2f_{\text{Fanning}}Lv^2}{Dg}$$

$$\text{Fanning friction factor, } f_{\text{Fanning}} = \frac{f}{4}$$

A chart that gives f versus Re for various values of ε/D, known as a *Moody, Darcy,* or *Stanton diagram*, is available in this section.

Minor Losses in Pipe Fittings, Contractions, and Expansions

Head losses also occur as the fluid flows through pipe fittings (i.e., elbows, valves, couplings, etc.) and sudden pipe contractions and expansions.

$$\frac{P_1}{\gamma} + z_1 + \frac{v_1^2}{2g} = \frac{P_2}{\gamma} + z_2 + \frac{v_2^2}{2g} + h_f + h_{f,\text{ fitting}}$$

$$\frac{P_1}{\rho g} + z_1 + \frac{v_1^2}{2g} = \frac{P_2}{\rho g} + z_2 + \frac{v_2^2}{2g} + h_f + h_{f,\text{ fitting}}, \text{ where}$$

$$h_{f,\text{ fitting}} = C\frac{v^2}{2g}, \text{ and } \frac{v^2}{2g} = 1 \text{ velocity head}$$

Specific fittings have characteristic values of C, which will be provided in the problem statement. A generally accepted *nominal value* for head loss in *well-streamlined gradual contractions* is

$$h_{f,\text{ fitting}} = 0.04\, v^2/2g$$

The *head loss* at either an *entrance* or *exit* of a pipe from or to a reservoir is also given by the $h_{f,\text{ fitting}}$ equation. Values for C for various cases are shown as follows.

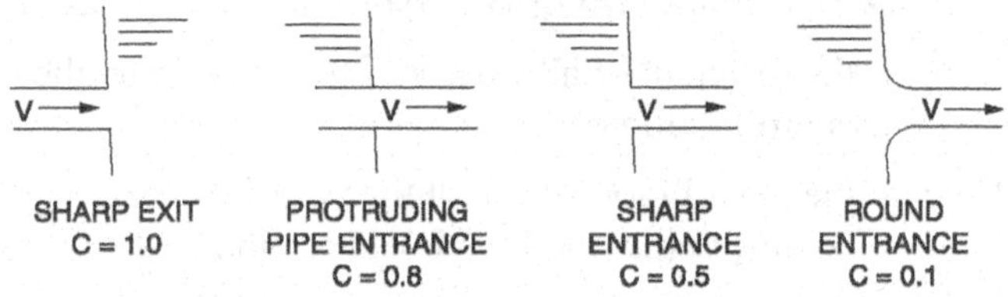

Bober, W., and R.A. Kenyon, *Fluid Mechanics*, Wiley, 1980. Diagrams reprinted by permission of William Bober and Richard A. Kenyon.

Pressure Drop for Laminar Flow

The equation for Q in terms of the pressure drop ΔP_f is the Hagen-Poiseuille equation. This relation is valid only for flow in the laminar region.

$$Q = \frac{\pi R^4 \Delta P_f}{8\mu L} = \frac{\pi D^4 \Delta P_f}{128\mu L}$$

Flow in Noncircular Conduits

Analysis of flow in conduits having a noncircular cross section uses the *hydraulic radius* R_H, or the *hydraulic diameter* D_H, as follows:

$$R_H = \frac{\text{cross-sectional area}}{\text{wetted perimeter}} = \frac{D_H}{4}$$

Drag Force

The *drag force* F_D on objects immersed in a large body of flowing fluid or objects moving through a stagnant fluid is

$$F_D = \frac{C_D \rho v^2 A}{2}, \text{ where}$$

C_D = the *drag coefficient,*

v = the velocity (m/s) of the flowing fluid or moving object, and

A = the *projected area* (m^2) of blunt objects such as spheres, ellipsoids, disks, and plates, cylinders, ellipses, and air foils with axes perpendicular to the flow.

ρ = fluid density

For flat plates placed parallel with the flow:

$C_D = 1.33/\text{Re}^{0.5}\ (10^4 < \text{Re} < 5 \times 10^5)$

$C_D = 0.031/\text{Re}^{1/7}\ (10^6 < \text{Re} < 10^9)$

The characteristic length in the Reynolds Number (Re) is the length of the plate parallel with the flow. For blunt objects, the characteristic length is the largest linear dimension (diameter of cylinder, sphere, disk, etc.) that is perpendicular to the flow.

CHARACTERISTICS OF SELECTED FLOW CONFIGURATIONS

Open-Channel Flow and/or Pipe Flow of Water

Manning's Equation

$$Q = (K/n)AR_H^{2/3}S^{1/2}$$

$$V = (K/n)R_H^{2/3}S^{1/2}$$

Q = discharge (ft^3/sec or m^3/s)

V = velocity (ft/sec or m/s)

K = 1.486 for USCS units, 1.0 for SI units

n = roughness coefficient

A = cross-sectional area of flow (ft^2 or m^2)

R_H = hydraulic radius = A/P (ft or m)

P = wetted perimeter (ft or m)

S = slope (ft/ft or m/m)

Hazen-Williams Equation

$$v = k_1 C R_H^{0.63} S^{0.54}$$

$$Q = k_1 C A R_H^{0.63} S^{0.54}$$

k_1 = 0.849 for SI units, 1.318 for USCS units

C = roughness coefficient, as tabulated in the Civil Engineering section. Other symbols are defined as before.

Flow Through a Packed Bed

A porous, fixed bed of solid particles can be characterized by

L = length of particle bed (m)

D_p = average particle diameter (m)

Φ_s = sphericity of particles, dimensionless (0–1)

ε = porosity or void fraction of the particle bed, dimensionless (0–1)

The Ergun equation can be used to estimate pressure loss through a packed bed under laminar and turbulent flow conditions.

$$\frac{\Delta P}{L} = \frac{150 v_o \mu (1-\varepsilon)^2}{\Phi_s^2 D_p^2 \varepsilon^3} + \frac{1.75 \rho v_o^2 (1-\varepsilon)}{\Phi_s D_p \varepsilon^3}$$

ΔP = pressure loss across packed bed (Pa)

v_o = superficial (flow through empty vessel) fluid velocity (m/s)

ρ = fluid density (kg/m³)

μ = fluid viscosity [kg/(m•s)]

Submerged Orifice operating under steady-flow conditions:

- •

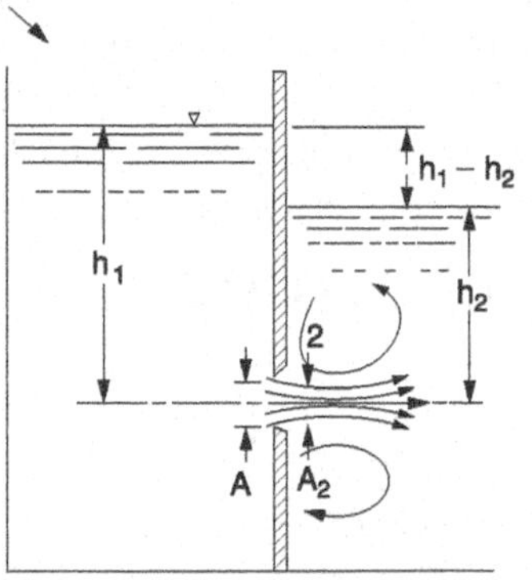

$$Q = A_2 v_2 = C_c C_v A\sqrt{2g(h_1 - h_2)}$$
$$= CA\sqrt{2g(h_1 - h_2)}$$

in which the product of C_c and C_v is defined as the *coefficient of discharge* of the orifice.

v_2 = velocity of fluid exiting orifice

Orifice Discharging Freely into Atmosphere

- •

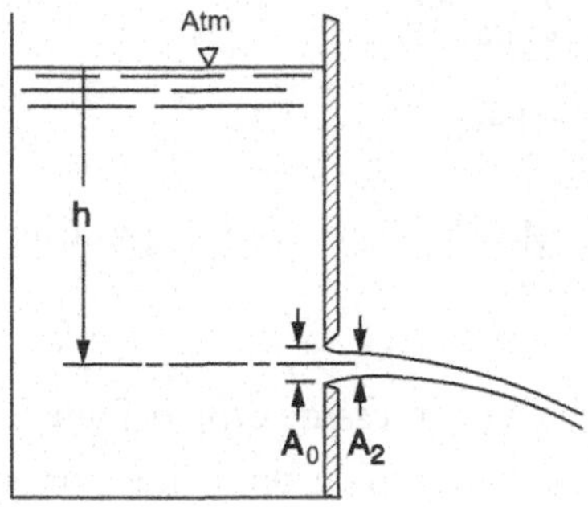

$$Q = CA_0\sqrt{2gh}$$

in which h is measured from the liquid surface to the centroid of the orifice opening.

Q = volumetric flow

A_0 = cross-sectional area of flow

g = acceleration of gravity

h = height of fluid above orifice

Multipath Pipeline Problems

- •

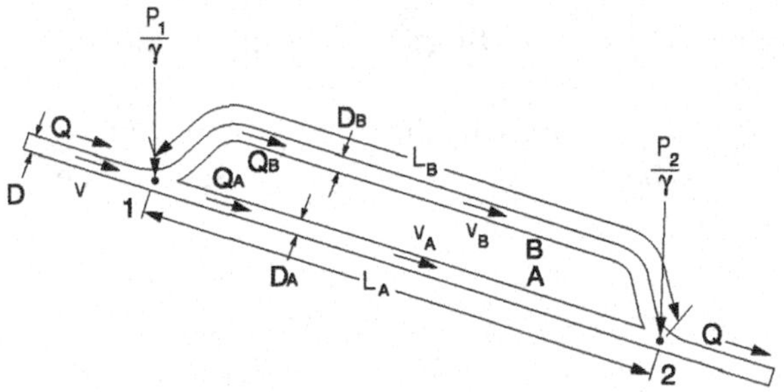

For pipes in parallel, the head loss is the same in each pipe.

$$h_L = f_A \frac{L_A}{D_A}\frac{v_A^2}{2g} = f_B \frac{L_B}{D_B}\frac{v_B^2}{2g}$$

$$(\pi D^2/4)v = (\pi D_A^2/4)v_A + (\pi D_B^2/4)v_B$$

The total flowrate Q is the sum of the flowrates in the parallel pipes.

THE IMPULSE-MOMENTUM PRINCIPLE

The resultant force in a given direction acting on the fluid equals the rate of change of momentum of the fluid.

$$\Sigma F = \Sigma Q_2 \rho_2 v_2 - \Sigma Q_1 \rho_1 v_1, \text{ where}$$

ΣF = the resultant of all external forces acting on the control volume

$\Sigma Q_1 \rho_1 v_1$ = the rate of momentum of the fluid flow entering the control volume in the same direction of the force

$\Sigma Q_2 \rho_2 v_2$ = the rate of momentum of the fluid flow leaving the control volume in the same direction of the force

- • Vennard, J.K., *Elementary Fluid Mechanics*, 6th ed., J.K. Vennard, 1954.

Pipe Bends, Enlargements, and Contractions

The force exerted by a flowing fluid on a bend, enlargement, or contraction in a pipeline may be computed using the impulse-momentum principle.

•

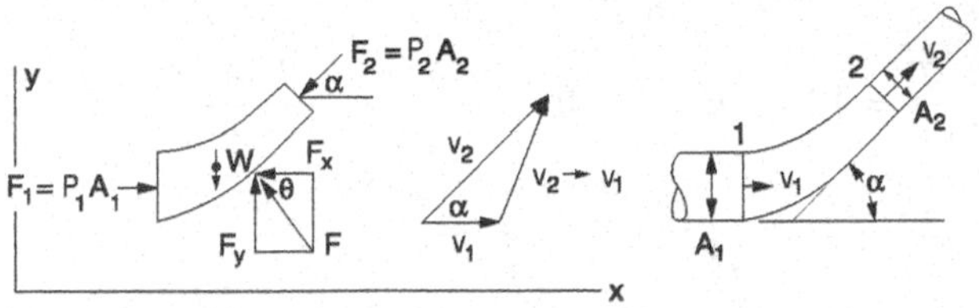

$P_1A_1 - P_2A_2\cos\alpha - F_x = Q\rho\,(v_2\cos\alpha - v_1)$

$F_y - W - P_2A_2\sin\alpha = Q\rho\,(v_2\sin\alpha - 0)$, where

F = the force exerted by the bend on the fluid (the force exerted by the fluid on the bend is equal in magnitude and opposite in sign), F_x and F_y are the x-component and y-component of the force $F = \sqrt{F_x^2 + F_y^2}$ and $\theta = \tan^{-1}\left(\frac{F_y}{F_x}\right)$.

P = the internal pressure in the pipe line

A = the cross-sectional area of the pipe line

W = the weight of the fluid

v = the velocity of the fluid flow

α = the angle the pipe bend makes with the horizontal

ρ = the density of the fluid

Q = fluid volumetric flowrate

Jet Propulsion

•

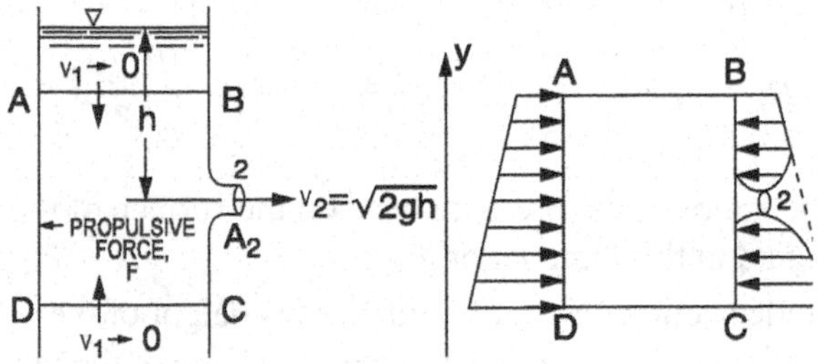

$F = Q\rho(v_2 - 0)$

$F = 2\gamma hA_2$, where

F = the propulsive force

γ = the specific weight of the fluid

h = the height of the fluid above the outlet

A_2 = the area of the nozzle tip

Q = $A_2\sqrt{2gh}$

v_2 = $\sqrt{2gh}$

Deflectors and Blades

Fixed Blade

•

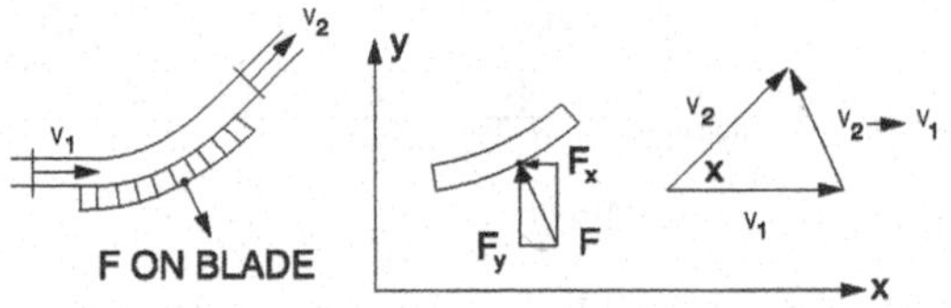

$-F_x = Q\rho(v_2\cos\alpha - v_1)$

$F_y = Q\rho(v_2\sin\alpha - 0)$

Moving Blade

•

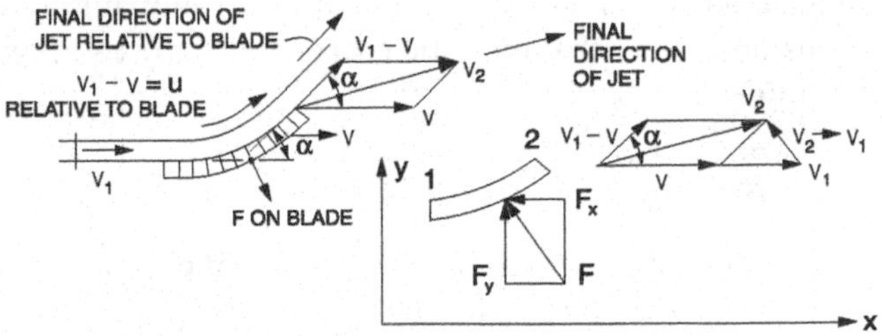

$-F_x = Q\rho(v_{2x} - v_{1x})$

$= -Q\rho(v_1 - v)(1 - \cos\alpha)$

$F_y = Q\rho(v_{2y} - v_{1y})$

$= +Q\rho(v_1 - v)\sin\alpha$, where

v = the velocity of the blade.

Impulse Turbine

•

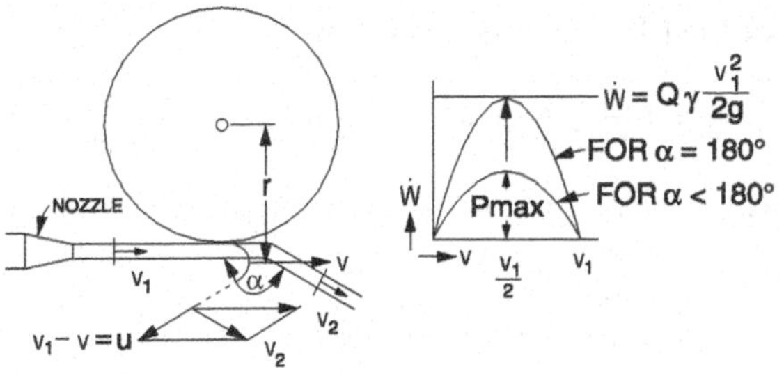

$\dot{W} = Q\rho(v_1 - v)(1 - \cos\alpha)v$, where

$\dot{W}$ = power of the turbine.

$\dot{W}_{max} = Q\rho\left(v_1^2/4\right)(1 - \cos\alpha)$

When $\alpha = 180°$,

$\dot{W}_{max} = \left(Q\rho v_1^2\right)/2 = \left(Q\gamma v_1^2\right)/2g$

COMPRESSIBLE FLOW

Mach Number

The local *speed of sound* in an ideal gas is given by:

$c = \sqrt{kRT}$, where

$c \equiv$ local speed of sound

$k \equiv$ ratio of specific heats $= \frac{c_p}{c_v}$

$R \equiv$ specific gas constant $= \bar{R}$/(molecular weight)

$T \equiv$ absolute temperature

example: speed of sound in dry air at 1 atm 20°C is 343.2 m/s.

This shows that the acoustic velocity in an ideal gas depends only on its temperature. The *Mach number* (Ma) is the ratio of the fluid velocity to the speed of sound.

$\mathrm{Ma} \equiv \frac{V}{c}$

$V \equiv$ mean fluid velocity

• Vennard, J.K., *Elementary Fluid Mechanics*, 6th ed., J.K. Vennard, 1954.

Isentropic Flow Relationships

In an ideal gas for an isentropic process, the following relationships exist between static properties at any two points in the flow.

$$\frac{P_2}{P_1} = \left(\frac{T_2}{T_1}\right)^{\frac{k}{(k-1)}} = \left(\frac{\rho_2}{\rho_1}\right)^{k}$$

The stagnation temperature, T_0, at a point in the flow is related to the static temperature as follows:

$$T_0 = T + \frac{V^2}{2 \cdot c_p}$$

Energy relation between two points:

$$h_1 + \frac{V_1^2}{2} = h_2 + \frac{V_2^2}{2}$$

The relationship between the static and stagnation properties (T_0, P_0, and ρ_0) at any point in the flow can be expressed as a function of the Mach number as follows:

$$\frac{T_0}{T} = 1 + \frac{k-1}{2} \cdot \text{Ma}^2$$

$$\frac{P_0}{P} = \left(\frac{T_0}{T}\right)^{\frac{k}{(k-1)}} = \left(1 + \frac{k-1}{2} \cdot \text{Ma}^2\right)^{\frac{k}{(k-1)}}$$

$$\frac{\rho_0}{\rho} = \left(\frac{T_0}{T}\right)^{\frac{1}{(k-1)}} = \left(1 + \frac{k-1}{2} \cdot \text{Ma}^2\right)^{\frac{1}{(k-1)}}$$

Compressible flows are often accelerated or decelerated through a nozzle or diffuser. For subsonic flows, the velocity decreases as the flow cross-sectional area increases and vice versa. For supersonic flows, the velocity increases as the flow cross-sectional area increases and decreases as the flow cross-sectional area decreases. The point at which the Mach number is sonic is called the throat and its area is represented by the variable, A^*. The following area ratio holds for any Mach number.

$$\frac{A}{A^*} = \frac{1}{\text{Ma}}\left[\frac{1 + \frac{1}{2}(k-1)\text{Ma}^2}{\frac{1}{2}(k+1)}\right]^{\frac{(k+1)}{2(k-1)}}$$

where

A ≡ area [length^2]

A^* ≡ area at the sonic point (Ma = 1.0)

Normal Shock Relationships

A normal shock wave is a physical mechanism that slows a flow from supersonic to subsonic. It occurs over an infinitesimal distance. The flow upstream of a normal shock wave is always supersonic and the flow downstream is always subsonic as depicted in the figure.

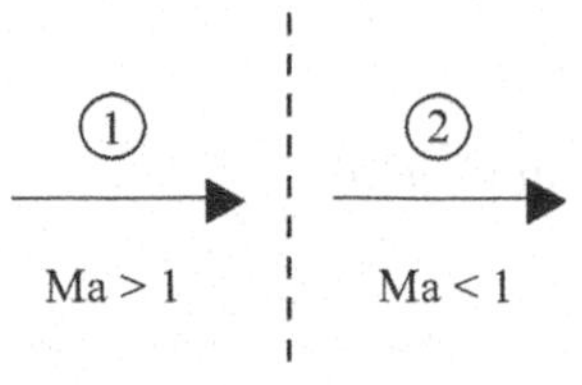

NORMAL SHOCK

The following equations relate downstream flow conditions to upstream flow conditions for a normal shock wave.

$$\text{Ma}_2 = \sqrt{\frac{(k-1)\text{Ma}_1^2 + 2}{2k\,\text{Ma}_1^2 - (k-1)}}$$

$$\frac{T_2}{T_1} = \left[2 + (k-1)\text{Ma}_1^2\right]\frac{2k\,\text{Ma}_1^2 - (k-1)}{(k+1)^2\text{Ma}_1^2}$$

$$\frac{P_2}{P_1} = \frac{1}{k+1}\left[2k\,\text{Ma}_1^2 - (k-1)\right]$$

$$\frac{\rho_2}{\rho_1} = \frac{V_1}{V_2} = \frac{(k+1)\text{Ma}_1^2}{(k-1)\text{Ma}_1^2 + 2}$$

$$T_{01} = T_{02}$$

FLUID FLOW MACHINERY

Centrifugal Pump Characteristics

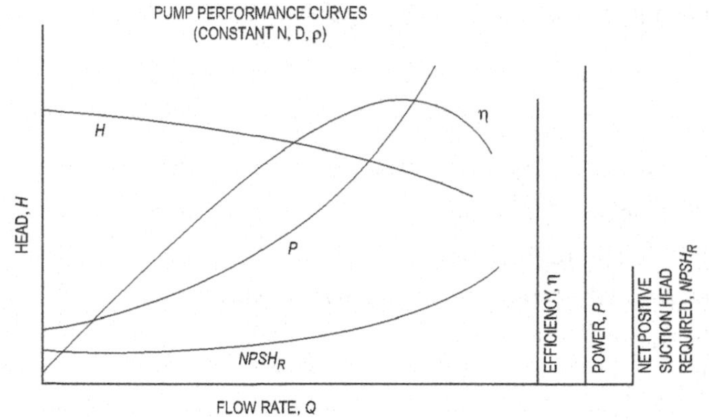

Net Positive Suction Head Available ($NPSH_A$)

$$NPSH_A = H_{pa} + H_s - \Sigma h_L - H_{vp} = \frac{P_{\text{inlet}}}{\rho g} + \frac{v_{\text{inlet}}^2}{2g} - \frac{P_{\text{vapor}}}{\rho g}$$

where,

H_{pa} = the atmospheric pressure head on the surface of the liquid in the sump (ft or m)

H_s = static suction head of liquid. This is height of the surface of the liquid above the centerline of the pump impeller (ft or m)

Σh_L = total friction losses in the suction line (ft or m)

H_{vp} = the vapor pressure head of the liquid at the operating temperature (ft or m)

v = fluid velocity at pump inlet

P_{vapor} = fluid vapor pressure at pump inlet

ρ = fluid density

g = acceleration due to gravity

Fluid power $\dot{W}_{\text{fluid}} = \rho g H Q$

Pump (brake) power $\dot{W} = \dfrac{\rho g H Q}{\eta_{\text{pump}}}$

Purchased power $\dot{W}_{\text{purchased}} = \dfrac{\dot{W}}{\eta_{\text{motor}}}$

η_{pump} = pump efficiency (0 to 1)

η_{motor} = motor efficiency (0 to 1)

H = head increase provided by pump

Pump Power Equation

$\dot{W} = Q\gamma h/\eta_t = Q\rho gh/\eta_t$, where

Q = volumetric flow (m^3/s or cfs)

h = head (m or ft) the fluid has to be lifted

η_t = total efficiency $(\eta_{pump} \times \eta_{motor})$

$\dot{W}$ = power ($kg{\bullet}m^2/sec^3$ or ft-lbf/sec)

Fan Characteristics

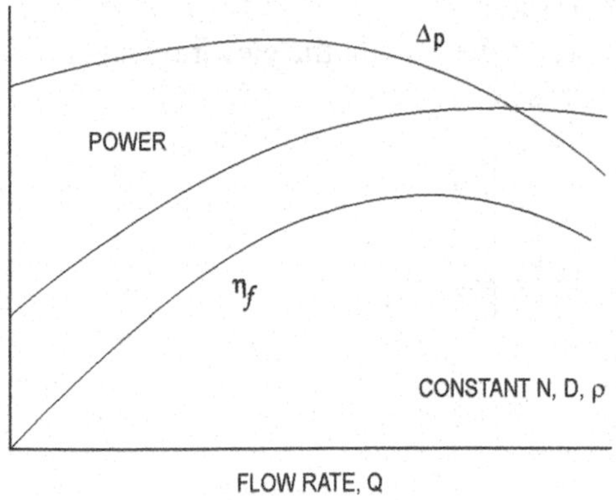

Typical *Backward Curved* Fans

$\dot{W} = \dfrac{\Delta PQ}{\eta_f}$, where

$\dot{W}$ = fan power

ΔP = pressure rise

η_f = fan efficiency

Compressors

Compressors consume power to add energy to the working fluid. This energy addition results in an increase in fluid pressure (head).

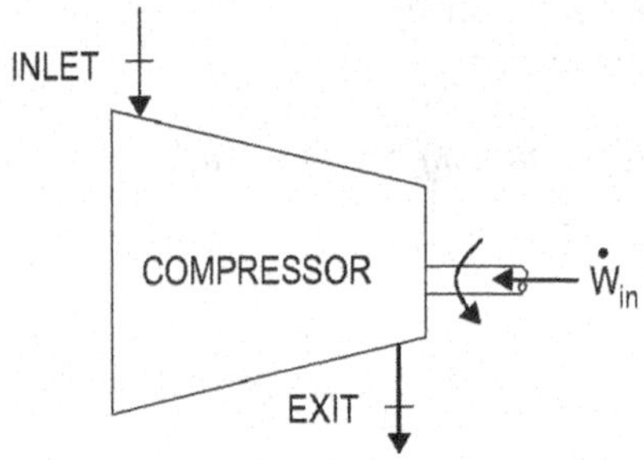

For an adiabatic compressor with $\Delta PE = 0$ and negligible ΔKE:

$$\dot{W}_{comp} = -\dot{m}(h_e - h_i)$$

For an ideal gas with constant specific heats:

$$\dot{W}_{comp} = -\dot{m}c_p(T_e - T_i)$$

Per unit mass:

$$w_{comp} = -c_p(T_e - T_i)$$

Compressor Isentropic Efficiency

$$\eta_C = \frac{w_s}{w_a} = \frac{T_{es} - T_i}{T_e - T_i} \text{ where,}$$

$w_a \equiv$ actual compressor work per unit mass

$w_s \equiv$ isentropic compressor work per unit mass

$T_{es} \equiv$ isentropic exit temperature

For a compressor where ΔKE is included:

$$\dot{W}_{comp} = -\dot{m}\left(h_e - h_i + \frac{V_e^2 - V_i^2}{2}\right) = -\dot{m}\left(c_p(T_e - T_i) + \frac{V_e^2 - V_i^2}{2}\right)$$

Adiabatic Compression

$$\dot{W}_{comp} = \frac{\dot{m}\, P_i\, k}{(k-1)\rho_i\,\eta_c}\left[\left(\frac{P_e}{P_i}\right)^{1-1/k} - 1\right]$$

$\dot{W}_{comp}$ = fluid or gas power (W)

P_i = inlet or suction pressure (N/m^2)

P_e = exit or discharge pressure (N/m^2)

k = ratio of specific heats = c_p/c_v

ρ_i = inlet gas density (kg/m^3)

η_c = isentropic compressor efficiency

Isothermal Compression

$$\dot{W}_{comp} = \frac{\bar{R}T_i}{M\eta_c}\ln\frac{P_e}{P_i}(\dot{m})$$

$\dot{W}_{comp}$, P_i, P_e, and η_c as defined for adiabatic compression

$\bar{R}$ = universal gas constant

T_i = inlet temperature of gas (K)

M = molecular weight of gas (kg/kmol)

♦ **Blowers**

$$P_W = \frac{WRT_1}{Cne}\left[\left(\frac{P_2}{P_1}\right)^{0.283} - 1\right]$$

C = 29.7 (constant for SI unit conversion)
= 550 ft-lb/(sec-hp) (U.S. Customary Units)

P_W = power requirement (hp)

W = weight of flow of air (lb/sec)

R = engineering gas constant for air = 53.3 ft-lb/(lb air-°R)

T_1 = absolute inlet temperature (°R)

P_1 = absolute inlet pressure (lbf/in^2)

P_2 = absolute outlet pressure (lbf/in^2)

n = $(k - 1)/k = 0.283$ for air

e = efficiency (usually $0.70 < e < 0.90$)

♦ Metcalf and Eddy, *Wastewater Engineering: Treatment, Disposal, and Reuse*, 3rd ed., McGraw-Hill, 1991.

Turbines

Turbines produce power by extracting energy from a working fluid. The energy loss shows up as a decrease in fluid pressure (head).

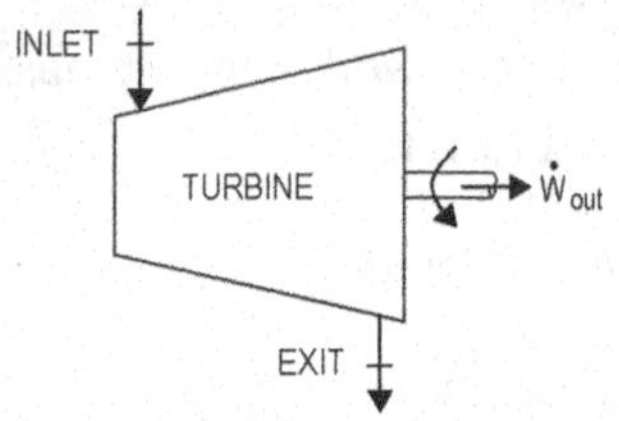

For an adiabatic turbine with $\Delta PE = 0$ and negligible ΔKE:

$$\dot{W}_{turb} = \dot{m}(h_i - h_e)$$

For an ideal gas with constant specific heats:

$$\dot{W}_{turb} = \dot{m}c_p(T_i - T_e)$$

Per unit mass:

$$w_{turb} = c_p(T_i - T_e)$$

Turbine Isentropic Efficiency

$$\eta_T = \frac{w_a}{w_s} = \frac{T_i - T_e}{T_i - T_{es}}$$

For a turbine where ΔKE is included:

$$\dot{W}_{turb} = \dot{m}\left(h_i - h_e + \frac{V_i^2 - V_e^2}{2}\right)$$

$$= \dot{m}\left(c_p(T_i - T_e) + \frac{V_i^2 - V_e^2}{2}\right)$$

Performance of Components

Fans, Pumps, and Compressors

Scaling Laws; Affinity Laws

$$\left(\frac{Q}{ND^3}\right)_2 = \left(\frac{Q}{ND^3}\right)_1$$

$$\left(\frac{\dot{m}}{\rho ND^3}\right)_2 = \left(\frac{\dot{m}}{\rho ND^3}\right)_1$$

$$\left(\frac{H}{N^2D^2}\right)_2 = \left(\frac{H}{N^2D^2}\right)_1$$

$$\left(\frac{P}{\rho N^2D^2}\right)_2 = \left(\frac{P}{\rho N^2D^2}\right)_1$$

$$\left(\frac{\dot{W}}{\rho N^3D^5}\right)_2 = \left(\frac{\dot{W}}{\rho N^3D^5}\right)_1$$

where

Q = volumetric flowrate

$\dot{m}$ = mass flowrate

H = head

P = pressure rise

$\dot{W}$ = power

ρ = fluid density

N = rotational speed

D = impeller diameter

Subscripts 1 and 2 refer to different but similar machines or to different operating conditions of the same machine.

FLUID FLOW MEASUREMENT

The Pitot Tube – From the stagnation pressure equation for an *incompressible fluid,*

$$v = \sqrt{(2/\rho)(P_0 - P_s)} = \sqrt{2g(P_0 - P_s)/\gamma}, \text{ where}$$

v = the velocity of the fluid

P_0 = the stagnation pressure

P_s = the static pressure of the fluid at the elevation where the measurement is taken

•

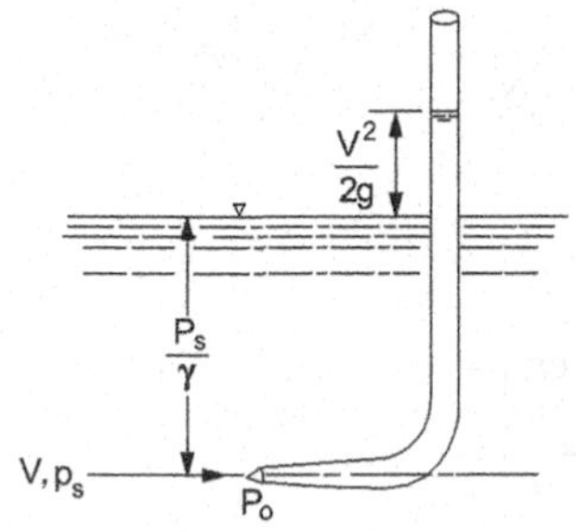

For a *compressible fluid,* use the above incompressible fluid equation if the Mach number ≤ 0.3.

Venturi Meters

$$Q = \frac{C_v A_2}{\sqrt{1 - (A_2/A_1)^2}} \sqrt{2g\left(\frac{P_1}{\gamma} + z_1 - \frac{P_2}{\gamma} - z_2\right)}, \text{ where}$$

Q = volumetric flowrate

C_v = the coefficient of velocity

A = cross-sectional area of flow

P = pressure

$\gamma = \rho g$

z_1 = elevation of venturi entrance

z_2 = elevation of venturi throat

The above equation is for *incompressible fluids.*

•

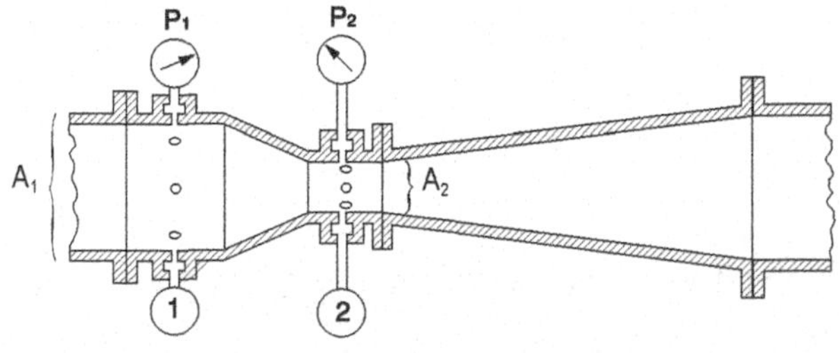

• Vennard, J.K., *Elementary Fluid Mechanics,* 6th ed., J.K. Vennard, 1954.

Orifices

The cross-sectional area at the vena contracta A_2 is characterized by a *coefficient of contraction* C_c and given by C_cA.

•

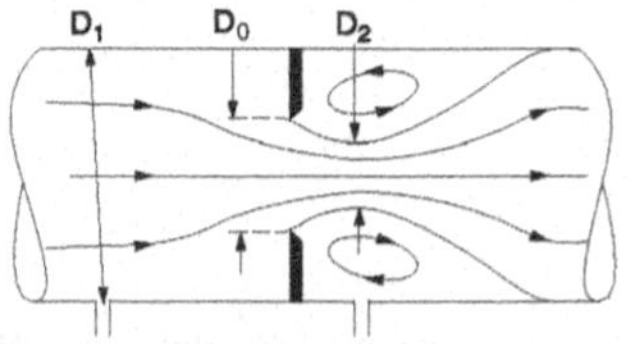

$$Q = CA_0\sqrt{2g\left(\frac{P_1}{\gamma} + z_1 - \frac{P_2}{\gamma} - z_2\right)}$$

where C, the *coefficient of the meter* (*orifice coefficient*), is given by

$$C = \frac{C_vC_c}{\sqrt{1 - C_c^2\left(A_0/A_1\right)^2}}$$

♦

ORIFICES AND THEIR NOMINAL COEFFICIENTS				
	SHARP EDGED	ROUNDED	SHORT TUBE	BORDA
	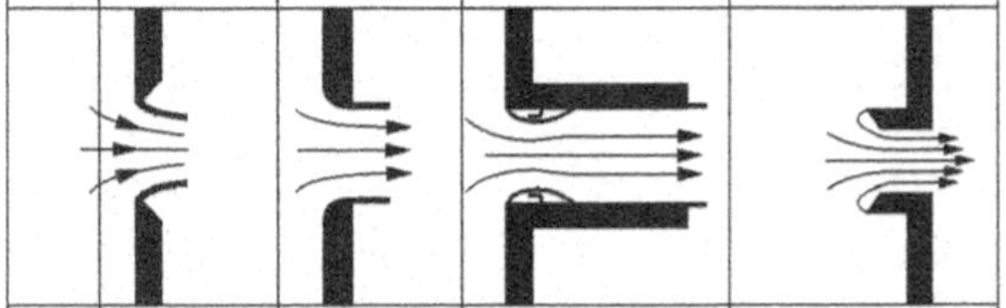			
C	0.61	0.98	0.80	0.51
C_c	0.62	1.00	1.00	0.52
C_v	0.98	0.98	0.80	0.98

For incompressible flow through a horizontal orifice meter installation

$$Q = CA_0\sqrt{\frac{2}{\rho}\left(P_1 - P_2\right)}$$

DIMENSIONAL HOMOGENEITY

Dimensional Analysis

A dimensionally homogeneous equation has the same dimensions on the left and right sides of the equation. Dimensional analysis involves the development of equations that relate dimensionless groups of variables to describe physical phemona.

Buckingham Pi Theorem: The *number of independent dimensionless groups* that may be employed to describe a phenomenon known to involve n variables is equal to the number $(n - \bar{r})$, where $\bar{r}$ is the number of basic dimensions (e.g., M, L, T) needed to express the variables dimensionally.

Similitude

In order to use a model to simulate the conditions of the prototype, the model must be *geometrically*, *kinematically*, and *dynamically similar* to the prototype system.

To obtain dynamic similarity between two flow pictures, all independent force ratios that can be written must be the same in both the model and the prototype. Thus, dynamic similarity between two flow pictures (when all possible forces are acting) is expressed in the five simultaneous equations below.

$$\left[\frac{F_I}{F_P}\right]_p = \left[\frac{F_I}{F_P}\right]_m = \left[\frac{\rho v^2}{P}\right]_p = \left[\frac{\rho v^2}{P}\right]_m$$

$$\left[\frac{F_I}{F_V}\right]_p = \left[\frac{F_I}{F_V}\right]_m = \left[\frac{vl\rho}{\mu}\right]_p = \left[\frac{vl\rho}{\mu}\right]_m = \left[\text{Re}\right]_p = \left[\text{Re}\right]_m$$

$$\left[\frac{F_I}{F_G}\right]_p = \left[\frac{F_I}{F_G}\right]_m = \left[\frac{v^2}{lg}\right]_p = \left[\frac{v^2}{lg}\right]_m = \left[\text{Fr}\right]_p = \left[\text{Fr}\right]_m$$

$$\left[\frac{F_I}{F_E}\right]_p = \left[\frac{F_I}{F_E}\right]_m = \left[\frac{\rho v^2}{E_v}\right]_p = \left[\frac{\rho v^2}{E_v}\right]_m = \left[\text{Ca}\right]_p = \left[\text{Ca}\right]_m$$

$$\left[\frac{F_I}{F_T}\right]_p = \left[\frac{F_I}{F_T}\right]_m = \left[\frac{\rho l v^2}{\sigma}\right]_p = \left[\frac{\rho l v^2}{\sigma}\right]_m = \left[\text{We}\right]_p = \left[\text{We}\right]_m$$

where the subscripts p and m stand for *prototype* and *model* respectively, and

F_I = inertia force
F_P = pressure force
F_V = viscous force
F_G = gravity force
F_E = elastic force
F_T = surface tension force
Re = Reynolds number
We = Weber number
Ca = Cauchy number
Fr = Froude number
l = characteristic length
v = velocity
ρ = density
σ = surface tension
E_v = bulk modulus
μ = dynamic viscosity
P = pressure
g = acceleration of gravity

• Vennard, J.K., *Elementary Fluid Mechanics*, 6th ed., J.K. Vennard, 1954.
♦ Bober, W., and R.A. Kenyon, *Fluid Mechanics*, Wiley, 1980. Diagrams reprinted by permission of William Bober and Richard A. Kenyon.

AERODYNAMICS

Airfoil Theory

The lift force on an airfoil, F_L, is given by

$$F_L = \frac{C_L \rho v^2 A_P}{2}$$

C_L = the lift coefficient
ρ = fluid density
v = velocity (m/s) of the undisturbed fluid and
A_P = the projected area of the airfoil as seen from above (plan area). This same area is used in defining the drag coefficient for an airfoil.

The lift coefficient, C_L, can be approximated by the equation

$C_L = 2\pi k_1 \sin(\alpha + \beta)$, which is valid for small values of α and β
k_1 = a constant of proportionality
α = angle of attack (angle between chord of airfoil and direction of flow)
β = negative of angle of attack for zero lift

The drag coefficient, C_D, may be approximated by

$$C_D = C_{D\infty} + \frac{C_L^2}{\pi AR}$$

$C_{D\infty}$ = infinite span drag coefficient

The aspect ratio, AR, is defined

$$AR = \frac{b^2}{A_p} = \frac{A_p}{c^2}$$

b = span length
A_p = plan area
c = chord length

The aerodynamic moment, M, is given by

$$M = \frac{C_M \rho v^2 A_p c}{2}$$

where the moment is taken about the front quarter point of the airfoil.

C_M = moment coefficient
ρ = fluid density
v = velocity

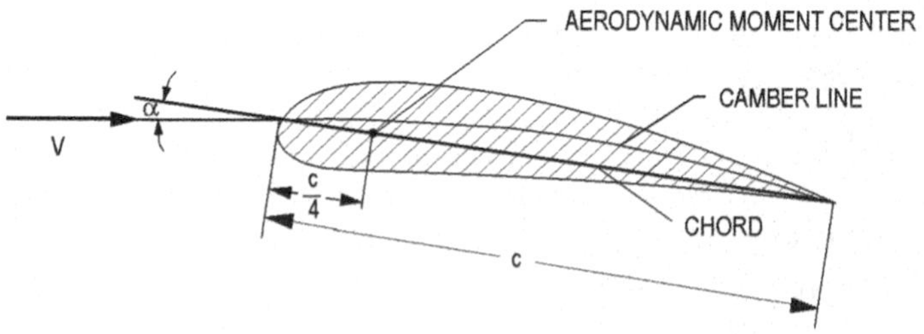

Properties of Water (SI Metric Units)

Temperature (°C)	Specific Weight γ (kN/m^3)	Density ρ (kg/m^3)	Absolute Dynamic Viscosity μ (Pa•s)	Kinematic Viscosity υ (m^2/s)	Vapor Pressure P_v (kPa)
0	9.805	999.8	0.001781	0.000001785	0.61
5	9.807	1000.0	0.001518	0.000001518	0.87
10	9.804	999.7	0.001307	0.000001306	1.23
15	9.798	999.1	0.001139	0.000001139	1.70
20	9.789	998.2	0.001002	0.000001003	2.34
25	9.777	997.0	0.000890	0.000000893	3.17
30	9.764	995.7	0.000798	0.000000800	4.24
40	9.730	992.2	0.000653	0.000000658	7.38
50	9.689	988.0	0.000547	0.000000553	12.33
60	9.642	983.2	0.000466	0.000000474	19.92
70	9.589	977.8	0.000404	0.000000413	31.16
80	9.530	971.8	0.000354	0.000000364	47.34
90	9.466	965.3	0.000315	0.000000326	70.10
100	9.399	958.4	0.000282	0.000000294	101.33

Properties of Water (English Units)

Temperature (°F)	Specific Weight γ (lbf/ft^3)	Mass Density ρ ($lbf\text{-}sec^2/ft^4$)	Absolute Dynamic Viscosity μ ($\times 10^{-5}$ $lbf\text{-}sec/ft^2$)	Kinematic Viscosity υ ($\times 10^{-5}$ ft^2/sec)	Vapor Pressure P_v (psi)
32	62.42	1.940	3.746	1.931	0.09
40	62.43	1.940	3.229	1.664	0.12
50	62.41	1.940	2.735	1.410	0.18
60	62.37	1.938	2.359	1.217	0.26
70	62.30	1.936	2.050	1.059	0.36
80	62.22	1.934	1.799	0.930	0.51
90	62.11	1.931	1.595	0.826	0.70
100	62.00	1.927	1.424	0.739	0.95
110	61.86	1.923	1.284	0.667	1.24
120	61.71	1.918	1.168	0.609	1.69
130	61.55	1.913	1.069	0.558	2.22
140	61.38	1.908	0.981	0.514	2.89
150	61.20	1.902	0.905	0.476	3.72
160	61.00	1.896	0.838	0.442	4.74
170	60.80	1.890	0.780	0.413	5.99
180	60.58	1.883	0.726	0.385	7.51
190	60.36	1.876	0.678	0.362	9.34
200	60.12	1.868	0.637	0.341	11.52
212	59.83	1.860	0.593	0.319	14.70

Vennard, John K., and Robert L. Street, *Elementary Fluid Mechanics*, 6th ed., New York: Wiley, 1982, p. 663. Reproduced with permission of John Wiley & Sons, Inc.

Moody (Stanton) Diagram

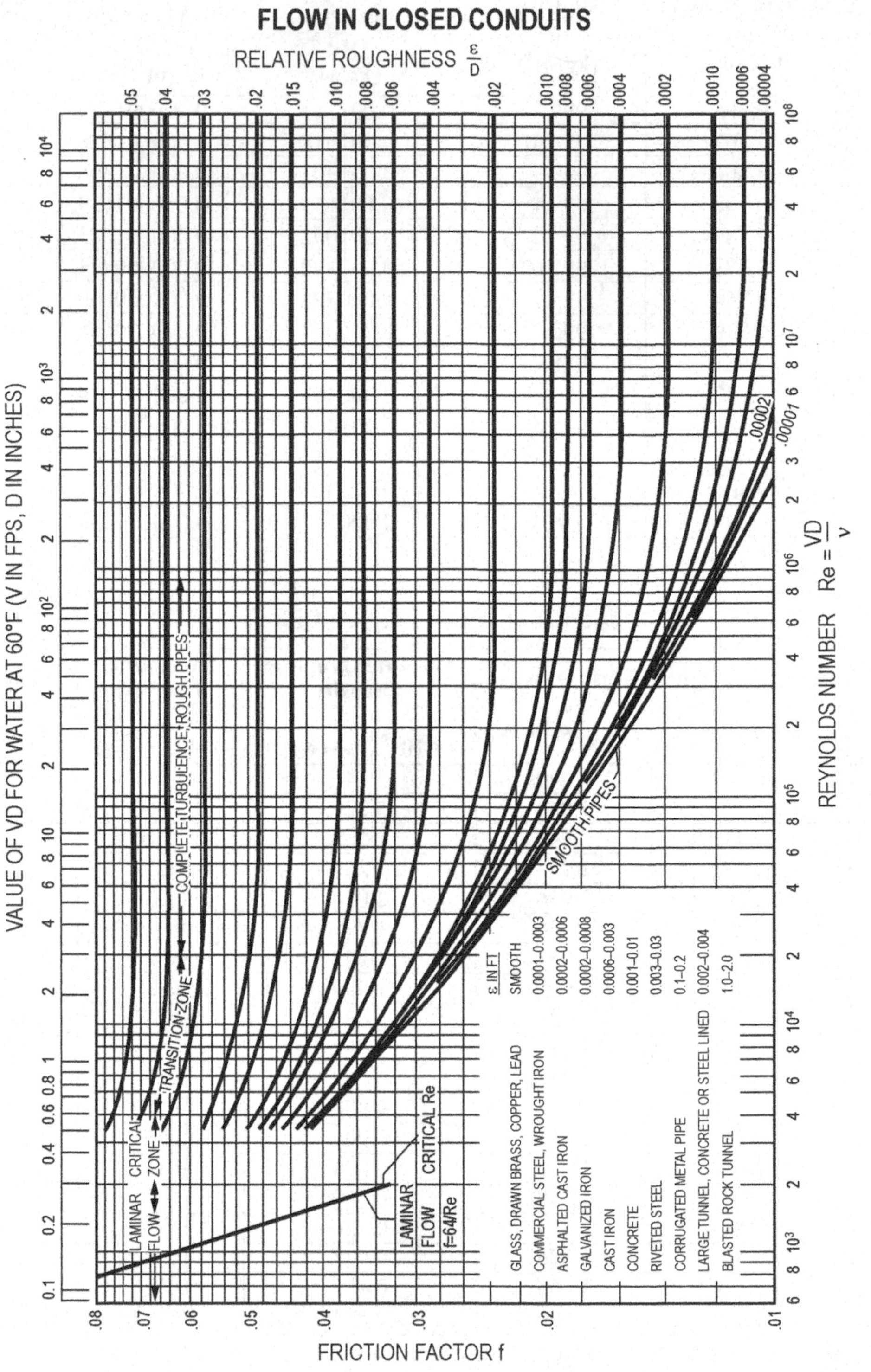

Chow, Ven Te, *Handbook of Applied Hydrology*, McGraw-Hill, 1964.

Drag Coefficient for Spheres, Disks, and Cylinders

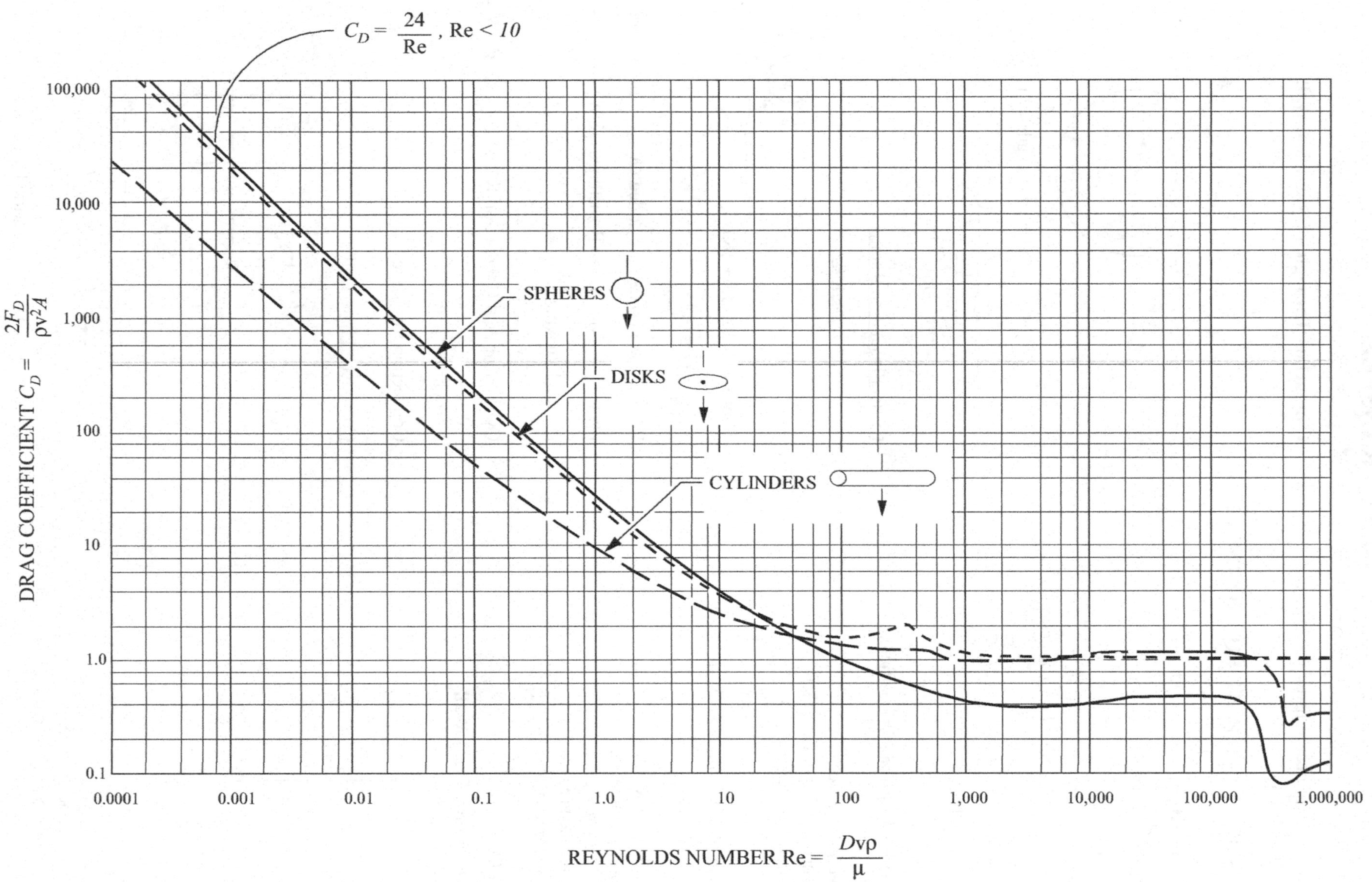

Note: Intermediate divisions are 2, 4, 6, and 8

HEAT TRANSFER

There are three modes of heat transfer: conduction, convection, and radiation.

BASIC HEAT-TRANSFER RATE EQUATIONS

Conduction

Fourier's Law of Conduction

$$\dot{Q} = -kA\frac{dT}{dx}, \text{ where}$$

$\dot{Q}$ = rate of heat transfer (W)
k = the thermal conductivity [W/(m•K)]
A = the surface area perpendicular to direction of heat transfer (m^2)

Convection

Newton's Law of Cooling

$$\dot{Q} = hA(T_w - T_\infty), \text{ where}$$

h = the convection heat-transfer coefficient of the fluid [W/(m^2•K)]
A = the convection surface area (m^2)
T_w = the wall surface temperature (K)
T_∞ = the bulk fluid temperature (K)

Radiation

The radiation emitted by a body is given by

$$\dot{Q} = \varepsilon\sigma AT^4, \text{ where}$$

ε = the emissivity of the body
σ = the Stefan-Boltzmann constant
= 5.67×10^{-8} W/(m^2•K^4)
A = the body surface area (m^2)
T = the absolute temperature (K)

CONDUCTION

Conduction Through a Plane Wall

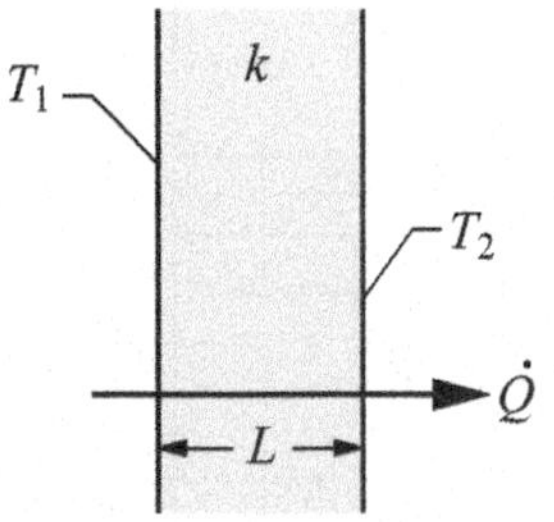

$$\dot{Q} = \frac{-kA(T_2 - T_1)}{L}, \text{ where}$$

A = wall surface area normal to heat flow (m^2)
L = wall thickness (m)
T_1 = temperature of one surface of the wall (K)
T_2 = temperature of the other surface of the wall (K)

Conduction Through a Cylindrical Wall

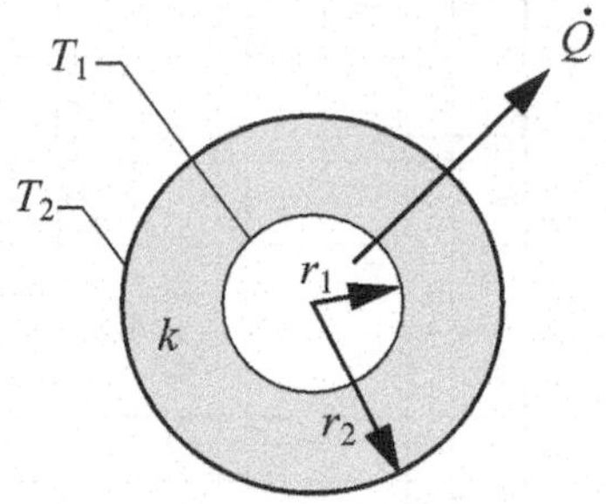

Cylinder (Length = L)

$$\dot{Q} = \frac{2\pi kL(T_1 - T_2)}{\ln\left(\frac{r_2}{r_1}\right)}$$

Critical Insulation Radius

$$r_{cr} = \frac{k_{insulation}}{h_\infty}$$

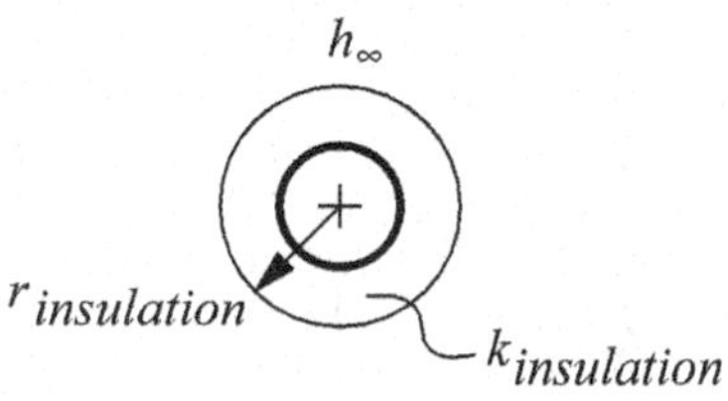

Thermal Resistance (R)

$$\dot{Q} = \frac{\Delta T}{R_{total}}$$

Resistances in series are added: $R_{total} = \Sigma R$, where

Plane Wall Conduction Resistance (K/W): $R = \frac{L}{kA}$, where

L = wall thickness

Cylindrical Wall Conduction Resistance (K/W): $R = \frac{\ln\left(\frac{r_2}{r_1}\right)}{2\pi kL}$, where
L = cylinder length

Convection Resistance (K/W) : $R = \frac{1}{hA}$

Composite Plane Wall

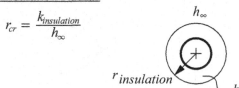

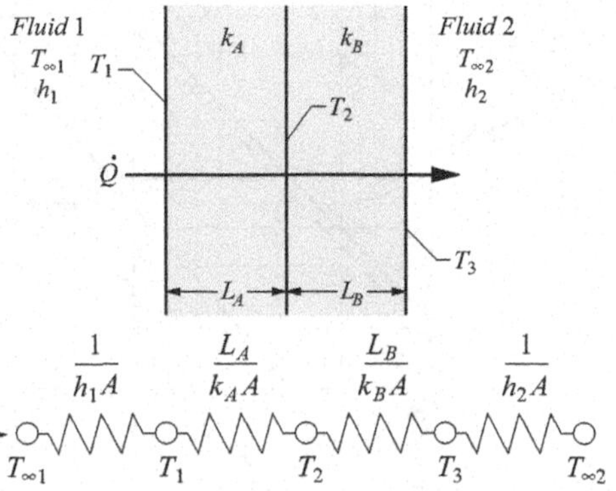

To evaluate surface or intermediate temperatures:

$$\dot{Q} = \frac{T_1 - T_2}{R_A} = \frac{T_2 - T_3}{R_B}$$

Transient Conduction Using the Lumped Capacitance Model

The lumped capacitance model is valid if

$$\text{Biot number, Bi} = \frac{hV}{kA_s} \ll 1, \text{ where}$$

h = the convection heat-transfer coefficient of the fluid [W/(m^2•K)]
V = the volume of the body (m^3)
k = thermal conductivity of the body [W/(m•K)]
A_s = the surface area of the body (m^2)

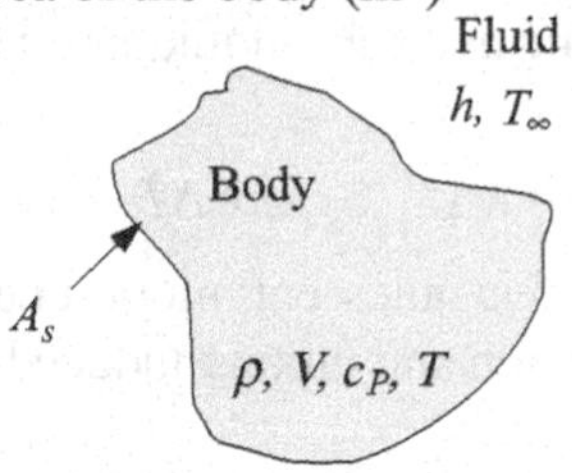

Constant Fluid Temperature

If the temperature may be considered uniform within the body at any time, the heat-transfer rate at the body surface is given by

$$\dot{Q} = hA_s(T - T_\infty) = -\rho V(c_P)\left(\frac{dT}{dt}\right), \text{ where}$$

T = the body temperature (K)
T_∞ = the fluid temperature (K)
ρ = the density of the body (kg/m^3)
c_P = the heat capacity of the body [J/(kg•K)]
t = time (s)

The temperature variation of the body with time is

$$T - T_\infty = (T_i - T_\infty)e^{-\beta t}, \text{ where}$$

$$\beta = \frac{hA_s}{\rho V c_P} \qquad \text{where } \beta = \frac{1}{\tau} \text{ and}$$

$$\tau = \text{time constant } (s)$$

The total heat transferred (Q_{total}) up to time t is

$$Q_{total} = \rho V c_P (T_i - T), \text{ where}$$

T_i = initial body temperature (K)

Fins

For a straight fin with uniform cross section (assuming negligible heat transfer from tip),

$$\dot{Q} = \sqrt{hPkA_c}\,(T_b - T_\infty)\tanh(mL_c), \text{ where}$$

h = the convection heat-transfer coefficient of the fluid [W/(m^2•K)]
P = perimeter of exposed fin cross section (m)
k = fin thermal conductivity [W/(m•K)]
A_c = fin cross-sectional area (m^2)
T_b = temperature at base of fin (K)
T_∞ = fluid temperature (K)

$$m = \sqrt{\frac{hP}{kA_c}}$$

$$L_c = L + \frac{A_c}{P}, \text{ corrected length of fin (m)}$$

Rectangular Fin

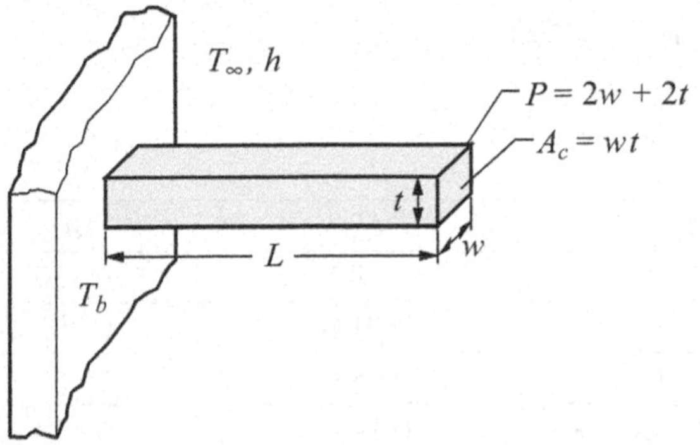

Pin Fin

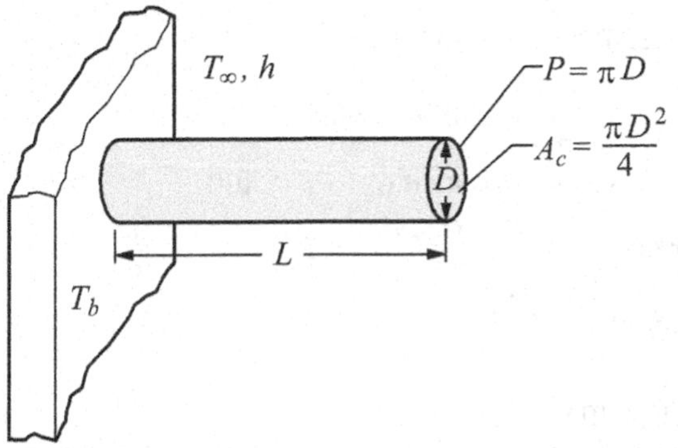

CONVECTION

Terms

D = diameter (m)
$\bar{h}$ = average convection heat-transfer coefficient of the fluid [W/(m^2•K)]
L = length (m)
$\overline{Nu}$ = average Nusselt number
Pr = Prandtl number = $\frac{c_P \mu}{k}$
u_m = mean velocity of fluid (m/s)
u_∞ = free stream velocity of fluid (m/s)
μ = dynamic viscosity of fluid [kg/(s•m)]
ρ = density of fluid (kg/m^3)

External Flow

In all cases, evaluate fluid properties at average temperature between that of the body and that of the flowing fluid.

Flat Plate of Length L in Parallel Flow

$$\text{Re}_L = \frac{\rho u_\infty L}{\mu}$$

$$\overline{Nu}_L = \frac{\bar{h}L}{k} = 0.6640\,\text{Re}_L^{1/2}\,\text{Pr}^{1/3} \qquad (\text{Re}_L < 10^5)$$

$$\overline{Nu}_L = \frac{\bar{h}L}{k} = 0.0366\,\text{Re}_L^{0.8}\,\text{Pr}^{1/3} \qquad (\text{Re}_L > 10^5)$$

Cylinder of Diameter D in Cross Flow

$$\mathrm{Re}_D = \frac{\rho u_\infty D}{\mu}$$

$$\overline{Nu}_D = \frac{\bar{h}D}{k} = C\,\mathrm{Re}_D^n\,\mathrm{Pr}^{1/3}, \text{ where}$$

Re_D	C	n
1 – 4	0.989	0.330
4 – 40	0.911	0.385
40 – 4,000	0.683	0.466
4,000 – 40,000	0.193	0.618
40,000 – 250,000	0.0266	0.805

Flow Over a Sphere of Diameter, D

$$\overline{Nu}_D = \frac{\bar{h}D}{k} = 2.0 + 0.60\,\mathrm{Re}_D^{1/2}\mathrm{Pr}^{1/3},$$

$$(1 < \mathrm{Re}_D < 70{,}000;\ 0.6 < \mathrm{Pr} < 400)$$

Internal Flow

$$\mathrm{Re}_D = \frac{\rho u_m D}{\mu}$$

Laminar Flow in Circular Tubes

For laminar flow ($\mathrm{Re}_D < 2300$), fully developed conditions

$Nu_D = 4.36$ (uniform heat flux)

$Nu_D = 3.66$ (constant surface temperature)

For laminar flow ($\mathrm{Re}_D < 2300$), combined entry length with constant surface temperature

$$Nu_D = 1.86\left(\frac{\mathrm{Re}_D\mathrm{Pr}}{\frac{L}{D}}\right)^{1/3}\left(\frac{\mu_b}{\mu_s}\right)^{0.14}, \text{ where}$$

L = length of tube (m)

D = tube diameter (m)

μ_b = dynamic viscosity of fluid [kg/(s•m)] at bulk temperature of fluid, T_b

μ_s = dynamic viscosity of fluid [kg/(s•m)] at inside surface temperature of the tube, T_s

Turbulent Flow in Circular Tubes

For turbulent flow ($\mathrm{Re}_D > 10^4$, $\mathrm{Pr} > 0.7$) for either uniform surface temperature or uniform heat flux condition, Sieder-Tate equation offers good approximation:

$$Nu_D = 0.023\,\mathrm{Re}_D^{0.8}\mathrm{Pr}^{1/3}\left(\frac{\mu_b}{\mu_s}\right)^{0.14}$$

Noncircular Ducts

In place of the diameter, D, use the equivalent (hydraulic) diameter (D_H) defined as

$$D_H = \frac{4 \times \text{cross-sectional area}}{\text{wetted perimeter}}$$

Circular Annulus ($D_o > D_i$)

In place of the diameter, D, use the equivalent (hydraulic) diameter (D_H) defined as

$$D_H = D_o - D_i$$

Liquid Metals ($0.003 < \mathrm{Pr} < 0.05$)

$$Nu_D = 6.3 + 0.0167\,\mathrm{Re}_D^{0.85}\mathrm{Pr}^{0.93} \text{ (uniform heat flux)}$$

$$Nu_D = 7.0 + 0.025\,\mathrm{Re}_D^{0.8}\mathrm{Pr}^{0.8} \text{ (constant wall temperature)}$$

Boiling

Evaporation occurring at a solid-liquid interface when $T_{\text{solid}} > T_{\text{sat, liquid}}$

$$q'' = h(T_s - T_{\text{sat}}) = h\Delta T_e, \text{ where } \Delta T_e = \text{excess temperature}$$

Pool Boiling – Liquid is quiescent; motion near solid surface is due to free convection and mixing induced by bubble growth and detachment.

Forced Convection Boiling – Fluid motion is induced by external means in addition to free convection and bubble-induced mixing.

Sub-Cooled Boiling – Temperature of liquid is below saturation temperature; bubbles forming at surface may condense in the liquid.

Saturated Boiling – Liquid temperature slightly exceeds the saturation temperature; bubbles forming at the surface are propelled through liquid by buoyancy forces.

♦

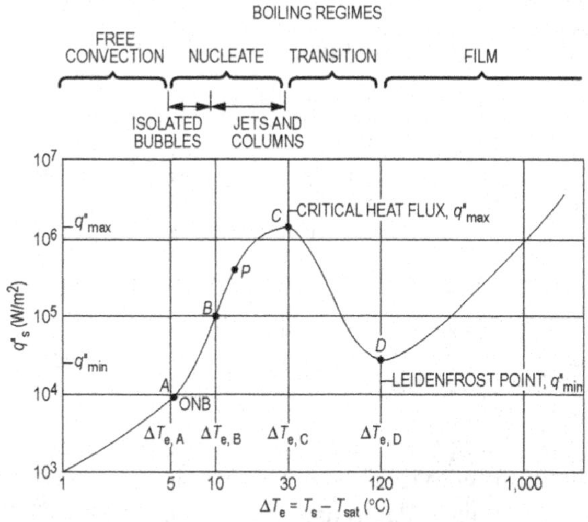

Typical boiling curve for water at one atmosphere: surface heat flux q''_s as a function of excess temperature, $\Delta T_e = T_s - T_{sat}$

Free Convection Boiling – Insufficient vapor is in contact with the liquid phase to cause boiling at the saturation temperature.

Nucleate Boiling – Isolated bubbles form at nucleation sites and separate from surface; vapor escapes as jets or columns.

♦ Incropera, Frank P. and David P. DeWitt, *Fundamentals of Heat and Mass Transfer*, 3rd ed., Wiley, 1990. Reproduced with permission of John Wiley & Sons, Inc.

♦ For nucleate boiling a widely used correlation was proposed in 1952 by Rohsenow:

$$\dot{q}_{\text{nucleate}} = \mu_l h_{fg} \left[\frac{g(\rho_l - \rho_v)}{\sigma}\right]^{1/2} \left[\frac{c_{pl}(T_s - T_{\text{sat}})}{C_{sf} h_{fg} \Pr_l^n}\right]^3$$

$\dot{q}_{\text{nucleate}}$ = nucleate boiling heat flux, W/m²
μ_l = viscosity of the liquid, kg/(m•s)
h_{fg} = enthalpy of vaporization, J/kg
g = gravitational acceleration, m/s²
ρ_l = density of the liquid, kg/m³
ρ_v = density of the vapor, kg/m³
σ = surface tension of liquid–vapor interface, N/m
c_{pl} = specific heat of the liquid, J/(kg•°C)
T_s = surface temperature of the heater,°C
T_{sat} = saturation temperature of the fluid,°C
C_{sf} = experimental constant that depends on surface–fluid combination
$\Pr_l$ = Prandtl number of the liquid
n = experimental constant that depends on the fluid

♦ Peak Heat Flux

The maximum (or critical) heat flux (CHF) in nucleate pool boiling:

$$\dot{q}_{\max} = C_{cr} h_{fg}\left[\sigma g \rho^2{}_v(\rho_l - \rho_v)\right]^{1/4}$$

C_{cr} is a constant whose value depends on the heater geometry, but generally is about 0.15.

The CHF is independent of the fluid–heating surface combination, as well as the viscosity, thermal conductivity, and specific heat of the liquid.

The CHF increases with pressure up to about one-third of the critical pressure, and then starts to decrease and becomes zero at the critical pressure.

The CHF is proportional to h_{fg}, and large maximum heat fluxes can be obtained using fluids with a large enthalpy of vaporization, such as water.

♦

Values of the coefficient C_{cr} for maximum heat flux (dimensionless parameter $L^* = L[g(\rho_l - \rho_v)/\sigma]^{1/2}$

Heater Geometry	C_{cr}	Charac. Dimension of Heater, L	Range of L^*
Large horizontal flat heater	0.149	Width or diameter	$L^* > 27$
Small horizontal flat heater[1]	$18.9\,K_1$	Width or diameter	$9 < L^* < 20$
Large horizontal cyclinder	0.12	Radius	$L^* > 1.2$
Small horizontal cyclinder	$0.12\,L^{*-0.25}$	Radius	$0.15 < L^* < 1.2$
Large sphere	0.11	Radius	$L^* > 4.26$
Small sphere	$0.227\,L^{*-0.5}$	Radius	$0.15 < L^* < 4.26$

[1] $K_1 = \sigma/[g(\rho_l - \rho_v)A_{\text{heater}}]$

♦ Minimum Heat Flux

Minimum heat flux, which occurs at the Leidenfrost point, is of practical interest since it represents the lower limit for the heat flux in the film boiling regime.

Zuber derived the following expression for the minimum heat flux for a large horizontal plate

$$\dot{q}_{\min} = 0.09\,\rho_v h_{fg}\left[\frac{\sigma g(\rho_l - \rho_v)}{(\rho_l + \rho_v)^2}\right]^{1/4}$$

The relation above can be in error by 50% or more.

Transition Boiling – Rapid bubble formation results in vapor film on surface and oscillation between film and nucleate boiling.

Film Boiling – Surface completely covered by vapor blanket; includes significant radiation through vapor film.

♦ Film Boiling

The heat flux for film boiling on a horizontal cylinder or sphere of diameter D is given by

$$\dot{q}_{\text{film}} = C_{\text{film}}\left[\frac{gk_v^3\,\rho_v(\rho_l - \rho_v)\left[h_{fg} + 0.4c_{pv}(T_s - T_{\text{sat}})\right]}{\mu_v D(T_s - T_{\text{sat}})}\right]^{1/4}(T_s - T_{\text{sat}})$$

$$C_{\text{film}} = \begin{Bmatrix} 0.62 \text{ for horizontal cylinders} \\ 0.67 \text{ for spheres} \end{Bmatrix}$$

Film Condensation of a Pure Vapor

<u>On a Vertical Surface</u>

$$\overline{Nu}_L = \frac{\bar{h}L}{k_l} = 0.943\left[\frac{\rho_l^2 g h_{fg} L^3}{\mu_l k_l (T_{\text{sat}} - T_s)}\right]^{0.25}, \text{ where}$$

ρ_l = density of liquid phase of fluid (kg/m³)
g = gravitational acceleration (9.81 m/s²)
h_{fg} = latent heat of vaporization [J/kg]
L = length of surface [m]
μ_l = dynamic viscosity of liquid phase of fluid [kg/(s•m)]
k_l = thermal conductivity of liquid phase of fluid [W/(m•K)]
T_{sat} = saturation temperature of fluid [K]
T_s = temperature of vertical surface [K]

Note: Evaluate all liquid properties at the average temperature between the saturated temperature, T_{sat}, and the surface temperature, T_s.

<u>Outside Horizontal Tubes</u>

$$\overline{Nu}_D = \frac{\bar{h}D}{k} = 0.729\left[\frac{\rho_l^2 g h_{fg} D^3}{\mu_l k_l (T_{\text{sat}} - T_s)}\right]^{0.25}, \text{ where}$$

D = tube outside diameter (m)

Note: Evaluate all liquid properties at the average temperature between the saturated temperature, T_{sat}, and the surface temperature, T_s.

♦ Cengel, Yunus A., and Afshin J. Ghajar, *Heat and Mass Transfer*, 4th ed., McGraw-Hill, 2011.

Natural (Free) Convection

Vertical Flat Plate in Large Body of Stationary Fluid

Equation also can apply to vertical cylinder of sufficiently large diameter in large body of stationary fluid.

$$\bar{h} = C\left(\frac{k}{L}\right)Ra_L^n, \text{ where}$$

L = the length of the plate (cylinder) in the vertical direction

Ra_L = Rayleigh Number = $\dfrac{g\beta(T_s - T_\infty)L^3}{\nu^2}Pr$

T_s = surface temperature (K)

T_∞ = fluid temperature (K)

β = coefficient of thermal expansion (1/K)

(For an ideal gas: $\beta = \dfrac{2}{T_s + T_\infty}$ with T in absolute temperature)

ν = kinematic viscosity (m^2/s)

Range of Ra_L	C	n
$10^4 - 10^9$	0.59	1/4
$10^9 - 10^{13}$	0.10	1/3

Long Horizontal Cylinder in Large Body of Stationary Fluid

$$\bar{h} = C\left(\frac{k}{D}\right)Ra_D^n, \text{ where}$$

$$Ra_D = \frac{g\beta(T_s - T_\infty)D^3}{\nu^2}Pr$$

Ra_D	C	n
$10^{-3} - 10^2$	1.02	0.148
$10^2 - 10^4$	0.850	0.188
$10^4 - 10^7$	0.480	0.250
$10^7 - 10^{12}$	0.125	0.333

Heat Exchangers

The rate of heat transfer in a heat exchanger is

$$\dot{Q} = UAF\Delta T_{lm}, \text{ where}$$

A = any convenient reference area (m^2)

F = correction factor for log mean temperature difference for more complex heat exchangers (shell and tube arrangements with several tube or shell passes or cross-flow exchangers with mixed and unmixed flow); otherwise $F = 1$.

U = overall heat-transfer coefficient based on area A and the log mean temperature difference [$W/(m^2 \bullet K)$]

ΔT_{lm} = log mean temperature difference (K)

Overall Heat-Transfer Coefficient for Concentric Tube and Shell-and-Tube Heat Exchangers

$$\frac{1}{UA} = \frac{1}{h_i A_i} + \frac{R_{fi}}{A_i} + \frac{\ln\left(\frac{D_o}{D_i}\right)}{2\pi kL} + \frac{R_{fo}}{A_o} + \frac{1}{h_o A_o}, \text{ where}$$

A_i = inside area of tubes (m^2)

A_o = outside area of tubes (m^2)

D_i = inside diameter of tubes (m)

D_o = outside diameter of tubes (m)

h_i = convection heat-transfer coefficient for inside of tubes [$W/(m^2 \bullet K)$]

h_o = convection heat-transfer coefficient for outside of tubes [$W/(m^2 \bullet K)$]

k = thermal conductivity of tube material [$W/(m \bullet K)$]

R_{fi} = fouling factor for inside of tube [$(m^2 \bullet K)/W$]

R_{fo} = fouling factor for outside of tube [$(m^2 \bullet K)/W$]

Log Mean Temperature Difference (LMTD)

For *counterflow* in tubular heat exchangers

$$\Delta T_{lm} = \frac{(T_{Ho} - T_{Ci}) - (T_{Hi} - T_{Co})}{\ln\left(\frac{T_{Ho} - T_{Ci}}{T_{Hi} - T_{Co}}\right)}$$

For *parallel flow* in tubular heat exchangers

$$\Delta T_{lm} = \frac{(T_{Ho} - T_{Co}) - (T_{Hi} - T_{Ci})}{\ln\left(\frac{T_{Ho} - T_{Co}}{T_{Hi} - T_{Ci}}\right)}, \text{ where}$$

ΔT_{lm} = log mean temperature difference (K)

T_{Hi} = inlet temperature of the hot fluid (K)

T_{Ho} = outlet temperature of the hot fluid (K)

T_{Ci} = inlet temperature of the cold fluid (K)

T_{Co} = outlet temperature of the cold fluid (K)

Heat Exchanger Effectiveness, ε

$$\varepsilon = \frac{\dot{Q}}{\dot{Q}_{max}} = \frac{\text{actual heat transfer rate}}{\text{maximum possible heat transfer rate}}$$

$$\varepsilon = \frac{C_H(T_{Hi} - T_{Ho})}{C_{min}(T_{Hi} - T_{Ci})} \quad \text{or} \quad \varepsilon = \frac{C_C(T_{Co} - T_{Ci})}{C_{min}(T_{Hi} - T_{Ci})}$$

where

$C = \dot{m}c_P$ = heat capacity rate (W/K)

C_{min} = smaller of C_C or C_H

Number of Transfer Units (*NTU*)

$$NTU = \frac{UA}{C_{min}}$$

Effectiveness-NTU Relations

$$C_r = \frac{C_{min}}{C_{max}} = \text{heat capacity ratio}$$

For *parallel flow concentric tube* heat exchanger

$$\varepsilon = \frac{1 - \exp[-NTU(1 + C_r)]}{1 + C_r}$$

$$NTU = -\frac{\ln[1 - \varepsilon(1 + C_r)]}{1 + C_r}$$

For *counterflow concentric tube* heat exchanger

$$\varepsilon = \frac{1 - \exp[-NTU(1 - C_r)]}{1 - C_r\exp[-NTU(1 - C_r)]} \qquad (C_r < 1)$$

$$\varepsilon = \frac{NTU}{1 + NTU} \qquad (C_r = 1)$$

$$NTU = \frac{1}{C_r - 1}\ln\left(\frac{\varepsilon - 1}{\varepsilon C_r - 1}\right) \qquad (C_r < 1)$$

$$NTU = \frac{\varepsilon}{1 - \varepsilon} \qquad (C_r = 1)$$

RADIATION

Types of Bodies

Any Body

For any body, $\alpha + \rho + \tau = 1$, where

α = absorptivity (ratio of energy absorbed to incident energy)
ρ = reflectivity (ratio of energy reflected to incident energy)
τ = transmissivity (ratio of energy transmitted to incident energy)

Opaque Body

For an opaque body: $\alpha + \rho = 1$

Gray Body

A gray body is one for which

$$\alpha = \varepsilon, (0 < \alpha < 1; 0 < \varepsilon < 1), \text{ where}$$

ε = the emissivity of the body

For a gray body: $\varepsilon + \rho = 1$

Real bodies are frequently approximated as gray bodies.

Black body

A black body is defined as one that absorbs all energy incident upon it. It also emits radiation at the maximum rate for a body of a particular size at a particular temperature. For such a body

$$\alpha = \varepsilon = 1$$

Shape Factor (View Factor, Configuration Factor) Relations

Reciprocity Relations

$$A_iF_{ij} = A_jF_{ji}, \text{ where}$$

A_i = surface area (m^2) of surface *i*
F_{ij} = shape factor (view factor, configuration factor); fraction of the radiation leaving surface *i* that is intercepted by surface *j*; $0 \le F_{ij} \le 1$

Summation Rule for *N* Surfaces

$$\sum_{j=1}^{N} F_{ij} = 1$$

Net Energy Exchange by Radiation between Two Bodies

Body Small Compared to its Surroundings

$$\dot{Q}_{12} = \varepsilon\sigma A\left(T_1^4 - T_2^4\right), \text{ where}$$

$\dot{Q}_{12}$ = the net heat-transfer rate from the body (W)
ε = the emissivity of the body
σ = the Stefan-Boltzmann constant [$\sigma = 5.67 \times 10^{-8}$ W/(m^2•K^4)]
A = the body surface area (m^2)
T_1 = the absolute temperature (K) of the body surface
T_2 = the absolute temperature (K) of the surroundings

Net Energy Exchange by Radiation between Two Black Bodies

The net energy exchange by radiation between two black bodies that see each other is given by

$$\dot{Q}_{12} = A_1 F_{12} \sigma \left(T_1^4 - T_2^4\right)$$

Net Energy Exchange by Radiation between Two Diffuse-Gray Surfaces that Form an Enclosure

Generalized Cases

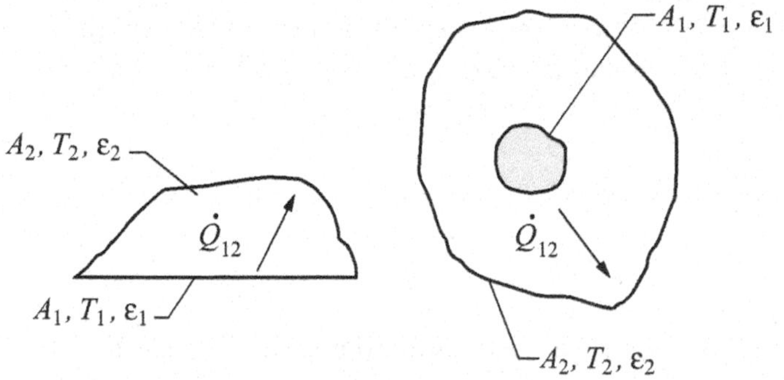

$$\dot{Q}_{12} = \frac{\sigma\left(T_1^4 - T_2^4\right)}{\dfrac{1-\varepsilon_1}{\varepsilon_1 A_1} + \dfrac{1}{A_1 F_{12}} + \dfrac{1-\varepsilon_2}{\varepsilon_2 A_2}}$$

One-Dimensional Geometry with Thin Low-Emissivity Shield Inserted between Two Parallel Plates

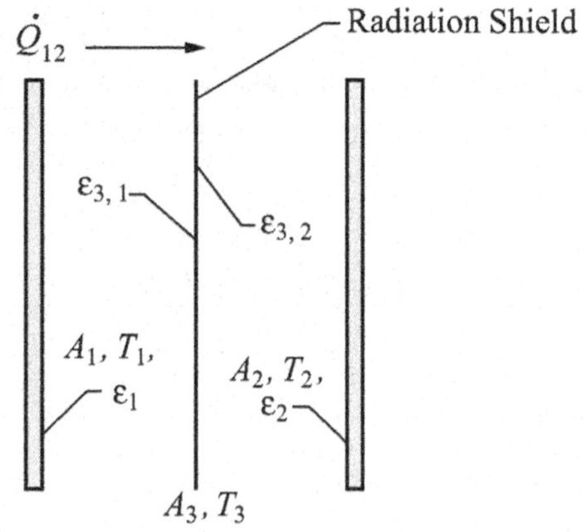

$$\dot{Q}_{12} = \frac{\sigma\left(T_1^4 - T_2^4\right)}{\dfrac{1-\varepsilon_1}{\varepsilon_1 A_1} + \dfrac{1}{A_1 F_{13}} + \dfrac{1-\varepsilon_{3,1}}{\varepsilon_{3,1} A_3} + \dfrac{1-\varepsilon_{3,2}}{\varepsilon_{3,2} A_3} + \dfrac{1}{A_3 F_{32}} + \dfrac{1-\varepsilon_2}{\varepsilon_2 A_2}}$$

Reradiating Surface

Reradiating Surfaces are considered to be insulated or adiabatic $\left(\dot{Q}_R = 0\right)$.

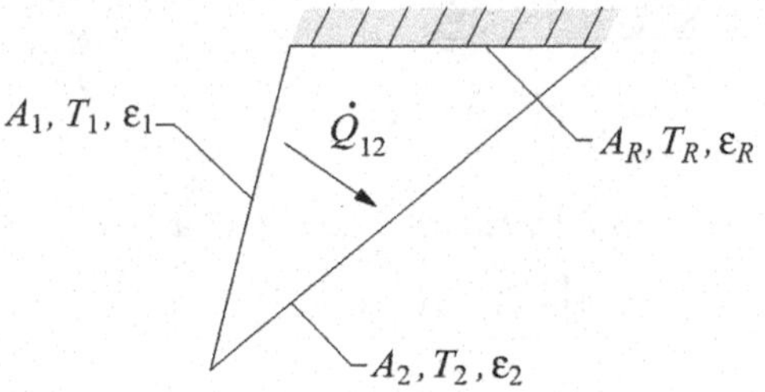

$$\dot{Q}_{12} = \frac{\sigma\left(T_1^4 - T_2^4\right)}{\dfrac{1-\varepsilon_1}{\varepsilon_1 A_1} + \dfrac{1}{A_1 F_{12} + \left[\left(\dfrac{1}{A_1 F_{1R}}\right) + \left(\dfrac{1}{A_2 F_{2R}}\right)\right]^{-1}} + \dfrac{1-\varepsilon_2}{\varepsilon_2 A_2}}$$

INSTRUMENTATION, MEASUREMENT, AND CONTROLS

MEASUREMENT

Definitions

Transducer – a device used to convert a physical parameter such as temperature, pressure, flow, light intensity, etc. into an electrical signal (also called a *sensor*).

Transducer Sensitivity – the ratio of change in electrical signal magnitude to the change in magnitude of the physical parameter being measured.

Resistance Temperature Detector (RTD) – a device used to relate change in resistance to change in temperature. Typically made from platinum, the controlling equation for an RTD is given by:

Sensitivity [RTD] = $\frac{\Delta R}{\Delta T}$

$$R_T = R_0\left[1 + \alpha\left(T - T_0\right)\right], \text{ where}$$

R_T = resistance of the RTD at temperature T (in °C)

R_0 = resistance of the RTD at the reference temperature T_0 (usually 0°C)

α = temperature coefficient of the RTD

The following graph shows tolerance values as a function of temperature for 100-Ω RTDs.

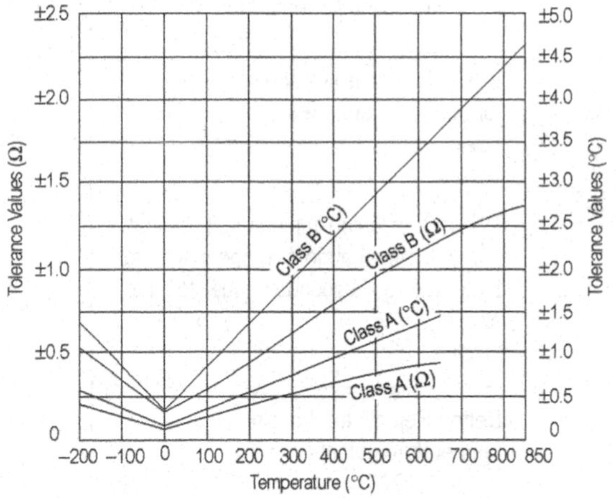

From Tempco Manufactured Products, as posted on www.tempco.com, July 2013.

Thermocouple (*TC*) – a device used to relate change in voltage to change in temperature. A thermocouple consists of two dissimilar conductors in contact, which produce a voltage when heated.

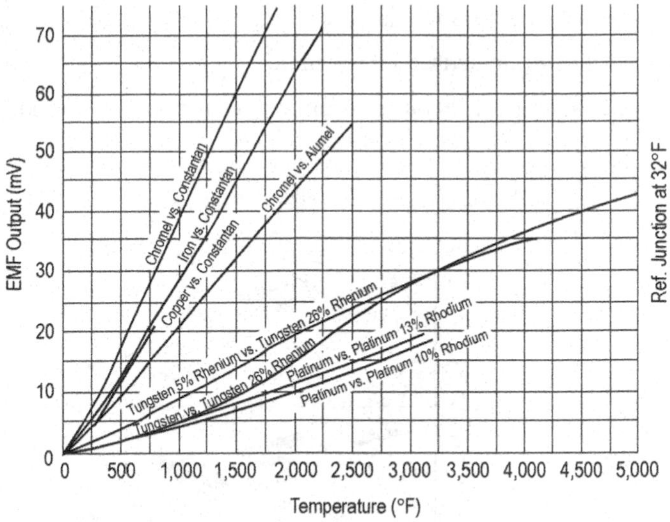

From Convectronics Inc., as posted on www.convectronics.com, July 2013.

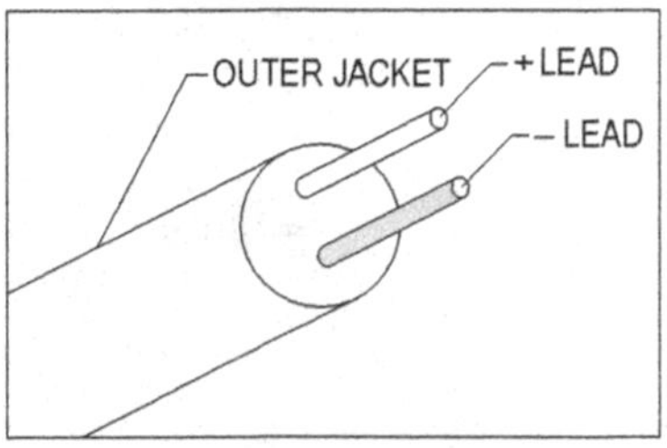

Typical Thermocouple (TC) Cable

From Convectronics Inc., as posted on www.convectronics.com, July 2013.

Accuracy refers to the closeness of a measured value to the standard or known value.
Precesion refers to the closeness of two or more measurements to each other.

ANSI Code	Alloy Combination and Color		Outer Jacket Color		Maximum Thermocouple Temperature Range	Environment
	+ Lead	– Lead	Thermocouple Leads	Extension Cable		
J	IRON Fe (magnetic) White	CONSTANTAN COPPER-NICKEL Cu-Ni Red	Brown	Black	–346 to 2,193°F –210 to 1,200°C	Reducing, Vacuum, Inert. Limited use in Oxidizing at High Temperatures. Not Recommended for Low Temperatures
K	NICKELCHROMIUM Ni-Cr Yellow	NICKEL-ALUMINUM Ni-Al (magnetic) Red	Brown	Yellow	–454 to 2,501°F –270 to 1,372°C	Clean Oxidizing and Inert. Limited Use in Vacuum or Reducing.
T	COPPER Cu Blue	CONSTANTAN COPPER-NICKEL Cu-Ni Red	Brown	Blue	–454 to 752°F –270 to 400°C	Mild Oxidizing, Reducing Vacuum or Inert. Good where moisture is present.
E	NICKELCHROMIUM Ni-Cr Purple	CONSTANTAN COPPER-NICKEL Cu-Ni Red	Brown	Purple	–454 to 1,832°F –270 to 1,000°C	Oxidizing or Inert. Limited Use in Vacuum or Reducing.

From Omega Engineering, as posted on www.omega.com, July 2013.

Strain Gauge – a device whose electrical resistance varies in proportion to the amount of strain in the device.

Gauge Factor (*GF*) – the ratio of fractional change in electrical resistance to the fractional change in length (strain):

$$GF = \frac{\Delta R/R}{\Delta L/L} = \frac{\Delta R/R}{\varepsilon}, \text{ where}$$

R = nominal resistance of the strain gauge at nominal length L

ΔR = change in resistance due the change in length ΔL

ε = normal strain sensed by the gauge

The gauge factor for metallic strain gauges is typically around 2.

Strain	Gauge Setup	Bridge Type	Sensitivity mV/V @ 1,000 $\mu\varepsilon$	Details
Axial	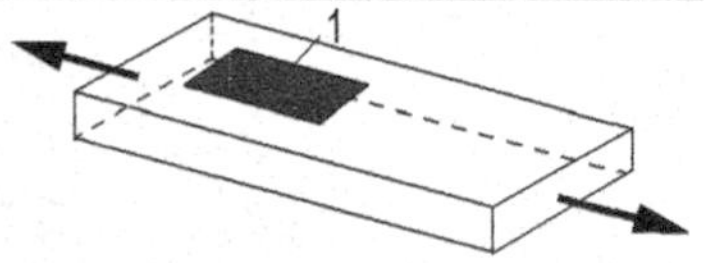	1/4	0.5	Good: Simplest to implement, but must use a dummy gauge if compensating for temperature. Also responds to bending strain.
	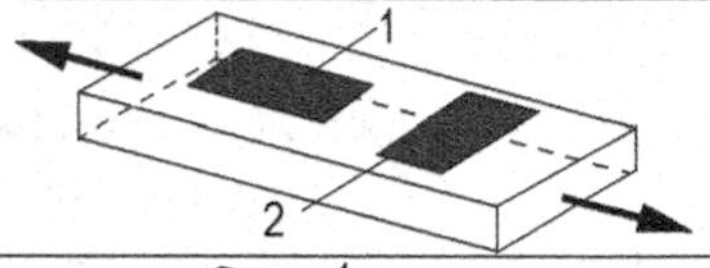	1/2	0.65	Better: Temperature compensated, but it is sensitive to bending strain.
	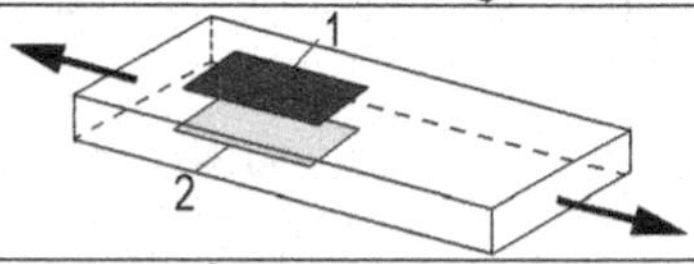	1/2	1.0	Better: Rejects bending strain, but not temperature. Must use dummy gauges if compensating for temperature
	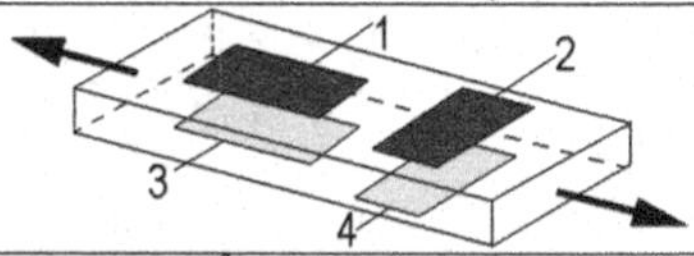	Full	1.3	Best: More sensitive and compensates for both temperature and bending strain.
Bending	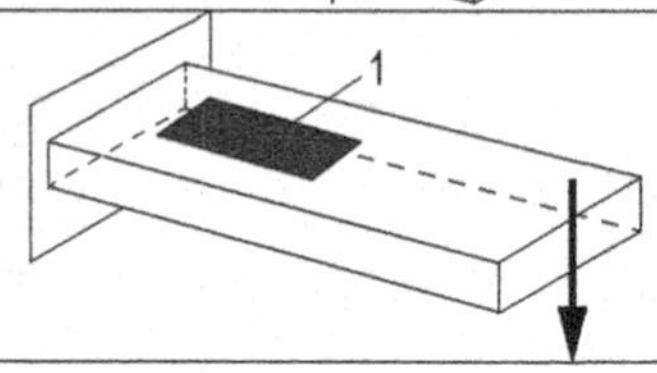	1/4	0.5	Good: Simplest to implement, but must use a dummy gauge if compensating for temperature. Responds equally to axial strain.
	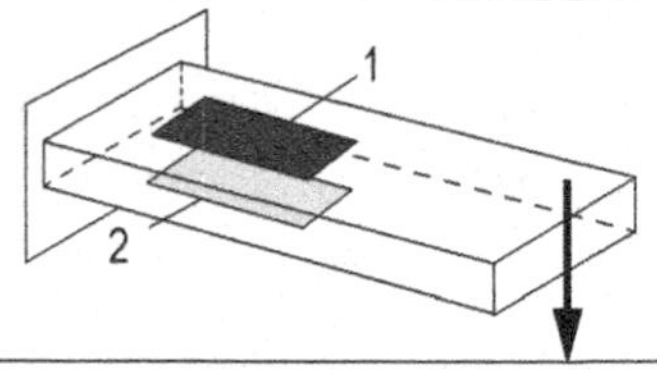	1/2	1.0	Better: Rejects axial strain and is temperature compensated.
	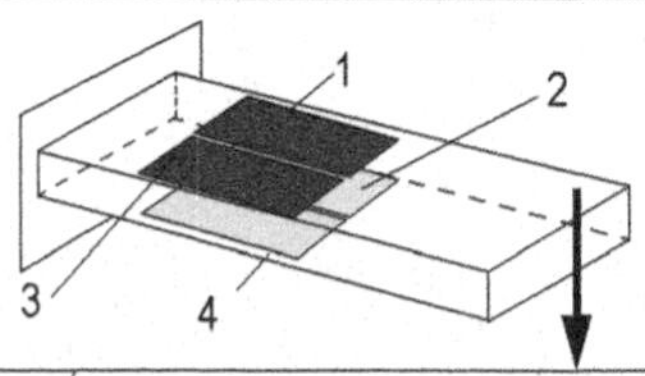	Full	2.0	Best: Rejects axial strain and is temperature compensated. Most sensitive to bending strain.
Torsional and Shear	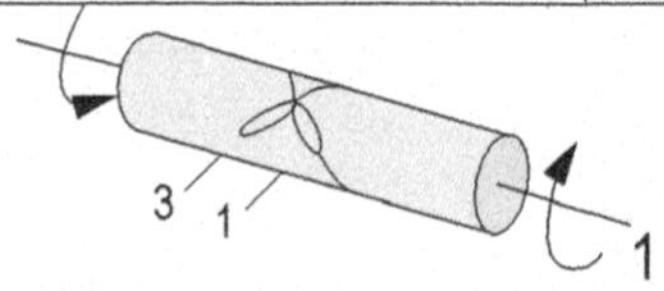	1/2	1.0	Good: Gauges must be mounted at 45 degrees from centerline.
	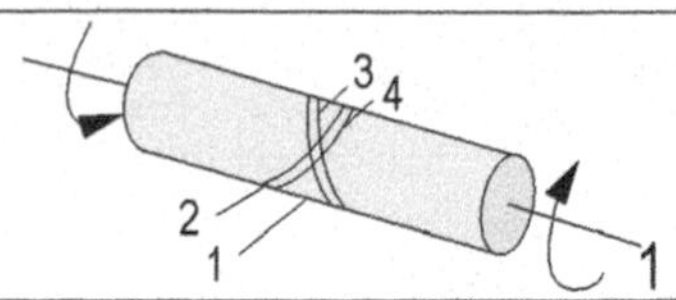	Full	2.0	Best: Most sensitive full-bridge version of previous setup. Rejects both axial and bending strains.

From National Instruments Corporation, as posted on www.ni.com, July 2013.

Wheatstone Bridge – an electrical circuit used to measure changes in resistance.

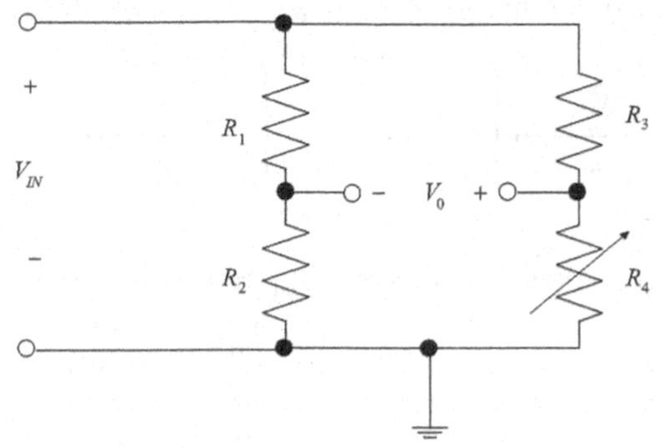

WHEATSTONE BRIDGE

If $\frac{R_1}{R_2} = \frac{R_3}{R_4}$ then $V_0 = 0$ V and the bridge is said to be *balanced*.

If $R_1 = R_2 = R_3 = R$ and $R_4 = R + \Delta R$, where $\Delta R \ll R$, then

$V_0 \approx \frac{\Delta R}{4R} \cdot V_{IN}$.

Piezoelectric Transducers – often comprised of a special ceramic that converts the electrical energy to mechanical energy, or electrical voltage to mechanical force, and vice versa. When an electric field is applied to the material, it will change dimension. Conversely, when a mechanical force is applied to the material, an electric field is produced. Piezoelectric transducers can have multiple layers and many different geometries.

A simple piezoelectric transducer generates a voltage that is proportional to the change in its ceramic's volume, or will change volume proportional to the applied voltage. Dimensional changes are usually very small and can be predominantly in one dimension.

♦ *Pressure Sensors* – can alternatively be called pressure transducers, pressure transmitters, pressure senders, pressure indicators, piezometers, and manometers.

♦

Pressure Relative Measurement Types	Comparison
Absolute	Relative to 0 Pa, the pressure in a vacuum
Gauge	Relative to local atmospheric pressure
Vacuum	Relative to either absolute vacuum (0 Pa) or local atmospheric pressure
Differential	Relative to another pressurized container
Sealed	Relative to sea level pressure

• *pH Sensor* – a typical pH meter consists of a special measuring probe connected to an electronic meter that measures and displays the pH reading.

$$E_{el} = E^0 - S(pH_a - pH_i)$$

E_{el} = electrode potential
E^0 = zero potential
S = slope (mV per pH unit)
pH_a = pH value of the measured solution
pH_i = pH value of the internal buffer

♦ From National Instruments Corporation, as posted on www.ni.com, July 2013.
• From Alliance Technical Sales, Inc., as posted on www.alliancets.com, July 2013.

Examples of Common Chemical Sensors

Sensor Type	Principle	Materials	Analyte
Semiconducting oxide sensor	Conductivity impedance	SnO_2, TiO_2, ZnO_2, WO_3, polymers	O_2, H_2, CO, SO_x, NO_x, combustible hydrocarbons, alcohol, H_2S, NH_3
Electrochemical sensor (liquid electrolyte)	Amperiometric	composite Pt, Au catalyst	H_2, O_2, O_3, CO, H_2S, SO_2, NO_x, NH_3, glucose, hydrazine
Ion-selective electrode (ISE)	Potentiometric	glass, LaF_3, CaF_2	pH, K^+, Na^+, Cl^-, Ca^2, Mg^{2+}, F^-, Ag^+
Solid electrode sensor	Amperiometric Potentiometric	YSZ, H^+-conductor YSZ, β-alumina, Nasicon, Nafion	O_2, H_2, CO, combustible hydrocarbons, O_2, H_2, CO_2, CO, NO_x, SO_x, H_2S, Cl_2 H_2O, combustible hydrocarbons
Piezoelectric sensor	Mechanical w/ polymer film	quartz	combustible hydrocarbons, VOCs
Catalytic combustion sensor	Calorimetric	Pt/Al_2O_3, Pt-wire	H_2, CO, combustible hydrocarbons
Pyroelectric sensor	Calorimetric	Pyroelectric + film	Vapors
Optical sensors	Colorimetric fluorescence	optical fiber/indicator dye	Acids, bases, combustible hydrocarbons, biologicals

SAMPLING

When a continuous-time or analog signal is sampled using a discrete-time method, certain basic concepts should be considered. The sampling rate or frequency is given by

$$f_s = \frac{1}{\Delta t}$$

Nyquist's (Shannon's) sampling theorem states that in order to accurately reconstruct the analog signal from the discrete sample points, the sample rate must be larger than twice the highest frequency contained in the measured signal. Denoting this frequency, which is called the Nyquist frequency, as f_N, the sampling theorem requires that

$$f_s > 2f_N$$

When the above condition is not met, the higher frequencies in the measured signal will not be accurately represented and will appear as lower frequencies in the sampled data. These are known as alias frequencies.

Analog-to-Digital Conversion

When converting an analog signal to digital form, the resolution of the conversion is an important factor. For a measured analog signal over the nominal range $[V_L, V_H]$, where V_L is the low end of the voltage range and V_H is the nominal high end of the voltage range, the voltage resolution is given by

$$\varepsilon_V = \frac{V_H - V_L}{2^n}$$

where n is the number of conversion bits of the A/D converter with typical values of 4, 8, 10, 12, or 16. This number is a key design parameter. After converting an analog signal, the A/D converter produces an integer number of n bits. Call this number N. Note that the range of N is $[0, 2^n - 1]$. When calculating the discrete voltage, V, using the reading, N, from the A/D converter the following equation is used.

$$V = \varepsilon_V N + V_L$$

Note that with this strategy, the highest measurable voltage is one voltage resolution less than V_H, or $V_H - \varepsilon_V$.

Signal Conditioning

Signal conditioning of the measured analog signal is often required to prevent alias frequencies from being measured, and to reduce measurement errors.

MEASUREMENT UNCERTAINTY

Suppose that a calculated result R depends on measurements whose values are $x_1 \pm w_1$, $x_2 \pm w_2$, $x_3 \pm w_3$, etc., where $R = f(x_1, x_2, x_3, \ldots x_n)$, x_i is the measured value, and w_i is the uncertainty in that value. The uncertainty in R, w_R, can be estimated using the Kline-McClintock equation:

$$w_R = \sqrt{\left(w_1 \frac{\partial f}{\partial x_1}\right)^2 + \left(w_2 \frac{\partial f}{\partial x_2}\right)^2 + \cdots + \left(w_n \frac{\partial f}{\partial x_n}\right)^2}$$

CONTROL SYSTEMS

The linear time-invariant transfer function model represented by the block diagram

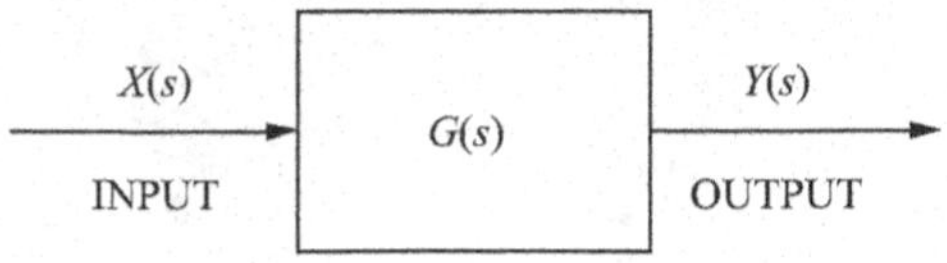

can be expressed as the ratio of two polynomials in the form

$$\frac{Y(s)}{X(s)} = G(s) = \frac{N(s)}{D(s)} = K\frac{\prod_{m=1}^{M}(s - z_m)}{\prod_{n=1}^{N}(s - p_n)}$$

where the M zeros, z_m, and the N poles, p_n, are the roots of the numerator polynomial, $N(s)$, and the denominator polynomial, $D(s)$, respectively.

One classical negative feedback control system model block diagram is

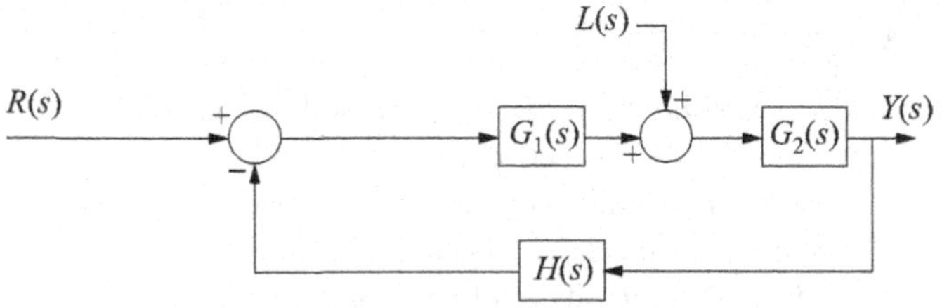

where $G_1(s)$ is a controller or compensator, $G_2(s)$ represents a plant model, and $H(s)$ represents the measurement dynamics. $Y(s)$ represents the controlled variable, $R(s)$ represents the reference input, and $L(s)$ represents a disturbance. $Y(s)$ is related to $R(s)$ and $L(s)$ by

$$Y(s) = \frac{G_1(s)G_2(s)}{1 + G_1(s)G_2(s)H(s)}R(s) + \frac{G_2(s)}{1 + G_1(s)G_2(s)H(s)}L(s)$$

$G_1(s)\,G_2(s)\,H(s)$ is the open-loop transfer function. The closed-loop characteristic equation is

$$1 + G_1(s)\,G_2(s)\,H(s) = 0$$

System performance studies normally include

1. Steady-state analysis using constant inputs based on the Final Value Theorem. If all poles of a $G(s)$ function have negative real parts, then

$$\text{dc gain} = \lim_{s \to 0} G(s)$$

Note that $G(s)$ could refer to either an open-loop or a closed-loop transfer function.

For the unity feedback control system model

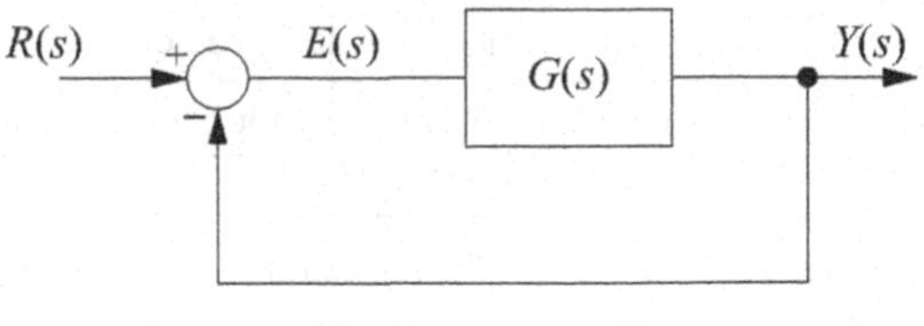

with the open-loop transfer function defined by

$$G(s) = \frac{K_B}{s^T} \times \frac{\prod_{m=1}^{M} (1 + s/\omega_m)}{\prod_{n=1}^{N} (1 + s/\omega_n)}$$

The following steady-state error analysis table can be constructed where T denotes the type of system, i.e., type 0, type 1, etc.

Steady-State Error e_{ss}			
Input \ Type	$T = 0$	$T = 1$	$T = 2$
Unit Step	$1/(K_B + 1)$	0	0
Ramp	∞	$1/K_B$	0
Acceleration	∞	∞	$1/K_B$

2. Frequency response evaluations to determine dynamic performance and stability. For example, relative stability can be quantified in terms of

   a. Gain margin (GM), which is the additional gain required to produce instability in the unity gain feedback control system. If at $\omega = \omega_{180}$,

      $\angle G(j\omega_{180}) = -180°$; then

      $$\text{GM} = -20\log_{10}\left(\left|G(j\omega_{180})\right|\right)$$

   b. Phase margin (PM), which is the additional phase required to produce instability. Thus,

      $$\text{PM} = 180° + \angle G(j\omega_{0\text{dB}})$$

      where $\omega_{0\text{dB}}$ is the ω that satisfies $\left|G(j\omega)\right| = 1$.

3. Transient responses are obtained by using Laplace transforms or computer solutions with numerical integration.

Common Compensator/Controller forms are

PID Controller $G_C(s) = K\left(1 + \frac{1}{T_I s} + T_D s\right)$

Lag or Lead Compensator $G_C(s) = K\left(\frac{1 + sT_1}{1 + sT_2}\right)$ depending on the ratio of T_1/T_2.

Routh Test

For the characteristic equation

$$a_n s^n + a_{n-1} s^{n-1} + a_{n-2} s^{n-2} + \ldots + a_0 = 0$$

the coefficients are arranged into the first two rows of an array. Additional rows are computed. The array and coefficient computations are defined by:

$$\begin{matrix} a_n & a_{n-2} & a_{n-4} & \cdots & \cdots & \cdots \\ a_{n-1} & a_{n-3} & a_{n-5} & \cdots & \cdots & \cdots \\ b_1 & b_2 & b_3 & \cdots & \cdots & \cdots \\ c_1 & c_2 & c_3 & \cdots & \cdots & \cdots \end{matrix}$$

where

$$b_1 = \frac{a_{n-1}a_{n-2} - a_n a_{n-3}}{a_{n-1}} \qquad c_1 = \frac{a_{n-3}b_1 - a_{n-1}b_2}{b_1}$$

$$b_2 = \frac{a_{n-1}a_{n-4} - a_n a_{n-5}}{a_{n-1}} \qquad c_2 = \frac{a_{n-5}b_1 - a_{n-1}b_3}{b_1}$$

$$b_i = \frac{a_{n-1}a_{n-2i} - a_n a_{n-2i-1}}{a_{n-1}}$$

$$c_i = \frac{b_1 a_{n-2i-1} - a_{n-1}b_{i+1}}{b_1}$$

The necessary and sufficient conditions for all the roots of the equation to have negative real parts is that all the elements in the first column be of the same sign and nonzero.

First-Order Control System Models

The transfer function model for a first-order system is

$$\frac{Y(s)}{R(s)} = \frac{K}{\tau s + 1}, \text{ where}$$

K = steady-state gain

τ = time constant

The step response of a first-order system to a step input of magnitude M is

$$y(t) = y_0 e^{-t/\tau} + KM\left(1 - e^{-t/\tau}\right)$$

In the chemical process industry, y_0 is typically taken to be zero, and $y(t)$ is referred to as a deviation variable.

For systems with time delay (dead time or transport lag) θ, the transfer function is

$$\frac{Y(s)}{R(s)} = \frac{Ke^{-\theta s}}{\tau s + 1}$$

The step response for $t \geq \theta$ to a step of magnitude M is

$$y(t) = \left[y_0 e^{-(t-\theta)/\tau} + KM\left(1 - e^{-(t-\theta)/\tau}\right)\right]u(t - \theta), \text{ where}$$

$u(t)$ is the unit step function.

Second-Order Control System Models

One standard second-order control system model is

$$\frac{Y(s)}{R(s)} = \frac{K\omega_n^2}{s^2 + 2\zeta\omega_n s + \omega_n^2}, \text{ where}$$

K = steady-state gain

ζ = the damping ratio

ω_n = the undamped natural ($\zeta = 0$) frequency

$\omega_d = \omega_n\sqrt{1 - \zeta^2}$, the damped natural frequency

$\omega_r = \omega_n\sqrt{1 - 2\zeta^2}$, the damped resonant frequency

If the damping ratio ζ is less than unity, the system is said to be underdamped; if ζ is equal to unity, it is said to be critically damped; and if ζ is greater than unity, the system is said to be overdamped.

For a unit step input to a normalized underdamped second-order control system, the time required to reach a peak value t_p and the value of that peak M_p are given by

$$t_p = \pi / \left(\omega_n\sqrt{1 - \zeta^2}\right)$$

$$M_p = 1 + e^{-\pi\zeta/\sqrt{1 - \zeta^2}}$$

The percent overshoot (% OS) of the response is given by

$$\%\,OS = 100e^{-\pi\zeta/\sqrt{1 - \zeta^2}}$$

For an underdamped second-order system, the logarithmic decrement is

$$\delta = \frac{1}{m}\ln\left(\frac{x_k}{x_{k+m}}\right) = \frac{2\pi\zeta}{\sqrt{1 - \zeta^2}}$$

where x_k and x_{k+m} are the amplitudes of oscillation at cycles k and $k + m$, respectively. The period of oscillation τ is related to ω_d by

$$\omega_d\tau = 2\pi$$

The time required for the output of a second-order system to settle to within 2% of its final value (2% settling time) is defined to be

$$T_s = \frac{4}{\zeta\omega_n}$$

An alternative form commonly employed in the chemical process industry is

$$\frac{Y(s)}{R(s)} = \frac{K}{\tau^2 s^2 + 2\zeta\tau s + 1}, \text{ where}$$

K = steady-state gain

ζ = the damping ratio

τ = the inverse natural frequency

Root Locus

The root locus is the locus of points in the complex s-plane satisfying

$$1 + K\frac{(s - z_1)(s - z_2)\ldots(s - z_m)}{(s - p_1)(s - p_2)\ldots(s - p_n)} = 0 \qquad m \le n$$

as K is varied. The p_i and z_j are the open-loop poles and zeros, respectively. When K is increased from zero, the locus has the following properties.

1. Locus branches exist on the real axis to the left of an odd number of open-loop poles and/or zeros.
2. The locus originates at the open-loop poles $p_1, \ldots, p_n$ and terminates at the zeros $z_1, \ldots, z_m$. If $m < n$ then $(n - m)$ branches terminate at infinity at asymptote angles

   $$\alpha = \frac{(2k + 1)180°}{n - m} \qquad k = 0, \pm 1, \pm 2, \pm 3, \ldots$$

   with the real axis.
3. The intersection of the real axis with the asymptotes is called the asymptote centroid and is given by

   $$\sigma_A = \frac{\sum_{i=1}^{n} \operatorname{Re}(p_i) - \sum_{i=1}^{m} \operatorname{Re}(z_i)}{n - m}$$

4. If the locus crosses the imaginary (ω) axis, the values of K and ω are given by letting $s = j\omega$ in the defining equation.

State-Variable Control System Models

One common state-variable model for dynamic systems has the form

$\dot{\mathbf{x}}(t) = \mathbf{Ax}(t) + \mathbf{Bu}(t)$ (state equation)

$\mathbf{y}(t) = \mathbf{Cx}(t) + \mathbf{Du}(t)$ (output equation)

where

$\mathbf{x}(t)$ = N by 1 state vector (N state variables)

$\mathbf{u}(t)$ = R by 1 input vector (R inputs)

$\mathbf{y}(t)$ = M by 1 output vector (M outputs)

$\mathbf{A}$ = system matrix

$\mathbf{B}$ = input distribution matrix

$\mathbf{C}$ = output matrix

$\mathbf{D}$ = feed-through matrix

The orders of the matrices are defined via variable definitions.

State-variable models are used to handle multiple inputs and multiple outputs. Furthermore, state-variable models can be formulated for open-loop system components or the complete closed-loop system.

The Laplace transform of the time-invariant state equation is

$$s\mathbf{X}(s) - \mathbf{x}(0) = \mathbf{AX}(s) + \mathbf{BU}(s)$$

from which

$$\mathbf{X}(s) = \Phi(s)\,\mathbf{x}(0) + \Phi(s)\,\mathbf{BU}(s)$$

where the Laplace transform of the state transition matrix is

$$\Phi(s) = [s\mathbf{I} - \mathbf{A}]^{-1}.$$

The state-transition matrix

$$\Phi(t) = L^{-1}\{\Phi(s)\}$$

(also defined as $e^{\mathbf{A}t}$) can be used to write

$$\mathbf{x}(t) = \Phi(t)\,\mathbf{x}(0) + \int_0^t \Phi(t-\tau)\,\mathbf{Bu}(\tau)\,d\tau$$

The output can be obtained with the output equation; e.g., the Laplace transform output is

$$\mathbf{Y}(s) = \{\mathbf{C}\Phi(s)\,\mathbf{B} + \mathbf{D}\}\mathbf{U}(s) + \mathbf{C}\Phi(s)\,\mathbf{x}(0)$$

The latter term represents the output(s) due to initial conditions, whereas the former term represents the output(s) due to the $\mathbf{U}(s)$ inputs and gives rise to transfer function definitions.

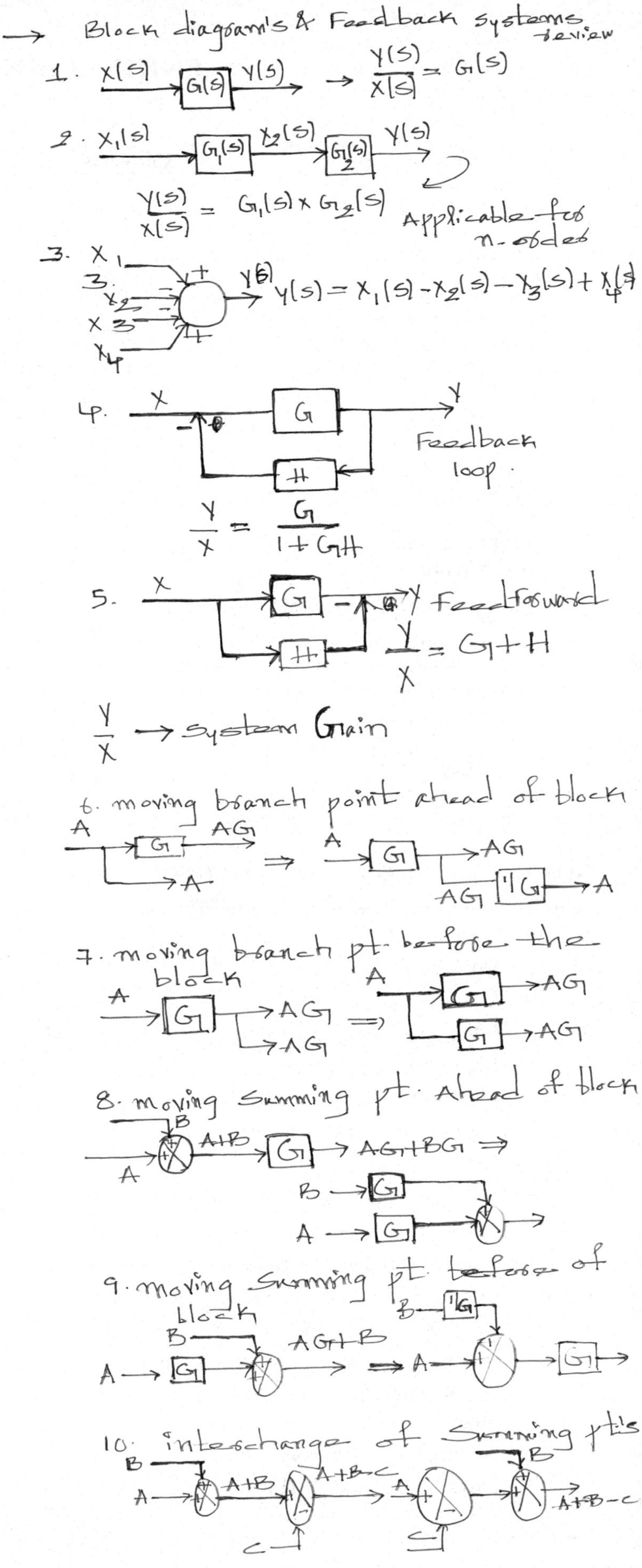

ENGINEERING ECONOMICS

Factor Name	Converts	Symbol	Formula
Single Payment Compound Amount	to F given P	$(F/P, i\%, n)$	$(1+i)^n$
Single Payment Present Worth	to P given F	$(P/F, i\%, n)$	$(1+i)^{-n}$
Uniform Series Sinking Fund	to A given F	$(A/F, i\%, n)$	$\frac{i}{(1+i)^n-1}$
Capital Recovery	to A given P	$(A/P, i\%, n)$	$\frac{i(1+i)^n}{(1+i)^n-1}$
Uniform Series Compound Amount	to F given A	$(F/A, i\%, n)$	$\frac{(1+i)^n-1}{i}$
Uniform Series Present Worth	to P given A	$(P/A, i\%, n)$	$\frac{(1+i)^n-1}{i(1+i)^n}$
Uniform Gradient Present Worth	to P given G	$(P/G, i\%, n)$	$\frac{(1+i)^n-1}{i^2(1+i)^n}-\frac{n}{i(1+i)^n}$
Uniform Gradient † Future Worth	to F given G	$(F/G, i\%, n)$	$\frac{(1+i)^n-1}{i^2}-\frac{n}{i}$
Uniform Gradient Uniform Series	to A given G	$(A/G, i\%, n)$	$\frac{1}{i}-\frac{n}{(1+i)^n-1}$

NOMENCLATURE AND DEFINITIONS

A............Uniform amount per interest period

B............Benefit

BV..........Book value

C............Cost

d.............Inflation adjusted interest rate per interest period

D_j...........Depreciation in year j

EV..........Expected value

F.............Future worth, value, or amount

f..............General inflation rate per interest period

G............Uniform gradient amount per interest period

i..............Interest rate per interest period

i_e.............Annual effective interest rate

MARR....Minimum acceptable/attractive rate of return

m............Number of compounding periods per year

n..............Number of compounding periods; or the expected life of an asset

P.............Present worth, value, or amount

r..............Nominal annual interest rate

S_n...........Expected salvage value in year n

Subscripts

j .... at time j

n.......... at time n

†.......... $F/G = (F/A - n)/i = (F/A) \times (A/G)$

Risk

Risk is the chance of an outcome other than what is planned to occur or expected in the analysis.

NON-ANNUAL COMPOUNDING

$$i_e = \left(1 + \frac{r}{m}\right)^m - 1$$

BREAK-EVEN ANALYSIS

By altering the value of any one of the variables in a situation, holding all of the other values constant, it is possible to find a value for that variable that makes the two alternatives equally economical. This value is the break-even point.

Break-even analysis is used to describe the percentage of capacity of operation for a manufacturing plant at which income will just cover expenses.

The payback period is the period of time required for the profit or other benefits of an investment to equal the cost of the investment.

INFLATION

To account for inflation, the dollars are deflated by the general inflation rate per interest period *f*, and then they are shifted over the time scale using the interest rate per interest period *i*. Use an inflation adjusted interest rate per interest period *d* for computing present worth values *P*.
The formula for *d* is $d = i + f + (i \times f)$

DEPRECIATION

Straight Line

$$D_j = \frac{C - S_n}{n}$$

Modified Accelerated Cost Recovery System (MACRS)

$$D_j = (\text{factor})\ C$$

A table of MACRS factors is provided below.

BOOK VALUE

$BV = \text{initial cost} - \Sigma D_j$

TAXATION

Income taxes are paid at a specific rate on taxable income. Taxable income is total income less depreciation and ordinary expenses. Expenses do not include capital items, which should be depreciated.

CAPITALIZED COSTS

Capitalized costs are present worth values using an assumed perpetual period of time.

$$\text{Capitalized Costs} = P = \frac{A}{i}$$

BONDS

Bond value equals the present worth of the payments the purchaser (or holder of the bond) receives during the life of the bond at some interest rate *i*.

Bond yield equals the computed interest rate of the bond value when compared with the bond cost.

RATE-OF-RETURN

The minimum acceptable rate-of-return (MARR) is that interest rate that one is willing to accept, or the rate one desires to earn on investments. The rate-of-return on an investment is the interest rate that makes the benefits and costs equal.

BENEFIT-COST ANALYSIS

In a benefit-cost analysis, the benefits *B* of a project should exceed the estimated costs *C*.

$$B - C \geq 0, \text{ or } B/C \geq 1$$

MODIFIED ACCELERATED COST RECOVERY SYSTEM (MACRS)

MACRS FACTORS				
	Recovery Period (Years)			
Year	3	5	7	10
	Recovery Rate (Percent)			
1	33.33	20.00	14.29	10.00
2	44.45	32.00	24.49	18.00
3	14.81	19.20	17.49	14.40
4	7.41	11.52	12.49	11.52
5		11.52	8.93	9.22
6		5.76	8.92	7.37
7			8.93	6.55
8			4.46	6.55
9				6.56
10				6.55
11				3.28

ECONOMIC DECISION TREES

The following symbols are used to model decisions with decision trees:

Decision node 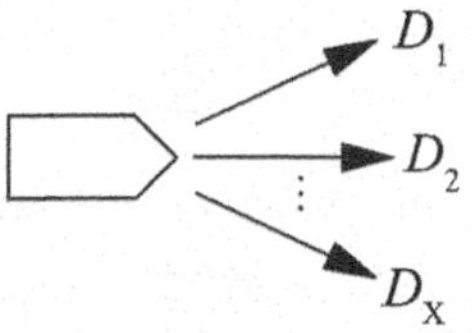

Decision maker chooses 1 of the available paths.

Chance node 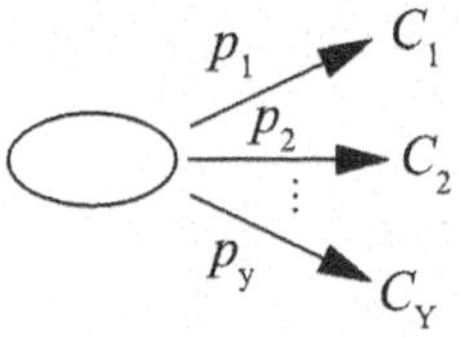

Represents a probabilistic (chance) event. Each possible outcome (C_1, C_2,..., C_y) has a probability (p_1, p_2,..., p_y) associated with it.

Outcome node 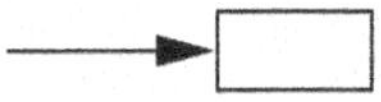

Shows result for a particular path through the decision tree.

Expected Value: $EV = (C_1)(p_1) + (C_2)(p_2) + \ldots$

Interest Rate Tables

Factor Table - $i = 0.50\%$

n	P/F	P/A	P/G	F/P	F/A	A/P	A/F	A/G
1	0.9950	0.9950	0.0000	1.0050	1.0000	1.0050	1.0000	0.0000
2	0.9901	1.9851	0.9901	1.0100	2.0050	0.5038	0.4988	0.4988
3	0.9851	2.9702	2.9604	1.0151	3.0150	0.3367	0.3317	0.9967
4	0.9802	3.9505	5.9011	1.0202	4.0301	0.2531	0.2481	1.4938
5	0.9754	4.9259	9.8026	1.0253	5.0503	0.2030	0.1980	1.9900
6	0.9705	5.8964	14.6552	1.0304	6.0755	0.1696	0.1646	2.4855
7	0.9657	6.8621	20.4493	1.0355	7.1059	0.1457	0.1407	2.9801
8	0.9609	7.8230	27.1755	1.0407	8.1414	0.1278	0.1228	3.4738
9	0.9561	8.7791	34.8244	1.0459	9.1821	0.1139	0.1089	3.9668
10	0.9513	9.7304	43.3865	1.0511	10.2280	0.1028	0.0978	4.4589
11	0.9466	10.6770	52.8526	1.0564	11.2792	0.0937	0.0887	4.9501
12	0.9419	11.6189	63.2136	1.0617	12.3356	0.0861	0.0811	5.4406
13	0.9372	12.5562	74.4602	1.0670	13.3972	0.0796	0.0746	5.9302
14	0.9326	13.4887	86.5835	1.0723	14.4642	0.0741	0.0691	6.4190
15	0.9279	14.4166	99.5743	1.0777	15.5365	0.0694	0.0644	6.9069
16	0.9233	15.3399	113.4238	1.0831	16.6142	0.0652	0.0602	7.3940
17	0.9187	16.2586	128.1231	1.0885	17.6973	0.0615	0.0565	7.8803
18	0.9141	17.1728	143.6634	1.0939	18.7858	0.0582	0.0532	8.3658
19	0.9096	18.0824	160.0360	1.0994	19.8797	0.0553	0.0503	8.8504
20	0.9051	18.9874	177.2322	1.1049	20.9791	0.0527	0.0477	9.3342
21	0.9006	19.8880	195.2434	1.1104	22.0840	0.0503	0.0453	9.8172
22	0.8961	20.7841	214.0611	1.1160	23.1944	0.0481	0.0431	10.2993
23	0.8916	21.6757	233.6768	1.1216	24.3104	0.0461	0.0411	10.7806
24	0.8872	22.5629	254.0820	1.1272	25.4320	0.0443	0.0393	11.2611
25	0.8828	23.4456	275.2686	1.1328	26.5591	0.0427	0.0377	11.7407
30	0.8610	27.7941	392.6324	1.1614	32.2800	0.0360	0.0310	14.1265
40	0.8191	36.1722	681.3347	1.2208	44.1588	0.0276	0.0226	18.8359
50	0.7793	44.1428	1,035.6966	1.2832	56.6452	0.0227	0.0177	23.4624
60	0.7414	51.7256	1,448.6458	1.3489	69.7700	0.0193	0.0143	28.0064
100	0.6073	78.5426	3,562.7934	1.6467	129.3337	0.0127	0.0077	45.3613

Factor Table - $i = 1.00\%$

n	P/F	P/A	P/G	F/P	F/A	A/P	A/F	A/G
1	0.9901	0.9901	0.0000	1.0100	1.0000	1.0100	1.0000	0.0000
2	0.9803	1.9704	0.9803	1.0201	2.0100	0.5075	0.4975	0.4975
3	0.9706	2.9410	2.9215	1.0303	3.0301	0.3400	0.3300	0.9934
4	0.9610	3.9020	5.8044	1.0406	4.0604	0.2563	0.2463	1.4876
5	0.9515	4.8534	9.6103	1.0510	5.1010	0.2060	0.1960	1.9801
6	0.9420	5.7955	14.3205	1.0615	6.1520	0.1725	0.1625	2.4710
7	0.9327	6.7282	19.9168	1.0721	7.2135	0.1486	0.1386	2.9602
8	0.9235	7.6517	26.3812	1.0829	8.2857	0.1307	0.1207	3.4478
9	0.9143	8.5650	33.6959	1.0937	9.3685	0.1167	0.1067	3.9337
10	0.9053	9.4713	41.8435	1.1046	10.4622	0.1056	0.0956	4.4179
11	0.8963	10.3676	50.8067	1.1157	11.5668	0.0965	0.0865	4.9005
12	0.8874	11.2551	60.5687	1.1268	12.6825	0.0888	0.0788	5.3815
13	0.8787	12.1337	71.1126	1.1381	13.8093	0.0824	0.0724	5.8607
14	0.8700	13.0037	82.4221	1.1495	14.9474	0.0769	0.0669	6.3384
15	0.8613	13.8651	94.4810	1.1610	16.0969	0.0721	0.0621	6.8143
16	0.8528	14.7179	107.2734	1.1726	17.2579	0.0679	0.0579	7.2886
17	0.8444	15.5623	120.7834	1.1843	18.4304	0.0643	0.0543	7.7613
18	0.8360	16.3983	134.9957	1.1961	19.6147	0.0610	0.0510	8.2323
19	0.8277	17.2260	149.8950	1.2081	20.8109	0.0581	0.0481	8.7017
20	0.8195	18.0456	165.4664	1.2202	22.0190	0.0554	0.0454	9.1694
21	0.8114	18.8570	181.6950	1.2324	23.2392	0.0530	0.0430	9.6354
22	0.8034	19.6604	198.5663	1.2447	24.4716	0.0509	0.0409	10.0998
23	0.7954	20.4558	216.0660	1.2572	25.7163	0.0489	0.0389	10.5626
24	0.7876	21.2434	234.1800	1.2697	26.9735	0.0471	0.0371	11.0237
25	0.7798	22.0232	252.8945	1.2824	28.2432	0.0454	0.0354	11.4831
30	0.7419	25.8077	355.0021	1.3478	34.7849	0.0387	0.0277	13.7557
40	0.6717	32.8347	596.8561	1.4889	48.8864	0.0305	0.0205	18.1776
50	0.6080	39.1961	879.4176	1.6446	64.4632	0.0255	0.0155	22.4363
60	0.5504	44.9550	1,192.8061	1.8167	81.6697	0.0222	0.0122	26.5333
100	0.3697	63.0289	2,605.7758	2.7048	170.4814	0.0159	0.0059	41.3426

Interest Rate Tables
Factor Table - $i = 1.50\%$

n	P/F	P/A	P/G	F/P	F/A	A/P	A/F	A/G
1	0.9852	0.9852	0.0000	1.0150	1.0000	1.0150	1.0000	0.0000
2	0.9707	1.9559	0.9707	1.0302	2.0150	0.5113	0.4963	0.4963
3	0.9563	2.9122	2.8833	1.0457	3.0452	0.3434	0.3284	0.9901
4	0.9422	3.8544	5.7098	1.0614	4.0909	0.2594	0.2444	1.4814
5	**0.9283**	**4.7826**	**9.4229**	**1.0773**	**5.1523**	**0.2091**	**0.1941**	**1.9702**
6	0.9145	5.6972	13.9956	1.0934	6.2296	0.1755	0.1605	2.4566
7	0.9010	6.5982	19.4018	1.1098	7.3230	0.1516	0.1366	2.9405
8	0.8877	7.4859	26.6157	1.1265	8.4328	0.1336	0.1186	3.4219
9	0.8746	8.3605	32.6125	1.1434	9.5593	0.1196	0.1046	3.9008
10	**0.8617**	**9.2222**	**40.3675**	**1.1605**	**10.7027**	**0.1084**	**0.0934**	**4.3772**
11	0.8489	10.0711	48.8568	1.1779	11.8633	0.0993	0.0843	4.8512
12	0.8364	10.9075	58.0571	1.1956	13.0412	0.0917	0.0767	5.3227
13	0.8240	11.7315	67.9454	1.2136	14.2368	0.0852	0.0702	5.7917
14	0.8118	12.5434	78.4994	1.2318	15.4504	0.0797	0.0647	6.2582
15	**0.7999**	**13.3432**	**89.6974**	**1.2502**	**16.6821**	**0.0749**	**0.0599**	**6.7223**
16	0.7880	14.1313	101.5178	1.2690	17.9324	0.0708	0.0558	7.1839
17	0.7764	14.9076	113.9400	1.2880	19.2014	0.0671	0.0521	7.6431
18	0.7649	15.6726	126.9435	1.3073	20.4894	0.0638	0.0488	8.0997
19	0.7536	16.4262	140.5084	1.3270	21.7967	0.0609	0.0459	8.5539
20	**0.7425**	**17.1686**	**154.6154**	**1.3469**	**23.1237**	**0.0582**	**0.0432**	**9.0057**
21	0.7315	17.9001	169.2453	1.3671	24.4705	0.0559	0.0409	9.4550
22	0.7207	18.6208	184.3798	1.3876	25.8376	0.0537	0.0387	9.9018
23	0.7100	19.3309	200.0006	1.4084	27.2251	0.0517	0.0367	10.3462
24	0.6995	20.0304	216.0901	1.4295	28.6335	0.0499	0.0349	10.7881
25	**0.6892**	**20.7196**	**232.6310**	**1.4509**	**30.0630**	**0.0483**	**0.0333**	**11.2276**
30	0.6398	24.0158	321.5310	1.5631	37.5387	0.0416	0.0266	13.3883
40	0.5513	29.9158	524.3568	1.8140	54.2679	0.0334	0.0184	17.5277
50	0.4750	34.9997	749.9636	2.1052	73.6828	0.0286	0.0136	21.4277
60	0.4093	39.3803	988.1674	2.4432	96.2147	0.0254	0.0104	25.0930
100	**0.2256**	**51.6247**	**1,937.4506**	**4.4320**	**228.8030**	**0.0194**	**0.0044**	**37.5295**

Factor Table - $i = 2.00\%$

n	P/F	P/A	P/G	F/P	F/A	A/P	A/F	A/G
1	0.9804	0.9804	0.0000	1.0200	1.0000	1.0200	1.0000	0.0000
2	0.9612	1.9416	0.9612	1.0404	2.0200	0.5150	0.4950	0.4950
3	0.9423	2.8839	2.8458	1.0612	3.0604	0.3468	0.3268	0.9868
4	0.9238	3.8077	5.6173	1.0824	4.1216	0.2626	0.2426	1.4752
5	**0.9057**	**4.7135**	**9.2403**	**1.1041**	**5.2040**	**0.2122**	**0.1922**	**1.9604**
6	0.8880	5.6014	13.6801	1.1262	6.3081	0.1785	0.1585	2.4423
7	0.8706	6.4720	18.9035	1.1487	7.4343	0.1545	0.1345	2.9208
8	0.8535	7.3255	24.8779	1.1717	8.5830	0.1365	0.1165	3.3961
9	0.8368	8.1622	31.5720	1.1951	9.7546	0.1225	0.1025	3.8681
10	**0.8203**	**8.9826**	**38.9551**	**1.2190**	**10.9497**	**0.1113**	**0.0913**	**4.3367**
11	0.8043	9.7868	46.9977	1.2434	12.1687	0.1022	0.0822	4.8021
12	0.7885	10.5753	55.6712	1.2682	13.4121	0.0946	0.0746	5.2642
13	0.7730	11.3484	64.9475	1.2936	14.6803	0.0881	0.0681	5.7231
14	0.7579	12.1062	74.7999	1.3195	15.9739	0.0826	0.0626	6.1786
15	**0.7430**	**12.8493**	**85.2021**	**1.3459**	**17.2934**	**0.0778**	**0.0578**	**6.6309**
16	0.7284	13.5777	96.1288	1.3728	18.6393	0.0737	0.0537	7.0799
17	0.7142	14.2919	107.5554	1.4002	20.0121	0.0700	0.0500	7.5256
18	0.7002	14.9920	119.4581	1.4282	21.4123	0.0667	0.0467	7.9681
19	0.6864	15.6785	131.8139	1.4568	22.8406	0.0638	0.0438	8.4073
20	**0.6730**	**16.3514**	**144.6003**	**1.4859**	**24.2974**	**0.0612**	**0.0412**	**8.8433**
21	0.6598	17.0112	157.7959	1.5157	25.7833	0.0588	0.0388	9.2760
22	0.6468	17.6580	171.3795	1.5460	27.2990	0.0566	0.0366	9.7055
23	0.6342	18.2922	185.3309	1.5769	28.8450	0.0547	0.0347	10.1317
24	0.6217	18.9139	199.6305	1.6084	30.4219	0.0529	0.0329	10.5547
25	**0.6095**	**19.5235**	**214.2592**	**1.6406**	**32.0303**	**0.0512**	**0.0312**	**10.9745**
30	0.5521	22.3965	291.7164	1.8114	40.5681	0.0446	0.0246	13.0251
40	0.4529	27.3555	461.9931	2.2080	60.4020	0.0366	0.0166	16.8885
50	0.3715	31.4236	642.3606	2.6916	84.5794	0.0318	0.0118	20.4420
60	0.3048	34.7609	823.6975	3.2810	114.0515	0.0288	0.0088	23.6961
100	**0.1380**	**43.0984**	**1,464.7527**	**7.2446**	**312.2323**	**0.0232**	**0.0032**	**33.9863**

Interest Rate Tables

Factor Table - $i = 4.00\%$

n	P/F	P/A	P/G	F/P	F/A	A/P	A/F	A/G
1	0.9615	0.9615	0.0000	1.0400	1.0000	1.0400	1.0000	0.0000
2	0.9246	1.8861	0.9246	1.0816	2.0400	0.5302	0.4902	0.4902
3	0.8890	2.7751	2.7025	1.1249	3.1216	0.3603	0.3203	0.9739
4	0.8548	3.6299	5.2670	1.1699	4.2465	0.2755	0.2355	1.4510
5	**0.8219**	**4.4518**	**8.5547**	**1.2167**	**5.4163**	**0.2246**	**0.1846**	**1.9216**
6	0.7903	5.2421	12.5062	1.2653	6.6330	0.1908	0.1508	2.3857
7	0.7599	6.0021	17.0657	1.3159	7.8983	0.1666	0.1266	2.8433
8	0.7307	6.7327	22.1806	1.3686	9.2142	0.1485	0.1085	3.2944
9	0.7026	7.4353	27.8013	1.4233	10.5828	0.1345	0.0945	3.7391
10	**0.6756**	**8.1109**	**33.8814**	**1.4802**	**12.0061**	**0.1233**	**0.0833**	**4.1773**
11	0.6496	8.7605	40.3772	1.5395	13.4864	0.1141	0.0741	4.6090
12	0.6246	9.3851	47.2477	1.6010	15.0258	0.1066	0.0666	5.0343
13	0.6006	9.9856	54.4546	1.6651	16.6268	0.1001	0.0601	5.4533
14	0.5775	10.5631	61.9618	1.7317	18.2919	0.0947	0.0547	5.8659
15	**0.5553**	**11.1184**	**69.7355**	**1.8009**	**20.0236**	**0.0899**	**0.0499**	**6.2721**
16	0.5339	11.6523	77.7441	1.8730	21.8245	0.0858	0.0458	6.6720
17	0.5134	12.1657	85.9581	1.9479	23.6975	0.0822	0.0422	7.0656
18	0.4936	12.6593	94.3498	2.0258	25.6454	0.0790	0.0390	7.4530
19	0.4746	13.1339	102.8933	2.1068	27.6712	0.0761	0.0361	7.8342
20	**0.4564**	**13.5903**	**111.5647**	**2.1911**	**29.7781**	**0.0736**	**0.0336**	**8.2091**
21	0.4388	14.0292	120.3414	2.2788	31.9692	0.0713	0.0313	8.5779
22	0.4220	14.4511	129.2024	2.3699	34.2480	0.0692	0.0292	8.9407
23	0.4057	14.8568	138.1284	2.4647	36.6179	0.0673	0.0273	9.2973
24	0.3901	15.2470	147.1012	2.5633	39.0826	0.0656	0.0256	9.6479
25	**0.3751**	**15.6221**	**156.1040**	**2.6658**	**41.6459**	**0.0640**	**0.0240**	**9.9925**
30	0.3083	17.2920	201.0618	3.2434	56.0849	0.0578	0.0178	11.6274
40	0.2083	19.7928	286.5303	4.8010	95.0255	0.0505	0.0105	14.4765
50	0.1407	21.4822	361.1638	7.1067	152.6671	0.0466	0.0066	16.8122
60	0.0951	22.6235	422.9966	10.5196	237.9907	0.0442	0.0042	18.6972
100	**0.0198**	**24.5050**	**563.1249**	**50.5049**	**1,237.6237**	**0.0408**	**0.0008**	**22.9800**

Factor Table - $i = 6.00\%$

n	P/F	P/A	P/G	F/P	F/A	A/P	A/F	A/G
1	0.9434	0.9434	0.0000	1.0600	1.0000	1.0600	1.0000	0.0000
2	0.8900	1.8334	0.8900	1.1236	2.0600	0.5454	0.4854	0.4854
3	0.8396	2.6730	2.5692	1.1910	3.1836	0.3741	0.3141	0.9612
4	0.7921	3.4651	4.9455	1.2625	4.3746	0.2886	0.2286	1.4272
5	**0.7473**	**4.2124**	**7.9345**	**1.3382**	**5.6371**	**0.2374**	**0.1774**	**1.8836**
6	0.7050	4.9173	11.4594	1.4185	6.9753	0.2034	0.1434	2.3304
7	0.6651	5.5824	15.4497	1.5036	8.3938	0.1791	0.1191	2.7676
8	0.6274	6.2098	19.8416	1.5938	9.8975	0.1610	0.1010	3.1952
9	0.5919	6.8017	24.5768	1.6895	11.4913	0.1470	0.0870	3.6133
10	**0.5584**	**7.3601**	**29.6023**	**1.7908**	**13.1808**	**0.1359**	**0.0759**	**4.0220**
11	0.5268	7.8869	34.8702	1.8983	14.9716	0.1268	0.0668	4.4213
12	0.4970	8.3838	40.3369	2.0122	16.8699	0.1193	0.0593	4.8113
13	0.4688	8.8527	45.9629	2.1329	18.8821	0.1130	0.0530	5.1920
14	0.4423	9.2950	51.7128	2.2609	21.0151	0.1076	0.0476	5.5635
15	**0.4173**	**9.7122**	**57.5546**	**2.3966**	**23.2760**	**0.1030**	**0.0430**	**5.9260**
16	0.3936	10.1059	63.4592	2.5404	25.6725	0.0990	0.0390	6.2794
17	0.3714	10.4773	69.4011	2.6928	28.2129	0.0954	0.0354	6.6240
18	0.3505	10.8276	75.3569	2.8543	30.9057	0.0924	0.0324	6.9597
19	0.3305	11.1581	81.3062	3.0256	33.7600	0.0896	0.0296	7.2867
20	**0.3118**	**11.4699**	**87.2304**	**3.2071**	**36.7856**	**0.0872**	**0.0272**	**7.6051**
21	0.2942	11.7641	93.1136	3.3996	39.9927	0.0850	0.0250	7.9151
22	0.2775	12.0416	98.9412	3.6035	43.3923	0.0830	0.0230	8.2166
23	0.2618	12.3034	104.7007	3.8197	46.9958	0.0813	0.0213	8.5099
24	0.2470	12.5504	110.3812	4.0489	50.8156	0.0797	0.0197	8.7951
25	**0.2330**	**12.7834**	**115.9732**	**4.2919**	**54.8645**	**0.0782**	**0.0182**	**9.0722**
30	0.1741	13.7648	142.3588	5.7435	79.0582	0.0726	0.0126	10.3422
40	0.0972	15.0463	185.9568	10.2857	154.7620	0.0665	0.0065	12.3590
50	0.0543	15.7619	217.4574	18.4202	290.3359	0.0634	0.0034	13.7964
60	0.0303	16.1614	239.0428	32.9877	533.1282	0.0619	0.0019	14.7909
100	**0.0029**	**16.6175**	**272.0471**	**339.3021**	**5,638.3681**	**0.0602**	**0.0002**	**16.3711**

Interest Rate Tables

Factor Table - $i = 8.00\%$

n	*P/F*	*P/A*	*P/G*	*F/P*	*F/A*	*A/P*	*A/F*	*A/G*
1	0.9259	0.9259	0.0000	1.0800	1.0000	1.0800	1.0000	0.0000
2	0.8573	1.7833	0.8573	1.1664	2.0800	0.5608	0.4808	0.4808
3	0.7938	2.5771	2.4450	1.2597	3.2464	0.3880	0.3080	0.9487
4	0.7350	3.3121	4.6501	1.3605	4.5061	0.3019	0.2219	1.4040
5	**0.6806**	**3.9927**	**7.3724**	**1.4693**	**5.8666**	**0.2505**	**0.1705**	**1.8465**
6	0.6302	4.6229	10.5233	1.5869	7.3359	0.2163	0.1363	2.2763
7	0.5835	5.2064	14.0242	1.7138	8.9228	0.1921	0.1121	2.6937
8	0.5403	5.7466	17.8061	1.8509	10.6366	0.1740	0.0940	3.0985
9	0.5002	6.2469	21.8081	1.9990	12.4876	0.1601	0.0801	3.4910
10	**0.4632**	**6.7101**	**25.9768**	**2.1589**	**14.4866**	**0.1490**	**0.0690**	**3.8713**
11	0.4289	7.1390	30.2657	2.3316	16.6455	0.1401	0.0601	4.2395
12	0.3971	7.5361	34.6339	2.5182	18.9771	0.1327	0.0527	4.5957
13	0.3677	7.9038	39.0463	2.7196	21.4953	0.1265	0.0465	4.9402
14	0.3405	8.2442	43.4723	2.9372	24.2149	0.1213	0.0413	5.2731
15	**0.3152**	**8.5595**	**47.8857**	**3.1722**	**27.1521**	**0.1168**	**0.0368**	**5.5945**
16	0.2919	8.8514	52.2640	3.4259	30.3243	0.1130	0.0330	5.9046
17	0.2703	9.1216	56.5883	3.7000	33.7502	0.1096	0.0296	6.2037
18	0.2502	9.3719	60.8426	3.9960	37.4502	0.1067	0.0267	6.4920
19	0.2317	9.6036	65.0134	4.3157	41.4463	0.1041	0.0241	6.7697
20	**0.2145**	**9.8181**	**69.0898**	**4.6610**	**45.7620**	**0.1019**	**0.0219**	**7.0369**
21	0.1987	10.0168	73.0629	5.0338	50.4229	0.0998	0.0198	7.2940
22	0.1839	10.2007	76.9257	5.4365	55.4568	0.0980	0.0180	7.5412
23	0.1703	10.3711	80.6726	5.8715	60.8933	0.0964	0.0164	7.7786
24	0.1577	10.5288	84.2997	6.3412	66.7648	0.0950	0.0150	8.0066
25	**0.1460**	**10.6748**	**87.8041**	**6.8485**	**73.1059**	**0.0937**	**0.0137**	**8.2254**
30	0.0994	11.2578	103.4558	10.0627	113.2832	0.0888	0.0088	9.1897
40	0.0460	11.9246	126.0422	21.7245	259.0565	0.0839	0.0039	10.5699
50	0.0213	12.2335	139.5928	46.9016	573.7702	0.0817	0.0017	11.4107
60	0.0099	12.3766	147.3000	101.2571	1,253.2133	0.0808	0.0008	11.9015
100	**0.0005**	**12.4943**	**155.6107**	**2,199.7613**	**27,484.5157**	**0.0800**		**12.4545**

Factor Table - $i = 10.00\%$

n	*P/F*	*P/A*	*P/G*	*F/P*	*F/A*	*A/P*	*A/F*	*A/G*
1	0.9091	0.9091	0.0000	1.1000	1.0000	1.1000	1.0000	0.0000
2	0.8264	1.7355	0.8264	1.2100	2.1000	0.5762	0.4762	0.4762
3	0.7513	2.4869	2.3291	1.3310	3.3100	0.4021	0.3021	0.9366
4	0.6830	3.1699	4.3781	1.4641	4.6410	0.3155	0.2155	1.3812
5	**0.6209**	**3.7908**	**6.8618**	**1.6105**	**6.1051**	**0.2638**	**0.1638**	**1.8101**
6	0.5645	4.3553	9.6842	1.7716	7.7156	0.2296	0.1296	2.2236
7	0.5132	4.8684	12.7631	1.9487	9.4872	0.2054	0.1054	2.6216
8	0.4665	5.3349	16.0287	2.1436	11.4359	0.1874	0.0874	3.0045
9	0.4241	5.7590	19.4215	2.3579	13.5735	0.1736	0.0736	3.3724
10	**0.3855**	**6.1446**	**22.8913**	**2.5937**	**15.9374**	**0.1627**	**0.0627**	**3.7255**
11	0.3505	6.4951	26.3962	2.8531	18.5312	0.1540	0.0540	4.0641
12	0.3186	6.8137	29.9012	3.1384	21.3843	0.1468	0.0468	4.3884
13	0.2897	7.1034	33.3772	3.4523	24.5227	0.1408	0.0408	4.6988
14	0.2633	7.3667	36.8005	3.7975	27.9750	0.1357	0.0357	4.9955
15	**0.2394**	**7.6061**	**40.1520**	**4.1772**	**31.7725**	**0.1315**	**0.0315**	**5.2789**
16	0.2176	7.8237	43.4164	4.5950	35.9497	0.1278	0.0278	5.5493
17	0.1978	8.0216	46.5819	5.0545	40.5447	0.1247	0.0247	5.8071
18	0.1799	8.2014	49.6395	5.5599	45.5992	0.1219	0.0219	6.0526
19	0.1635	8.3649	52.5827	6.1159	51.1591	0.1195	0.0195	6.2861
20	**0.1486**	**8.5136**	**55.4069**	**6.7275**	**57.2750**	**0.1175**	**0.0175**	**6.5081**
21	0.1351	8.6487	58.1095	7.4002	64.0025	0.1156	0.0156	6.7189
22	0.1228	8.7715	60.6893	8.1403	71.4027	0.1140	0.0140	6.9189
23	0.1117	8.8832	63.1462	8.9543	79.5430	0.1126	0.0126	7.1085
24	0.1015	8.9847	65.4813	9.8497	88.4973	0.1113	0.0113	7.2881
25	**0.0923**	**9.0770**	**67.6964**	**10.8347**	**98.3471**	**0.1102**	**0.0102**	**7.4580**
30	0.0573	9.4269	77.0766	17.4494	164.4940	0.1061	0.0061	8.1762
40	0.0221	9.7791	88.9525	45.2593	442.5926	0.1023	0.0023	9.0962
50	0.0085	9.9148	94.8889	117.3909	1,163.9085	0.1009	0.0009	9.5704
60	0.0033	9.9672	97.7010	304.4816	3,034.8164	0.1003	0.0003	9.8023
100	**0.0001**	**9.9993**	**99.9202**	**13,780.6123**	**137,796.1234**	**0.1000**		**9.9927**

Interest Rate Tables

Factor Table - $i = 12.00\%$

n	P/F	P/A	P/G	F/P	F/A	A/P	A/F	A/G
1	0.8929	0.8929	0.0000	1.1200	1.0000	1.1200	1.0000	0.0000
2	0.7972	1.6901	0.7972	1.2544	2.1200	0.5917	0.4717	0.4717
3	0.7118	2.4018	2.2208	1.4049	3.3744	0.4163	0.2963	0.9246
4	0.6355	3.0373	4.1273	1.5735	4.7793	0.3292	0.2092	1.3589
5	**0.5674**	**3.6048**	**6.3970**	**1.7623**	**6.3528**	**0.2774**	**0.1574**	**1.7746**
6	0.5066	4.1114	8.9302	1.9738	8.1152	0.2432	0.1232	2.1720
7	0.4523	4.5638	11.6443	2.2107	10.0890	0.2191	0.0991	2.5515
8	0.4039	4.9676	14.4714	2.4760	12.2997	0.2013	0.0813	2.9131
9	0.3606	5.3282	17.3563	2.7731	14.7757	0.1877	0.0677	3.2574
10	**0.3220**	**5.6502**	**20.2541**	**3.1058**	**17.5487**	**0.1770**	**0.0570**	**3.5847**
11	0.2875	5.9377	23.1288	3.4785	20.6546	0.1684	0.0484	3.8953
12	0.2567	6.1944	25.9523	3.8960	24.1331	0.1614	0.0414	4.1897
13	0.2292	6.4235	28.7024	4.3635	28.0291	0.1557	0.0357	4.4683
14	0.2046	6.6282	31.3624	4.8871	32.3926	0.1509	0.0309	4.7317
15	**0.1827**	**6.8109**	**33.9202**	**5.4736**	**37.2797**	**0.1468**	**0.0268**	**4.9803**
16	0.1631	6.9740	36.3670	6.1304	42.7533	0.1434	0.0234	5.2147
17	0.1456	7.1196	38.6973	6.8660	48.8837	0.1405	0.0205	5.4353
18	0.1300	7.2497	40.9080	7.6900	55.7497	0.1379	0.0179	5.6427
19	0.1161	7.3658	42.9979	8.6128	63.4397	0.1358	0.0158	5.8375
20	**0.1037**	**7.4694**	**44.9676**	**9.6463**	**72.0524**	**0.1339**	**0.0139**	**6.0202**
21	0.0926	7.5620	46.8188	10.8038	81.6987	0.1322	0.0122	6.1913
22	0.0826	7.6446	48.5543	12.1003	92.5026	0.1308	0.0108	6.3514
23	0.0738	7.7184	50.1776	13.5523	104.6029	0.1296	0.0096	6.5010
24	0.0659	7.7843	51.6929	15.1786	118.1552	0.1285	0.0085	6.6406
25	**0.0588**	**7.8431**	**53.1046**	**17.0001**	**133.3339**	**0.1275**	**0.0075**	**6.7708**
30	0.0334	8.0552	58.7821	29.9599	241.3327	0.1241	0.0041	7.2974
40	0.0107	8.2438	65.1159	93.0510	767.0914	0.1213	0.0013	7.8988
50	0.0035	8.3045	67.7624	289.0022	2,400.0182	0.1204	0.0004	8.1597
60	0.0011	8.3240	68.8100	897.5969	7,471.6411	0.1201	0.0001	8.2664
100		**8.3332**	**69.4336**	**83,522.2657**	**696,010.5477**	**0.1200**		**8.3321**

Factor Table - $i = 18.00\%$

n	P/F	P/A	P/G	F/P	F/A	A/P	A/F	A/G
1	0.8475	0.8475	0.0000	1.1800	1.0000	1.1800	1.0000	0.0000
2	0.7182	1.5656	0.7182	1.3924	2.1800	0.6387	0.4587	0.4587
3	0.6086	2.1743	1.9354	1.6430	3.5724	0.4599	0.2799	0.8902
4	0.5158	2.6901	3.4828	1.9388	5.2154	0.3717	0.1917	1.2947
5	**0.4371**	**3.1272**	**5.2312**	**2.2878**	**7.1542**	**0.3198**	**0.1398**	**1.6728**
6	0.3704	3.4976	7.0834	2.6996	9.4423	0.2859	0.1059	2.0252
7	0.3139	3.8115	8.9670	3.1855	12.1415	0.2624	0.0824	2.3526
8	0.2660	4.0776	10.8292	3.7589	15.3270	0.2452	0.0652	2.6558
9	0.2255	4.3030	12.6329	4.4355	19.0859	0.2324	0.0524	2.9358
10	**0.1911**	**4.4941**	**14.3525**	**5.2338**	**23.5213**	**0.2225**	**0.0425**	**3.1936**
11	0.1619	4.6560	15.9716	6.1759	28.7551	0.2148	0.0348	3.4303
12	0.1372	4.7932	17.4811	7.2876	34.9311	0.2086	0.0286	3.6470
13	0.1163	4.9095	18.8765	8.5994	42.2187	0.2037	0.0237	3.8449
14	0.0985	5.0081	20.1576	10.1472	50.8180	0.1997	0.0197	4.0250
15	**0.0835**	**5.0916**	**21.3269**	**11.9737**	**60.9653**	**0.1964**	**0.0164**	**4.1887**
16	0.0708	5.1624	22.3885	14.1290	72.9390	0.1937	0.0137	4.3369
17	0.0600	5.2223	23.3482	16.6722	87.0680	0.1915	0.0115	4.4708
18	0.0508	5.2732	24.2123	19.6731	103.7403	0.1896	0.0096	4.5916
19	0.0431	5.3162	24.9877	23.2144	123.4135	0.1881	0.0081	4.7003
20	**0.0365**	**5.3527**	**25.6813**	**27.3930**	**146.6280**	**0.1868**	**0.0068**	**4.7978**
21	0.0309	5.3837	26.3000	32.3238	174.0210	0.1857	0.0057	4.8851
22	0.0262	5.4099	26.8506	38.1421	206.3448	0.1848	0.0048	4.9632
23	0.0222	5.4321	27.3394	45.0076	244.4868	0.1841	0.0041	5.0329
24	0.0188	5.4509	27.7725	53.1090	289.4944	0.1835	0.0035	5.0950
25	**0.0159**	**5.4669**	**28.1555**	**62.6686**	**342.6035**	**0.1829**	**0.0029**	**5.1502**
30	0.0070	5.5168	29.4864	143.3706	790.9480	0.1813	0.0013	5.3448
40	0.0013	5.5482	30.5269	750.3783	4,163.2130	0.1802	0.0002	5.5022
50	0.0003	5.5541	30.7856	3,927.3569	21,813.0937	0.1800		5.5428
60	0.0001	5.5553	30.8465	20,555.1400	114,189.6665	0.1800		5.5526
100		**5.5556**	**30.8642**	**15,424,131.91**	**85,689,616.17**	**0.1800**		**5.5555**

CHEMICAL ENGINEERING

CHEMICAL REACTION ENGINEERING

Nomenclature

A chemical reaction may be expressed by the general equation

$$aA + bB \leftrightarrow cC + dD.$$

The rate of reaction of any component is defined as the moles of that component formed per unit time per unit volume.

$$-r_A = -\frac{1}{V}\frac{dN_A}{dt} \quad \text{(negative because A disappears)}$$

$$-r_A = \frac{-dC_A}{dt} \quad \text{if } V \text{ is constant}$$

The rate of reaction is frequently expressed by

$$-r_A = kf_r(C_A, C_B, \ldots), \text{ where}$$

k = reaction rate constant

C_I = concentration of component I

In the conversion of A, the fractional conversion X_A is defined as the moles of A reacted per mole of A fed.

$$X_A = (C_{A0} - C_A)/C_{A0} \quad \text{if } V \text{ is constant}$$

The Arrhenius equation gives the dependence of k on temperature

$$k = Ae^{-E_a/\bar{R}T}, \text{ where}$$

A = pre-exponential or frequency factor

E_a = activation energy (J/mol, cal/mol)

T = temperature (K)

$\bar{R}$ = gas law constant = 8.314 J/(mol•K)

For values of rate constant (k_i) at two temperatures (T_i),

$$E_a = \frac{RT_1T_2}{(T_1 - T_2)}\ln\left(\frac{k_1}{k_2}\right)$$

Reaction Order

If $-r_A = kC_A^xC_B^y$

the reaction is x order with respect to reactant A and y order with respect to reactant B. The overall order is

$$n = x + y$$

Batch Reactor, Constant Volume

For a well-mixed, constant-volume batch reactor

$$-r_A = -dC_A/dt$$

$$t = -C_{A0}\int_0^{X_A} dX_A/(-r_A)$$

Zero-Order Irreversible Reaction

$$-r_A = kC_A^0 = k(1)$$

$$-dC_A/dt = k \quad \text{or}$$

$$C_A = C_{A0} - kt$$

$$dX_A/dt = k/C_{A0} \quad \text{or}$$

$$C_{A0}X_A = kt$$

First-Order Irreversible Reaction

$$-r_A = kC_A$$

$$-dC_A/dt = kC_A \quad \text{or}$$

$$\ln(C_A/C_{A0}) = -kt$$

$$dX_A/dt = k(1 - X_A) \quad \text{or}$$

$$\ln(1 - X_A) = -kt$$

Second-Order Irreversible Reaction

$$-r_A = kC_A^2$$

$$-dC_A/dt = kC_A^2 \quad \text{or}$$

$$1/C_A - 1/C_{A0} = kt$$

$$dX_A/dt = kC_{A0}(1 - X_A)^2 \quad \text{or}$$

$$X_A/[C_{A0}(1 - X_A)] = kt$$

First-Order Reversible Reactions

$$A \underset{k_2}{\overset{k_1}{\rightleftarrows}} R$$

$$-r_A = -\frac{dC_A}{dt} = k_1C_A - k_2C_R$$

$$K_c = k_1/k_2 = \hat{C}_R/\hat{C}_A$$

$$M = C_{R_0}/C_{A_0}$$

$$\frac{dX_A}{dt} = \frac{k_1(M+1)}{M+\hat{X}_A}(\hat{X}_A - X_A)$$

$$-\ln\left(1 - \frac{X_A}{\hat{X}_A}\right) = -\ln\frac{C_A - \hat{C}_A}{C_{A_0} - \hat{C}_A}$$

$$= \frac{(M+1)}{(M+\hat{X}_A)}k_1t$$

$\hat{X}_A$ is the equilibrium conversion.

Reactions of Shifting Order

$$-r_A = \frac{k_1 C_A}{1 + k_2 C_A}$$

$$\ln\left(\frac{C_{A_o}}{C_A}\right) + k_2\left(C_{A_o} - C_A\right) = k_1 t$$

$$\frac{\ln\left(C_{A_o}/C_A\right)}{C_{A_o} - C_A} = -k_2 + \frac{k_1 t}{C_{A_o} - C_A}$$

This form of the rate equation is used for elementary enzyme-catalyzed reactions and for elementary surfaced-catalyzed reactions.

Batch Reactor, Variable Volume

If the volume of the reacting mass varies with the conversion (such as a variable-volume batch reactor) according to

$$V = V_{X_A = 0}\left(1 + \varepsilon_A X_A\right)$$

(i.e., at constant pressure), where

$$\varepsilon_A = \frac{V_{X_A = 1} - V_{X_A = 0}}{V_{X_A = 0}} = \frac{\Delta V}{V_{X_A = 0}}$$

then at any time

$$C_A = C_{A0}\left[\frac{1 - X_A}{1 + \varepsilon_A X_A}\right]$$

and

$$t = -C_{A0}\int_0^{X_A} dX_A/\left[\left(1 + \varepsilon_A X_A\right)\left(-r_A\right)\right]$$

For a first-order irreversible reaction,

$$kt = -\ln\left(1 - X_A\right) = -\ln\left(1 - \frac{\Delta V}{\varepsilon_A V_{XA = 0}}\right)$$

Flow Reactors, Steady State

Space-time τ is defined as the reactor volume divided by the inlet volumetric feed rate. Space-velocity SV is the reciprocal of space-time, $SV = 1/\tau$.

Plug-Flow Reactor (PFR)

$$\tau = \frac{C_{A0} V_{PFR}}{F_{A0}} = C_{A0}\int_0^{X_A} \frac{dX_A}{\left(-r_A\right)}, \text{ where}$$

F_{A0} = moles of A fed per unit time.

Continuous-Stirred Tank Reactor (CSTR)

For a constant-volume, well-mixed CSTR

$$\frac{\tau}{C_{A0}} = \frac{V_{CSTR}}{F_{A0}} = \frac{X_A}{-r_A}, \text{ where}$$

$-r_A$ is evaluated at exit stream conditions.

Continuous-Stirred Tank Reactors in Series

With a first-order reaction $A \rightarrow R$, no change in volume.

$$\tau_{N\text{-reactors}} = N\tau_{\text{individual}}$$

$$= \frac{N}{k}\left[\left(\frac{C_{A0}}{C_{AN}}\right)^{1/N} - 1\right], \text{ where}$$

N = number of CSTRs (equal volume) in series, and

C_{AN} = concentration of A leaving the Nth CSTR.

Two Irreversible Reactions in Parallel

$$A \overset{k_D}{\rightarrow} D(\text{desired})$$

$$A \overset{k_U}{\rightarrow} U(\text{undesired})$$

$$-r_A = -dc_A/dt = k_D C_A^x + k_U C_A^y$$

$$r_D = dc_D/dt = k_D C_A^x$$

$$r_U = dc_U/dt = k_U C_A^y$$

Y_D = instantaneous fractional yield of D

$$= dC_D/\left(-dC_A\right)$$

$\overline{Y}_D$ = overall fractional yield of D

$$= N_{D_f}/\left(N_{A_0} - N_{A_f}\right)$$

where N_{A_f} and N_{D_f} are measured at the outlet of the flow reactor.

$\overline{S}_{DU}$ = overall selectivity to D

$$= N_{D_f}/N_{U_f}$$

Two First-Order Irreversible Reactions in Series

$$A \overset{k_D}{\rightarrow} D \overset{k_U}{\rightarrow} U$$

$$r_A = -dC_A/dt = k_D C_A$$

$$r_D = dC_D/dt = k_D C_A - k_U C_D$$

$$r_U = dC_U/dt = k_U C_D$$

Yield and selectivity definitions are identical to those for two irreversible reactions in parallel. The optimal yield of D in a PFR is

$$\frac{C_{D,max}}{C_{A_0}} = \left(\frac{k_D}{k_U}\right)^{k_U/\left(k_U - k_D\right)}$$

at time

$$\tau_{\max} = \frac{1}{k_{\text{log mean}}} = \frac{\ln\left(k_U/k_D\right)}{\left(k_U - k_D\right)}$$

The optimal yield of D in a CSTR is

$$\frac{C_{D,\max}}{C_{A_0}} = \frac{1}{\left[\left(k_U/k_D\right)^{1/2} + 1\right]^2}$$

at time

$$\tau_{\max} = 1/\sqrt{k_D k_U}$$

MASS TRANSFER

Diffusion

Molecular Diffusion

$$\text{Gas: } N_A = \frac{p_A}{P}\left(N_A + N_B\right) - \frac{D_m}{\overline{R}T}\frac{\partial p_A}{\partial z}$$

$$\text{Liquid: } N_A = x_A\left(N_A + N_B\right) - CD_m\frac{\partial x_A}{\partial z}$$

where,

N_i = molar flux of component i

P = pressure

p_i = partial pressure of component i

D_m = mass diffusivity

$\overline{R}$ = universal gas constant

T = temperature

z = length

Unidirectional Diffusion of a Gas A Through a Second Stagnant Gas B (N_b= 0)

$$N_A = \frac{D_m P}{\overline{R}T\left(p_B\right)_{lm}} \times \frac{\left(p_{A2} - p_{A1}\right)}{z_2 - z_1}$$

in which $(p_B)_{lm}$ is the log mean of p_{B2} and p_{B1}

$$\left(p_{BM}\right)_{lm} = \frac{p_{B2} - p_{B1}}{\ln\left(\frac{p_{B2}}{p_{B1}}\right)}$$

N_i = diffusive flux [mole/(time × area)] of component i through area A, in z direction

D_m = mass diffusivity

p_I = partial pressure of species I

C = concentration (mole/volume)

$(z_2 - z_1)$ = diffusion flow path length

Equimolar Counter-Diffusion (Gases)

($N_B = -N_A$)

$$N_A = D_m/(\overline{R}T) \times \left[(p_{A1} - p_{A2})/(\Delta z)\right]$$

$$N_A = D_m(C_{A1} - C_{A2})/\Delta z$$

Convection

Two-Film Theory (for Equimolar Counter-Diffusion)

$$\begin{aligned} N_A &= k'_G(p_{AG} - p_{Ai}) \\ &= k'_L(C_{Ai} - C_{AL}) \\ &= K'_G(p_{AG} - p_A{}^*) \\ &= K'_L(C_A{}^* - C_{AL}) \end{aligned}$$

where,

N_A = molar flux of component A

k'_G = gas phase mass-transfer coefficient

k'_L = liquid phase mass-transfer coefficient

K'_G = overall gas phase mass-transfer coefficient

K'_L = overall liquid phase mass-transfer coefficient

p_{AG} = partial pressure in component A in the bulk gas phase

p_{Ai} = partial pressure at component A at the gas-liquid interface

C_{Ai} = concentration (mole/volume) of component A in the liquid phase at the gas-liquid interface

C_{AL} = concentration of component A in the bulk liquid phase

$p_A{}^*$ = partial pressure of component A in equilibrium with C_{AL}

$C_A{}^*$ = concentration of component A in equilibrium with the bulk gas vapor composition of A

Overall Coefficients

$$1/K'_G = 1/k'_G + H/k'_L$$

$$1/K'_L = 1/Hk'_G + 1/k'_L$$

H = Henry's Law constant where $p_A{}^* = H\,C_{AL}$ and $C_A{}^* = p_{AG}/H$

Dimensionless Group Equation (Sherwood)

For the turbulent flow inside a tube the Sherwood number

$$\text{Sh} = \left(\frac{k_m D}{D_m}\right) = 0.023\left(\frac{DV\rho}{\mu}\right)^{0.8}\left(\frac{\mu}{\rho D_m}\right)^{1/3}$$

where,

D = inside diameter

D_m = diffusion coefficient

V = average velocity in the tube

ρ = fluid density

μ = fluid viscosity

k_m = mass-transfer coefficient

Distillation

Definitions:

α = relative volatility

B = molar bottoms-product rate

D = molar overhead-product rate

F = molar feed rate

L = molar liquid downflow rate

R_D = ratio of reflux to overhead product

V = molar vapor upflow rate

W = total moles in still pot

x = mole fraction of the more volatile component in the liquid phase

y = mole fraction of the more volatile component in the vapor phase

Subscripts:

B = bottoms product

D = overhead product

F = feed

m = any plate in stripping section of column

$m+1$ = plate below plate m

n = any plate in rectifying section of column

$n+1$ = plate below plate n

o = original charge in still pot

Flash (or equilibrium) Distillation

Component material balance:

$$Fz_F = yV + xL$$

Overall material balance:

$$F = V + L$$

Differential (Simple or Rayleigh) Distillation

$$\ln\left(\frac{W}{W_o}\right) = \int_{x_o}^{x} \frac{dx}{y - x}$$

When the relative volatility α is constant,

$$y = \alpha x/[1 + (\alpha - 1)x]$$

can be substituted to give

$$\ln\left(\frac{W}{W_o}\right) = \frac{1}{(\alpha - 1)}\ln\left[\frac{x(1 - x_o)}{x_o(1 - x)}\right] + \ln\left[\frac{1 - x_o}{1 - x}\right]$$

For binary system following Raoult's Law

$$\alpha = (y/x)_a/(y/x)_b = p_a/p_b, \text{ where}$$

p_i = partial pressure of component i.

Continuous Distillation (Binary System)

Constant molal overflow is assumed.

Equilibrium stages numbered from top.

Overall Material Balances

Total Material:

$$F = D + B$$

Component A:

$$Fz_F = Dx_D + Bx_B$$

Operating Lines

Rectifying section

Total Material:

$$V_{n+1} = L_n + D$$

Component A:

$$V_{n+1}y_{n+1} = L_n x_n + Dx_D$$

$$y_{n+1} = [L_n/(L_n + D)]\, x_n + Dx_D/(L_n + D)$$

Stripping section

Total Material:

$$L_m = V_{m+1} + B$$

Component A:

$$L_m x_m = V_{m+1}y_{m+1} + Bx_B$$

$$y_{m+1} = [L_m/(L_m - B)]\, x_m - Bx_B/(L_m - B)$$

Reflux ratio

Ratio of reflux to overhead product

$$R_D = L_R/D = (V_R - D)/D$$

Minimum reflux ratio is defined as that value which results in an infinite number of contact stages. For a binary system the equation of the operating line is

$$y = \frac{R_{\min}}{R_{\min} + 1}x + \frac{x_D}{R_{\min} + 1}$$

Feed condition line

slope = $q/(q - 1)$, where

$$q = \frac{\text{heat to convert one mol of feed to saturated vapor}}{\text{molar heat of vaporization}}$$

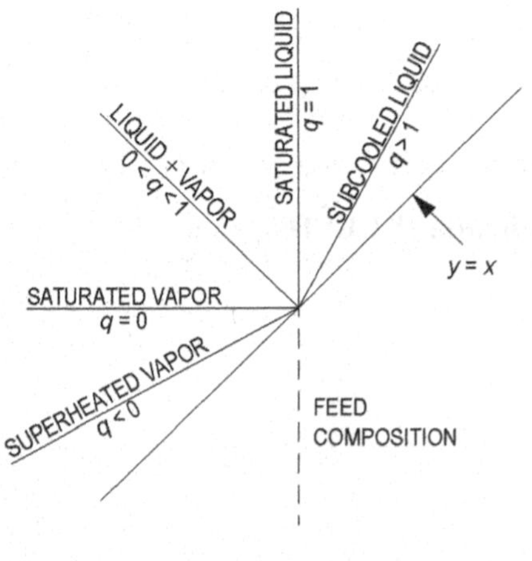

q-LINE SLOPES

Murphree plate efficiency

$$E_{ME} = (y_n - y_{n+1})/(y_n^* - y_{n+1}), \text{ where}$$

y_n = concentration of vapor above equilibrium stage n

y_{n+1} = concentration of vapor entering from equilibrium stage below n

y_n^* = concentration of vapor in equilibrium with liquid leaving equilibrium stage n

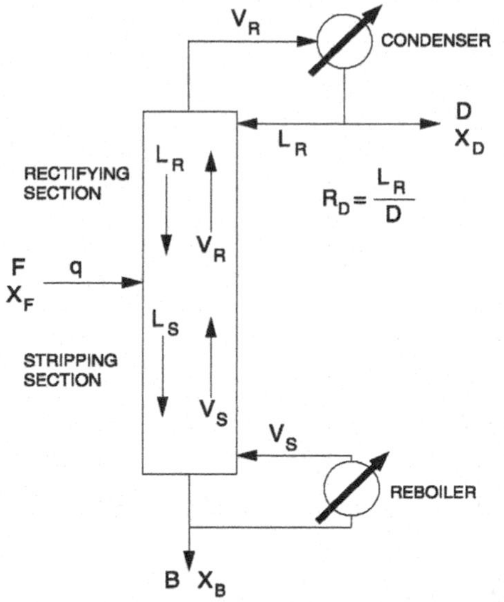

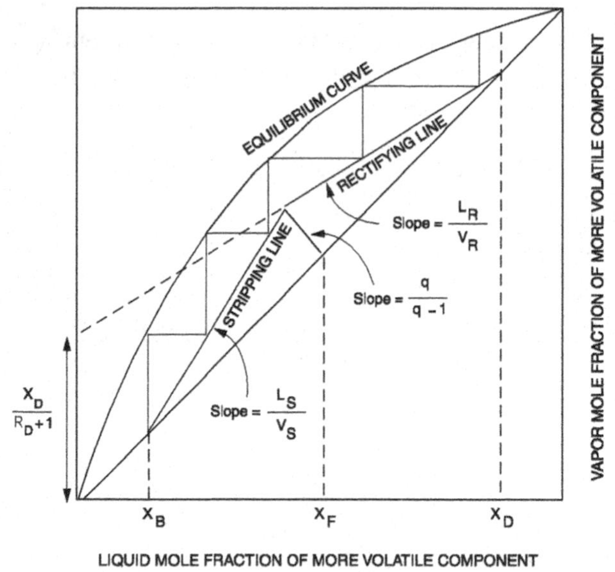

Vapor-Liquid Equilibrium (VLE) Diagram

Absorption (packed columns)

Continuous Contact Columns

$$Z = NTU_G \bullet HTU_G = NTU_L \bullet HTU_L = N_{EQ} \bullet HETP$$

Z = column height

NTU_G = number of transfer units (gas phase)

NTU_L = number of transfer units (liquid phase)

N_{EQ} = number of equilibrium stages

HTU_G = height of transfer unit (gas phase)

HTU_L = height of transfer unit (liquid phase)

$HETP$ = height equivalent to theoretical plate (stage)

$$HTU_G = \frac{G}{K'_G a} \qquad HTU_L = \frac{L}{K'_L a}$$

G = gas phase mass velocity (mass or moles/flow area • time)

L = liquid phase mass velocity (mass or moles/flow area • time)

K'_G = overall gas phase mass-transfer coefficient (mass or moles/mass-transfer area • time)

K'_L = overall liquid phase mass-transfer coefficient (mass or moles/mass-transfer area • time)

a = mass-transfer area/volume of column (length^{-1})

$$NTU_G = \int_{y_1}^{y_2} \frac{dy}{(y - y^*)} \qquad NTU_L = \int_{x_1}^{x_2} \frac{dx}{(x^* - x)}$$

y = gas phase solute mole fraction

x = liquid phase solute mole fraction

y^* = K • x, where K = equilibrium constant

x^* = y/K, where K = equilibrium constant

y_2, x_2 = mole fractions at the lean end of column

y_1, x_1 = mole fractions at the rich end of column

For dilute solutions (constant G/L and constant K value for entire column):

$$NTU_G = \frac{y_1 - y_2}{(y - y^*)_{LM}}$$

$$(y - y^*)_{LM} = \frac{(y_1 - y_1^*) - (y_2 - y_2^*)}{\ln\left(\frac{y_1 - y_1^*}{y_2 - y_2^*}\right)}$$

For a chemically reacting system—absorbed solute reacts in the liquid phase—the preceding relation simplifies to:

$$NTU_G = \ln\left(\frac{y_1}{y_2}\right)$$

TRANSPORT PHENOMENA–MOMENTUM, HEAT, AND MASS-TRANSFER ANALOGY

For the equations which apply to *turbulent flow in circular tubes*, the following definitions apply:

Nu = Nusselt Number $\left[\frac{hD}{k}\right]$

Pr = Prandtl Number $(c_p\mu/k)$

Re = Reynolds Number $(DV\rho/\mu)$

Sc = Schmidt Number $[\mu/(\rho D_m)]$

Sh = Sherwood Number $(k_m D/D_m)$

St = Stanton Number $[h/(c_p G)]$

c_m = concentration (mol/m^3)

c_p = heat capacity of fluid [J/(kg•K)]

D = tube inside diameter (m)

D_m = diffusion coefficient (m^2/s)

$(dc_m/dy)_w$ = concentration gradient at the wall (mol/m^4)

$(dT/dy)_w$ = temperature gradient at the wall (K/m)

$(dv/dy)_w$ = velocity gradient at the wall (s^{-1})

f = Moody friction factor

G = mass velocity [kg/(m^2•s)]

h = heat-transfer coefficient at the wall [W/(m^2•K)]

k = thermal conductivity of fluid [W/(m•K)]

k_m = mass-transfer coefficient (m/s)

L = length over which pressure drop occurs (m)

$(N/A)_w$ = inward mass-transfer flux at the wall [mol/(m^2•s)]

$(\dot{Q}/A)_w$ = inward heat-transfer flux at the wall (W/m^2)

y = distance measured from inner wall toward centerline (m)

Δc_m = concentration difference between wall and bulk fluid (mol/m^3)

ΔT = temperature difference between wall and bulk fluid (K)

μ = absolute dynamic viscosity (N•s/m^2)

τ_w = shear stress (momentum flux) at the tube wall (N/m^2)

Definitions already introduced also apply.

Rate of Transfer as a Function of Gradients at the Wall

Momentum Transfer

$$\tau_w = -\mu\left(\frac{dv}{dy}\right)_w = -\frac{f\rho V^2}{8} = \left(\frac{D}{4}\right)\left(-\frac{\Delta p}{L}\right)_f$$

Heat Transfer

$$\left(\frac{\dot{Q}}{A}\right)_w = -k\left(\frac{dT}{dy}\right)_w$$

Mass Transfer in Dilute Solutions

$$\left(\frac{N}{A}\right)_w = -D_m\left(\frac{dc_m}{dy}\right)_w$$

<u>Rate of Transfer in Terms of Coefficients</u>

Momentum Transfer

$$\tau_w = \frac{f\rho V^2}{8}$$

Heat Transfer

$$\left(\frac{\dot{Q}}{A}\right)_w = h\Delta T$$

Mass Transfer

$$\left(\frac{N}{A}\right)_w = k_m \Delta c_m$$

<u>Use of Friction Factor (f) to Predict Heat-Transfer and Mass-Transfer Coefficients (Turbulent Flow)</u>

Heat Transfer

$$j_H = \left(\frac{\text{Nu}}{\text{RePr}}\right)\text{Pr}^{2/3} = \frac{f}{8}$$

Mass Transfer

$$j_M = \left(\frac{\text{Sh}}{\text{ReSc}}\right)\text{Sc}^{2/3} = \frac{f}{8}$$

COST ESTIMATION

Cost Indexes

Cost indexes are used to update historical cost data to the present. If a purchase cost is available for an item of equipment in year M, the equivalent current cost would be found by:

$$\text{Current \$} = (\text{Cost in year } M)\left(\frac{\text{Current Index}}{\text{Index in year } M}\right)$$

Capital Cost Estimation

	Lang factors	
Type of plant	**Fixed capital investment**	**Total capital investment**
Solid processing	4.0	4.7
Solid-fluid processing	4.3	5.0
Fluid processing	5.0	6.0

From Green, Don W., and Robert H. Perry, *Perry's Chemical Engineers' Handbook,* 8th ed., McGraw-Hill, 2008.
Adapted from M. S. Peters, K. D. Timmerhaus, and R. West, *Plant Design and Economics for Chemical Engineers*, 5th ed., McGraw-Hill, 2004.

Component	Range
Direct costs	
Purchased equipment-delivered (including fabricated equipment and process machinery such as pumps and compressors)	100
Purchased-equipment installation	39–47
Instrumentation and controls (installed)	9–18
Piping (installed)	16–66
Electrical (installed)	10–11
Buildings (including services)	18–29
Yard improvements	10–13
Service facilities (installed)	40–70
Land (if purchase is required)	6
Total direct plant cost	264–346
Indirect costs	
Engineering and supervision	32–33
Construction expenses	34–41
Total direct and indirect plant costs	336–420
Contractor's fee (about 5% of direct and indirect plant costs)	17–21
Contingency (about 10% of direct and indirect plant costs)	36–42
Fixed-capital investment	387–483
Working capital (about 15% of total capital investment)	68–86
Total capital investment	455–569

Scaling of Equipment Costs

The cost of Unit A at one capacity related to the cost of a similar Unit B with X times the capacity of Unit A is approximately X^n times the cost of Unit B.

$$\text{Cost of Unit A} = \text{Cost of Unit B}\left(\frac{\text{Capacity of Unit A}}{\text{Capacity of Unit B}}\right)^n$$

Typical Exponents (*n*) for Equipment Cost vs. Capacity

Equipment	Size range	Exponent
Dryer, drum, single vacuum	10 to 10^2 ft^2	0.76
Dryer, drum, single atmospheric	10 to 10^2 ft^2	0.40
Fan, centrifugal	10^3 to 10^4 ft^3/min	0.44
Fan, centrifugal	2×10^4 to 7×10^4 ft^3/min	1.17
Heat exchanger, shell and tube, floating head, c.s.	100 to 400 ft^2	0.60
Heat exchanger, shell and tube, fixed sheet, c.s.	100 to 400 ft^2	0.44
Motor, squirrel cage, induction, 440 volts, explosion proof	5 to 20 hp	0.69
Motor, squirrel cage, induction, 440 volts, explosion proof	20 to 200 hp	0.99
Tray, bubble cup, c.s.	3- to 10-ft diameter	1.20
Tray, sieve, c.s.	3- to 10-ft diameter	0.86

CIVIL ENGINEERING

GEOTECHNICAL

Phase Relationships

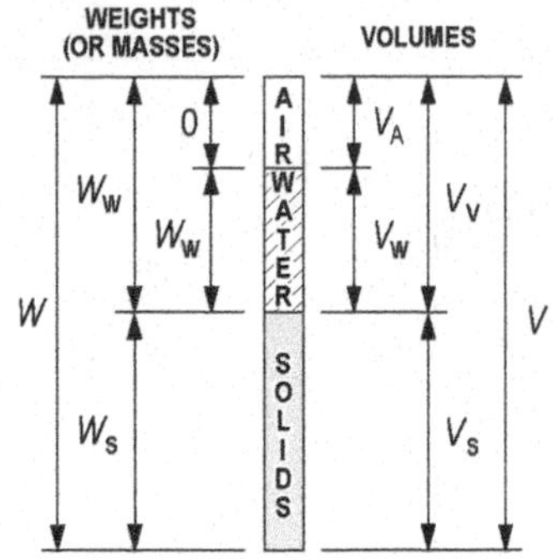

PHASE DIAGRAM

Volume of voids
$V_V = V_A + V_W$

Total unit weight
$\gamma = W/V$

Saturated unit weight
$\gamma_{sat} = (G_s + e)\,\gamma_w/(1 + e)$
$\gamma_W = 62.4\ \text{lb/ft}^3$ or $9.81\ \text{kN/m}^3$

Effective (submerged) unit weight
$\gamma' = \gamma_{sat} - \gamma_W$

Unit weight of solids
$\gamma_S = W_S/V_S$

Dry unit weight
$\gamma_D = W_S/V$

Water content (%)
$\omega = (W_W/W_S) \times 100$

Specific gravity of soil solids
$G_S = (W_S/V_S)/\gamma_W$

Void ratio
$e = V_V/V_S$

Porosity
$n = V_V/V = e/(1+e)$

Degree of saturation (%)
$S = (V_W/V_V) \times 100$
$S = \omega\, G_S/e$

Relative density
$D_r = [(e_{max} - e)/(e_{max} - e_{min})] \times 100$
$= [(\gamma_{D\ field} - \gamma_{D\ min})/(\gamma_{D\ max} - \gamma_{D\ min})][\gamma_{D\ max}/\gamma_{D\ field}] \times 100$

Relative compaction (%)
$RC = (\gamma_{D\ field}/\gamma_{D\ max}) \times 100$

Plasticity index
PI = LL – PL
LL= liquid limit
PL = Plastic limit

Coefficient of uniformity
$C_U = D_{60}/D_{10}$

Coefficient of concavity (or curvature)
$C_C = (D_{30})^2/(D_{10} \times D_{60})$

Hydraulic conductivity (also coefficient of permeability)
From constant head test: $k = Q/(iAt_e)$
$i = dh/dL$
Q = total quantity of water

From falling head test: $k = 2.303[(aL)/(At_e)]\log_{10}(h_1/h_2)$
A = cross-sectional area of test specimen perpendicular to flow
a = cross-sectional area of reservoir tube
t_e = elapsed time
h_1 = head at time $t = 0$
h_2 = head at time $t = t_e$
L = length of soil column
Discharge velocity, $v = ki$

Flow Nets

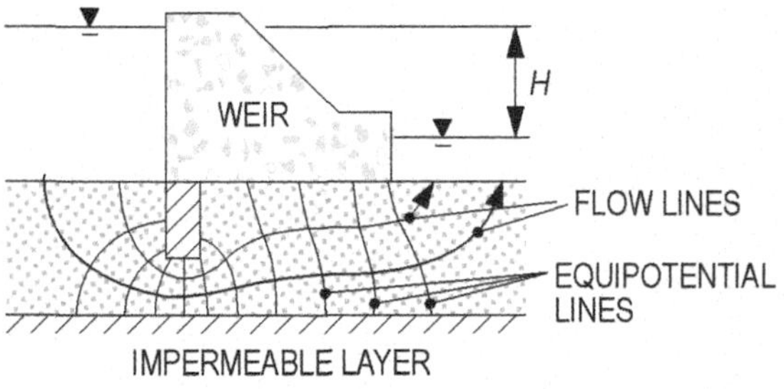

FLOW NET

$Q = kH\,(N_f/N_d)$, where
Q = flow per unit time
N_f = number of flow channels
N_d = number of equipotential drops
H = total hydraulic head differential

Factor of safety against seepage liquefaction
$FS_s = i_c/i_e$
$i_c = (\gamma_{sat} - \gamma_W)/\gamma_W$
i_e = seepage exit gradient

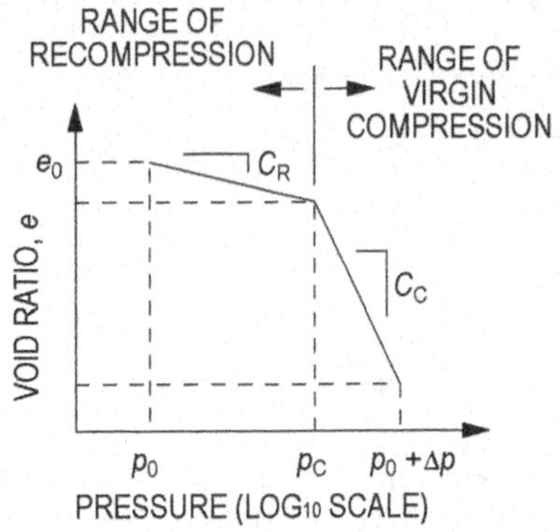

SOIL CONSOLIDATION CURVE

e_0 = initial void ratio (prior to consolidation)
Δe = change in void ratio
p_0 = initial effective consolidation stress, σ'_0
p_c = past maximum consolidation stress, σ'_c
Δp = induced change in consolidation stress at center of consolidating stratum
$\Delta p = I\, q_s$
I = Stress influence value at center of consolidating stratum
q_s = applied surface stress causing consolidation

If $(p_o \text{ and } p_o + \Delta p) < p_c$, then $\Delta H = \frac{H_o}{1+e_o}\left[C_R \log \frac{p_o + \Delta p}{p_o}\right]$

If $(p_o \text{ and } p_o + \Delta p) > p_c$, then $\Delta H = \frac{H_o}{1+e_o}\left[C_C \log \frac{p_o + \Delta p}{p_o}\right]$

If $p_o < p_c < (p_o + \Delta p)$, then $\Delta H = \frac{H_o}{1+e_o}\left[C_R \log \frac{p_c}{p_o} + C_C \log \frac{p_o + \Delta p}{p_c}\right]$

where: ΔH = change in thickness of soil layer

Compression index
In virgin compression range: $C_C = \Delta e / \Delta \log p$
By correlation to liquid limit: $C_C = 0.009\ (\text{LL} - 10)$

Recompression index
In recompression range: $C_R = \Delta e / \Delta \log p$
By correlation to compression index, C_C: $C_R = C_C/6$

Ultimate consolidation settlement in soil layer
$S_{ULT} = \varepsilon_v H_S$
H_S = thickness of soil layer
$\varepsilon_v = \Delta e_{TOT}/(1 + e_0)$
Δe_{TOT} = total change in void ratio due to recompression and virgin compression

Approximate settlement (at time $t = t_C$)
$S_T = U_{AV} S_{ULT}$
U_{AV} = average degree of consolidation
t_C = elapsed time since application of consolidation load

Variation of time factor with degree of consolidation*

U (%)	T_v	U (%)	T_v	U (%)	T_v
0	0	34	0.0907	68	0.377
1	0.00008	35	0.0962	69	0.390
2	0.0003	36	0.102	70	0.403
3	0.00071	37	0.107	71	0.417
4	0.00126	38	0.113	72	0.431
5	0.00196	39	0.119	73	0.446
6	0.00283	40	0.126	74	0.461
7	0.00385	41	0.132	75	0.477
8	0.00502	42	0.138	76	0.493
9	0.00636	43	0.145	77	0.511
10	0.00785	44	0.152	78	0.529
11	0.0095	45	0.159	79	0.547
12	0.0113	46	0.166	80	0.567
13	0.0133	47	0.173	81	0.588
14	0.0154	48	0.181	82	0.610
15	0.0177	49	0.188	83	0.633
16	0.0201	50	0.197	84	0.658
17	0.0227	51	0.204	85	0.684
18	0.0254	52	0.212	86	0.712
19	0.0283	53	0.221	87	0.742
20	0.0314	54	0.230	88	0.774
21	0.0346	55	0.239	89	0.809
22	0.0380	56	0.248	90	0.848
23	0.0415	57	0.257	91	0.891
24	0.0452	58	0.267	92	0.938
25	0.0491	59	0.276	93	0.993
26	0.0531	60	0.286	94	1.055
27	0.0572	61	0.297	95	1.129
28	0.0615	62	0.307	96	1.219
29	0.0660	63	0.318	97	1.336
30	0.0707	64	0.329	98	1.500
31	0.0754	65	0.340	99	1.781
32	0.0803	66	0.352	100	∞
33	0.0855	67	0.364		

*u_0 constant with depth.

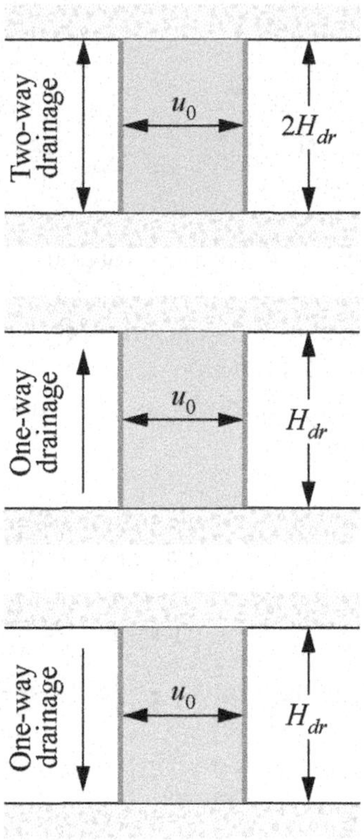

Different types of drainage with u_0 constant

where time factor is $T_v = \dfrac{c_v t}{H_{dr}^2}$

Das, Braja M., *Fundamentals of Geotechnical Engineering*, Cengage Learning (formerly Brooks/Cole), 2000.

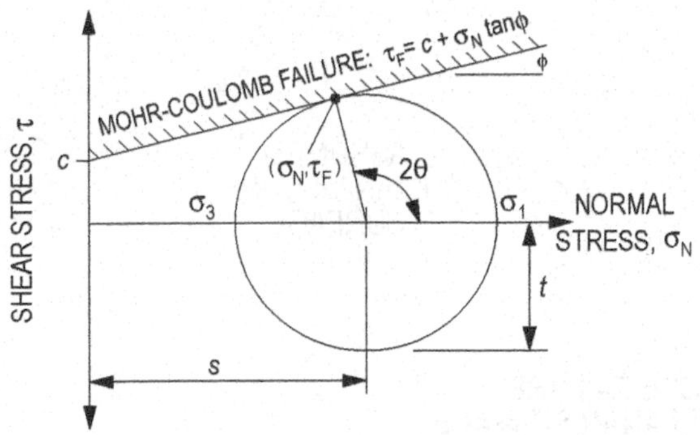

s = mean normal stress
t = maximum shear stress
σ_1 = major principal stress
σ_3 = minor principal stress
θ = orientation angle between plane of existing normal stress and plane of major principal stress

Total normal stress
$\sigma_N = P/A$
P = normal force
A = cross-sectional area over which force acts

Effective stress
$\sigma' = \sigma - u$
$u = h_u \gamma_w$
h_u = uplift or pressure head

Shear stress
$\tau = T/A$
T = shearing force

Shear stress at failure
$\tau_F = c + \sigma_N \tan\phi$
c = cohesion
ϕ = angle of internal friction

Horizontal Stress Profiles and Forces

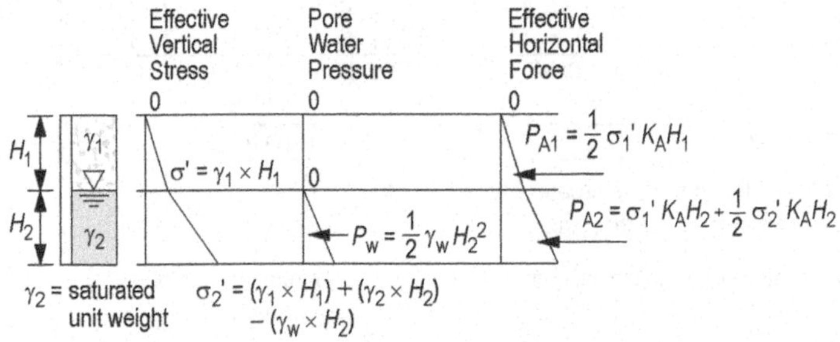

Active forces on retaining wall per unit wall length (as shown):
K_A = Rankine active earth pressure coefficient (smooth wall, $c = 0$, level backfill) = $\tan^2(45° - \phi/2)$

Passive forces on retaining wall per unit wall length (similar to the active forces shown):

K_P = Rankine passive earth pressure coefficient (smooth wall, $c = 0$, level backfill) = $\tan^2(45° + \phi/2)$

At rest forces on wall per unit length of wall

K_0 = at rest earth pressure coefficient (smooth wall, C = 0, level backfill)

$K_0 \approx 1 - \sin\phi$ for normally consolidated soil

$K_0 = (1 - \sin\phi)\,OCR^{\sin\phi}$ for overconsolidated soil where

OCR = overconsolidation ratio

Vertical Stress Profiles

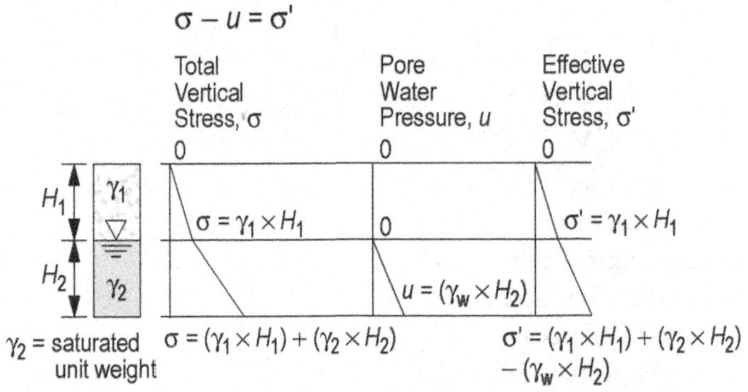

Vertical Stress Profiles with Surcharge

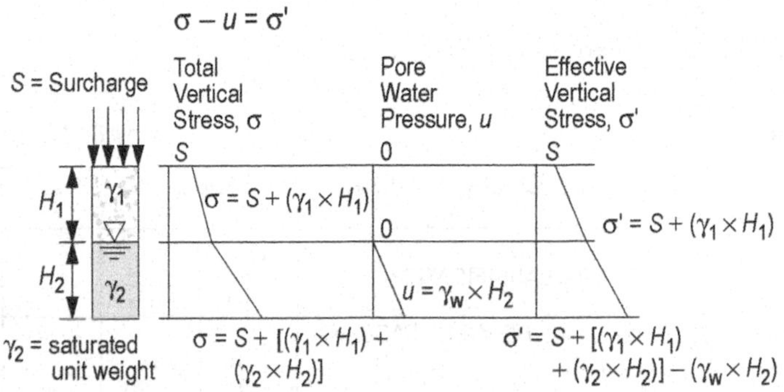

Ultimate Bearing Capacity

$$q_{ULT} = cN_c + \gamma' D_f N_q + \frac{1}{2}\gamma' B N_\gamma$$

N_c = bearing capacity factor for cohesion
N_q = bearing capacity factor for depth
N_γ = bearing capacity factor for unit weight
D_f = depth of footing below ground surface
B = width of strip footing

Retaining Walls

$$FS_{\text{overturning}} = \frac{\Sigma M_R}{M_O}$$

$$FS_{\text{sliding}} = \frac{\Sigma F_R}{\Sigma F_D}$$

$$FS_{\text{sliding}} = \frac{(\Sigma V)\tan\delta + BC_a + P_p}{P_a\cos\alpha}$$

$$FS_{\text{bearing capacity}} = \frac{q_{ULT}}{q_{toe}}$$

$$q_{toe} = \frac{\Sigma V}{B}\left(1 + \frac{6e}{B}\right)$$

$$e = \frac{B}{2} - \left(\frac{\Sigma M_R - M_O}{\Sigma V}\right)$$

e = eccentricity
B = width of base
M_R = resisting moment
M_O = overturning moment
F_R = resisting forces
F_D = driving forces
V = vertical forces
$\delta = k_1\phi_2$
$C_a = k_2C_2$
k_1 and k_2 are given, ranging from 1/2 to 2/3

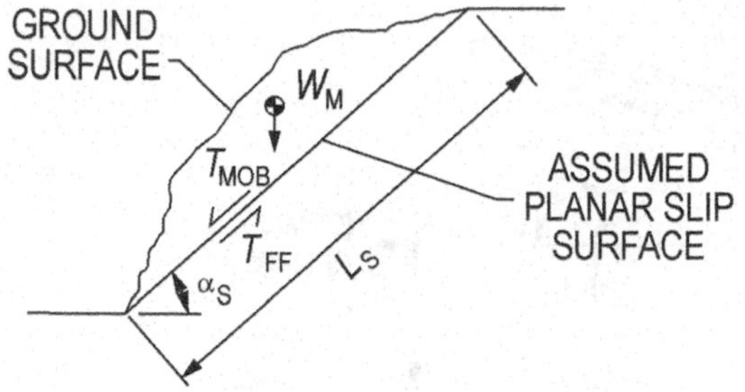

SLOPE FAILURE ALONG PLANAR SURFACE

FS = factor of safety against slope instability
= T_{FF}/T_{MOB}
T_{FF} = available shearing resistance along slip surface
= $cL_S + W_M\cos\alpha_S\tan\phi$
T_{MOB} = mobilized shear force along slip surface
= $W_M\sin\alpha_S$
L_S = length of assumed planar slip surface
W_M = weight of soil above slip surface
α_S = angle of assumed slip surface with respect to horizontal

♦ **AASHTO Soil Classification**

GENERAL CLASSIFICATION	GRANULAR MATERIALS (35% OR LESS PASSING 0.075 SIEVE)							SILT-CLAY MATERIALS (MORE THAN 35% PASSING 0.075 SIEVE)			
GROUP CLASSIFICATION	A-1		A-3	A-2				A-4	A-5	A-6	A-7-5 A-7-6
	A-1-a	A-1-b		A-2-4	A-2-5	A-2-6	A-2-7				
SIEVE ANALYSIS, PERCENT PASSING:											
2.00 mm (No. 10)	≤ 50	–	–	–	–	–	–	–	–	–	–
0.425 mm (No. 40)	≤ 30	≤ 50	≥ 51	–	–	–	–	–	–	–	–
0.075 mm (No. 200)	≤ 15	≤ 25	≤ 10	≤ 35	≤ 35	≤ 35	≤ 35	≥ 36	≥ 36	≥ 36	≥ 36
CHARACTERISTICS OF FRACTION PASSING 0.425 SIEVE (No. 40):											
LIQUID LIMIT	–		–	≤ 40	≥ 41	≤ 40	≥ 41	≤ 40	≥ 41	≤ 40	≥ 41
PLASTICITY INDEX *	6 max		NP	≤ 10	≤ 10	≥ 11	≥ 11	≤ 10	≤ 10	≥ 11	≥ 11
USUAL TYPES OF CONSTITUENT MATERIALS	STONE FRAGM'TS, GRAVEL, SAND		FINE SAND	SILTY OR CLAYEY GRAVEL AND SAND				SILTY SOILS		CLAYEY SOILS	
GENERAL RATING AS A SUBGRADE	EXCELLENT TO GOOD							FAIR TO POOR			

*Plasticity index of A-7-5 subgroup is equal to or less than LL – 30. Plasticity index of A-7-6 subgroup is greater than LL – 30.
NP = Non-plastic (use "0"). Symbol "–" means that the particular sieve analysis is not considered for that classification.

If the soil classification is A4-A7, then calculate the group index (GI) as shown below and report with classification. The higher the GI, the less suitable the soil. Example: A-6 with GI = 15 is less suitable than A-6 with GI = 10.

$$GI = (F - 35)[0.2 + 0.005(LL - 40)] + 0.01(F - 15)(PI - 10)$$

where:
F = Percent passing No. 200 sieve, expressed as a whole number. This percentage is based only on the material passing the No. 200 sieve.
LL = Liquid limit
PI = Plasticity index

If the computed value of GI < 0, then use GI = 0.

ASTM D2487-11 Standard Practice for Classification of Soils for Engineering Purposes (Unified Soil Classification System)

Criteria for Assigning Group Symbols and Group Names Using Laboratory Tests[A]				Soil Classification	
				Group Symbol	Group Name[B]
COARSE-GRAINED SOILS More than 50% retained on No. 200 sieve	Gravels (more than 50% of coarse fraction retained on No. 4 sieve)	Clean Gravels (Less than 5% fines[C])	$Cu \geq 4$ and $1 \leq Cc \leq 3$[D]	GW	Well-graded gravel[E]
			$Cu < 4$ and/or $[Cc < 1$ or $Cc > 3]$[D]	GP	Poorly graded gravel[E]
		Gravels with Fines (More than 12% fines[C])	Fines classify as ML or MH	GM	Silty gravel[E, F, G]
			Fines classify as CL or CH	GC	Clayey gravel[E, F, G]
	Sands (50% or more of coarse fraction passes No. 4 sieve)	Clean Sands (Less than 5% fines[H])	$Cu \geq 6$ and $1 \leq Cc \leq 3$[D]	SW	Well-graded sand[I]
			$Cu < 6$ and/or $[Cc < 1$ or $Cc > 3]$[D]	SP	Poorly graded sand[I]
		Sands with Fines (More than 12% fines[H])	Fines classify as ML or MH	SM	Silty sand[F, G, I]
			Fines classify as CL or CH	SC	Clayey sand[F, G, I]
FINE-GRAINED SOILS 50% or more passes the No. 200 sieve	Silts and Clays Liquid limit less than 50	inorganic	PI > 7 and plots on or above "A" line[J]	CL	Lean clay[K, L, M]
			PI < 4 or plots below "A" line[J]	ML	Silt[K,L, M]
		organic	Liquid limit - oven dried/Liquid&#10 < 0.75	OL	Organic clay[K, L, M, N] Organic silt[K, L, M, O]
	Silts and Clays Liquid limit 50 or more	inorganic	PI plots on or above "A" line	CH	Fat clay[K, L, M]
			PI plots below "A" line	MH	Elastic silt[K, L, M]
		organic	Liquid limit - oven dried/Liquid&#10 < 0.75	OH	Organic clay[K, L, M, P] Organic silt[K, L, M, Q]
HIGHLY ORGANIC SOILS	Primarily organic matter, dark in color, and organic odor			PT	Peat

[A]Based on the material passing the 3-in. (75-mm) sieve.
[B]If field sample contained cobbles or boulders, or both, add "with cobbles or boulders, or both" to group name.
[C]Gravels with 5 to 12% fines require dual symbols:
- GW-GM well-graded gravel with silt
- GW-GC well-graded gravel with clay
- GP-GM poorly graded gravel with silt
- GP-GC poorly graded gravel with clay

[D] $Cu = D_{60}/D_{10}$ $\qquad Cc = \dfrac{(D_{30})^2}{D_{10} \times D_{60}}$

[E]If soil contains ≥ 15% sand, add "with sand" to group name.
[F]If fines classify as CL-ML, use dual symbol GC-GM, or SC-SM.
[G]If fines are organic, add "with organic fines" to group name.
[H]Sands with 5 to 12% fines require dual symbols:
- SW-SM well-graded sand with silt
- SW-SC well-graded sand with clay
- SP-SM poorly graded sand with silt
- SP-SC poorly graded sand with clay

[I]If soil contains ≥ 15% gravel, add "with gravel" to group name.
[J]If Atterberg limits plot in hatched area, soil is a CL-ML, silty clay.
[K]If soil contains 15 to 30% plus No. 200, add "with sand" or "with gravel", whichever is predominant.
[L]If soil contains ≥ 30% plus No. 200, predominantly sand, add "sandy" to group name.
[M]If soil contains ≥ 30% plus No. 200, predominantly gravel, add "gravelly" to group name.
[N]PI ≥ 4 and plots on or above "A" line.
[O]PI < 4 or plots below "A" line.
[P]PI plots on or above "A" line.
[Q]PI plots below "A" line.

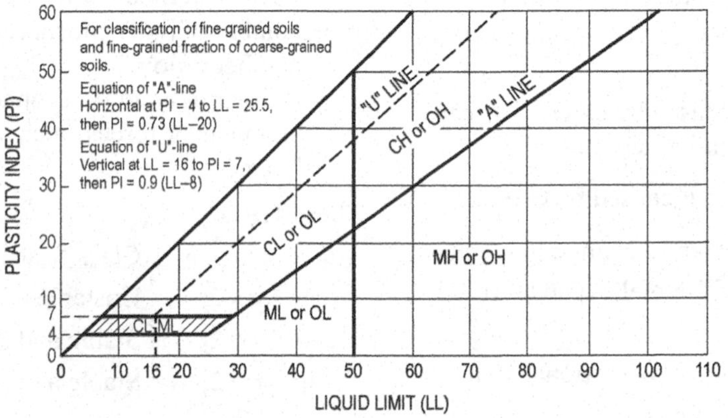

STRUCTURAL ANALYSIS

Influence Lines for Beams and Trusses

An influence line shows the variation of an effect (reaction, shear and moment in beams, bar force in a truss) caused by moving a unit load across the structure. An influence line is used to determine the position of a moveable set of loads that causes the maximum value of the effect.

Moving Concentrated Load Sets

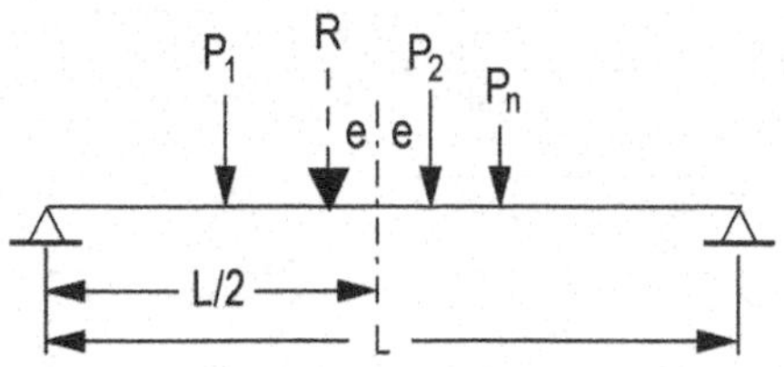

The **absolute maximum moment** produced in a beam by a set of "n" moving loads occurs when the resultant "R" of the load set and an adjacent load are equal distance from the centerline of the beam. In general, two possible load set positions must be considered, one for each adjacent load.

Beam Stiffness and Moment Carryover

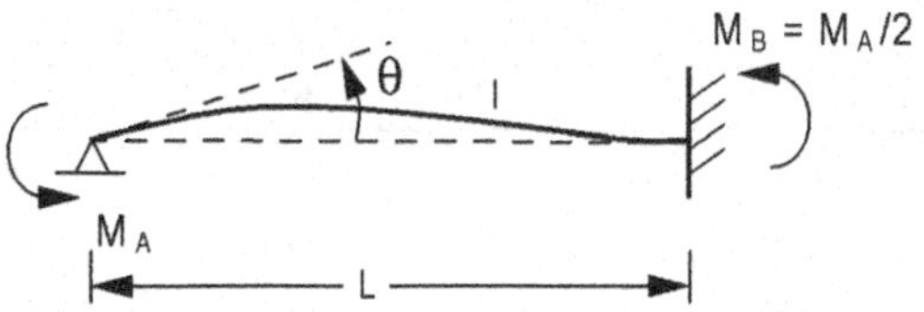

$$\theta = \frac{ML}{4EI} \Rightarrow M = \left(\frac{4EI}{L}\right)\theta = k_{AB}\,\theta$$

k_{AB} = stiffness $\qquad M_B = M_A/2$ = carryover

Truss Deflection by Unit Load Method

The displacement of a truss joint caused by external effects (truss loads, member temperature change, member misfit) is found by applying a unit load at the point that corresponds to the desired displacement.

$$\Delta_{\text{joint}} = \sum_{i=1}^{\text{members}} f_i (\Delta L)_i$$

where: Δ_{joint} = joint displacement at point of application of unit load (+ in direction of unit load)

f_i = force in member "i" caused by unit load (+ tension)

$(\Delta L)_i$ = change in length caused by external effect (+ for increase in member length):

= $\left(\frac{FL}{AE}\right)_i$ for bar force F caused by external load

= $\alpha L_i (\Delta T)_i$ for temperature change in member (α = coefficient of thermal expansion)

= member misfit

L, A = member length and cross-sectional area

E = member elastic modulus

Frame Deflection by Unit Load Method

The displacement of any point on a frame caused by external loads is found by applying a unit load at that point that corresponds to the desired displacement:

$$\Delta = \sum_{i=1}^{\text{members}} \int_{x=0}^{x=L_i} \frac{m_i M_i}{EI_i} dx$$

where: Δ = displacement at point of application of unit load (+ in direction of unit load)

m_i = moment equation in member "i" caused by the unit load

M_i = moment equation in member "i" caused by loads applied to frame

L_i = length of member "i"

I_i = moment of inertia of member "i"

If either the real loads or the unit load cause no moment in a member, that member can be omitted from the summation.

Member Fixed-End Moments (Magnitudes)

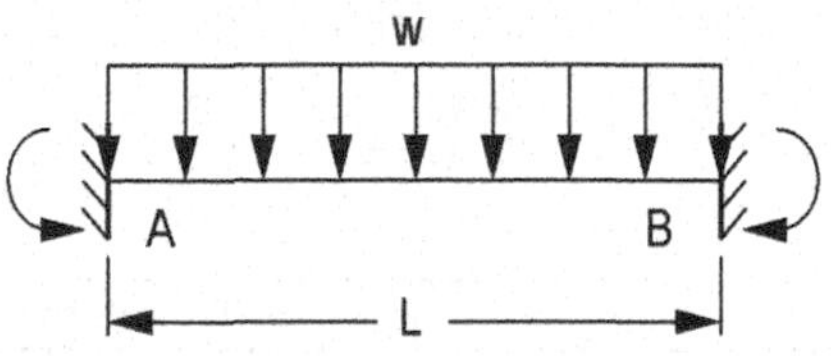

$$FEM_{AB} = FEM_{BA} = \frac{wL^2}{12}$$

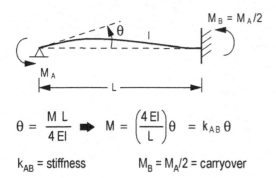

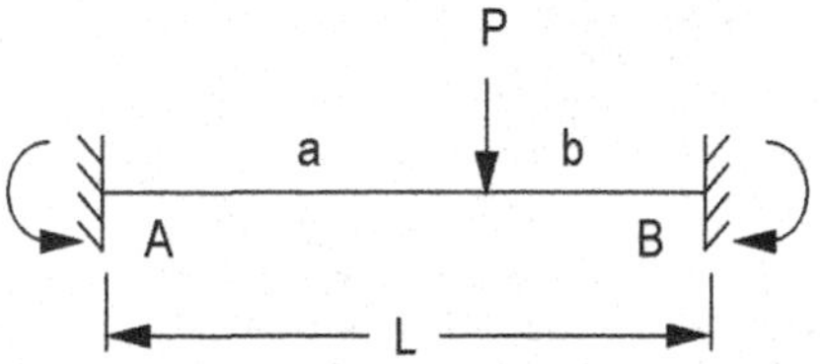

$$FEM_{AB} = \frac{Pab^2}{L^2} \qquad FEM_{BA} = \frac{Pa^2b}{L^2}$$

STABILITY, DETERMINACY, AND CLASSIFICATION OF STRUCTURES

m = number of members

r = number of independent reaction components

j = number of joints

c = number of condition equations based on known internal moments or forces, such as internal moment of zero at a hinge

Plane Truss

Static Analysis	Classification
m + r < 2j	Unstable
m + r = 2j	Stable and statically determinate
m + r > 2j	Stable and statically indeterminate

Plane Frame

Static Analysis	Classification
$3m + r < 3j + c$	Unstable
$3m + r = 3j + c$	Stable and statically determinate
$3m + r > 3j + c$	Stable and statically indeterminate

Stability also requires an appropriate arrangement of members and reaction components.

STRUCTURAL DESIGN

Live Load Reduction

The effect on a building member of nominal occupancy live loads may often be reduced based on the loaded floor area supported by the member. A typical model used for computing reduced live load (as found in ASCE 7 and many building codes) is:

For members supporting one floor

$$L_{\text{reduced}} = L_{\text{nominal}}\left(0.25 + \frac{15}{\sqrt{K_u A_T}}\right) \ge 0.5 L_{\text{nominal}}$$

For members supporting two or more floors

$$L_{\text{reduced}} = L_{\text{nominal}}\left(0.25 + \frac{15}{\sqrt{K_{LL} A_T}}\right) \ge 0.4 L_{\text{nominal}}$$

where: L_{nominal} = nominal live load given in ASCE 7 or a building code

A_T = the cumulative floor tributary area supported by the member

$K_{LL}A_T$ = area of influence supported by the member

K_{LL} = ratio of area of influence to the tributary area supported by the member:

$K_{LL} = 4$ (typical columns)

$K_{LL} = 2$ (typical beams and girders)

Load Combinations using Strength Design (LRFD, USD)

Nominal loads used in following combinations:

D = dead loads
E = earthquake loads
L = live loads (floor)
L_r = live loads (roof)
R = rain load
S = snow load
W = wind load

Load factors λ: λ_D (dead load), λ_L (live load), etc.

Basic combinations $L_r/S/R$ = largest of L_r, S, R

L or $0.8W$ = larger of L, $0.8W$

$1.4D$

$1.2D + 1.6L + 0.5\,(L_r/S/R)$

$1.2D + 1.6(L_r/S/R) + (L \text{ or } 0.8W)$

$1.2D + 1.6W + L + 0.5(L_r/S/R)$

$1.2D + 1.0E + L + 0.2S$

$0.9D + 1.6W$

$0.9D + 1.0E$

DESIGN OF REINFORCED CONCRETE COMPONENTS (ACI 318-11)

U.S. Customary units

Definitions

a = depth of equivalent rectangular stress block, in.

A_g = gross area of column, in^2

A_s = area of tension reinforcement, in^2

A_{st} = total area of longitudinal reinforcement, in^2

A_v = area of shear reinforcement within a distance s, in.

b = width of compression face of member, in.

β_1 = ratio of depth of rectangular stress block, a, to depth to neutral axis, c

$$= 0.85 \ge 0.85 - 0.05\left(\frac{f'_c - 4{,}000}{1{,}000}\right) \ge 0.65$$

c = distance from extreme compression fiber to neutral axis, in.

d = distance from extreme compression fiber to centroid of nonprestressed tension reinforcement, in.

d_t = distance from extreme compression fiber to extreme tension steel, in.

E_c = modulus of elasticity $= 33w_c^{1.5}\sqrt{f'_c}$, psi

ε_t = net tensile strain in extreme tension steel at nominal strength

f'_c = compressive strength of concrete, psi

f_y = yield strength of steel reinforcement, psi

M_n = nominal moment strength at section, in.-lb

ϕM_n = design moment strength at section, in.-lb

M_u = factored moment at section, in.-lb

P_n = nominal axial load strength at given eccentricity, lb

ϕP_n = design axial load strength at given eccentricity, lb

P_u = factored axial force at section, lb

ρ_g = ratio of total reinforcement area to cross-sectional area of column $= A_{st}/A_g$

s = spacing of shear ties measured along longitudinal axis of member, in.

V_c = nominal shear strength provided by concrete, lb

V_n = nominal shear strength at section, lb

ϕV_n = design shear strength at section, lb

V_s = nominal shear strength provided by reinforcement, lb

V_u = factored shear force at section, lb

ASTM STANDARD REINFORCING BARS

BAR SIZE	DIAMETER, IN.	AREA, IN^2	WEIGHT, LB/FT
#3	0.375	0.11	0.376
#4	0.500	0.20	0.668
#5	0.625	0.31	1.043
#6	0.750	0.44	1.502
#7	0.875	0.60	2.044
#8	1.000	0.79	2.670
#9	1.128	1.00	3.400
#10	1.270	1.27	4.303
#11	1.410	1.56	5.313
#14	1.693	2.25	7.650
#18	2.257	4.00	13.60

UNIFIED DESIGN PROVISIONS

Internal Forces and Strains

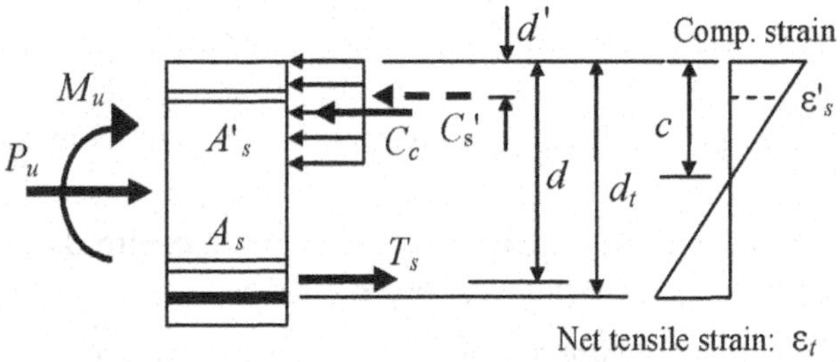

Strain Conditions

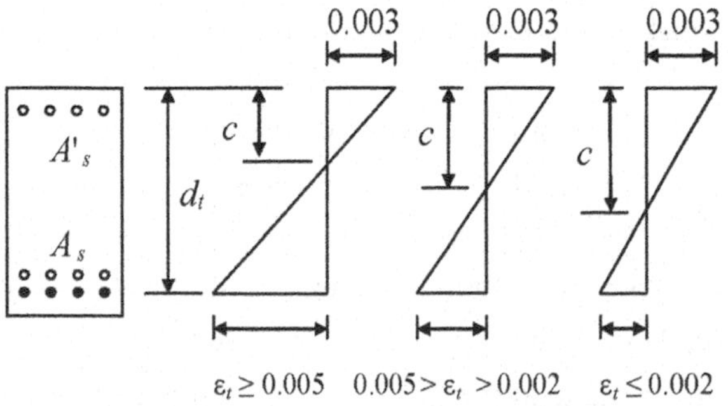

Tension-controlled section: $c \leq 0.375\, d_t$

Transition section

Compression-controlled section: $c \geq 0.6\, d_t$

RESISTANCE FACTORS, ϕ

Tension-controlled sections ($\varepsilon_t \geq 0.005$):	$\phi = 0.9$
Compression-controlled sections ($\varepsilon_t \leq 0.002$):	
Members with tied reinforcement	$\phi = 0.65$
Transition sections ($0.002 < \varepsilon_t < 0.005$):	
Members with tied reinforcement	$\phi = 0.48 + 83\varepsilon_t$
Shear and torsion	$\phi = 0.75$
Bearing on concrete	$\phi = 0.65$

BEAMS—FLEXURE

$$\phi M_n \geq M_u$$

For All Beams

Net tensile strain: $a = \beta_1 c$

$$\varepsilon_t = \frac{0.003(d_t - c)}{c} = \frac{0.003(\beta_1 d_t - a)}{a}$$

Design moment strength: ϕM_n

where: $\phi = 0.9\ [\varepsilon_t \geq 0.005]$

$\phi = 0.48 + 83\varepsilon_t\ [0.002 \leq \varepsilon_t < 0.005]$

$\phi = 0.65\ [\varepsilon_t < 0.002]$

Singly-Reinforced Beams

$$a = \frac{A_s f_y}{0.85 f'_c b}$$

$$M_n = 0.85 f'_c a b\left(d - \frac{a}{2}\right) = A_s f_y\left(d - \frac{a}{2}\right)$$

BEAMS—SHEAR

$$\phi V_n \geq V_u$$

Nominal shear strength:

$$V_n = V_c + V_s$$

$$V_c = 2\, b_w d\sqrt{f'_c}$$

$$V_s = \frac{A_v f_y d}{s} \text{(may not exceed } 8\, b_w d\sqrt{f'_c}\text{)}$$

Required and maximum-permitted stirrup spacing, s

$V_u \le \frac{\phi V_c}{2}$: No stirrups required

$V_u > \frac{\phi V_c}{2}$: Use the following table (A_v given)

	$\frac{\phi V_c}{2} < V_u \le \phi V_c$	$V_u > \phi V_c$
Required spacing	Smaller of: $s = \frac{A_v f_y}{50 b_w}$ $s = \frac{A_v f_y}{0.75 b_w \sqrt{f_c'}}$	$V_s = \frac{V_u}{\phi} - V_c$ $s = \frac{A_v f_y d}{V_s}$
Maximum permitted spacing	Smaller of: $s = \frac{d}{2}$ OR $s = 24"$	$V_s \le 4\, b_w d \sqrt{f_c'}$ Smaller of: $s = \frac{d}{2}$ OR $s = 24"$
		$V_s > 4\, b_w d \sqrt{f_c'}$ Smaller of: $s = \frac{d}{4}$ $s = 12"$

SHORT COLUMNS

Limits for Main Reinforcements

$\rho_g = \frac{A_{st}}{A_g}$

$0.01 \le \rho_g \le 0.08$

Design Column Strength, Tied Columns

$\phi = 0.65$

$\phi P_n = 0.80\phi\,[0.85\, f_c'\,(A_g - A_{st}) + A_{st}\, f_y]$

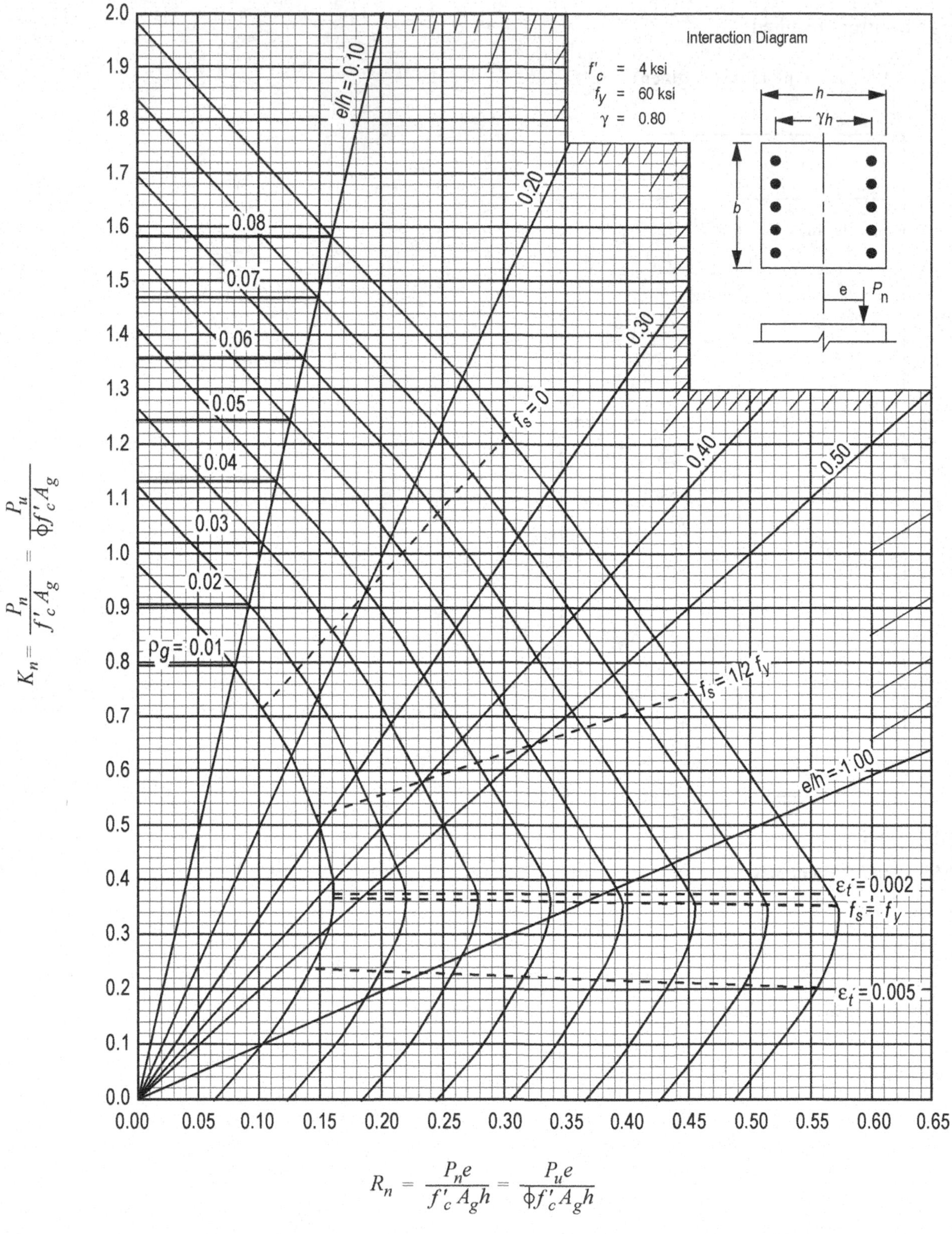

$$K_n = \frac{P_n}{f'_c A_g} = \frac{P_u}{\phi f'_c A_g}$$

$$R_n = \frac{P_n e}{f'_c A_g h} = \frac{P_u e}{\phi f'_c A_g h}$$

GRAPH A.11

Column strength interaction diagram for rectangular section with bars on end faces and $\gamma = 0.80$ (for instructional use only).

Nilson, Arthur H., David Darwin, and Charles W. Dolan, *Design of Concrete Structures*, 13th ed., McGraw-Hill, 2004.

DESIGN OF STEEL COMPONENTS (ANSI/AISC 360-10) LRFD, E = 29,000 ksi

BEAMS

For doubly symmetric compact I-shaped members bent about their major axis, the *design flexural strength* $\phi_b M_n$ is determined with $\phi_b = 0.90$ as follows:

Yielding

$M_n = M_p = F_y Z_x$

where

F_y = specified minimum yield stress

Z_x = plastic section modulus about the x-axis

Lateral-Torsional Buckling

Based on bracing where L_b is the length between points that are either braced against lateral displacement of the compression flange or braced against twist of the cross section with respect to the length limits L_p and L_r:

When $L_b \le L_p$, the limit state of lateral-torsional buckling does not apply.

When $L_p < L_b \le L_r$

$$M_n = C_b\left[M_p - (M_p - 0.7\,F_y S_x)\left(\frac{L_b - L_p}{L_r - L_p}\right)\right] \le M_p$$

where

$$C_b = \frac{12.5 M_{max}}{2.5 M_{max} + 3M_A + 4M_B + 3M_C}$$

M_{max} = absolute value of maximum moment in the unbraced segment

M_A = absolute value of maximum moment at quarter point of the unbraced segment

M_B = absolute value of maximum moment at centerline of the unbraced segment

M_C = absolute value of maximum moment at three-quarter of the unbraced segment

Shear

The *design shear strength* $\phi_v V_n$ is determined with $\phi_v = 1.00$ for webs of rolled I-shaped members and is determined as follows:

$V_n = 0.6\,F_y A_w C_v$

$C_v = 1.0$

A_w = area of web, the overall depth times the web thickness, dt_w, in^2 (mm^2)

COLUMNS

The *design compressive strength* $\phi_c P_n$ is determined with $\phi_c = 0.90$ for flexural buckling of members without slender elements and is determined as follows:

$P_n = F_{cr} A_g$

where the critical stress F_{cr} is determined as follows:

(a) When $\frac{KL}{r} \le 4.71\sqrt{\frac{E}{F_y}}$, $F_{cr} = \left[0.658^{\frac{F_y}{F_e}}\right]F_y$

(b) When $\frac{KL}{r} > 4.71\sqrt{\frac{E}{F_y}}$, $F_{cr} = 0.877\,F_e$

where

KL/r is the effective slenderness ratio based on the column effective length (KL) and radius of gyration (r)

KL is determined from AISC Table C-A-7.1 or AISC Figures C-A-7.1 and C-A-7.2 in the Civil Engineering section of this handbook.

F_e is the elastic buckling stress = $\pi^2 E/(KL/r)^2$

VALUES OF C_b FOR SIMPLY SUPPORTED BEAMS

LOAD	LATERAL BRACING ALONG SPAN	C_b
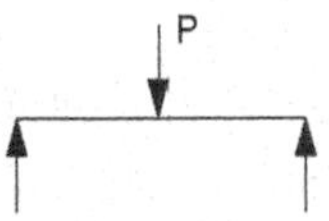	NONE LOAD AT MIDPOINT	1.32
	AT LOAD POINT	1.67 1.67
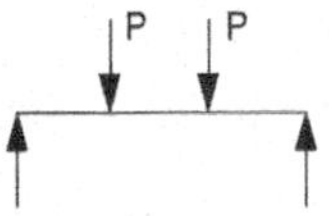	NONE LOADS AT THIRD POINTS	1.14
	AT LOAD POINTS LOADS SYMMETRICALLY PLACED	1.67 1.00 1.67
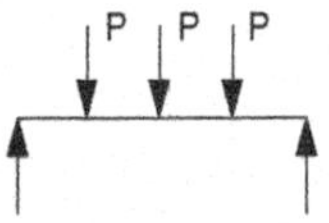	NONE LOADS AT QUARTER POINTS	1.14
	AT LOAD POINTS LOADS AT QUARTER POINTS	1.67 1.11 1.11 1.67
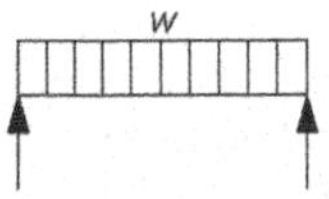	NONE	1.14
	AT MIDPOINT	1.30 1.30
	AT THIRD POINTS	1.45 1.01 1.45
	AT QUARTER POINTS	1.52 1.06 1.06 1.52
	AT FIFTH POINTS	1.56 1.12 1.00 1.12 1.56

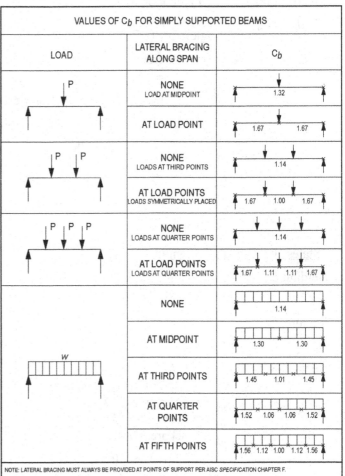

NOTE: LATERAL BRACING MUST ALWAYS BE PROVIDED AT POINTS OF SUPPORT PER AISC *SPECIFICATION* CHAPTER F.

Adapted from *Steel Construction Manual*, 14th ed., AISC, 2011.

TENSION MEMBERS

Flat bars or angles, bolted or welded

Definitions

Bolt diameter: d_b

Nominal hole diameter: $d_h = d_b + 1/16"$

Gross width of member: b_g

Member thickness: t

Connection eccentricity: $\bar{x}$

Gross area: $A_g = b_g\, t$ (use tabulated areas for angles)

Net area (parallel holes): $A_n = \left[b_g - \Sigma\left(d_h + \frac{1}{16}"\right)\right]t$

Net area (staggered holes):

$$A_n = \left[b_g - \Sigma\left(d_h + \frac{1}{16}"\right) + \Sigma\frac{s^2}{4g}\right]t$$

s = longitudinal spacing of consecutive holes

g = transverse spacing between lines of holes

Effective area (bolted members):

$$A_e = UA_n \quad \begin{cases} \text{Flat bars: } U = 1.0 \\ \text{Angles: } U = 1 - \bar{x}/L \end{cases}$$

Effective area (welded members):

$$A_e = UA_n \quad \begin{cases} \text{Flat bars or angles with transverse welds: } U = 1.0 \\ \text{Flat bars of width "}w\text{", longitudinal welds of length "}L\text{" only:} \\ \quad U = 1.0 \ (L \ge 2w) \\ \quad U = 0.87 \ (2w > L \ge 1.5w) \\ \quad U = 0.75 \ (1.5w > L > w) \\ \text{Angles with longitudinal welds only} \\ \quad U = 1 - \bar{x}/L \end{cases}$$

Limit States and Available Strengths

Yielding:

$\phi_y = 0.90$

$\phi T_n = \phi_y F_y A_g$

Fracture:

$\phi_f = 0.75$

$\phi T_n = \phi_f F_u A_e$

Block shear:

$\phi = 0.75$

$U_{bs} = 1.0$ (flat bars and angles)

A_{gv} = gross area for shear

A_{nv} = net area for shear

A_{nt} = net area for tension

$$\phi T_n = \underset{\text{smaller}}{} \begin{cases} 0.75\, F_u\left[0.6A_{nv} + U_{bs}A_{nt}\right] \\ 0.75\left[0.6F_yA_{gv} + U_{bs}F_uA_{nt}\right] \end{cases}$$

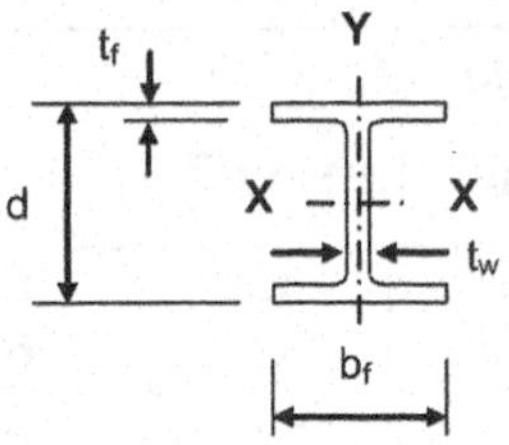

Table 1-1: W Shapes Dimensions and Properties

Shape	Area	Depth	Web	Flange		Axis X-X				Axis Y-Y	
	A	d	t_w	b_f	t_f	I	S	r	Z	I	r
	$In.^2$	In.	In.	In.	In.	$In.^4$	$In.^3$	In.	$In.^3$	$In.^4$	In.
W24X68	20.1	23.7	0.415	8.97	0.585	1830	154	9.55	177	70.4	1.87
W24X62	18.2	23.7	0.430	7.04	0.590	1550	131	9.23	153	34.5	1.38
W24X55	16.3	23.6	0.395	7.01	0.505	1350	114	9.11	134	29.1	1.34
W21X73	21.5	21.2	0.455	8.30	0.740	1600	151	8.64	172	70.6	1.81
W21X68	20.0	21.1	0.430	8.27	0.685	1480	140	8.60	160	64.7	1.80
W21X62	18.3	21.0	0.400	8.24	0.615	1330	127	8.54	144	57.5	1.77
W21X55	16.2	20.8	0.375	8.22	0.522	1140	110	8.40	126	48.4	1.73
W21X57	16.7	21.1	0.405	6.56	0.650	1170	111	8.36	129	30.6	1.35
W21X50	14.7	20.8	0.380	6.53	0.535	984	94.5	8.18	110	24.9	1.30
W21X48	14.1	20.6	0.350	8.14	0.430	959	93.0	8.24	107	38.7	1.66
W21X44	13.0	20.7	0.350	6.50	0.450	843	81.6	8.06	95.4	20.7	1.26
W18X71	20.8	18.5	0.495	7.64	0.810	1170	127	7.50	146	60.3	1.70
W18X65	19.1	18.4	0.450	7.59	0.750	1070	117	7.49	133	54.8	1.69
W18X60	17.6	18.2	0.415	7.56	0.695	984	108	7.47	123	50.1	1.68
W18X55	16.2	18.1	0.390	7.53	0.630	890	98.3	7.41	112	44.9	1.67
W18X50	14.7	18.0	0.355	7.50	0.570	800	88.9	7.38	101	40.1	1.65
W18X46	13.5	18.1	0.360	6.06	0.605	712	78.8	7.25	90.7	22.5	1.29
W18X40	11.8	17.9	0.315	6.02	0.525	612	68.4	7.21	78.4	19.1	1.27
W16X67	19.7	16.3	0.395	10.2	0.67	954	117	6.96	130	119	2.46
W16X57	16.8	16.4	0.430	7.12	0.715	758	92.2	6.72	105	43.1	1.60
W16X50	14.7	16.3	0.380	7.07	0.630	659	81.0	6.68	92.0	37.2	1.59
W16X45	13.3	16.1	0.345	7.04	0.565	586	72.7	6.65	82.3	32.8	1.57
W16X40	11.8	16.0	0.305	7.00	0.505	518	64.7	6.63	73.0	28.9	1.57
W16X36	10.6	15.9	0.295	6.99	0.430	448	56.5	6.51	64.0	24.5	1.52
W14X74	21.8	14.2	0.450	10.1	0.785	795	112	6.04	126	134	2.48
W14X68	20.0	14.0	0.415	10.0	0.720	722	103	6.01	115	121	2.46
W14X61	17.9	13.9	0.375	9.99	0.645	640	92.1	5.98	102	107	2.45
W14X53	15.6	13.9	0.370	8.06	0.660	541	77.8	5.89	87.1	57.7	1.92
W14X48	14.1	13.8	0.340	8.03	0.595	484	70.2	5.85	78.4	51.4	1.91
W12X79	23.2	12.4	0.470	12.1	0.735	662	107	5.34	119	216	3.05
W12X72	21.1	12.3	0.430	12.0	0.670	597	97.4	5.31	108	195	3.04
W12X65	19.1	12.1	0.390	12.0	0.605	533	87.9	5.28	96.8	174	3.02
W12X58	17.0	12.2	0.360	10.0	0.640	475	78.0	5.28	86.4	107	2.51
W12X53	15.6	12.1	0.345	9.99	0.575	425	70.6	5.23	77.9	95.8	2.48
W12X50	14.6	12.2	0.370	8.08	0.640	391	64.2	5.18	71.9	56.3	1.96
W12X45	13.1	12.1	0.335	8.05	0.575	348	57.7	5.15	64.2	50.0	1.95
W12X40	11.7	11.9	0.295	8.01	0.515	307	51.5	5.13	57.0	44.1	1.94
W10x60	17.6	10.2	0.420	10.1	0.680	341	66.7	4.39	74.6	116	2.57
W10x54	15.8	10.1	0.370	10.0	0.615	303	60.0	4.37	66.6	103	2.56
W10x49	14.4	10.0	0.340	10.0	0.560	272	54.6	4.35	60.4	93.4	2.54
W10x45	13.3	10.1	0.350	8.02	0.620	248	49.1	4.32	54.9	53.4	2.01
W10x39	11.5	9.92	0.315	7.99	0.530	209	42.1	4.27	46.8	45.0	1.98

Adapted from *Steel Construction Manual*, 14th ed., AISC, 2011.

Z_x

AISC Table 3-2
W Shapes – Selection by Z_x

F_y = 50 ksi
ϕ_b = 0.90
ϕ_v = 1.00

Shape	Z_x in.^{3}	$\phi_b M_{px}$ kip-ft	$\phi_b M_{rx}$ kip-ft	$\phi_b BF$ kips	L_p ft.	L_r ft.	I_x in.^{4}	$\phi_v V_{nx}$ kips
W24 x 55	**134**	**503**	**299**	**22.2**	**4.73**	**13.9**	**1350**	**251**
W18 x 65	133	499	307	14.9	5.97	18.8	1070	248
W12 x 87	132	495	310	5.76	10.8	43.0	740	194
W16 x 67	130	488	307	10.4	8.69	26.1	954	194
W10 x 100	130	488	294	4.01	9.36	57.7	623	226
W21 x 57	129	484	291	20.1	4.77	14.3	1170	256
W21 x 55	**126**	**473**	**289**	**16.3**	**6.11**	**17.4**	**1140**	**234**
W14 x 74	126	473	294	8.03	8.76	31.0	795	191
W18 x 60	123	461	284	14.5	5.93	18.2	984	227
W12 x 79	119	446	281	5.67	10.8	39.9	662	175
W14 x 68	115	431	270	7.81	8.69	29.3	722	175
W10 x 88	113	424	259	3.95	9.29	51.1	534	197
W18 x 55	**112**	**420**	**258**	**13.9**	**5.90**	**17.5**	**890**	**212**
W21 x 50	**110**	**413**	**248**	**18.3**	**4.59**	**13.6**	**984**	**237**
W12 x 72	108	405	256	5.59	10.7	37.4	597	158
W21 x 48	**107**	**398**	**244**	**14.7**	**6.09**	**16.6**	**959**	**217**
W16 x 57	105	394	242	12.0	5.56	18.3	758	212
W14 x 61	102	383	242	7.46	8.65	27.5	640	156
W18 x 50	101	379	233	13.1	5.83	17.0	800	192
W10 x 77	97.6	366	225	3.90	9.18	45.2	455	169
W12 x 65	96.8	356	231	5.41	11.9	35.1	533	142
W21 x 44	**95.4**	**358**	**214**	**16.8**	**4.45**	**13.0**	**843**	**217**
W16 x 50	92.0	345	213	11.4	5.62	17.2	659	185
W18 x 46	90.7	340	207	14.6	4.56	13.7	712	195
W14 x 53	87.1	327	204	7.93	6.78	22.2	541	155
W12 x 58	86.4	324	205	5.66	8.87	29.9	475	132
W10 x 68	85.3	320	199	3.86	9.15	40.6	394	147
W16 x 45	82.3	309	191	10.8	5.55	16.5	586	167
W18 x 40	**78.4**	**294**	**180**	**13.3**	**4.49**	**13.1**	**612**	**169**
W14 x 48	78.4	294	184	7.66	6.75	21.1	484	141
W12 x 53	77.9	292	185	5.48	8.76	28.2	425	125
W10 x 60	74.6	280	175	3.80	9.08	36.6	341	129
W16 x 40	**73.0**	**274**	**170**	**10.1**	**5.55**	**15.9**	**518**	**146**
W12 x 50	71.9	270	169	5.97	6.92	23.9	391	135
W8 x 67	70.1	263	159	2.60	7.49	47.7	272	154
W14 x 43	69.6	261	164	7.24	6.68	20.0	428	125
W10 x 54	66.6	250	158	3.74	9.04	33.7	303	112
W18 x 35	**66.5**	**249**	**151**	**12.3**	**4.31**	**12.4**	**510**	**159**
W12 x 45	64.2	241	151	5.75	6.89	22.4	348	121
W16 x 36	64.0	240	148	9.31	5.37	15.2	448	140
W14 x 38	61.5	231	143	8.10	5.47	16.2	385	131
W10 x 49	60.4	227	143	3.67	8.97	31.6	272	102
W8 x 58	59.8	224	137	2.56	7.42	41.7	228	134
W12 x 40	57.0	214	135	5.50	6.85	21.1	307	106
W10 x 45	54.9	206	129	3.89	7.10	26.9	248	106

$M_{rx} = (0.7F_y)S_x$ $\qquad$ $BF = \dfrac{M_{px} - M_{rx}}{L_r - L_p}$

Adapted from *Steel Construction Manual*, 14th ed., AISC, 2011.

$C_b = 1$	F_y = 50 ksi ϕ = 0.90	TABLE 3-10 (CONTINUED) W SHAPES
ϕM_n kip-ft LRFD	AVAILABLE MOMENT VS. UNBRACED LENGTH	

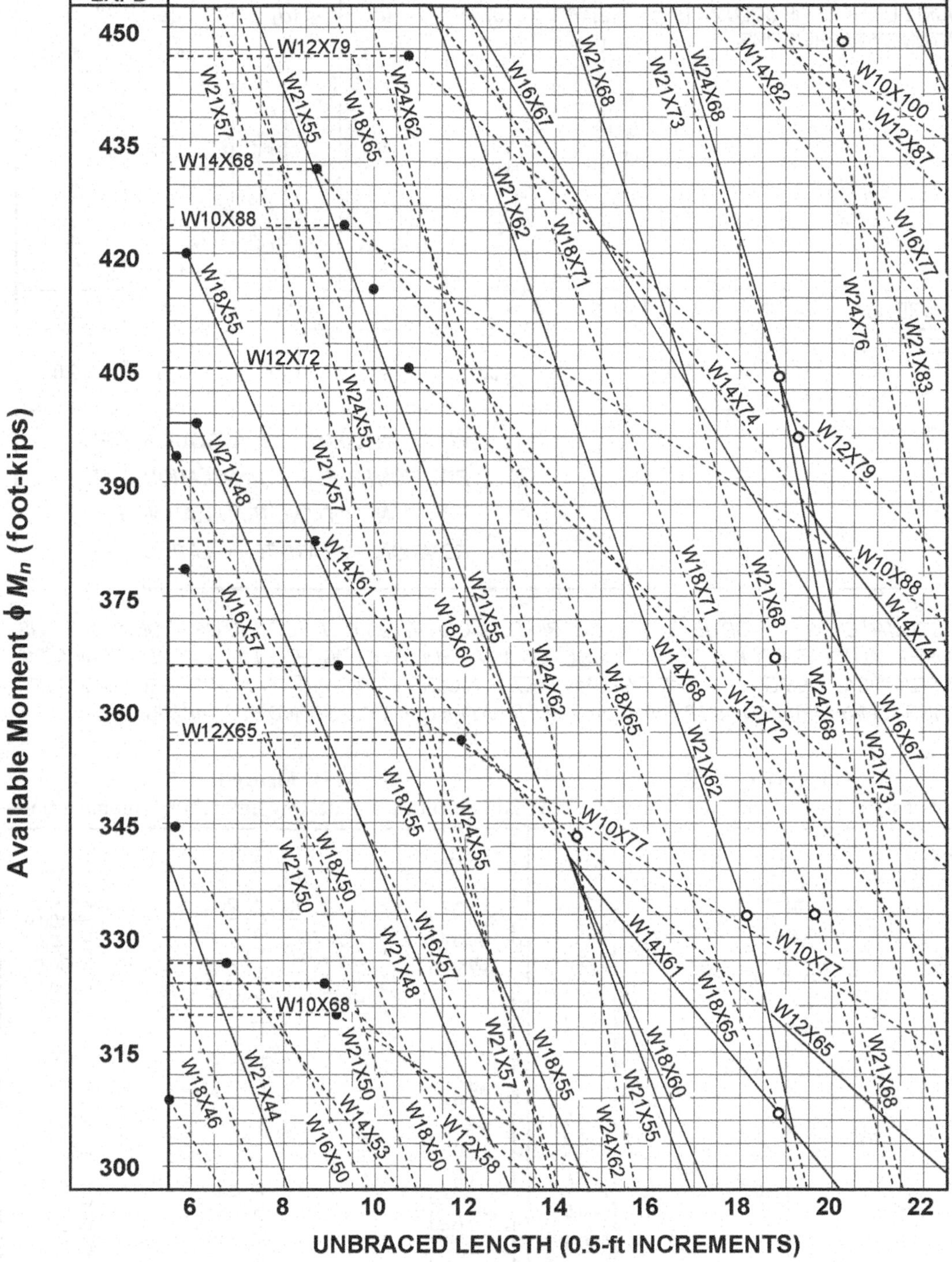

Steel Construction Manual, 14th ed., AISC, 2011.

TABLE C-A-7.1 APPROXIMATE VALUES OF EFFECTIVE LENGTH FACTOR, *K*						
BUCKLED SHAPE OF COLUMN IS SHOWN BY DASHED LINE.	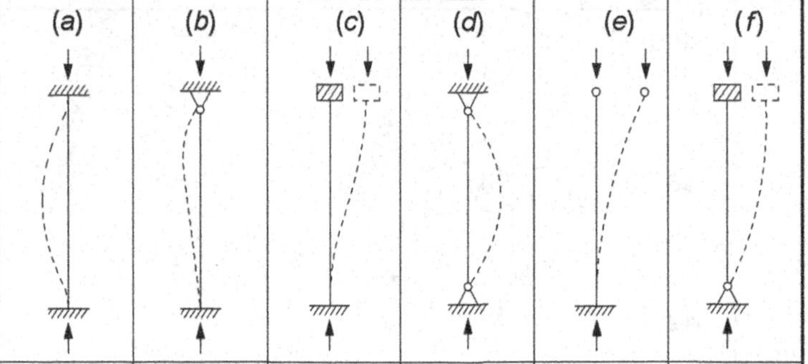					
	(*a*)	(*b*)	(*c*)	(*d*)	(*e*)	(*f*)
THEORETICAL *K* VALUE	0.5	0.7	1.0	1.0	2.0	2.0
RECOMMENDED DESIGN VALUE WHEN IDEAL CONDITIONS ARE APPROXIMATED	0.65	0.80	1.2	1.0	2.10	2.0
END CONDITION CODE	*ROTATION FIXED AND TRANSLATION FIXED* *ROTATION FREE AND TRANSLATION FIXED* *ROTATION FIXED AND TRANSLATION FREE* *ROTATION FREE AND TRANSLATION FREE*					

FOR COLUMN ENDS SUPPORTED BY, BUT NOT RIGIDLY CONNECTED TO, A FOOTING OR FOUNDATION, G IS THEORETICALLY INFINITY BUT UNLESS DESIGNED AS A TRUE FRICTION-FREE PIN, MAY BE TAKEN AS 10 FOR PRACTICAL DESIGNS. IF THE COLUMN END IS RIGIDLY ATTACHED TO A PROPERLY DESIGNED FOOTING, G MAY BE TAKEN AS 1.0. SMALLER VALUES MAY BE USED IF JUSTIFIED BY ANALYSIS.

AISC Figure C-A-7.1

Alignment chart, sidesway inhibited (braced frame)

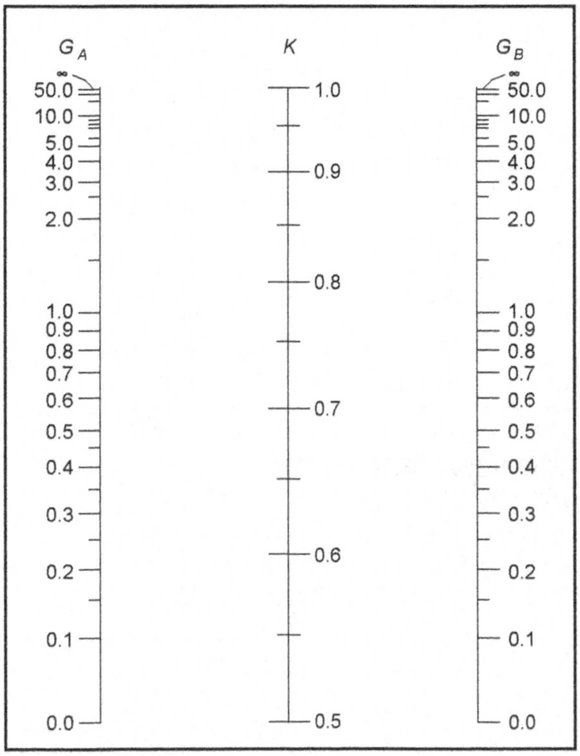

AISC Figure C-A-7.2

Alignment chart, sidesway uninhibited (moment frame)

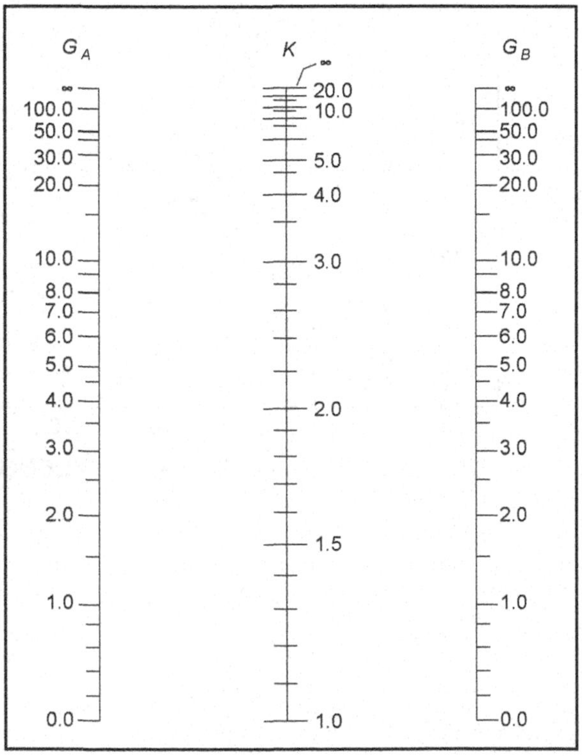

Steel Construction Manual, 14th ed., AISC, 2011.

AISC Table 4-22

Available Critical Stress $\phi_c F_{cr}$ for Compression Members

F_y = 50 ksi ϕ_c = 0.90

$\frac{KL}{r}$	ϕF_{cr} ksi	$\frac{KL}{r}$	ϕF_{cr} ksi	$\frac{KL}{r}$	ϕF_{cr} ksi	$\frac{KL}{r}$	ϕF_{cr} ksi	$\frac{KL}{r}$	ϕF_{cr} ksi
1	45.0	41	39.8	81	27.9	121	15.4	161	8.72
2	45.0	42	39.5	82	27.5	122	15.2	162	8.61
3	45.0	43	39.3	83	27.2	123	14.9	163	8.50
4	44.9	44	39.1	84	26.9	124	14.7	164	8.40
5	44.9	45	38.8	85	26.5	125	14.5	165	8.30
6	44.9	46	38.5	86	26.2	126	14.2	166	8.20
7	44.8	47	38.3	87	25.9	127	14.0	167	8.10
8	44.8	48	38.0	88	25.5	128	13.8	168	8.00
9	44.7	49	37.7	89	25.2	129	13.6	169	7.89
10	44.7	50	37.5	90	24.9	130	13.4	170	7.82
11	44.6	51	37.2	91	24.6	131	13.2	171	7.73
12	44.5	52	36.9	92	24.2	132	13.0	172	7.64
13	44.4	53	36.7	93	23.9	133	12.8	173	7.55
14	44.4	54	36.4	94	23.6	134	12.6	174	7.46
15	44.3	55	36.1	95	23.3	135	12.4	175	7.38
16	44.2	56	35.8	96	22.9	136	12.2	176	7.29
17	44.1	57	35.5	97	22.6	137	12.0	177	7.21
18	43.9	58	35.2	98	22.3	138	11.9	178	7.13
19	43.8	59	34.9	99	22.0	139	11.7	179	7.05
20	43.7	60	34.6	100	21.7	140	11.5	180	6.97
21	43.6	61	34.3	101	21.3	141	11.4	181	6.90
22	43.4	62	34.0	102	21.0	142	11.2	182	6.82
23	43.3	63	33.7	103	20.7	143	11.0	183	6.75
24	43.1	64	33.4	104	20.4	144	10.9	184	6.67
25	43.0	65	33.0	105	20.1	145	10.7	185	6.60
26	42.8	66	32.7	106	19.8	146	10.6	186	6.53
27	42.7	67	32.4	107	19.5	147	10.5	187	6.46
28	42.5	68	32.1	108	19.2	148	10.3	188	6.39
29	42.3	69	31.8	109	18.9	149	10.2	189	6.32
30	42.1	70	31.4	110	18.6	150	10.0	190	6.26
31	41.9	71	31.1	111	18.3	151	9.91	191	6.19
32	41.8	72	30.8	112	18.0	152	9.78	192	6.13
33	41.6	73	30.5	113	17.7	153	9.65	193	6.06
34	41.4	74	30.2	114	17.4	154	9.53	194	6.00
35	41.2	75	29.8	115	17.1	155	9.40	195	5.94
36	40.9	76	29.5	116	16.8	156	9.28	196	5.88
37	40.7	77	29.2	117	16.5	157	9.17	197	5.82
38	40.5	78	28.8	118	16.2	158	9.05	198	5.76
39	40.3	79	28.5	119	16.0	159	8.94	199	5.70
40	40.0	80	28.2	120	15.7	160	8.82	200	5.65

Adapted from *Steel Construction Manual*, 14th ed., AISC, 2011.

Selected W14, W12, W10

AISC Table 4–1

Available Strength in Axial Compression, kips—W shapes

LRFD: ϕP_n

$F_y = 50$ ksi

$\phi c = 0.90$

Shape	W14					W12					W10				
wt/ft	74	68	61	53	48	58	53	50	45	40	60	54	49	45	39
0	980	899	806	702	636	767	701	657	590	526	794	712	649	597	516
6	922	844	757	633	573	722	659	595	534	475	750	672	612	543	469
7	901	826	740	610	552	707	644	574	516	458	734	658	599	525	452
8	878	804	721	585	529	689	628	551	495	439	717	643	585	505	435
9	853	781	700	557	504	670	610	526	472	419	698	625	569	483	415
10	826	755	677	528	477	649	590	499	448	397	677	607	551	460	395
11	797	728	652	497	449	627	569	471	422	375	655	586	533	435	373
12	766	700	626	465	420	603	547	443	396	351	631	565	513	410	351
13	734	670	599	433	391	578	525	413	370	328	606	543	493	384	328
14	701	639	572	401	361	553	501	384	343	304	581	520	471	358	305
15	667	608	543	369	332	527	477	354	317	280	555	496	450	332	282
16	632	576	515	338	304	500	452	326	291	257	528	472	428	306	260
17	598	544	486	308	276	473	427	297	265	234	501	448	405	281	238
18	563	512	457	278	250	446	402	270	241	212	474	423	383	256	216
19	528	480	428	250	224	420	378	244	217	191	447	399	360	233	195
20	494	448	400	226	202	393	353	220	196	172	420	375	338	210	176
22	428	387	345	186	167	342	306	182	162	142	367	327	295	174	146
24	365	329	293	157	140	293	261	153	136	120	317	282	254	146	122
26	311	281	250	133	120	249	222	130	116	102	270	241	216	124	104
28	268	242	215	115	103	215	192	112	99.8	88.0	233	208	186	107	90.0
30	234	211	187	100	89.9	187	167	97.7	87.0	76.6	203	181	162	93.4	78.4
32	205	185	165	88.1		165	147	82.9	76.4	67.3	179	159	143	82.1	68.9
34	182	164	146			146	130				158	141	126		
36	162	146	130			130	116				141	126	113		
38	146	131	117			117	104				127	113	101		
40	131	119	105			105	93.9				114	102	91.3		

Effective length KL (ft) with respect to least radius of gyration r_y

Heavy line indicates KL/r equal to or greater than 200

Adapted from *Steel Construction Manual*, 14th ed., AISC, 2011

HYDROLOGY/WATER RESOURCES

NRCS (SCS) Rainfall-Runoff

$$Q = \frac{(P - 0.2S)^2}{P + 0.8S}$$

$$S = \frac{1{,}000}{CN} - 10$$

$$CN = \frac{1{,}000}{S + 10}$$

P = precipitation (inches)

S = maximum basin retention (inches)

Q = runoff (inches)

CN = curve number

Rational Formula

$Q = CIA$, where

A = watershed area (acres)

C = runoff coefficient

I = rainfall intensity (in./hr)

Q = peak discharge (cfs)

Darcy's Law

$Q = -KA(dh/dx)$, where

Q = discharge rate (ft^3/sec or m^3/s)

K = hydraulic conductivity (ft/sec or m/s)

h = hydraulic head (ft or m)

A = cross-sectional area of flow (ft^2 or m^2)

q = $-K(dh/dx)$

q = specific discharge (also called Darcy velocity or superficial velocity)

v $= q/n = \dfrac{-K}{n}\dfrac{dh}{dx}$

v = average seepage velocity

n = effective porosity

Unit hydrograph: The direct runoff hydrograph that would result from one unit of runoff occurring uniformly in space and time over a specified period of time.

Transmissivity, T: The product of hydraulic conductivity and thickness, b, of the aquifer (L^2T^{-1}).

Storativity or storage coefficient of an aquifer, S: The volume of water taken into or released from storage per unit surface area per unit change in potentiometric (piezometric) head.

Well Drawdown

Unconfined aquifer

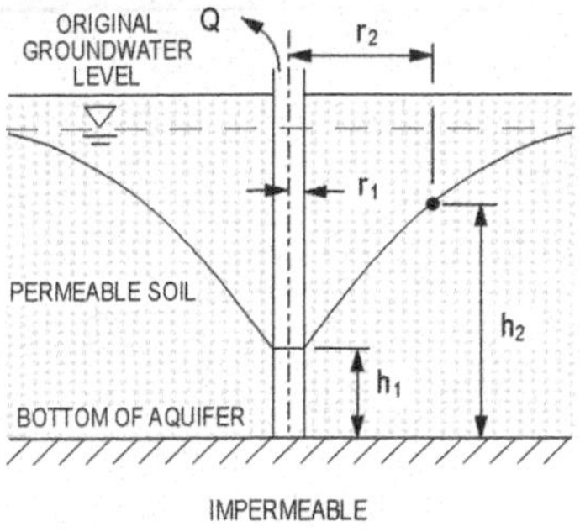

Dupuit's Formula

$$Q = \frac{\pi K\left(h_2^2 - h_1^2\right)}{\ln\left(\frac{r_2}{r_1}\right)}, \text{ where}$$

Q = flowrate of water drawn from well (cfs)

K = coefficient of permeability of soil
= hydraulic conductivity (ft/sec)

h_1 = height of water surface above bottom of aquifer at perimeter of well (ft)

h_2 = height of water surface above bottom of aquifer at distance r_2 from well centerline (ft)

r_1 = radius to water surface at perimeter of well, i.e., radius of well (ft)

r_2 = radius to water surface whose height is h_2 above bottom of aquifer (ft)

ln = natural logarithm

Q/D_w = specific capacity

D_w = well drawdown (ft)

Confined aquifer

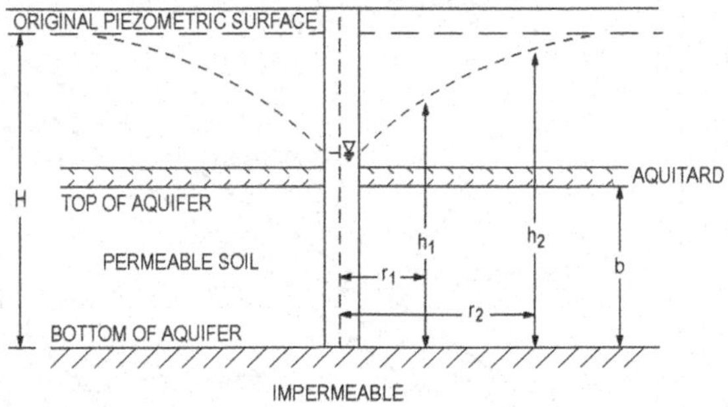

Theim Equation

$$Q = \frac{2\pi T\left(h_2 - h_1\right)}{\ln\left(\frac{r_2}{r_1}\right)}, \text{ where}$$

T = Kb = transmissivity (ft^2/sec)

b = thickness of confined aquifer (ft)

h_1, h_2 = heights of piezometric surface above bottom of aquifer (ft)

r_1, r_2 = radii from pumping well (ft)

ln = natural logarithm

H = height of peizometric surface prior to pumping (ft)

♦ Sewage Flow Ratio Curves

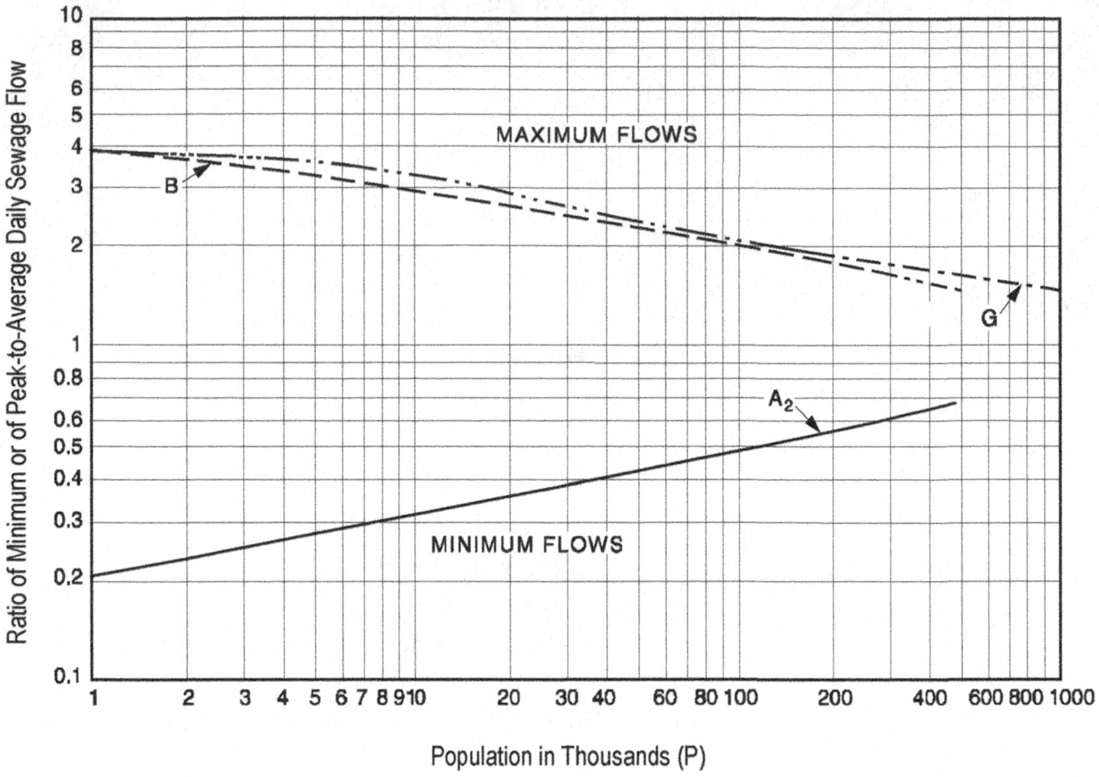

Curve A_2: $\frac{P^{0.2}}{5}$

Curve B: $\frac{14}{4+\sqrt{P}}+1$

Curve G: $\frac{18+\sqrt{P}}{4+\sqrt{P}}$

♦ Hydraulic-Elements Graph for Circular Sewers

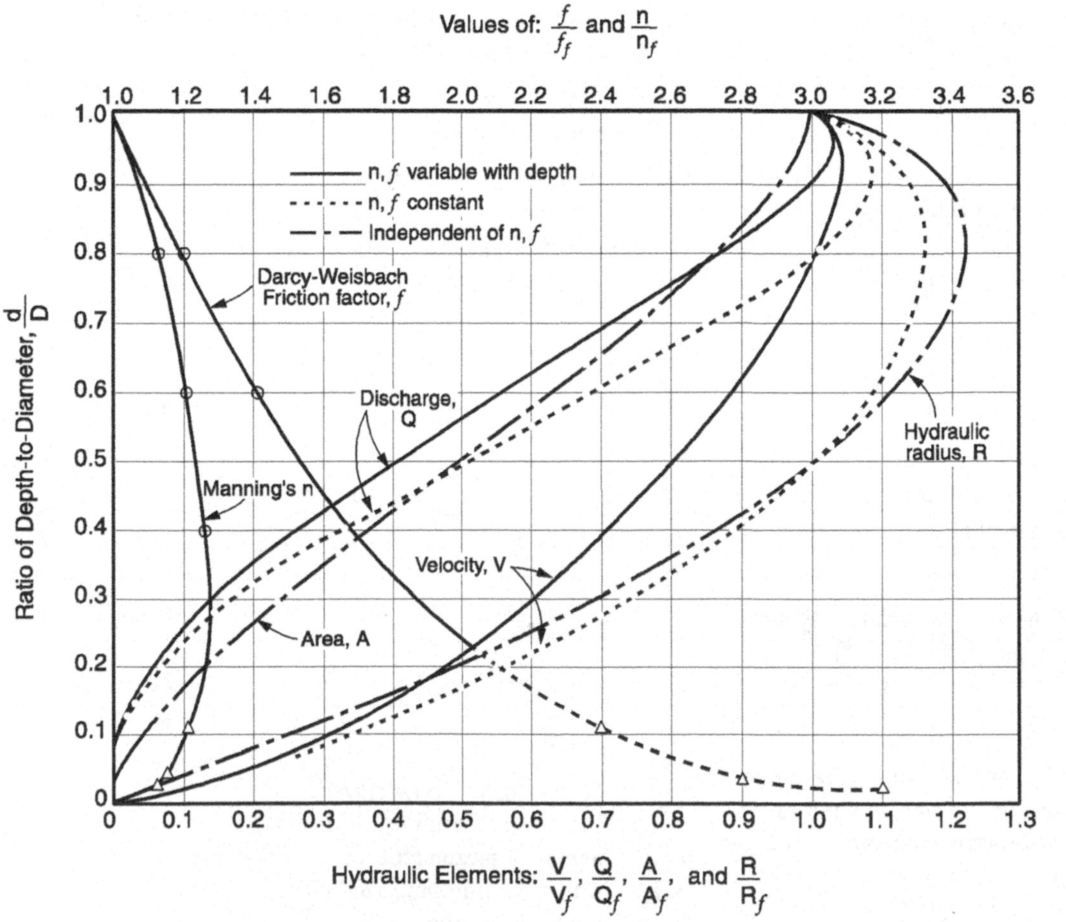

♦ *Design and Construction of Sanitary and Storm Sewers,* Water Pollution Control Federation and American Society of Civil Engineers, 1970. Reprinted with permission from ASCE.

Open-Channel Flow

Specific Energy

$$E = \alpha\frac{V^2}{2g} + y = \frac{\alpha Q^2}{2gA^2} + y, \text{ where}$$

E = specific energy

Q = discharge

V = velocity

y = depth of flow

A = cross-sectional area of flow

α = kinetic energy correction factor, usually 1.0

Critical Depth = that depth in a channel at minimum specific energy

$$\frac{Q^2}{g} = \frac{A^3}{T}$$

where Q and A are as defined above,

g = acceleration due to gravity

T = width of the water surface

For rectangular channels

$$y_c = \left(\frac{q^2}{g}\right)^{1/3}, \text{ where}$$

y_c = critical depth

q = unit discharge = Q/B

B = channel width

g = acceleration due to gravity

Froude Number = ratio of inertial forces to gravity forces

$$Fr = \frac{V}{\sqrt{gy_h}} = \sqrt{\frac{Q^2T}{gA^3}}, \text{ where}$$

y_h = hydraulic depth = A/T

Supercritical flow: $Fr > 1$

Subcritical flow: $Fr < 1$

Critical flow: $Fr = 1$

Specific Energy Diagram

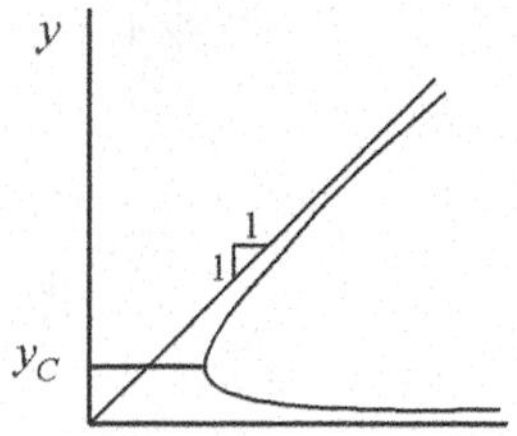

$$E = \frac{\alpha V^2}{2g} + y$$

Alternate depths: depths with the same specific energy

Uniform flow: a flow condition where depth and velocity do not change along a channel

Momentum Depth Diagram

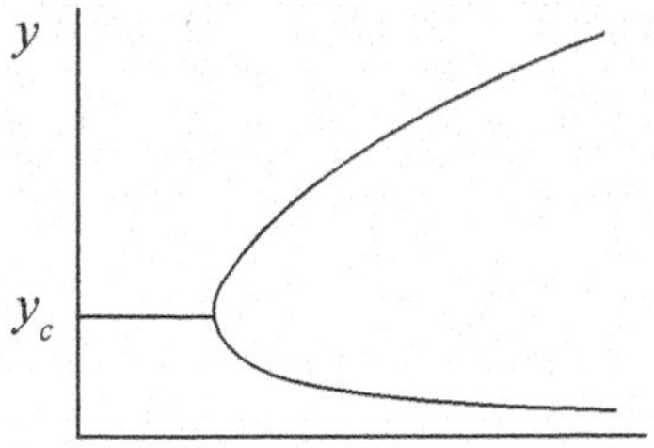

$$M = \frac{Q^2}{gA} + Ah_c, \text{ where}$$

h_c = vertical distance from liquid surface to centroid of area

Sequent (*conjugate*) *depths*: depths with the same momentum

Hydraulic Jump

$$y_2 = \frac{y_1}{2}\left(-1 + \sqrt{1 + 8Fr_1^2}\right), \text{ where}$$

y_1 = flow depth at upstream supercritical flow location

y_2 = flow depth at downstream subcritical flow location

Fr_1 = Froude number at upstream supercritical flow location

Manning's Equation

$$Q = (K/n)AR_H^{2/3}S^{1/2}$$

$$V = (K/n)R_H^{2/3}S^{1/2}$$

Q = discharge (ft^3/sec or m^3/s)

V = velocity (ft/sec or m/s)

K = 1.486 for USCS units, 1.0 for SI units

A = cross-sectional area of flow (ft^2 or m^2)

R_H = hydraulic radius = A/P (ft or m)

P = wetted perimeter (ft or m)

S = slope (ft/ft or m/m)

n = roughness coefficient

Weir Formulas

Rectangular

Free discharge suppressed

$$Q = CLH^{3/2}$$

Free discharge contracted

$$Q = C(L - 0.2H)H^{3/2}$$

V-Notch

$$Q = CH^{5/2}, \text{ where}$$

Q = discharge (cfs or m^3/s)

C = 3.33 for rectangular weir (USCS units)

C = 1.84 for rectangular weir (SI units)

C = 2.54 for 90° V-notch weir (USCS units)

C = 1.40 for 90° V-notch weir (SI units)

L = weir length (ft or m)

H = head (depth of discharge over weir) ft or m

Hazen-Williams Equation

$$V = k_1 C R_H^{0.63} S^{0.54}, \text{ where}$$

$$Q = k_1 C A R_H^{0.63} S^{0.54}$$

C = roughness coefficient
k_1 = 0.849 for SI units
k_1 = 1.318 for USCS units
R_H = hydraulic radius (ft or m)
S = slope of energy grade line
= h_f/L (ft/ft or m/m)
V = velocity (ft/sec or m/s)
Q = discharge (ft^3/sec or m^3/s)

Circular Pipe Head Loss Equation (Head Loss Expressed in Feet)

$$h_f = \frac{4.73\,L}{C^{1.852}\,D^{4.87}}\,Q^{1.852}, \text{ where}$$

h_f = head loss (ft)
L = pipe length (ft)
D = pipe diameter (ft)
Q = flow (cfs)
C = Hazen-Williams coefficient

Circular Pipe Head Loss Equation (Head Loss Expressed as Pressure)

U.S. Customary Units

$$P = \frac{4.52\,Q^{1.85}}{C^{1.85}\,D^{4.87}}, \text{ where}$$

P = pressure loss (psi per foot of pipe)
Q = flow (gpm)
D = pipe diameter (inches)
C = Hazen-Williams coefficient

SI Units

$$P = \frac{6.05\,Q^{1.85}}{C^{1.85}\,D^{4.87}} \times 10^5, \text{ where}$$

P = pressure loss (bars per meter of pipe)
Q = flow (liters/minute)
D = pipe diameter (mm)

Values of Hazen-Williams Coefficient C

Pipe Material	C
Ductile iron	140
Concrete (regardless of age)	130
Cast iron:	
New	130
5 yr old	120
20 yr old	100
Welded steel, new	120
Wood stave (regardless of age)	120
Vitrified clay	110
Riveted steel, new	110
Brick sewers	100
Asbestos-cement	140
Plastic	150

♦ Formula for Calculating Rated Capacity at 20 psi from Fire Hydrant

$$Q_R = Q_F \times (H_R/H_F)^{0.54}, \text{ where}$$

Q_R = rated capacity (gpm) at 20 psi
Q_F = total test flow
H_R = P_S − 20 psi
H_F = $P_S - P_R$
P_S = static pressure
P_R = residual pressure

♦ Fire Hydrant Discharging to Atmosphere

$$Q = 29.8\,D^2 C_d P^{1/2}, \text{ where}$$

Q = discharge (gpm)
D = outlet diameter (in.)
P = pressure detected by pitot gauge (psi)
C_d = hydrant coefficient based on hydrant outlet geometry

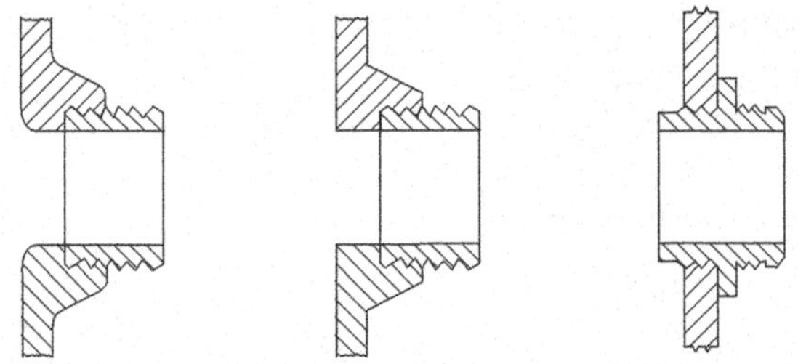

OUTLET SMOOTH AND WELL-ROUNDED COEF. 0.90 | OUTLET SQUARE AND SHARP COEF. 0.80 | OUTLET SQUARE AND PROJECTING INTO BARREL COEF. 0.70

• Fire Sprinkler Discharge

$$Q = KP^{1/2}, \text{ where}$$

Q = flow (gpm)
K = measure of the ease of getting water out of the orifice, related to size and shape of the orifice in units of gpm per (psi)$^{1/2}$
P = pressure (psi)

♦ NFPA Standard 291, *Recommended Practice for Fire Flow Testing and Marking of Hydrants*, Section 4.10.1.2
• Isman, Kenneth E., P.E., 2001, "Which Sprinkler to Choose?", *Fire Protection Engineering*, Winter 2001 (9), Society of Fire Protection Engineers, p. 28.

Sprinkler K Factors

Orifice Size	Name	K Factor
1/2"	Standard	5.6
17/32"	Large	8.0
5/8"	Extra large	11.2

TRANSPORTATION

U.S. Customary Units

a = deceleration rate (ft/sec^2)
A = absolute value of algebraic difference in grades (%)
e = superelevation (%)
f = side friction factor
$\pm G$ = percent grade divided by 100 (uphill grade "+")
h_1 = height of driver's eyes above the roadway surface (ft)
h_2 = height of object above the roadway surface (ft)
L = length of curve (ft)
L_s = spiral transition length (ft)
R = radius of curve (ft)
SSD = stopping sight distance (ft)
ISD = intersection sight distance (ft)
t_g = time gap for vehicle entering roadway (sec)
t = driver reaction time (sec)
V = design speed (mph)
v = vehicle approach speed (fps)
W = width of intersection, curb-to-curb (ft)
l = length of vehicle (ft)
y = length of yellow interval to nearest 0.1 sec (sec)
r = length of red clearance interval to nearest 0.1 sec (sec)

Vehicle Signal Change Interval

$$y = t + \frac{v}{2a \pm 64.4\,G}$$

$$r = \frac{W + l}{v}$$

Stopping Sight Distance

$$SSD = 1.47Vt + \frac{V^2}{30\left(\left(\frac{a}{32.2}\right) \pm G\right)}$$

$$ISD = 1.47\,V_{major}\,t_g$$

Peak Hour Factor (PHF)

$$PHF = \frac{\text{Hourly Volume}}{\text{Hourly Flow Rate}} = \frac{\text{Hourly Volume}}{4 \times \text{Peak 15-minute Volume}}$$

Queueing models are found in the Industrial Engineering section.

• SIGHT DISTANCE

Vertical Curves: Sight Distance Related to Curve Length

	$S \le L$	$S > L$
Crest Vertical Curve General equation:	$L = \frac{AS^2}{100(\sqrt{2h_1}+\sqrt{2h_2})^2}$	$L = 2S - \frac{200\left(\sqrt{h_1}+\sqrt{h_2}\right)^2}{A}$
Standard Criteria: $h_1 = 3.50$ ft and $h_2 = 2.0$ ft:	$L = \frac{AS^2}{2{,}158}$	$L = 2S - \frac{2{,}158}{A}$
Sag Vertical Curve (based on standard headlight criteria)	$L = \frac{AS^2}{400 + 3.5S}$	$L = 2S - \left(\frac{400 + 3.5S}{A}\right)$
Sag Vertical Curve (based on riding comfort)	$L = \frac{AV^2}{46.5}$	
Sag Vertical Curve (based on adequate sight distance under an overhead structure to see an object beyond a sag vertical curve)	$L = \frac{AS^2}{800\left(C - \frac{h_1+h_2}{2}\right)}$	$L = 2S - \frac{800}{A}\left(C - \frac{h_1+h_2}{2}\right)$
	C = vertical clearance for overhead structure (overpass) located within 200 feet of the midpoint of the curve	

• Compiled from AASHTO, *A Policy on Geometric Design of Highways and Streets*, 6th ed., 2011.

Horizontal Curves	
Side friction factor (based on superelevation)	$0.01e + f = \dfrac{V^2}{15R}$
Spiral Transition Length	$L_s = \dfrac{3.15V^3}{RC}$ C = rate of increase of lateral acceleration [use 1 ft/sec^3 unless otherwise stated]
Sight Distance (to see around obstruction)	$\text{HSO} = R\left[1 - \cos\left(\dfrac{28.65S}{R}\right)\right]$ HSO = Horizontal sight line offset

♦ Basic Freeway Segment Highway Capacity

FFS (mi/h)	Breakpoint (pc/h/ln)	Flow Rate Range: ≥ 0 ≤ Breakpoint	Flow Rate Range: > Breakpoint ≤ Capacity
75	1,000	75	$75 - 0.00001107\,(v_p - 1{,}000)^2$
70	1,200	70	$70 - 0.00001160\,(v_p - 1{,}200)^2$
65	1,400	65	$65 - 0.00001418\,(v_p - 1{,}400)^2$
60	1,600	60	$60 - 0.00001816\,(v_p - 1{,}600)^2$
55	1,800	55	$55 - 0.00002469\,(v_p - 1{,}800)^2$

Notes: FFS = free-flow speed, v_p = demand flow rate (pc/h/ln) under equivalent base conditions. Maximum flow rate for the equations is capacity: 2,400 pc/h/ln for 70- and 75-mph FFS; 2,350 pc/h/ln for 65-mph FFS; 2,300 pc/h/ln for 60-mph FFS; and 2,250 pc/h/ln for 55-mph FFS.

LOS	Density (pc/mi/ln)
A	≤11
B	>11 – 18
C	>18 – 26
D	>26 – 35
E	>35 – 45
F	Demand exceeds capacity >45

$$FFS = 75.4 - f_{LW} - f_{LC} - 3.22\ TRD^{0.84}$$

where

FFS = free flow speed of basic freeway segment (mi/h)

f_{LW} = adjustment for lane width (mi/h)

f_{LC} = adjustment for right-side lateral clearance (mi/h)

TRD = total ramp density (ramps/mi)

Average Lane Width (ft)	Reduction in FFS, f_{LW} (mi/h)
≥12	0.0
≥11 – 12	1.9
≥10 – 11	6.6

Right-Side Lateral Clearance (ft)	Lanes in One Direction: 2	3	4	≥5
≥6	0.0	0.0	0.0	0.0
5	0.6	0.4	0.2	0.1
4	1.2	0.8	0.4	0.2
3	1.8	1.2	0.6	0.3
2	2.4	1.6	0.8	0.4
1	3.0	2.0	1.0	0.5
0	3.6	2.4	1.2	0.6

♦ *HCM 2010: Highway Capacity Manual, Vol 2, Uninterrupted Flow*, Transportation Research Board of the National Academics, Washington, DC, 2010.

$$v_p = \frac{V}{PHF \times N \times f_{HV} \times f_p}$$

where

v_p = demand flowrate under equivalent base conditions (pc/h/ln)

V = demand volume under prevailing conditions (veh/h)

PHF = peak-hour factor

N = number of lanes in analysis direction

f_{HV} = adjustment factor for presence of heavy vehicles in traffic stream

f_p = adjustment factor for unfamiliar driver populations

$$f_{HV} = \frac{1}{1 + P_T(E_T - 1) + P_R(E_R - 1)}$$

where

f_{HV} = heavy-vehicle adjustment factor

P_T = proportion of trucks and buses in traffic stream

P_R = proportion of RVs in traffic stream

E_T = passenger-car equivalent (PCE) of one truck or bus in traffic stream

E_R = PCE of one RV in traffic stream

	PCE by Type of Terrain		
Vehicle	Level	Rolling	Mountainous
Trucks and buses, E_T	1.5	2.5	4.5
RVs, E_R	1.2	2.0	4.0

$$D = \frac{v_p}{S}$$

where

D = density(pc/mi/ln)

v_p = demand flowrate (pc/h/ln)

S = mean speed of traffic stream under base conditions (mi/h)

- **Traffic Flow Relationships**

Greenshields Model

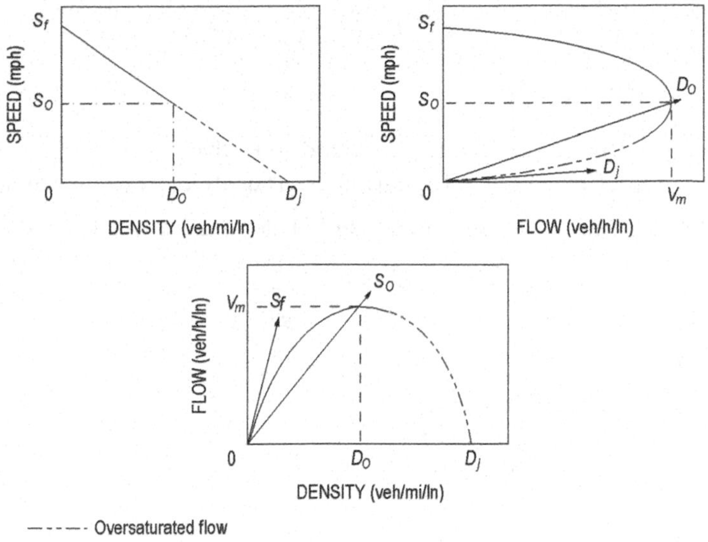

$$S = S_f - \frac{S_f}{D_j} D$$

$$V = S_f D - \frac{S_f}{D_j} D^2$$

$$V_m = \frac{D_j S_f}{4}$$

$$D_o = \frac{D_j}{2}$$

where

D = density (veh/mi)

S = speed (mi/hr)

V = flow (veh/hr)

V_m = maximum flow

D_o = optimum density (sometimes called critical density)

D_j = jam density

S_o = optimum speed (often called critical speed)

S_f = theoretical speed selected by the first driver entering a facility (i.e., under zero density and zero flowrate conditions)

Gravity Model

$$T_{ij} = P_i \left[\frac{A_j F_{ij} K_{ij}}{\sum_j A_j F_{ij} K_{ij}} \right]$$

where

T_{ij} = number of trips that are produced in zone i and attracted to zone j

P_i = total number of trips produced in zone i

A_j = number of trips attracted to zone j

F_{ij} = a friction factor that is an inverse function of travel time between zones i and j

K_i = socioeconomic adjustment factor for interchange ij

- AASHTO, *A Policy on Geometric Design of Highways and Streets*, 6th ed., 2011. Used by permission.

Logit Models

$$U_x = \sum_{i=1}^{n} a_i X_i$$

where

U_x = utility of mode x

n = number of attributes

X_i = attribute value (time, cost, and so forth)

a_i = coefficient value for attributes i (negative, since the values are disutilities)

If two modes, auto (A) and transit (T), are being considered, the probability of selecting the auto mode A can be written as

$$P(A) = \frac{e^{U_A}}{e^{U_A} + e^{U_T}} \qquad P(x) = \frac{e^{U_x}}{\sum_{e=1}^{n} e^{U_{xi}}}$$

Traffic Safety Equations

Crash Rates at Intersections

$$RMEV = \frac{A \times 1{,}000{,}000}{V}$$

where

$RMEV$ = crash rate per million entering vehicles

A = number of crashes, total or by type occurring in a single year at the location

V = $ADT \times 365$

ADT = average daily traffic entering intersection

Crash Rates for Roadway Segments

$$RMVM = \frac{A \times 100{,}000{,}000}{VMT}$$

where

$RMVM$ = crash rate per hundred million vehicle miles

A = number of crashes, total or by type at the study location, during a given period

VMT = vehicle miles of travel during the given period
= $ADT \times$ (number of days in study period) $\times$ (length of road)

ADT = average daily traffic on the roadway segment

Crash Reduction

$$\text{Crashes prevented} = N \times CR \frac{(ADT \text{ after improvement})}{(ADT \text{ before improvement})}$$

where

N = expected number of crashes if countermeasure is not implemented and if the traffic volume remains the same

CR = $CR_1 + (1 - CR_1)CR_2 + (1 - CR_1)(1 - CR_2)CR_3 + \ldots + (1 - CR_1)\ldots(1 - CR_{m-1})\,CR_m$
= overall crash reduction factor for multiple mutually exclusive improvements at a single site

CR_i = crash reduction factor for a specific countermeasure i

m = number of countermeasures at the site

Garber, Nicholas J., and Lester A. Hoel, *Traffic and Highway Engineering*, 4th ed., Cengage Learning, 2009.

AASHTO Structural Number Equation
$SN = a_1D_1 + a_2D_2 + \ldots + a_nD_n$, where
SN = structural number for the pavement
a_i = layer coefficient and D_i = thickness of layer (inches).

Gross Axle Load		Load Equivalency Factors		Gross Axle Load		Load Equivalency Factors	
kN	lb	Single Axles	Tandem Axles	kN	lb	Single Axles	Tandem Axles
4.45	1,000	0.00002		187.0	42,000	25.64	2.51
8.9	2,000	0.00018		195.7	44,000	31.00	3.00
17.8	4,000	0.00209		200.0	45,000	34.00	3.27
22.25	5,000	0.00500		204.5	46,000	37.24	3.55
26.7	**6,000**	**0.01043**		**213.5**	**48,000**	**44.50**	**4.17**
35.6	8,000	0.0343		222.4	50,000	52.88	4.86
44.5	10,000	0.0877	0.00688	231.3	52,000		5.63
53.4	12,000	0.189	0.0144	240.2	54,000		6.47
62.3	14,000	0.360	0.0270	244.6	55,000		6.93
66.7	**15,000**	**0.478**	**0.0360**	**249.0**	**56,000**		**7.41**
71.2	16,000	0.623	0.0472	258.0	58,000		8.45
80.0	18,000	1.000	0.0773	267.0	60,000		9.59
89.0	20,000	1.51	0.1206	275.8	62,000		10.84
97.8	22,000	2.18	0.180	284.5	64,000		12.22
106.8	**24,000**	**3.03**	**0.260**	**289.0**	**65,000**		**12.96**
111.2	25,000	3.53	0.308	293.5	66,000		13.73
115.6	26,000	4.09	0.364	302.5	68,000		15.38
124.5	28,000	5.39	0.495	311.5	70,000		17.19
133.5	30,000	6.97	0.658	320.0	72,000		19.16
142.3	**32,000**	**8.88**	**0.857**	**329.0**	**74,000**		**21.32**
151.2	34,000	11.18	1.095	333.5	75,000		22.47
155.7	35,000	12.50	1.23	338.0	76,000		23.66
160.0	36,000	13.93	1.38	347.0	78,000		26.22
169.0	38,000	17.20	1.70	356.0	80,000		28.99
178.0	**40,000**	**21.08**	**2.08**				

Note: kN converted to lb are within 0.1 percent of lb shown

Superpave

PERFORMANCE-GRADED (PG) BINDER GRADING SYSTEM

PERFORMANCE GRADE	PG 52							PG 58					PG 64				
	–10	–16	–22	–28	–34	–40	–46	–16	–22	–28	–34	–40	–16	–22	–28	–34	–40
AVERAGE 7-DAY MAXIMUM PAVEMENT DESIGN TEMPERATURE, °C[a]	<52							<58					<64				
MINIMUM PAVEMENT DESIGN TEMPERATURE, °C[a]	>–10	>–16	>–22	>–28	>–34	>–40	>–46	>–16	>–22	>–28	>–34	>–40	>–16	>–22	>–28	>–34	>–40
	ORIGINAL BINDER																
FLASH POINT TEMP, T48: MINIMUM °C	230																
VISCOSITY, ASTM D 4402:[b] MAXIMUM, 3 Pa-s (3,000 cP), TEST TEMP, °C	135																
DYNAMIC SHEAR, TP5:[c] $G^*/\sin\delta$, MINIMUM, 1.00 kPa TEST TEMPERATURE @ 10 rad/sec., °C	52							58					64				
	ROLLING THIN FILM OVEN (T240) OR THIN FILM OVEN (T179) RESIDUE																
MASS LOSS, MAXIMUM, %	1.00																
DYNAMIC SHEAR, TP5: $G^*/\sin\delta$, MINIMUM, 2.20 kPa TEST TEMP @ 10 rad/sec. °C	52							58					64				
	PRESSURE AGING VESSEL RESIDUE (PP1)																
PAV AGING TEMPERATURE, °C[d]	90							100					100				
DYNAMIC SHEAR, TP5: $G^*/\sin\delta$, MAXIMUM, 5,000 kPa TEST TEMP @ 10 rad/sec. °C	25	22	19	16	13	10	7	25	22	19	16	13	28	25	22	19	16
PHYSICAL HARDENING[e]	REPORT																
CREEP STIFFNESS, TP1:[f] S, MAXIMUM, 300 MPa M-VALUE, MINIMUM, 0.300 TEST TEMP, @ 60 sec., °C	0	–6	–12	–18	–24	–30	–36	–6	–12	–18	–24	–30	–6	–12	–18	–24	–30
DIRECT TENSION, TP3:[f] FAILURE STRAIN, MINIMUM, 1.0% TEST TEMP @ 1.0 mm/min, °C	0	–6	–12	–18	–24	–30	–36	–6	–12	–18	–24	–30	–6	–12	–18	–24	–30

Superpave Mixture Design: Compaction Requirements

TRAFFIC, MILLION ESALs	SUPERPAVE GYRATORY COMPACTION EFFORT — AVERAGE DESIGN HIGH AIR TEMPERATURE											
	< 39°C			39° – 40°C			41° – 42°C			42° – 43°C		
	N_{int}	N_{des}	N_{max}	N_{int}	N_{des}	N_{max}	N_{int}	N_{des}	N_{max}	N_{int}	N_{des}	N_{max}
< 0.3	7	68	104	7	74	114	7	78	121	7	82	127
< 1	7	76	117	7	83	129	7	88	138	8	93	146
< 3	7	86	134	8	95	150	8	100	158	8	105	167
< 10	8	96	152	8	106	169	8	113	181	9	119	192
< 30	8	109	174	9	121	195	9	128	208	9	135	220
< 100	9	126	204	9	139	228	9	146	240	10	153	253
≥ 100	9	142	233	10	158	262	10	165	275	10	177	288

VFA REQUIREMENTS @ 4% AIR VOIDS	
TRAFFIC, MILLION ESALs	DESIGN VFA (%)
< 0.3	70 – 80
< 1	65 – 78
< 3	65 – 78
< 10	65 – 75
< 30	65 – 75
< 100	65 – 75
≥ 100	65 – 75

VMA REQUIREMENTS @ 4% AIR VOIDS					
NOMINAL MAXIMUM AGGREGATE SIZE (mm)	9.5	12.5	19.0	25.0	37.5
MINIMUM VMA (%)	15	14	13	12	11

COMPACTION KEY			
SUPERPAVE GYRATORY COMPACTION	N_{int}	N_{des}	N_{max}
PERCENT OF Gmm	≤ 89%	96%	≤ 98%

Horizontal Curve Formulas

D = Degree of Curve, Arc Definition
PC = Point of Curve (also called *BC*)
PT = Point of Tangent (also called *EC*)
PI = Point of Intersection
I = Intersection Angle (also called Δ) Angle Between Two Tangents
L = Length of Curve, from *PC* to *PT*
T = Tangent Distance
E = External Distance
R = Radius
LC = Length of Long Chord
M = Length of Middle Ordinate
c = Length of Sub-Chord
d = Angle of Sub-Chord
l = Curve Length for Sub-Chord

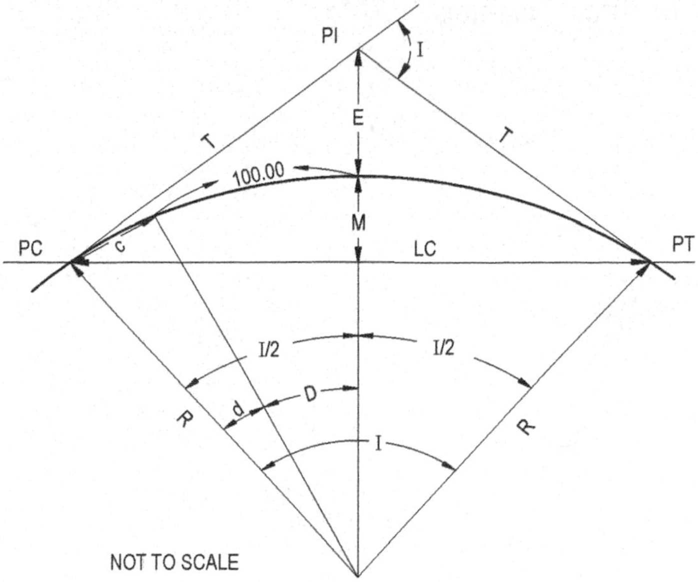

$$R = \frac{5729.58}{D}$$

$$R = \frac{LC}{2\sin(I/2)}$$

$$T = R\tan(I/2) = \frac{LC}{2\cos(I/2)}$$

$$L = RI\frac{\pi}{180} = \frac{I}{D}100$$

$$M = R\left[1 - \cos(I/2)\right]$$

$$\frac{R}{E+R} = \cos(I/2)$$

$$\frac{R-M}{R} = \cos(I/2)$$

$$c = 2R\sin(d/2)$$

$$l = Rd\left(\frac{\pi}{180}\right)$$

$$E = R\left[\frac{1}{\cos(I/2)} - 1\right]$$

Deflection angle per 100 feet of arc length equals $D/2$

LATITUDES AND DEPARTURES

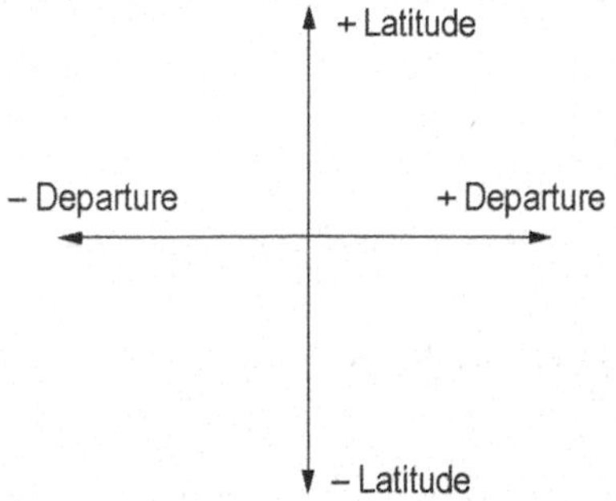

Vertical Curve Formulas

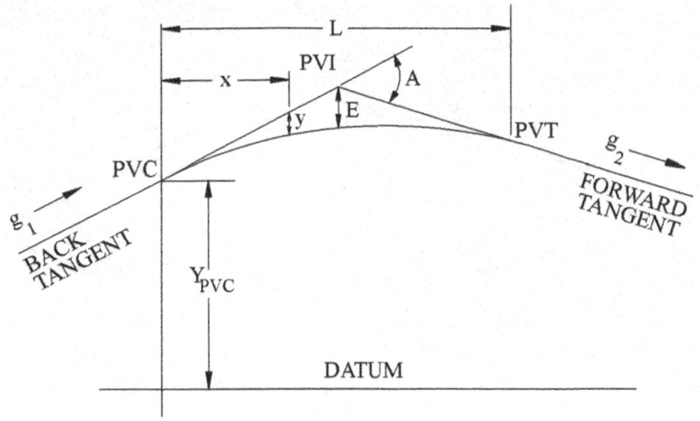

VERTICAL CURVE FORMULAS
NOT TO SCALE

$$y = ax^2$$

$$A = g_2 - g_1$$

$$a = \frac{A}{2L} = \frac{g_2 - g_1}{2L}$$

$$E = a\left(\frac{L}{2}\right)^2$$

$$r = \frac{g_2 - g_1}{L}$$

$$K = \frac{L}{A}$$

$$x_m = -\frac{g_1}{2a} = \frac{g_1 L}{g_1 - g_2}$$

PVC = Point of vertical curvature, or beginning of curve

PVI = Point of vertical intersection, or vertex

PVT = Point of vertical tangency, or end of curve

L = Length of curve

y = Tangent offset

x = Horizontal distance from PVC to point on curve

x_m = Horizontal distance to min/max elevation on curve

g_1 = Grade of back tangent

g_2 = Grade of forward tangent

A = Algebraic difference in grades

a = Parabola constant

E = Tangent offset at PVI

r = Rate of change of grade

K = Rate of vertical curvature

Tangent elevation = $Y_{\text{PVC}} + g_1 x = Y_{\text{PVI}} + g_2 (x - L/2)$

Curve elevation = $Y_{\text{PVC}} + g_1 x + ax^2 = Y_{\text{PVC}} + g_1 x + [(g_2 - g_1)/(2L)]x^2$

EARTHWORK FORMULAS

Average End Area Formula, $V = L(A_1 + A_2)/2$

Prismoidal Formula, $V = L\,(A_1 + 4A_m + A_2)/6$

where A_m = area of mid-section

L = distance between A_1 and A_2

Pyramid or Cone, $V = h$ (Area of Base)/3

AREA FORMULAS

Area by Coordinates: Area = $[X_A (Y_B - Y_N) + X_B (Y_C - Y_A) + X_C (Y_D - Y_B) + \ldots + X_N (Y_A - Y_{N-1})] / 2$

Trapezoidal Rule: Area = $w\left(\frac{h_1 + h_n}{2} + h_2 + h_3 + h_4 + \ldots + h_{n-1}\right)$

w = common interval

Simpson's 1/3 Rule: Area = $w\left[h_1 + 2\left(\sum_{k=3,5,\ldots}^{n-2} h_k\right) + 4\left(\sum_{k=2,4,\ldots}^{n-1} h_k\right) + h_n\right] \Big/ 3$

n must be odd number of measurements (only for Simpson's 1/3 Rule)

CONSTRUCTION

Construction project scheduling and analysis questions may be based on either the activity-on-node method or the activity-on-arrow method.

CPM PRECEDENCE RELATIONSHIPS

ACTIVITY-ON-NODE

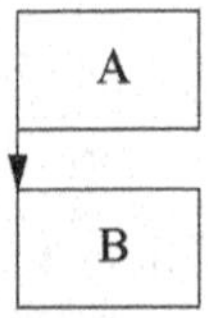

START-TO-START: START OF B DEPENDS ON THE START OF A

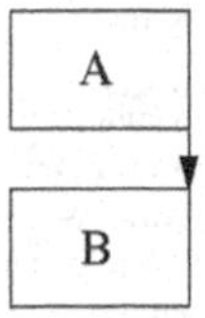

FINISH-TO-FINISH: FINISH OF B DEPENDS ON THE FINISH OF A

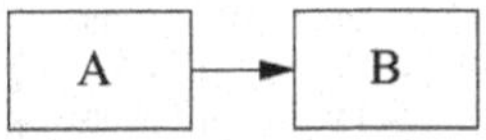

FINISH-TO-START: START OF B DEPENDS ON THE FINISH OF A

ACTIVITY-ON-ARROW ANNOTATION

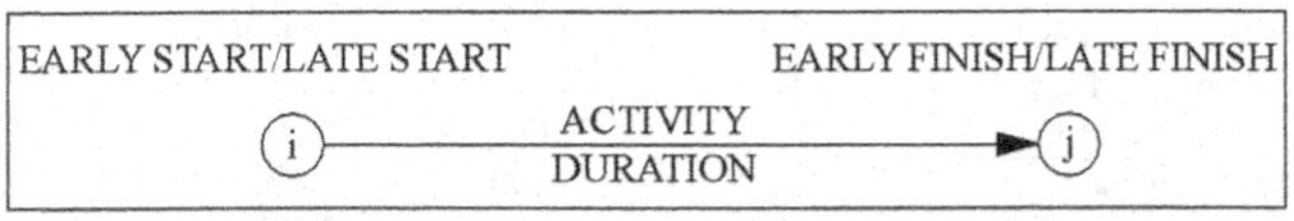

ACTIVITY-ON-NODE ANNOTATION

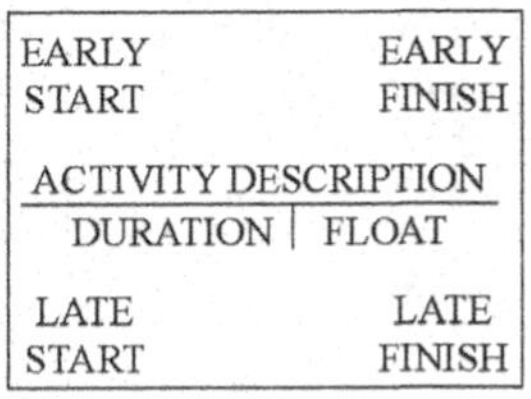

Nomenclature

ES = Early start = Latest EF of predecessors
EF = Early finish = ES + duration
LS = Late start = LF – duration
LF = Late finish = Earliest LS of successors
D = Duration
Float = LS – ES or LF – EF

Determining the Project Size Modifier

Square Foot Base Size							
Building Type	**Median Cost per S.F.**	**Typical Size Gross S.F.**	**Typical Range Gross S.F.**	**Building Type**	**Median Cost per S.F.**	**Typical Size Gross S.F.**	**Typical Range Gross S.F.**
Apartments, Low Rise	$ 54.05	21,000	9,700–37,200	Jails	$ 165.00	13,700	7,500–28,000
Apartments, Mid Rise	68.25	50,000	32,000–100,000	Libraries	97.30	12,000	7,000–31,000
Apartments, High Rise	78.30	310,000	100,000–650,000	Medical Clinics	93.15	7,200	4,200–15,700
Auditoriums	90.35	25,000	7,600–39,000	Medical Offices	87.50	6,000	4,000–15,000
Auto Sales	55.90	20,000	10,800–28,600	Motels	67.00	27,000	15,800–51,000
Banks	121.00	4,200	2,500–7,500	Nursing Homes	89.95	23,000	15,000–37,000
Churches	81.60	9,000	5,300–13,200	Offices, Low Rise	73.00	8,600	4,700–19,000
Clubs, Country	81.40	6,500	4,500–15,000	Offices, Mid Rise	76.65	52,000	31,300–83,100
Clubs, Social	79.15	10,000	6,000–13,500	Offices, High Rise	98.05	260,000	151,000–468,000
Clubs, YMCA	81.60	28,300	12,800–39,400	Police Stations	122.00	10,500	4,000–19,000
Colleges (Class)	107.00	50,000	23,500–98,500	Post Offices	90.40	12,400	6,800–30,000
Colleges (Science Lab)	156.00	45,600	16,600–80,000	Power Plants	678.00	7,500	1,000–20,000
College (Student Union)	119.00	33,400	16,000–85,000	Religious Education	74.85	9,000	6,000–12,000
Community Center	85.05	9,400	5,300–16,700	Research	127.00	19,000	6,300–45,000
Court Houses	116.00	32,400	17,800–106,000	Restaurants	110.00	4,400	2,800–6,000
Dept. Stores	50.50	90,000	44,000–122,000	Retail Stores	53.70	7,200	4,000–17,600
Dormitories, Low Rise	87.20	24,500	13,400–40,000	Schools, Elementary	78.20	41,000	24,500–55,000
Dormitories, Mid Rise	113.00	55,600	36,100–90,000	Schools, Jr. High	79.65	92,000	52,000–119,000
Factories	48.95	26,400	12,900–50,000	Schools, Sr. High	79.65	101,000	50,500–175,000
Fire Stations	85.45	5,800	4,000–8,700	Schools, Vocational	79.35	37,000	20,500–82,000
Fraternity Houses	84.10	12,500	8,200–14,800	Sports Arenas	66.45	15,000	5,000–40,000
Funeral Homes	94.00	7,800	4,500–11,000	Supermarkets	53.85	20,000	12,000–30,000
Garages, Commercial	59.70	9,300	5,000–13,600	Swimming Pools	125.00	13,000	7,800–22,000
Garages, Municipal	76.40	8,300	4,500–12,600	Telephone Exchange	145.00	4,500	1,200–10,600
Garages, Parking	31.30	163,000	76,400–225,300	Theaters	79.70	10,500	8,800–17,500
Gymnasiums	78.95	19,200	11,600–41,000	Town Halls	87.65	10,800	4,800–23,400
Hospitals	149.00	55,000	27,200–125,000	Warehouses	36.15	25,000	8,000–72,000
House (Elderly)	73.90	37,000	21,000–66,000	Warehouse and Office	41.75	25,000	8,000–72,000
Housing (Public)	68.45	36,000	14,400–74,400				
Ice Rinks	76.00	29,000	27,200–33,600				

Earned-Value Analysis

BCWS = Budgeted cost of work scheduled (Planned)

ACWP = Actual cost of work performed (Actual)

BCWP = Budgeted cost of work performed (Earned)

Variances

CV = BCWP – ACWP (Cost variance = Earned – Actual)

SV = BCWP – BCWS (Schedule variance = Earned – Planned)

Indices

$$\mathrm{CPI} = \frac{\mathrm{BCWP}}{\mathrm{ACWP}} \quad \left(\text{Cost Performance Index} = \frac{\text{Earned}}{\text{Actual}}\right)$$

$$\mathrm{SPI} = \frac{\mathrm{BCWP}}{\mathrm{BCWS}} \quad \left(\text{Schedule Performance Index} = \frac{\text{Earned}}{\text{Planned}}\right)$$

Forecasting

BAC = Original project estimate (Budget at completion)

$$\mathrm{ETC} = \frac{\mathrm{BAC} - \mathrm{BCWP}}{\mathrm{CPI}} \quad (\text{Estimate to complete})$$

$$\mathrm{EAC} = (\mathrm{ACWP} + \mathrm{ETC}) \quad (\text{Estimate at completion})$$

ENVIRONMENTAL ENGINEERING

AIR POLLUTION

Nomenclature

$$\frac{\mu g}{m^3} = \text{ppb} \times \frac{P(\text{MW})}{RT}$$

ppb = parts per billion

P = pressure (atm)

R = ideal gas law constant
= 0.0821 L•atm/(mole•K)

T = absolute temperature, K = 273.15 + °C

MW = molecular weight (g/mole)

Atmospheric Dispersion Modeling (Gaussian)

σ_y and σ_z as a function of downwind distance and stability class, see following figures.

$$C = \frac{Q}{2\pi u \sigma_y \sigma_z} \exp\left(-\frac{1}{2}\frac{y^2}{\sigma_y^2}\right)\left[\exp\left(-\frac{1}{2}\frac{(z-H)^2}{\sigma_z^2}\right) + \exp\left(-\frac{1}{2}\frac{(z+H)^2}{\sigma_z^2}\right)\right]$$

where

C = steady-state concentration at a point (x, y, z) ($\mu g/m^3$)

Q = emissions rate (μg/s)

σ_y = horizontal dispersion parameter (m)

σ_z = vertical dispersion parameter (m)

u = average wind speed at stack height (m/s)

y = horizontal distance from plume centerline (m)

z = vertical distance from ground level (m)

H = effective stack height (m) = $h + \Delta h$
where h = physical stack height

Δh = plume rise

x = downwind distance along plume centerline (m)

Maximum concentration at ground level and directly downwind from an elevated source.

$$C_{\max} = \frac{Q}{\pi u \sigma_y \sigma_z} \exp\left(-\frac{1}{2}\frac{(H^2)}{\sigma_z^2}\right)$$

where variables are as above except

$C_{\max}$ = maximum ground-level concentration

$\sigma_z = \frac{H}{\sqrt{2}}$ for neutral atmospheric conditions

Selected Properties of Air

Nitrogen (N_2) by volume		78.09%
Oxygen (O_2) by volume		20.94%
Argon (Ar) by volume		0.93%
Molecular weight of air		28.966 g/mol
Absolute viscosity, μ		
	at 80°F	0.045 lbm/(hr-ft)
	at 100°F	0.047 lbm/(hr-ft)
Density		
	at 80°F	0.0734 lbm/ft^3
	at 100°F	0.0708 lbm/ft^3

The dry adiabatic lapse rate Γ_{AD} is 0.98°C per 100 m (5.4°F per 1,000 ft). This is the rate at which dry air cools adiabatically with altitude.

$$\text{Lapse rate} = \Gamma = -\frac{\Delta T}{\Delta z}$$

where ΔT is the change in temperature and Δz is the change in elevation.

The actual (environmental) lapse rate Γ is compared to Γ_{AD} to determine stability as follows:

Lapse Rate	Stability Condition
$\Gamma > \Gamma_{AD}$	Unstable
$\Gamma = \Gamma_{AD}$	Neutral
$\Gamma < \Gamma_{AD}$	Stable

Atmospheric Stability Under Various Conditions

Surface Wind Speed[a] (m/s)	Day Solar Insolation Strong[b]	Moderate[c]	Slight[d]	Night Cloudiness[e] Cloudy (≥4/8)	Clear (≤3/8)
<2	A	A–B[f]	B	E	F
2–3	A–B	B	C	E	F
3–5	B	B–C	C	D	E
5–6	C	C–D	D	D	D
>6	C	D	D	D	D

Notes:

a. Surface wind speed is measured at 10 m above the ground.

b. Corresponds to clear summer day with sun higher than 60° above the horizon.

c. Corresponds to a summer day with a few broken clouds, or a clear day with sun 35-60° above the horizon.

d. Corresponds to a fall afternoon, or a cloudy summer day, or clear summer day with the sun 15-35°.

e. Cloudiness is defined as the fraction of sky covered by the clouds.

f. For A–B, B–C, or C–D conditions, average the values obtained for each.

* A = Very unstable D = Neutral
B = Moderately unstable E = Slightly stable
C = Slightly unstable F = Stable

Regardless of wind speed, Class D should be assumed for overcast conditions, day or night.

Turner, D.B., "Workbook of Atmospheric Dispersion Estimates: An Introduction to Dispersion Modeling," 2nd ed., Lewis Publishing/CRC Press, Florida, 1994.

♦

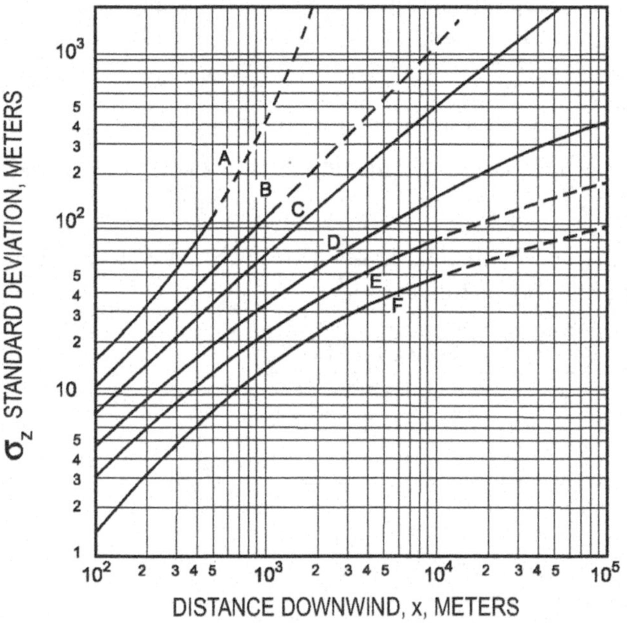

VERTICAL STANDARD DEVIATIONS OF A PLUME

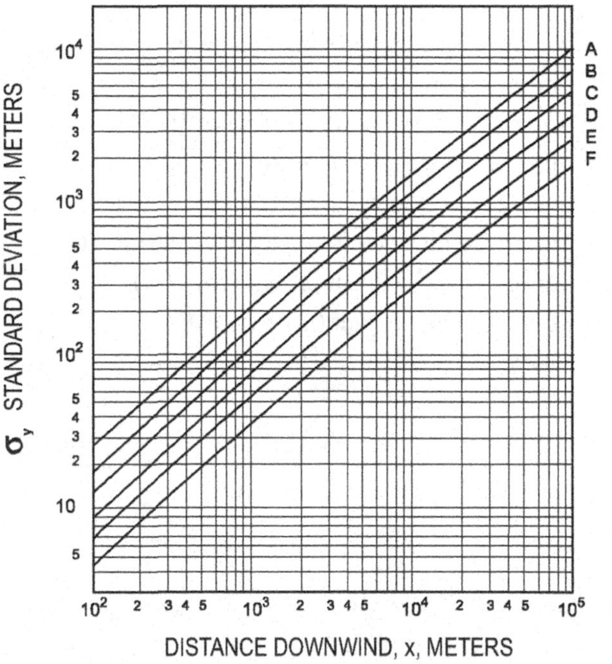

HORIZONTAL STANDARD DEVIATIONS OF A PLUME

A – EXTREMELY UNSTABLE
B – MODERATELY UNSTABLE
C – SLIGHTLY UNSTABLE
D – NEUTRAL
E – SLIGHTLY STABLE
F – MODERATELY STABLE

♦ Turner, D.B., "Workbook of Atmospheric Dispersion Estimates," U.S. Department of Health, Education, and Welfare, Washington, DC, 1970.

Downwind distance where the maximum concentration occurs, x_{max}, versus $(Cu/Q)_{max}$ as a function of stability class

♦

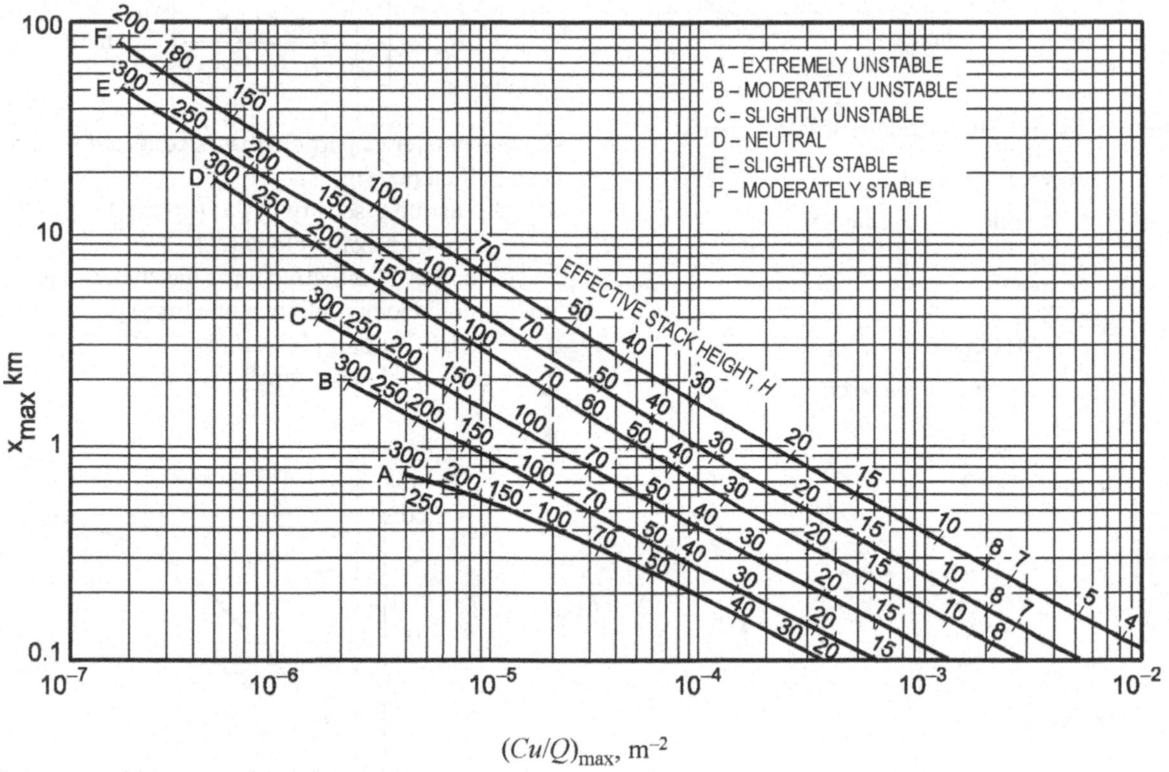

NOTES: Effective stack height shown on curves numerically.

x_{max} = distance along plume centerline to the point of maximum concentration

$$(Cu/Q)_{max} = e^{[a + b \ln H + c (\ln H)^2 + d(\ln H)^3]}$$

H = effective stack height, stack height + plume rise, m

♦

Values of Curve-Fit Constants for Estimating $(Cu/Q)_{max}$ from H as a Function of Atmospheric Stability

Stability	Constants			
	a	*b*	*c*	*d*
A	–1.0563	–2.7153	0.1261	0
B	–1.8060	–2.1912	0.0389	0
C	–1.9748	–1.9980	0	0
D	–2.5302	–1.5610	–0.0934	0
E	–1.4496	–2.5910	0.2181	–0.0343
F	–1.0488	–3.2252	0.4977	–0.0765

Adapted from Ranchoux, R.J.P., 1976.

♦ Turner, D.B., "Workbook of Atmospheric Dispersion Estimates: An Introduction to Dispersion Modeling," 2nd ed., Lewis Publishing/CRC Press, Florida, 1994.

Cyclone

Cyclone Collection (Particle Removal) Efficiency

$$\eta = \frac{1}{1 + \left(d_{pc}/d_p\right)^2}, \text{ where}$$

d_{pc} = diameter of particle collected with 50% efficiency

d_p = diameter of particle of interest

η = fractional particle collection efficiency

♦

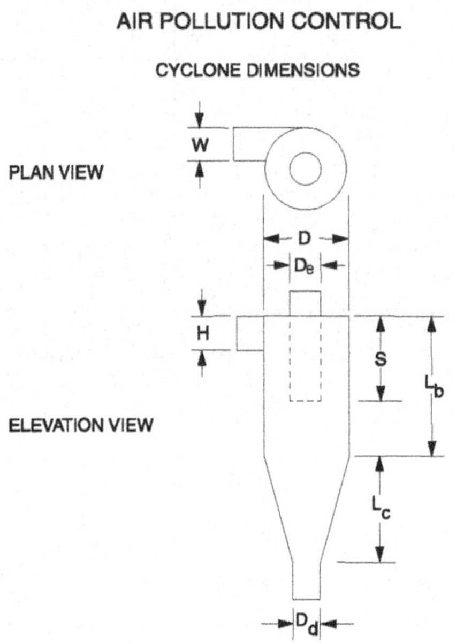

Cyclone Effective Number of Turns Approximation

$$N_e = \frac{1}{H}\left[L_b + \frac{L_c}{2}\right], \text{ where}$$

N_e = number of effective turns gas makes in cyclone

H = inlet height of cyclone (m)

L_b = length of body cyclone (m)

L_c = length of cone of cyclone (m)

Cyclone 50% Collection Efficiency for Particle Diameter

$$d_{pc} = \left[\frac{9\mu W}{2\pi N_e V_i\left(\rho_p - \rho_g\right)}\right]^{0.5}, \text{ where}$$

d_{pc} = diameter of particle that is collected with 50% efficiency (m)

μ = dynamic viscosity of gas (kg/m•s)

W = inlet width of cyclone (m)

N_e = number of effective turns gas makes in cyclone

V_i = inlet velocity into cyclone (m/s)

ρ_p = density of particle (kg/m^3)

ρ_g = density of gas (kg/m^3)

♦

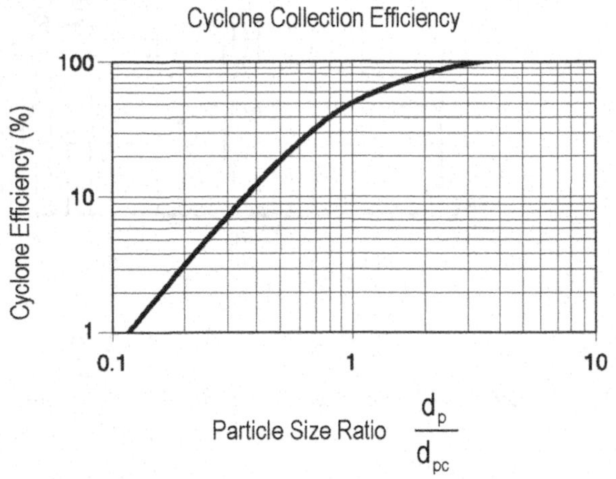

♦

Cyclone Ratio of Dimensions to Body Diameter

Dimension	High Efficiency	Conventional	High Throughput
Inlet height, H	0.44	0.50	0.80
Inlet width, W	0.21	0.25	0.35
Body length, L_b	1.40	1.75	1.70
Cone length, L_c	2.50	2.00	2.00
Vortex finder length, S	0.50	0.60	0.85
Gas exit diameter, D_e	0.40	0.50	0.75
Dust outlet diameter, D_d	0.40	0.40	0.40

♦ Adapted from Cooper, David C., and F.C. Alley, *Air Pollution Control: A Design Approach*, 2nd ed., Waveland Press, Illinois, 1986.

Baghouse

Air-to-Cloth Ratio for Baghouses

Dust	Shaker/Woven Reverse Air/Woven $[m^3/(min \cdot m^2)]$	Pulse Jet/Felt $[m^3/(min \cdot m^2)]$
alumina	0.8	2.4
asbestos	0.9	3.0
bauxite	0.8	2.4
carbon black	0.5	1.5
coal	0.8	2.4
cocoa	0.8	3.7
clay	0.8	2.7
cement	0.6	2.4
cosmetics	0.5	3.0
enamel frit	0.8	2.7
feeds, grain	1.1	4.3
feldspar	0.7	2.7
fertilizer	0.9	2.4
flour	0.9	3.7
fly ash	0.8	1.5
graphite	0.6	1.5
gypsum	0.6	3.0
iron ore	0.9	3.4
iron oxide	0.8	2.1
iron sulfate	0.6	1.8
lead oxide	0.6	1.8
leather dust	1.1	3.7
lime	0.8	3.0
limestone	0.8	2.4
mica	0.8	2.7
paint pigments	0.8	2.1
paper	1.1	3.0
plastics	0.8	2.1
quartz	0.9	2.7
rock dust	0.9	2.7
sand	0.8	3.0
sawdust (wood)	1.1	3.7
silica	0.8	2.1
slate	1.1	3.7
soap detergents	0.6	1.5
spices	0.8	3.0
starch	0.9	2.4
sugar	0.6	2.1
talc	0.8	3.0
tobacco	1.1	4.0

U.S. EPA OAQPS Control Cost Manual, 4th ed., EPA 450/3-90-006 (NTIS PB 90-169954), January 1990.

Electrostatic Precipitator Efficiency

Deutsch-Anderson equation:

$$\eta = 1 - e^{(-WA/Q)}$$

where

η = fractional collection efficiency

W = terminal drift velocity

A = total collection area

Q = volumetric gas flowrate

Note that any consistent set of units can be used for W, A, and Q (for example, ft/min, ft^2, and ft^3/min).

Incineration

$$DRE = \frac{W_{in} - W_{out}}{W_{in}} \times 100\%$$

where

DRE = destruction and removal efficiency (%)

W_{in} = mass feed rate of a particular POHC (kg/h or lb/h)

W_{out} = mass emission rate of the same POHC (kg/h or lb/h)

POHC = principal organic hazardous contaminant

$$CE = \frac{CO_2}{CO_2 + CO} \times 100\%$$

CO_2 = volume concentration (dry) of CO_2 (parts per million, volume, ppm_v)

CO = volume concentration (dry) of CO (ppm_v)

CE = combustion efficiency

Kiln Formula

$$t = \frac{2.28\ L/D}{SN}$$

where

t = mean residence time, min

L/D = internal length-to-diameter ratio

S = kiln rake slope, in./ft of length

N = rotational speed, rev/min

Energy Content of Waste

Typical Waste Values	Moisture, %	Energy, Btu/lb
Food Waste	70	2,000
Paper	6	7,200
Cardboard	5	7,000
Plastics	2	14,000
Wood	20	8,000
Glass	2	60
Bi-metallic Cans	3	300

FATE AND TRANSPORT

Mass Calculations

Mass balance: $\frac{dM}{dt} = \frac{dM_{in}}{dt} + \frac{dM_{out}}{dt} \pm r$

$M = CQ = CV$

Continuity equation = $Q = vA$

M = mass

M_{in} = mass in

M_{out} = mass out

r = reaction rate = kC^n

k = reaction rate constant (1/time)

n = order of reaction

C = concentration

Q = flowrate

V = volume

v = velocity

A = cross-sectional area of flow

M (lb/day) = C (mg/L) × Q (MGD) × 8.34 [lb-L/(mg-MG)]

where:

MGD = million gallons per day

MG = million gallons

Microbial Kinetics

BOD Exertion

$$BOD_t = L_o\left(1 - e^{-kt}\right)$$

where

k = BOD decay rate constant (base e, $days^{-1}$)

L_o = ultimate BOD (mg/L)

t = time (days)

BOD_t = the amount of BOD exerted at time t (mg/L)

Stream Modeling

Streeter Phelps

$$D = \frac{k_d L_a}{k_r - k_d}\left[\exp\left(-k_d t\right) - \exp\left(-k_r t\right)\right] + D_a \exp\left(-k_r t\right)$$

$$t_c = \frac{1}{k_r - k_d}\ln\left[\frac{k_r}{k_d}\left(1 - D_a\frac{\left(k_r - k_d\right)}{k_d L_a}\right)\right]$$

$$DO = DO_{sat} - D$$

where

D = dissolved oxygen deficit (mg/L)

DO = dissolved oxygen concentration (mg/L)

D_a = initial dissolved oxygen deficit in mixing zone (mg/L)

DO_{sat} = saturated dissolved oxygen concentration (mg/L)

k_d = deoxygenation rate constant, base e ($days^{-1}$)

k_r = reaeration rate constant, base e ($days^{-1}$)

L_a = initial ultimate BOD in mixing zone (mg/L)

t = time (days)

t_c = time at which minimum dissolved oxygen occurs (days)

Davis and Cornwell, *Introduction to Environmental Engineering*, 4th ed., McGraw-Hill, 2008.

Monod Kinetics—Substrate Limited Growth

Continuous flow systems where growth is limited by one substrate (chemostat):

$$\mu = \frac{Yk_m S}{K_s + S} - k_d = \mu_{max}\frac{S}{K_s + S} - k_d$$

Multiple Limiting Substrates

$$\frac{\mu}{\mu_{max}} = \left[\mu_1(S_1)\right]\left[\mu_2(S_2)\right]\left[\mu_3(S_3)\right]\ldots\left[\mu_n(S_n)\right]$$

$$\text{where } \mu_i = \frac{S_i}{K_{si} + S_i} \text{ for } i = 1 \text{ to } n$$

Non-steady State Continuous Flow

$$\frac{dx}{dt} = Dx_0 + \left(\mu - k_d - D\right)x$$

Steady State Continuous Flow

$$\mu = D \text{ with } k_d << \mu$$

Product production at steady state, single substrate limiting

$$X_1 = Y_{P/S}\left(S_0 - S_i\right)$$

k_d = microbial death rate or endogenous decay rate constant ($time^{-1}$)

k_m = maximum growth rate constant ($time^{-1}$)

K_s = saturation constant or half-velocity constant [= concentration at $\mu_{max}/2$]

S = concentration of substrate in solution (mass/unit volume)

Y = yield coefficient [(mass/L product)/(mass/L food used)]

μ = specific growth rate ($time^{-1}$)

μ_{max} = maximum specific growth rate ($time^{-1}$) = Yk_m

Kinetic Temperature Corrections

$k_T = k_{20}\,(\theta)^{T-20}$

BOD (k): $\theta = 1.135$ (T = 4 – 20°C)

$\theta = 1.056$ (T = 20 – 30°C)

Reaeration (k_r) $\theta = 1.024$

♦ Monod growth rate constant as a function of limiting food concentration.

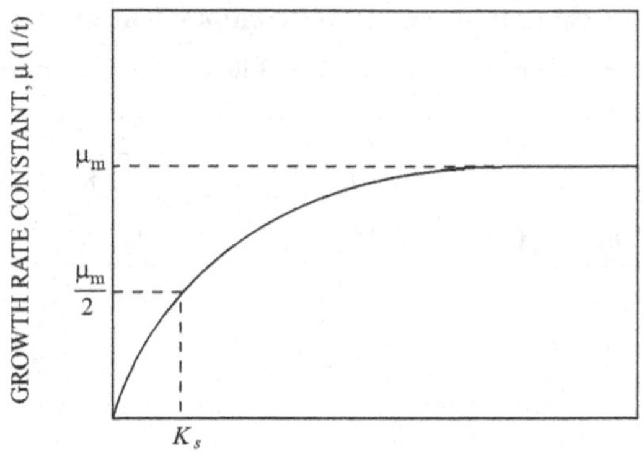

X_1 = product (mg/L)
V_r = volume (L)
D = dilution rate (flow f /reactor volume V_r; hr^{-1})
f = flowrate (L/hr)
μ_i = growth rate with one or multiple limiting substrates (hr^{-1})
S_i = substrate i concentration (mass/unit volume)
S_0 = initial substrate concentration (mass/unit volume)
$Y_{P/S}$ = product yield per unit of substrate (mass/mass)
p = product concentration (mass/unit volume)
x = cell concentration (mass/unit volume)
x_0 = initial cell concentration (mass/unit volume)
t = time (time)

Partition Coefficients

♦ Bioconcentration Factor BCF

The amount of a chemical to accumulate in aquatic organisms.

$$BCF = C_{org}/C$$

where

C_{org} = equilibrium concentration in organism (mg/kg or ppm)
C = concentration in water (ppm)

♦ Octanol-Water Partition Coefficient

The ratio of a chemical's concentration in the octanol phase to its concentration in the aqueous phase of a two-phase octanol-water system.

$$K_{ow} = C_o/C_w$$

where

C_o = concentration of chemical in octanol phase (mg/L or μg/L)
C_w = concentration of chemical in aqueous phase (mg/L or μg/L)

♦ Organic Carbon Partition Coefficient K_{oc}

$$K_{oc} = C_{soil}/C_{water}$$

where

C_{soil} = concentration of chemical in organic carbon component of soil (μg adsorbed/kg organic C, or ppb)
C_{water} = concentration of chemical in water (ppb or μg/kg)

• Retardation Factor R

$$R = 1 + (\rho_b/n_e)K_d$$

where

ρ_b = bulk density (mass/length³)
n_e = effective porosity of the media at saturation
K_d = partition or distribution coefficient
$= K_{OC}f_{OC}$

■ Soil-Water Partition Coefficient $K_{sw} = K_p$

$$K_{sw} = X/C$$

where

X = concentration of chemical in soil (ppb or μg/kg)
C = concentration of chemical in water (ppb or μg/kg)

$$K_{sw} = K_{oc}f_{oc}$$

K_p = partition coefficient
f_{oc} = fraction of organic carbon in the soil (dimensionless)

♦ Davis, M.L., and D. Cornwell, *Introduction to Environmental Engineering*, 3rd ed., McGraw-Hill, 1998.

• USEPA 402-R-99-004B, 1999, Understanding variations in partition coefficient, K_d, values.

■ LaGrega, Michael D., et al, *Hazardous Waste Management*, 2nd ed, McGraw-Hill, 2001.

Steady-State Reactor Parameters (Constant Density Systems)

Comparison of Steady-State Retention Times (θ) for Decay Reactions of Different Order[a]

		Equations for Mean Retention Times (θ)		
Reaction Order	**r**	**Ideal Batch**	**Ideal Plug Flow**	**Ideal CMFR**
Zero[b]	$-k$	$\frac{(C_o - C_t)}{k}$	$\frac{(C_o - C_t)}{k}$	$\frac{(C_o - C_t)}{k}$
First	$-kC$	$\frac{\ln(C_o/C_t)}{k}$	$\frac{\ln(C_o/C_t)}{k}$	$\frac{(C_o/C_t) - 1}{k}$
Second	$-kC^2$	$\frac{(C_o/C_t) - 1}{kC_o}$	$\frac{(C_o/C_t) - 1}{kC_o}$	$\frac{(C_o/C_t) - 1}{kC_t}$

[a]C_o = initial concentration or influent concentration; C_t = final condition or effluent concentration.

[b]Expressions are valid for $k\theta \le C_o$; otherwise $C_t = 0$.

Comparison of Steady-State Performance for Decay Reactions of Different Order[a]

		Equations for C_t		
Reaction Order	**r**	**Ideal Batch**	**Ideal Plug Flow**	**Ideal CMFR**
Zero[b] $t \le C_o/k$	$-k$	$C_o - kt$	$C_o - k\theta$	$C_o - k\theta$
$t > C_o/k$		0		
First	$-kC$	$C_o[\exp(-kt)]$	$C_o[\exp(-k\theta)]$	$\frac{C_o}{1 + k\theta}$
Second	$-kC^2$	$\frac{C_o}{1 + ktC_o}$	$\frac{C_o}{1 + k\theta C_o}$	$\frac{(4k\theta C_o + 1)^{1/2} - 1}{2k\theta}$

[a]C_o = initial concentration or influent concentration; C_t = final condition or effluent concentration.

[b]Time conditions are for ideal batch reactor only.

Davis, M.L., and S.J. Masten, *Principles of Environmental Engineering and Science*, 2nd ed., McGraw-Hill, 2004.

LANDFILL

Break-Through Time for Leachate to Penetrate a Clay Liner

$$t = \frac{d^2 \eta}{K(d + h)}$$

where

t = breakthrough time (yr)

d = thickness of clay liner (ft)

η = porosity

K = hydraulic conductivity (ft/yr)

h = hydraulic head (ft)

Typical porosity values for clays with a coefficient of permeability in the range of 10^{-6} to 10^{-8} cm/s vary from 0.1 to 0.3.

Effect of Overburden Pressure

$$SW_p = SW_i + \frac{p}{a + bp}$$

where

SW_p = specific weight of the waste material at pressure p (lb/yd^3) (typical 1,750 to 2,150)

SW_i = initial compacted specific weight of waste (lb/yd^3) (typical 1,000)

p = overburden pressure (lb/in^2)

a = empirical constant (yd^3/in^2)

b = empirical constant (yd^3/lb)

Tchobanoglous and F. Kreith, Handbook of Solid Waste Management, 2nd ed., McGraw-Hill, 2002.

Gas Flux

$$N_A = \frac{D\eta^{4/3}\left(C_{A_{atm}} - C_{A_{fill}}\right)}{L}$$

where

N_A = gas flux of compound A, $[g/(cm^2 \bullet s)][lb \bullet mol/(ft^2 \bullet d)]$

$C_{A_{atm}}$ = concentration of compound A at the surface of the landfill cover, g/cm^3 ($lb \bullet mol/ft^3$)

$C_{A_{fill}}$ = concentration of compound A at the bottom of the landfill cover, g/cm^3 ($lb \bullet mol/ft^3$)

L = depth of the landfill cover, cm (ft)

Typical values for the coefficient of diffusion for methane and carbon dioxide are 0.20 cm^2/s (18.6 ft^2/d) and 0.13 cm^2/s (12.1 ft^2/d), respectively.

D = diffusion coefficient, cm^2/s (ft^2/d)

η_{gas} = gas-filled porosity, cm^3/cm^3 (ft^3/ft^3)

η = porosity, cm^3/cm^3 (ft^3/ft^3)

Soil Landfill Cover Water Balance

$$\Delta S_{LC} = P - R - ET - PER_{sw}$$

where

ΔS_{LC} = change in the amount of water held in storage in a unit volume of landfill cover (in.)

P = amount of precipitation per unit area (in.)

R = amount of runoff per unit area (in.)

ET = amount of water lost through evapotranspiration per unit area (in.)

PER_{sw} = amount of water percolating through the unit area of landfill cover into compacted solid waste (in.)

Tchobanoglous and Kreith, *Handbook of Solid Waste Management*, 2nd ed., McGraw-Hill, 2002.

POPULATION MODELING

Population Projection Equations

Linear Projection = Algebraic Projection

$$P_t = P_0 + k\Delta t$$

where

P_t = population at time t

P_0 = population at time zero

k = growth rate

Δt = elapsed time in years relative to time zero

Log Growth = Exponential Growth = Geometric Growth

$$P_t = P_0 e^{k\Delta t}$$

$$\ln P_t = \ln P_0 + k\Delta t$$

where

P_t = population at time t

P_0 = population at time zero

k = growth rate

Δt = elapsed time in years relative to time zero

Percent Growth

$$P_t = P_0(1 + k)^n$$

where

P_t = population at time t

P_0 = population at time zero

k = growth rate

n = number of periods

Ratio and Correlation Growth

$$\frac{P_2}{P_{2R}} = \frac{P_1}{P_{1R}} = k$$

where

P_2 = projected population

P_{2R} = projected population of a larger region

P_1 = population at last census

P_{1R} = population of larger region at last census

k = growth ratio constant

Decreasing-Rate-of-Increase Growth

$$P_t = P_0 + (S - P_0)(1 - e^{-k(t-t_0)})$$

where

P_t = population at time t

P_0 = population at time zero

k = growth rate constant

S = saturation population

t, t_0 = future time, initial time

RADIATION

Effective Half-Life

Effective half-life, τ_e, is the combined radioactive and biological half-life.

$$\frac{1}{\tau_e} = \frac{1}{\tau_r} + \frac{1}{\tau_b}$$

where

τ_r = radioactive half-life

τ_b = biological half-life

Half-Life

$$N = N_0 e^{-0.693\, t/\tau}$$

where

N_0 = original number of atoms

N = final number of atoms

t = time

τ = half-life

Flux at distance 2 = (Flux at distance 1) $(r_1/r_2)^2$

where r_1 and r_2 are distances from source.

The half-life of a biologically degraded contaminant assuming a first-order rate constant is given by:

$$t_{1/2} = \frac{0.693}{k}$$

k = rate constant ($time^{-1}$)

$t_{1/2}$ = half-life (time)

Daughter Product Activity

$$N_2 = \frac{\lambda_1 N_{10}}{\lambda_2 - \lambda_1}\left(e^{-\lambda_1 t} - e^{-\lambda_2 t}\right)$$

where $\lambda_{1,2}$ = decay constants ($time^{-1}$)

N_{10} = initial activity (curies) of parent nuclei

t = time

Daughter Product Maximum Activity Time

$$t' = \frac{\ln\lambda_2 - \ln\lambda_1}{\lambda_2 - \lambda_1}$$

Inverse Square Law

$$\frac{I_1}{I_2} = \frac{(R_2)^2}{(R_1)^2}$$

where $I_{1,2}$ = Radiation intensity at locations 1 and 2

$R_{1,2}$ = Distance from the source at locations 1 and 2

SAMPLING AND MONITORING

Data Quality Objectives (DQO) for Sampling Soils and Solids

Investigation Type	Confidence Level (1–α) (%)	Power (1–β) (%)	Minimum Detectable Relative Difference (%)
Preliminary site investigation	70–80	90–95	10–30
Emergency clean-up	80–90	90–95	10–20
Planned removal and remedial response operations	90–95	90–95	10–20

Confidence level: 1– (Probability of a Type I error) = 1 – α = size probability of not making a Type I error.
Power = 1– (Probability of a Type II error) = 1 – β = probability of not making a Type II error.

EPA Document "EPA/600/8–89/046" *Soil Sampling Quality Assurance User's Guide*, Chapter 7.

CV = $(100 * s)/\bar{x}$

CV = coefficient of variation

s = standard deviation of sample

$\bar{x}$ = sample average

Minimum Detectable Relative Difference = Relative increase over background [100 $(\mu_S - \mu_B)/\mu_B$ to be detectable with a probability $(1 - \beta)$

μ_S = mean of pollutant concentration of the site of the contamination

μ_B = mean of pollutant concentration of the site before contamination or the noncontaminated area (background)

Number of Samples Required in a One-Sided One-Sample t-Test to Achieve a Minimum Detectable Relative Difference at Confidence Level (1–α) and Power (1–β)

Coefficient of Variation (%)	Power (%)	Confidence Level (%)	Minimum Detectable Relative Difference (%)				
			5	10	20	30	40
15	95	99	145	39	12	7	5
		95	99	26	8	5	3
		90	78	21	6	3	3
		80	57	15	4	2	2
	90	99	120	32	11	6	5
		95	79	21	7	4	3
		90	60	16	5	3	2
		80	41	11	3	2	1
	80	99	94	26	9	6	5
		95	58	16	5	3	3
		90	42	11	4	2	2
		80	26	7	2	2	1
25	95	99	397	102	28	14	9
		95	272	69	19	9	6
		90	216	55	15	7	5
		80	155	40	11	5	3
	90	99	329	85	24	12	8
		95	272	70	19	9	6
		90	166	42	12	6	4
		80	114	29	8	4	3
	80	99	254	66	19	10	7
		95	156	41	12	6	4
		90	114	30	8	4	3
		80	72	19	5	3	2
35	95	99	775	196	42	25	15
		95	532	134	35	17	10
		90	421	106	28	13	8
		80	304	77	20	9	6
	90	99	641	163	43	21	13
		95	421	107	28	14	8
		90	323	82	21	10	6
		80	222	56	15	7	4
	80	99	495	126	34	17	11
		95	305	78	21	10	7
		90	222	57	15	7	5
		80	140	36	10	5	3

WASTEWATER TREATMENT AND TECHNOLOGIES

Activated Sludge

$$X_A = \frac{\theta_c Y(S_0 - S_e)}{\theta(1 + k_d \theta_c)}$$

Steady-State Mass Balance around Secondary Clarifier:

$$(Q_0 + Q_R)X_A = Q_e X_e + Q_R X_r + Q_w X_w$$

$$\theta_c = \text{Solids residence time} = \frac{V(X_A)}{Q_w X_w + Q_e X_e}$$

$$\text{Sludge volume/day: } Q_s = \frac{M(100)}{\rho_s(\%\ \text{solids})}$$

$$\text{SVI} = \frac{\text{Sludge volume after settling (mL/L)} * 1{,}000}{\text{MLSS (mg/L)}}$$

k_d = microbial death ratio; kinetic constant; day^{-1}; typical range 0.1–0.01, typical domestic wastewater value = 0.05 day^{-1}

S_e = effluent BOD or COD concentration (kg/m^3)

S_0 = influent BOD or COD concentration (kg/m^3)

X_A = biomass concentration in aeration tank (MLSS or MLVSS kg/m^3)

Y = yield coefficient (kg biomass/kg BOD or COD consumed); range 0.4–1.2

θ = hydraulic residence time = V/Q

Solids loading rate = $Q X_A$

For activated sludge secondary clarifier $Q = Q_0 + Q_R$

Organic loading rate (volumetric) = $Q_0 S_0 / Vol$

Organic loading rate (F:M) = $Q_0 S_0 / (Vol\ X_A)$

Organic loading rate (surface area) = $Q_0 S_0 / A_M$

ρ_s = density of solids

A = surface area of unit

A_M = surface area of media in fixed-film reactor

A_x = cross-sectional area of channel

M = sludge production rate (dry weight basis)

Q_0 = influent flowrate

Q_e = effluent flowrate

Q_w = waste sludge flowrate

ρ_s = wet sludge density

R = recycle ratio = Q_R/Q_0

Q_R = recycle flowrate = $Q_0 R$

X_e = effluent suspended solids concentration

X_w = waste sludge suspended solids concentration

V = aeration basin volume

Q = flowrate

X_r = recycled sludge suspended solids concentration

Design and Operational Parameters for Activated-Sludge Treatment of Municipal Wastewater

Type of Process	Mean cell residence time (θ_c, d)	Food-to-mass ratio [(kg BOD_5/(day·kg MLSS)]	Volumetric loading (kg BOD_5/m^3)	Hydraulic residence time in aeration basin (θ, h)	Mixed liquor suspended solids (MLSS, mg/L)	Recycle ratio (Q_r/Q)	Flow regime*	BOD_5 removal efficiency (%)	Air supplied (m^3/kg BOD_5)
Tapered aeration	5–15	0.2–0.4	0.3–0.6	4–8	1,500–3,000	0.25–0.5	PF	85–95	45–90
Conventional	4–15	0.2–0.4	0.3–0.6	4–8	1,500–3,000	0.25–0.5	PF	85–95	45–90
Step aeration	4–15	0.2–0.4	0.6–1.0	3–5	2,000–3,500	0.25–0.75	PF	85–95	45–90
Completely mixed	4–15	0.2–0.4	0.8–2.0	3–5	3,000–6,000	0.25–1.0	CM	85–95	45–90
Contact stabilization	4–15	0.2–0.6	1.0–1.2			0.25–1.0			45–90
Contact basin				0.5–1.0	1,000–3,000		PF	80–90	
Stabilization basin				4–6	4,000–10,000		PF		
High-rate aeration	4–15	0.4–1.5	1.6–16	0.5–2.0	4,000–10,000	1.0–5.0	CM	75–90	25–45
Pure oxygen	8–20	0.2–1.0	1.6–4	1–3	6,000–8,000	0.25–0.5	CM	85–95	
Extended aeration	20–30	0.05–0.15	0.16–0.40	18–24	3,000–6,000	0.75–1.50	CM	75–90	90–125

*PF = plug flow, CM = completely mixed.

Metcalf and Eddy, *Wastewater Engineering: Treatment, Disposal, and Reuse*, 3rd ed., McGraw-Hill, 1991.

Facultative Pond

BOD Loading Total System ≤ 35 pounds BOD_5/(acre-day)
Minimum = 3 ponds
Depth = 3–8 ft
Minimum t = 90–120 days

Biotower

Fixed-Film Equation without Recycle

$$\frac{S_e}{S_0} = e^{-kD/q^n}$$

Fixed-Film Equation with Recycle

$$\frac{S_e}{S_a} = \frac{e^{-kD/q^n}}{(1+R) - R\left(e^{-kD/q^n}\right)}$$

where

S_e = effluent BOD_5 (mg/L)
S_0 = influent BOD_5 (mg/L)
R = recycle ratio = Q_R/Q_0
Q_R = recycle flowrate

$$S_a = \frac{S_o + RS_e}{1+R}$$

D = depth of biotower media (m)
q = hydraulic loading [$m^3/(m^2 \bullet min)$]
= $(Q_0 + RQ_0)/A_{plan}$ (with recycle)
k = treatability constant; functions of wastewater and medium (min^{-1}); range 0.01–0.1; for municipal wastewater and modular plastic media 0.06 min^{-1} @ 20°C
k_T = $k_{20}(1.035)^{T-20}$
n = coefficient relating to media characteristics; modular plastic, n = 0.5

♦ Aerobic Digestion

Design criteria for aerobic digesters[a]

Parameter	Value
Sludge retention time, d	
At 20°C	40
At 15°C	60
Solids loading, lb volatile solids/ft^3•d	0.1–0.3
Oxygen requirements, lb O_2/lb solids destroyed	
Cell tissue	~2.3
BOD_5 in primary sludge	1.6–1.9
Energy requirements for mixing	
Mechanical aerators, hp/10^3 ft^3	0.7–1.50
Diffused-air mixing, $ft^3/10^3$ ft^3•min	20–40
Dissolved-oxygen residual in liquid, mg/L	1–2
Reduction in volatile suspended solids (VSS), %	40–50

Tank Volume

$$V = \frac{Q_i(X_i + FS_i)}{X_d(k_d P_v + 1/\theta_c)}$$

where

V = volume of aerobic digester (ft^3)
Q_i = influent average flowrate to digester (ft^3/d)
X_i = influent suspended solids (mg/L)
F = fraction of the influent BOD_5 consisting of raw primary sludge (expressed as a decimal)
S_i = influent BOD_5 (mg/L)
X_d = digester suspended solids (mg/L); typically $X_d = (0.7)X_i$
k_d = reaction-rate constant (d^{-1})
P_v = volatile fraction of digester suspended solids (expressed as a decimal)
θ_c = solids residence time (sludge age) (d)

$F\,S_i$ can be neglected if primary sludge is not included on the sludge flow to the digester.

♦

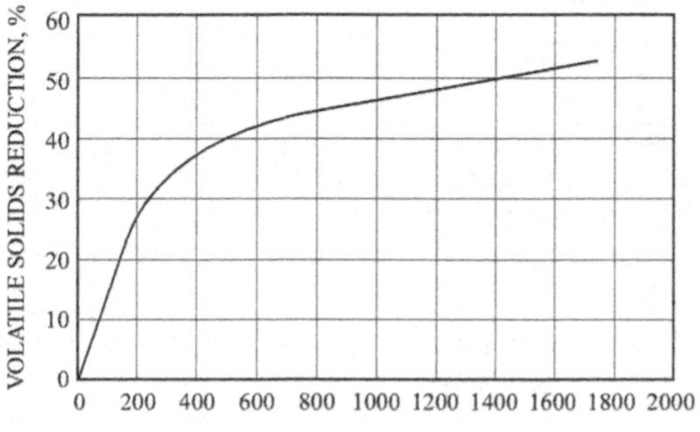

VOLATILE SOLIDS REDUCTION IN AN AEROBIC DIGESTER AS A FUNCTION OF DIGESTER LIQUID TEMPERATURE AND DIGESTER SLUDGE AGE

• Anaerobic Digestion

Design parameters for anaerobic digesters

Parameter	Standard-rate	High-rate
Solids residence time, d	30–90	10–20
Volatile solids loading, $kg/m^3/d$	0.5–1.6	1.6–6.4
Digested solids concentration, %	4–6	4–6
Volatile solids reduction, %	35–50	45–55
Gas production (m^3/kg VSS added)	0.5–0.55	0.6–0.65
Methane content, %	65	65

Standard Rate

$$\text{Reactor Volume} = \frac{V_1 + V_2}{2} t_r + V_2 t_s$$

High Rate

First stage

$$\text{Reactor Volume} = V_1 t_r$$

Second Stage

$$\text{Reactor Volume} = \frac{V_1 + V_2}{2} t_t + V_2 t_s$$

where

V_1 = raw sludge input (volume/day)
V_2 = digested sludge accumulation (volume/day)
t_r = time to react in a high-rate digester = time to react and thicken in a standard-rate digester
t_t = time to thicken in a high-rate digester
t_s = storage time

♦ Tchobanoglous, G., and Metcalf and Eddy, *Wastewater Engineering: Treatment, Disposal, and Reuse*, 4th ed., McGraw-Hill, 2003.
• Peavy, HS, D.R. Rowe, and G. Tchobanoglous, *Environmental Engineering*, McGraw-Hill, 1985.

WATER TREATMENT TECHNOLOGIES

Activated Carbon Adsorption

Freundlich Isotherm

$$\frac{x}{m} = X = KC_e^{1/n}$$

where

x = mass of solute adsorbed

m = mass of adsorbent

X = mass ratio of the solid phase—that is, the mass of adsorbed solute per mass of adsorbent

C_e = equilibrium concentration of solute, mass/volume

K, n = experimental constants

Linearized Form

$$\ln \frac{x}{m} = \frac{1}{n} \ln C_e + \ln K$$

For linear isotherm, $n = 1$

Langmuir Isotherm

$$\frac{x}{m} = X = \frac{aKC_e}{1 + KC_e}$$

where

a = mass of adsorbed solute required to saturate completely a unit mass of adsorbent

K = experimental constant

Linearized Form

$$\frac{m}{x} = \frac{1}{a} + \frac{1}{aK} \frac{1}{C_e}$$

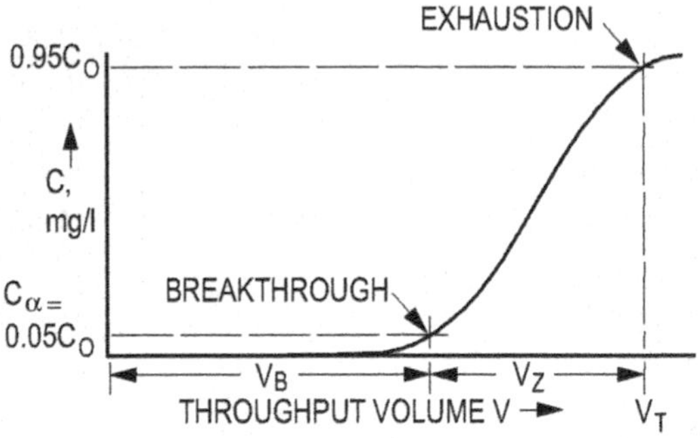

Depth of Sorption Zone

$$Z_s = Z\left[\frac{V_Z}{V_T - 0.5V_Z}\right]$$

where

V_Z = $V_T - V_B$

Z_S = depth of sorption zone

Z = total carbon depth

V_T = total volume treated at exhaustion ($C = 0.95\ C_0$)

V_B = total volume at breakthrough ($C = C_\alpha = 0.05\ C_0$)

C_0 = concentration of contaminant in influent

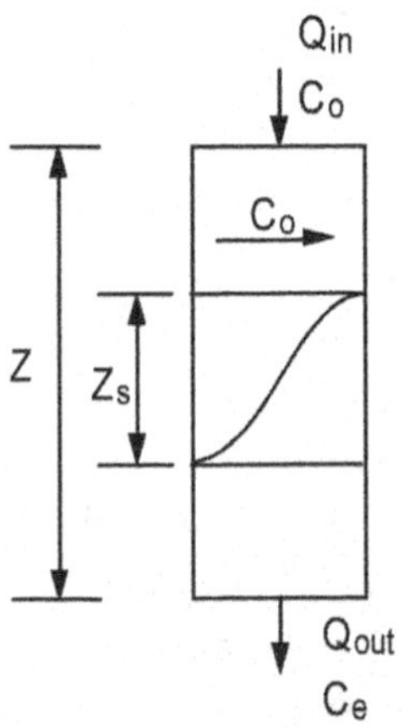

Air Stripping

$P_i = HC_i$ = Henry's Law

P_i = partial pressure of component i, atm

H = Henry's Law constant, atm-m^3/kmol

C_i = concentration of component i in solvent, kmol/m^3

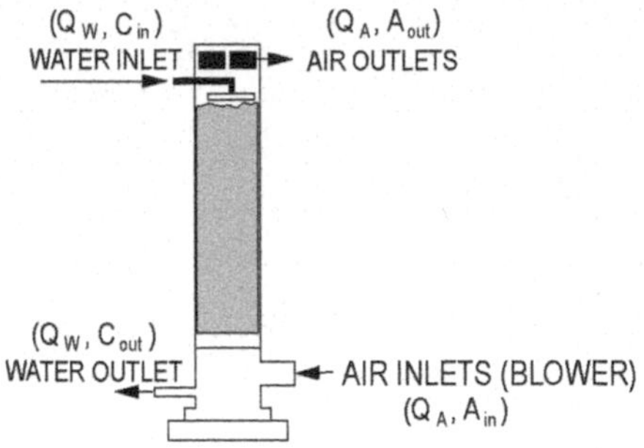

$$A_{out} = H'C_{in}$$

$$Q_W \bullet C_{in} = Q_A H'C_{in}$$

$$Q_W = Q_A H'$$

$$H'(Q_A / Q_W) = 1$$

where

A_{out} = concentration in the effluent air (kmol/m^3); in this formulation of the equation A_{in} and C_{out} are assumed to be negligible for simplicity.

Q_W = water flowrate (m^3/s)

Q_A = air flowrate (m^3/s)

A_{in} = concentration of contaminant in air (kmol/m^3)

C_{out} = concentration of contaminants in effluent water (kmol/m^3)

C_{in} = concentration of contaminants in influent water (kmol/m^3)

Stripper Packing Height = Z

$$Z = \text{HTU} \times \text{NTU}$$

Assuming rapid equilibrium:

$$\text{NTU} = \left(\frac{R_S}{R_S - 1}\right) \ln\left(\frac{\left(C_{in}/C_{out}\right)\left(R_S - 1\right) + 1}{R_S}\right)$$

where

NTU = number of transfer units

H = Henry's Law constant

H' = H/RT = dimensionless Henry's Law constant

T = temperature in units consistent with K

R = universal gas constant, atm•m^3/(kmol•K)

R_S = stripping factor $H'(Q_A/Q_W)$

C_{in} = concentration in the influent water (kmol/m^3)

C_{out} = concentration in the effluent water (kmol/m^3)

HTU = Height of Transfer Units = $\frac{L}{M_W K_L a}$, where

L = liquid molar loading rate [kmol/(s•m^2)]

M_W = molar density of water (55.6 kmol/m^3) = 3.47 lbmol/ft^3

$K_L a$ = overall transfer rate constant (s^{-1})

Clarifier

Overflow rate = Hydraulic loading rate = $v_o = Q/A_{surface}$

v_o = critical settling velocity

= terminal settling velocity of smallest particle that is 100% removed

Weir loading = weir overflow rate, WOR = Q/Weir Length

Horizontal velocity = approach velocity = v_h

$= Q/A_{cross\text{-}section} = Q/A_x$

Hydraulic residence time = $V/Q = \theta$

where

Q = flowrate

A_x = cross-sectional area

A = surface area, plan view

V = tank volume

Typical Primary Clarifier Efficiency Percent Removal

	Overflow rates			
	1,200 (gpd/ft^2) 48.9 (m/d)	1,000 (gpd/ft^2) 40.7 (m/d)	800 (gpd/ft^2) 32.6 (m/d)	600 (gpd/ft^2) 24.4 (m/d)
Suspended Solids	54%	58%	64%	68%
BOD_5	30%	32%	34%	36%

Weir Loadings

1. Water Treatment—weir overflow rates should not exceed 20,000 gpd/ft
2. Wastewater Treatment
   a. Flow ≤ 1 MGD: weir overflow rates should not exceed 10,000 gpd/ft
   b. Flow > 1 MGD: weir overflow rates should not exceed 15,000 gpd/ft

Horizontal Velocities

1. Water Treatment—horizontal velocities should not exceed 0.5 fpm
2. Wastewater Treatment—no specific requirements (use the same criteria as for water)

Dimensions

1. Rectangular Tanks
   a. Length:Width ratio = 3:1 to 5:1
   b. Basin width is determined by the scraper width (or multiples of the scraper width)
   c. Bottom slope is set at 1%
2. Circular Tanks
   a. Diameters up to 200 ft
   b. Diameters must match the dimensions of the sludge scraping mechanism
   c. Bottom slope is less than 8%

Design Criteria for Sedimentation Basins

Type of Basin	Overflow Rate Average (gpd/ft^2)	Overflow Rate Average (m^3/m^2•d)	Overflow Rate Peak (gpd/ft^2)	Overflow Rate Peak (m^3/m^2•d)	Solids Loading Rate Average (lb/ft^2–d)	Solids Loading Rate Average (kg/m^2•h)	Solids Loading Rate Peak (lb/ft^2–h)	Solids Loading Rate Peak (kg/m^2•h)	Hydraulic Residence Time (hr)	Depth (ft)
Water Treatment										
Clarification following coagulation and flocculation:										
Alum coagulation	350–550	14–22							4–8	12–16
Ferric coagulation	550–700	22–28							4–8	12–16
Upflow clarifiers										
Groundwater	1,500–2,200	61–90							1	
Surface water	1,000–1,500	41–61							4	
Clarification following lime-soda softening										
Conventional	550–1,000	22–41							2–4	
Upflow clarifiers										
Groundwater	1,000–2,500	41–102							1	
Surface water	1,000–1,800	41–73							4	
Wastewater Treatment										
Primary clarifiers	800–1,200	32–49	1,200–2,000	50–80					2	10–12
Settling basins following fixed film reactors	400–800	16–33							2	
Settling basins following air-activated sludge reactors										
All configurations EXCEPT extended aeration	400–700	16–28							2	12–15
Extended aeration	200–400	8–16	1,000–1,200	40–64	19–29	4–6	38	8	2	12–15
Settling basins following chemical flocculation reactors	800–1,200		600–800	24–32	5–24	1–5	34	7	2	

Settling Equations

General Spherical

$$v_t = \sqrt{\frac{4g(\rho_p - \rho_f)d}{3C_D\rho_f}}$$

where

C_D = drag coefficient
= 24/Re (Laminar; Re ≤ 1.0)
= 24/Re + 3/(Re$^{1/2}$) + 0.34 (Transitional)
= 0.4(Turbulent; Re ≥ 10^4)

Re = Reynolds number $\frac{v_t \rho d}{\mu}$

g = gravitational constant

ρ_p and ρ_f = density of particle and fluid respectively

d = diameter of sphere

μ = bulk viscosity of liquid = absolute viscosity

v_t = terminal settling velocity

Stokes' Law

$$v_t = \frac{g(\rho_p - \rho_f)d^2}{18\mu} = \frac{g\,\rho_f(S.G. - 1)d^2}{18\mu}$$

Approach velocity = horizontal velocity = Q/A_x

Hydraulic loading rate = Q/A

Hydraulic residence time = $V/Q = \theta$

where

Q = flowrate

A_x = cross-sectional area

A = surface area, plan view

V = tank volume

ρ_f = fluid mass density

$S.G.$ = specific gravity

Filtration Equations

Filter bay length:width ratio = 1.2:1 to 1.5:1

Effective size = d_{10}

Uniformity coefficient = d_{60}/d_{10}

d_x = diameter of particle class for which x% of sample is less than (units meters or feet)

Filter equations can be used with any consistent set of units.

Head Loss Through Clean Bed

Rose Equation

Monosized Media

$$h_f = \frac{1.067(v_s)^2 L C_D}{g\eta^4 d}$$

Multisized Media

$$h_f = \frac{1.067(v_s)^2 L}{g\eta^4}\Sigma\frac{C_{D_{ij}}x_{ij}}{d_{ij}}$$

Carmen-Kozeny Equation

Monosized Media

$$h_f = \frac{f'L(1-\eta)v_s^2}{\eta^3 g d}$$

Multisized Media

$$h_f = \frac{L(1-\eta)v_s^2}{\eta^3 g}\Sigma\frac{f'_{ij}x_{ij}}{d_{ij}}$$

$$f' = \text{friction factor} = 150\left(\frac{1-\eta}{\text{Re}}\right) + 1.75$$

where

h_f = head loss through the clean bed (m of H_2O)

L = depth of filter media (m)

η = porosity of bed = void volume/total volume

v_s = filtration rate = empty bed approach velocity = Q/A_{plan} (m/s)

g = gravitational acceleration (m/s^2)

Re = Reynolds number = $\frac{v_s \rho d}{\mu}$

d_{ij}, d = diameter of filter media particles; arithmetic average of adjacent screen openings (m)

i = filter media (sand, anthracite, garnet)

j = filter media particle size

x_{ij} = mass fraction of media retained between adjacent sieves

f'_{ij} = friction factors for each media fraction

C_D = drag coefficient as defined in settling velocity equations

Bed Expansion

Monosized

$$L_f = \frac{L_o(1-\eta_o)}{1 - \left(\frac{v_B}{v_t}\right)^{0.22}}$$

Multisized

$$L_f = L_o(1-\eta_o)\Sigma\frac{x_{ij}}{1 - \left(\frac{v_B}{v_{t,i,j}}\right)^{0.22}}$$

$$\eta_f = \left(\frac{v_B}{v_t}\right)^{0.22}$$

where

L_f = depth of fluidized filter media (m)

v_B = backwash velocity (m/s), Q_B/A_{plan}

Q_B = backwash flowrate

v_t = terminal setting velocity

η_f = porosity of fluidized bed

L_o = initial bed depth

η_o = initial bed porosity

Lime-Soda Softening Equations

1. Carbon dioxide removal
   $CO_2 + Ca(OH)_2 \rightarrow CaCO_3(s) + H_2O$
2. Calcium carbonate hardness removal
   $Ca(HCO_3)_2 + Ca(OH)_2 \rightarrow 2CaCO_3(s) + 2H_2O$
3. Calcium non-carbonate hardness removal
   $CaSO_4 + Na_2CO_3 \rightarrow CaCO_3(s) + 2Na^+ + SO_4^{-2}$
4. Magnesium carbonate hardness removal
   $Mg(HCO_3)_2 + 2Ca(OH)_2 \rightarrow 2CaCO_3(s) + Mg(OH)_2(s) + 2H_2O$
5. Magnesium non-carbonate hardness removal
   $MgSO_4 + Ca(OH)_2 + Na_2CO_3 \rightarrow CaCO_3(s) + Mg(OH)_2(s) + 2Na^+ + SO_4^{2-}$
6. Destruction of excess alkalinity
   $2HCO_3^- + Ca(OH)_2 \rightarrow CaCO_3(s) + CO_3^{2-} + 2H_2O$

7. Recarbonation

$$Ca^{2+} + 2OH^- + CO_2 \rightarrow CaCO_3(s) + H_2O$$

Molecular Formulas	Molecular Weight	*n* # Equiv per mole	Equivalent Weight
CO_3^{2-}	60.0	2	30.0
CO_2	44.0	2	22.0
$Ca(OH)_2$	74.1	2	37.1
$CaCO_3$	100.1	2	50.0
$Ca(HCO_3)_2$	162.1	2	81.1
$CaSO_4$	136.1	2	68.1
Ca^{2+}	40.1	2	20.0
H^+	1.0	1	1.0
HCO_3^-	61.0	1	61.0
$Mg(HCO_3)_2$	146.3	2	73.2
$Mg(OH)_2$	58.3	2	29.2
$MgSO_4$	120.4	2	60.2
Mg^{2+}	24.3	2	12.2
Na^+	23.0	1	23.0
Na_2CO_3	106.0	2	53.0
OH^-	17.0	1	17.0
SO_4^{2-}	96.1	2	48.0

Rapid Mix and Flocculator Design

$$G = \sqrt{\frac{P}{\mu V}} = \sqrt{\frac{\gamma H_L}{t\mu}}$$

$Gt = 10^4$ to 10^5

where

G = root mean square velocity gradient (mixing intensity) [ft/(sec-ft) or m/(s•m)]

P = power to the fluid (ft-lb/sec or N•m/s)

V = volume (ft^3 or m^3)

μ = dynamic viscosity [lb/(ft-sec) or Pa•s]

γ = specific weight of water (lb/ft^3 or N/m^3)

H_L = head loss (ft or m)

t = time (sec or s)

Reel and Paddle

$$P = \frac{C_D A_P \rho_f v_r^3}{2}$$

where

C_D = drag coefficient = 1.8 for flat blade with a L:W > 20:1

A_p = area of blade (m^2) perpendicular to the direction of travel through the water

ρ_f = density of H_2O (kg/m^3)

v_p = velocity of paddle (m/s)

v_r = relative or effective paddle velocity

= v_p• slip coefficient

slip coefficient = 0.5 to 0.75

Turbulent Flow Impeller Mixer

$$P = K_T(n)^3(D_i)^5\rho_f$$

where

K_T = impeller constant (see table)

n = rotational speed (rev/sec)

D_i = impeller diameter (m)

Values of the Impeller Constant K_T
(Assume Turbulent Flow)

Type of Impeller	K_T
Propeller, pitch of 1, 3 blades	0.32
Propeller, pitch of 2, 3 blades	1.00
Turbine, 6 flat blades, vaned disc	6.30
Turbine, 6 curved blades	4.80
Fan turbine, 6 blades at 45°	1.65
Shrouded turbine, 6 curved blades	1.08
Shrouded turbine, with stator, no baffles	1.12

Note: Constant assumes baffled tanks having four baffles at the tank wall with a width equal to 10% of the tank diameter.

Reprinted with permission from *Industrial & Engineering Chemistry*, "Mixing of Liquids in Chemical Processing," J. Henry Rushton, 1952, v. 44, no. 12. p. 2934, American Chemical Society.

Reverse Osmosis

Osmotic Pressure of Solutions of Electrolytes

$$\Pi = \phi v \frac{n}{V} RT$$

where

Π = osmotic pressure, Pa

ϕ = osmotic coefficient

v = number of ions formed from one molecule of electrolyte

n = number of moles of electrolyte

V = specific volume of solvent, $m^3/kmol$

R = universal gas constant, Pa • m^3/(kmol • K)

T = absolute temperature, K

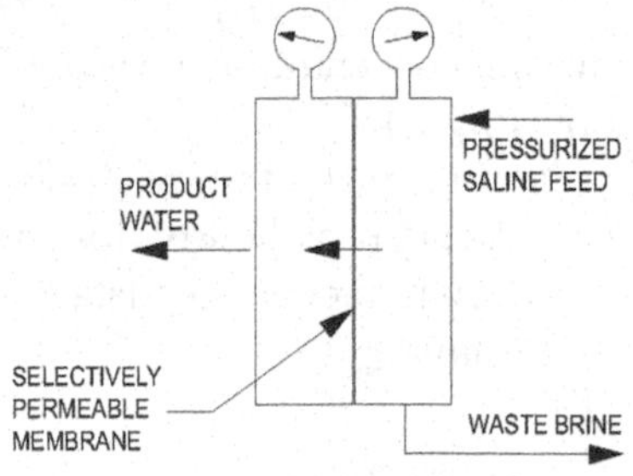

A CONTINUOUS-FLOW REVERSE OSMOSIS UNIT

Salt Flux through the Membrane

$$J_s = (D_s K_s / \Delta Z)(C_{in} - C_{out})$$

where

J_s = salt flux through the membrane [kmol/(m^2 • s)]

D_s = diffusivity of the solute in the membrane (m^2/s)

K_s = solute distribution coefficient (dimensionless)

C = concentration (kmol/m^3)

ΔZ = membrane thickness (m)

$$J_s = K_p (C_{in} - C_{out})$$

Kp = membrane solute mass-transfer coefficient

$$= \frac{D_s K_s}{\Delta Z} (L/t, m/s)$$

Water Flux

$$J_w = W_p (\Delta P - \Delta \pi)$$

where

J_w = water flux through the membrane [kmol/(m^2 • s)]

Wp = coefficient of water permeation, a characteristic of the particular membrane [kmol/(m^2 • s • Pa)]

ΔP = pressure differential across membrane = $P_{in} - P_{out}$ (Pa)

$\Delta\pi$ = osmotic pressure differential across membrane $\pi_{in} - \pi_{out}$ (Pa)

Ultrafiltration

$$J_w = \frac{\varepsilon r^2 \int \Delta P}{8 \mu \delta}$$

where

ε = membrane porosity

r = membrane pore size

ΔP = net transmembrane pressure

μ = viscosity

δ = membrane thickness

J_w = volumetric flux (m/s)

♦ Disinfection

Chlorine contact chamber length:width ratio = 20:1 to 50:1

$CT_{calc} = C \times t_{10}$

CT_{calc} = calculated CT value (mg • mm/L)

C = residual disinfectant concentration measured during peak hourly flow (mg/L)

t_{10} = time it takes 10% of the water to flow through the reactor measured during peak hourly flow (min)

= can be determined from traces study data or the following relationship

$t_{10(approx)} = \theta \times BF$

θ = hydraulic residence time (min)

BF = baffling factor

♦ Adapted from *Guidance Manual LT1ESWTR Disinfection Profiling and Benchmarking*, U.S. Environmental Protection Agency, 2003.

Baffling Factors

Baffling Condition	Baffling Factor	Baffling Description
Unbaffled (mixed flow)	0.1	None, agitated basin, very low length to width ratio, high inlet and outlet flow velocities.
Poor	0.3	Single or multiple unbaffled inlets and outlets, no intra-basin baffles.
Average	0.5	Baffled inlet or outlet with some intra-basin baffles.
Superior	0.7	Perforated inlet baffle, serpentine or perforated intra-basin baffles, outlet weir or perforated launders.
Perfect (plug flow)	1.0	Very high length to width ratio (pipeline flow), perforated inlet, outlet, and intra-basin baffles.

♦ *Guidance Manual LT1ESWTR Disinfection Profiling and Benchmarking*, U.S. Environmental Protection Agency, 2003.

♦ **Removal and Inactivation Requirements**

Microorganism	Required Log Reduction	Treatment
Giardia	3-log (99.9%)	Removal and/or inactivation
Virsuses	4-log (99.99%)	Removal and/or inactivation
Cryptosporidium	2-log (99%)	Removal

♦ **Typical Removal Credits and Inactivation Requirements for Various Treatment Technologies**

Process	Typical Log Removal Credits		Resulting Disinfection Log Inactivation Requirements	
	Giardia	Viruses	*Giardia*	Viruses
Conventional Treatment	2.5	2.0	0.5	2.0
Direct Filtration	2.0	1.0	1.0	3.0
Slow Sand Filtration	2.0	2.0	1.0	2.0
Diatomaceous Earth Filtration	2.0	1.0	1.0	3.0
Unfiltered	0	0	3.0	4.0

♦ *Guidance Manual LT1ESWTR Disinfection Profiling and Benchmarking*, U.S. Environmental Protection Agency, 2003.

CT Values* For 3-LOG Inactivation Of *Giardia* Cysts By Free Chlorine

Chlorine Concentration (mg/L)	Temperature <= 0.5°C							Temperature = 5°C							Temperature = 10°C						
	pH							pH							pH						
	<=6.0	6.5	7.0	7.5	8.0	8.5	9.0	<=6.0	6.5	7.0	7.5	8.0	8.5	9.0	<=6.0	6.5	7.0	7.5	8.0	8.5	9.0
<=0.4	137	163	195	237	277	329	390	97	117	139	166	198	236	279	73	88	104	125	149	177	209
0.6	141	168	200	239	286	342	407	100	120	143	171	204	244	291	75	90	107	128	153	183	218
0.8	145	172	205	246	295	354	422	103	122	146	175	210	252	301	78	92	110	131	158	189	226
1.0	148	176	210	253	304	365	437	105	125	149	179	216	260	312	79	94	112	134	162	195	234
1.2	152	180	215	259	313	376	451	107	127	152	183	221	267	320	80	95	114	137	166	200	240
1.4	155	184	221	266	321	387	464	109	130	155	187	227	274	329	82	98	116	140	170	206	247
1.6	157	189	226	273	329	397	477	111	132	158	192	232	281	337	83	99	119	144	174	211	253
1.8	162	193	231	279	338	407	489	114	135	162	196	238	287	345	86	101	122	147	179	215	259
2.0	165	197	236	286	346	417	500	116	138	165	200	243	294	353	87	104	124	150	182	221	265
2.2	169	201	242	297	353	426	511	118	140	169	204	248	300	361	89	105	127	153	186	225	271
2.4	172	205	247	298	361	435	522	120	143	172	209	253	306	368	90	107	129	157	190	230	276
2.6	175	209	252	304	368	444	533	122	146	175	213	258	312	375	92	110	131	160	194	234	281
2.8	178	213	257	310	375	452	543	124	148	178	217	263	318	382	93	111	134	163	197	239	287
3.0	181	217	261	316	382	460	552	126	151	182	221	268	324	389	95	113	137	166	201	243	292

Chlorine Concentration (mg/L)	Temperature = 15°C							Temperature = 20°C							Temperature = 25°C						
	pH							pH							pH						
	<=6.0	6.5	7.0	7.5	8.0	8.5	9.0	<=6.0	6.5	7.0	7.5	8.0	8.5	9.0	<=6.0	6.5	7.0	7.5	8.0	8.5	9.0
<=0.4	49	59	70	83	99	118	140	36	44	52	62	74	89	105	24	29	35	42	50	59	70
0.6	50	60	72	86	102	122	146	38	45	54	64	77	92	109	25	30	36	43	51	61	73
0.8	52	61	73	88	105	126	151	39	46	55	66	79	95	113	26	31	37	44	53	63	75
1.0	53	63	75	90	108	130	156	39	47	56	67	81	98	117	26	31	37	45	54	65	78
1.2	54	64	76	92	111	134	160	40	48	57	69	83	100	120	27	32	38	46	55	67	80
1.4	55	65	78	94	114	137	165	41	49	58	70	85	103	123	27	33	39	47	57	69	82
1.6	56	66	79	96	116	141	169	42	50	59	72	87	105	126	28	33	40	48	58	70	84
1.8	57	68	81	98	119	144	173	43	51	61	74	89	106	129	29	34	41	49	60	72	86
2.0	58	69	83	100	122	147	177	44	52	62	75	91	110	132	29	35	41	50	61	74	88
2.2	59	70	85	102	124	150	181	44	53	63	77	93	113	135	30	35	42	51	62	75	90
2.4	60	72	86	105	127	153	184	45	54	65	78	95	115	138	30	36	43	52	63	77	92
2.6	61	73	88	107	129	156	188	46	55	66	80	97	117	141	31	37	44	53	65	78	94
2.8	62	74	89	109	132	159	191	47	56	67	81	99	119	143	31	37	45	54	66	80	96
3.0	63	76	91	111	134	162	195	47	57	68	83	101	122	146	32	38	46	55	67	81	97

*Although units did not appear in the original tables, units are min-mg/L

CT VALUES* FOR 4-LOG INACTIVATION OF VIRUSES BY FREE CHLORINE

	pH	
Temperature (°C)	6–9	10
0.5	12	90
5	8	60
10	6	45
15	4	30
20	3	22
25	2	15

*Although units did not appear in the original tables, units are min-mg/L

♦ *Guidance Manual LT1ESWTR Disinfection Profiling and Benchmarking*, U.S. Environmental Protection Agency, 2003.

ELECTRICAL AND COMPUTER ENGINEERING

UNITS

The basic electrical units are coulombs for charge, volts for voltage, amperes for current, and ohms for resistance and impedance.

ELECTROSTATICS

$$\mathbf{F}_2 = \frac{Q_1 Q_2}{4\pi\varepsilon r^2}\mathbf{a}_{r12}, \text{ where}$$

$\mathbf{F}_2$ = the force on charge 2 due to charge 1

Q_i = the ith point charge

r = the distance between charges 1 and 2

$\mathbf{a}_{r12}$ = a unit vector directed from 1 to 2

ε = the permittivity of the medium

For free space or air:

$$\varepsilon = \varepsilon_0 = 8.85 \times 10^{-12} \text{ farads/meter}$$

Electrostatic Fields

Electric field intensity **E** (volts/meter) at point 2 due to a point charge Q_1 at point 1 is

$$\mathbf{E} = \frac{Q_1}{4\pi\varepsilon r^2}\mathbf{a}_{r12}$$

For a line charge of density ρ_L coulombs/meter on the z-axis, the radial electric field is

$$\mathbf{E}_L = \frac{\rho_L}{2\pi\varepsilon r}\mathbf{a}_r$$

For a sheet charge of density ρ_s coulombs/meter^2 in the x-y plane:

$$\mathbf{E}_s = \frac{\rho_s}{2\varepsilon}\mathbf{a}_z, z > 0$$

Gauss' law states that the integral of the electric flux density $\mathbf{D} = \varepsilon\mathbf{E}$ over a closed surface is equal to the charge enclosed or

$$Q_{encl} = \oiint_S \varepsilon\mathbf{E} \cdot d\mathbf{S}$$

The force on a point charge Q in an electric field with intensity **E** is $\mathbf{F} = Q\mathbf{E}$.

The work done by an external agent in moving a charge Q in an electric field from point p_1 to point p_2 is

$$W = -Q\int_{p_1}^{p_2}\mathbf{E} \cdot d\mathbf{l}$$

The energy W_E stored in an electric field **E** is

$$W_E = (1/2)\iiint_V \varepsilon|\mathbf{E}|^2 dV$$

Voltage

The potential difference V between two points is the work per unit charge required to move the charge between the points.

For two parallel plates with potential difference V, separated by distance d, the strength of the E field between the plates is

$$E = \frac{V}{d}$$

directed from the + plate to the – plate.

Current

Electric current $i(t)$ through a surface is defined as the rate of charge transport through that surface or

$i(t) = dq(t)/dt$, which is a function of time t

since $q(t)$ denotes instantaneous charge.

A constant current $i(t)$ is written as I, and the vector current density in amperes/m^2 is defined as **J**.

Magnetic Fields

For a current-carrying wire on the z-axis

$$\mathbf{H} = \frac{\mathbf{B}}{\mu} = \frac{I\mathbf{a}_\phi}{2\pi r}, \text{ where}$$

H = the magnetic field strength (amperes/meter)

B = the magnetic flux density (tesla)

$\mathbf{a}_\phi$ = the unit vector in positive ϕ direction in cylindrical coordinates

I = the current

μ = the permeability of the medium

For air: $\mu = \mu_o = 4\pi \times 10^{-7}$ H/m

Force on a current-carrying conductor in a uniform magnetic field is

$$\mathbf{F} = I\mathbf{L} \times \mathbf{B}, \text{ where}$$

L = the length vector of a conductor

The energy stored W_H in a magnetic field **H** is

$$W_H = (1/2)\iiint_V \mu|\mathbf{H}|^2 dv$$

Induced Voltage

Faraday's Law states for a coil of N turns enclosing flux ϕ:

$v = -N\, d\phi/dt$, where

v = the induced voltage

ϕ = the average flux (webers) enclosed by each turn

$\phi = \int_S \mathbf{B} \cdot d\mathbf{S}$

Resistivity

For a conductor of length L, electrical resistivity ρ, and cross-sectional area A, the resistance is

$$R = \frac{\rho L}{A}$$

For metallic conductors, the resistivity and resistance vary linearly with changes in temperature according to the following relationships:

$$\rho = \rho_0\left[1 + \alpha\left(T - T_0\right)\right], \text{and}$$
$$R = R_0\left[1 + \alpha\left(T - T_0\right)\right], \text{where}$$

ρ_0 = resistivity at T_0

R_0 = resistance at T_0

α = temperature coefficient

Ohm's Law: $V = IR$; $v(t) = i(t)\,R$

Resistors in Series and Parallel

For series connections, the current in all resistors is the same and the equivalent resistance for n resistors in series is

$$R_S = R_1 + R_2 + \ldots + R_n$$

For parallel connections of resistors, the voltage drop across each resistor is the same and the equivalent resistance for n resistors in parallel is

$$R_P = 1/(1/R_1 + 1/R_2 + \ldots + 1/R_n)$$

For two resistors R_1 and R_2 in parallel

$$R_P = \frac{R_1 R_2}{R_1 + R_2}$$

Power Absorbed by a Resistive Element

$$P = VI = \frac{V^2}{R} = I^2 R$$

Kirchhoff's Laws

Kirchhoff's voltage law for a closed path is expressed by

$$\Sigma V_{\text{rises}} = \Sigma V_{\text{drops}}$$

Kirchhoff's current law for a closed surface is

$$\Sigma I_{\text{in}} = \Sigma I_{\text{out}}$$

SOURCE EQUIVALENTS

For an arbitrary circuit

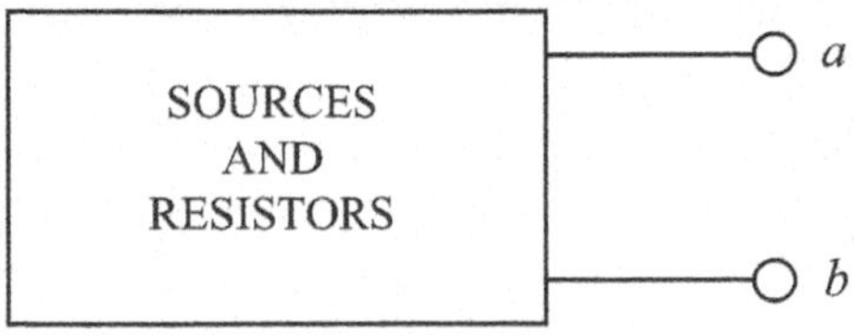

The Thévenin equivalent is

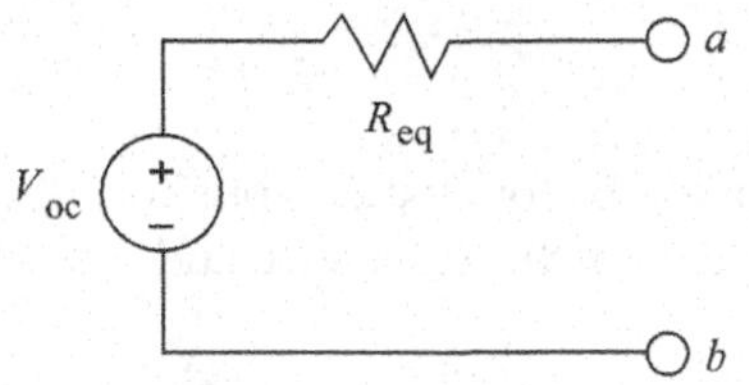

$$R_{\text{eq}} = \frac{V_{\text{oc}}}{I_{\text{sc}}}$$

The open circuit voltage V_{oc} is $V_a - V_b$, and the short circuit current is I_{sc} from a to b.

The Norton equivalent circuit is

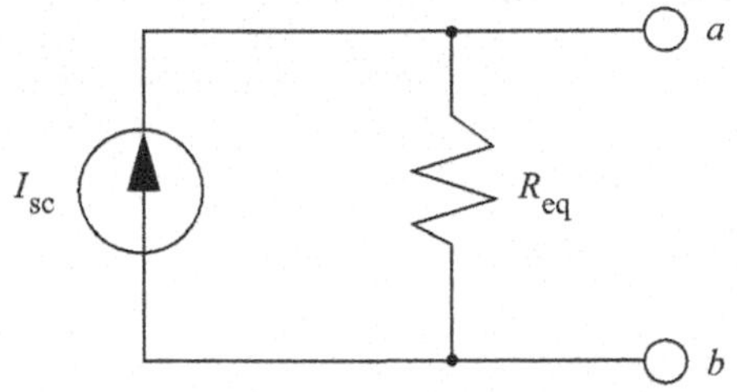

where I_{sc} and R_{eq} are defined above.

A load resistor R_L connected across terminals a and b will draw maximum power when $R_L = R_{\text{eq}}$.

CAPACITORS AND INDUCTORS

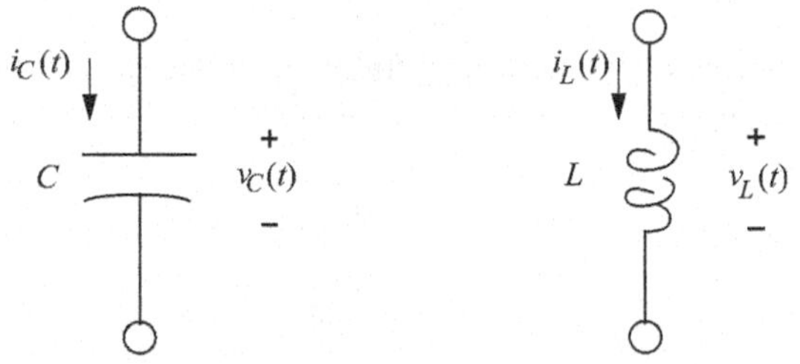

The charge $q_C(t)$ and voltage $v_C(t)$ relationship for a capacitor C in farads is

$$C = q_C(t)/v_C(t) \qquad \text{or} \qquad q_C(t) = Cv_C(t)$$

A parallel plate capacitor of area A with plates separated a distance d by an insulator with a permittivity ε has a capacitance

$$C = \frac{\varepsilon A}{d}$$

ε is often given as $\varepsilon = \varepsilon_r(\varepsilon_o)$ where ε_r is the relative permittivity or dielectric constant and $\varepsilon_o = 8.85 \times 10^{-12}$ F/m.

The current-voltage relationships for a capacitor are

$$v_C(t) = v_C(0) + \frac{1}{C}\int_0^t i_C(\tau)\,d\tau$$

and $i_C(t) = C\,(dv_C/dt)$

The energy stored in a capacitor is expressed in joules and given by

$$\text{Energy} = Cv_C^2/2 = q_C^2/2C = q_C v_C/2$$

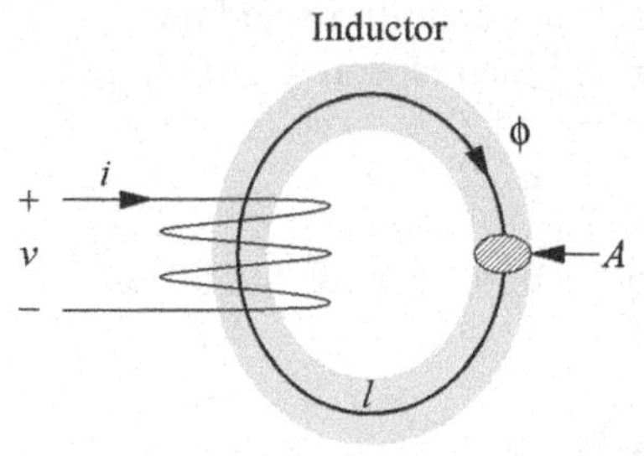

The inductance L (henrys) of a coil of N turns wound on a core with cross-sectional area A (m^2), permeability μ and flux ϕ with a mean path of l (m) is given as:

$$L = N^2 \mu A/l = N^2/\Re$$
$$N\phi = Li$$

where $\Re$ = reluctance = $l/\mu A$ (H^{-1}). μ is sometimes given as $\mu = \mu_r \cdot \mu_o$ where μ_r is the relative permeability and $\mu_o = 4\pi \times 10^{-7}$ H/m.

Using Faraday's law, the voltage-current relations for an inductor are

$$v_L(t) = L\,(di_L/dt)$$

$$i_L(t) = i_L(0) + \frac{1}{L}\int_0^t v_L(\tau)\,d\tau, \text{where}$$

v_L = inductor voltage

L = inductance (henrys)

i_L = inductor current (amperes)

The energy stored in an inductor is expressed in joules and given by

$$\text{Energy} = Li_L^2/2$$

Capacitors and Inductors in Parallel and Series

Capacitors in Parallel

$$C_P = C_1 + C_2 + \ldots + C_n$$

Capacitors in Series

$$C_S = \frac{1}{1/C_1 + 1/C_2 + \ldots + 1/C_n}$$

Inductors in Parallel

$$L_P = \frac{1}{1/L_1 + 1/L_2 + \ldots + 1/L_n}$$

Inductors in Series

$$L_S = L_1 + L_2 + \ldots + L_n$$

AC CIRCUITS

For a sinusoidal voltage or current of frequency f (Hz) and period T (seconds),

$$f = 1/T = \omega/(2\pi), \text{ where}$$

ω = the angular frequency in radians/s

Average Value

For a periodic waveform (either voltage or current) with period T,

$$X_{ave} = (1/T)\int_0^T x(t)\,dt$$

The average value of a full-wave rectified sinusoid is

$$X_{ave} = (2X_{max})/\pi$$

and half this for half-wave rectification, where

X_{max} = the peak amplitude of the sinusoid.

Effective or RMS Values

For a periodic waveform with period T, the rms or effective value is

$$X_{eff} = X_{rms} = \left[(1/T)\int_0^T x^2(t)\,dt\right]^{1/2}$$

For a sinusoidal waveform and full-wave rectified sine wave,

$$X_{eff} = X_{rms} = X_{max}/\sqrt{2}$$

For a half-wave rectified sine wave,

$$X_{eff} = X_{rms} = X_{max}/2$$

For a periodic signal,

$$X_{rms} = \sqrt{X_{dc}^2 + \sum_{n=1}^{\infty} X_n^2} \text{ where}$$

X_{dc} is the dc component of $x(t)$

X_n is the rms value of the nth harmonic

Sine-Cosine Relations and Trigonometric Identities

$$\cos(\omega t) = \sin(\omega t + \pi/2) = -\sin(\omega t - \pi/2)$$

$$\sin(\omega t) = \cos(\omega t - \pi/2) = -\cos(\omega t + \pi/2)$$

Other trigonometric identities for sinusoids are given in the section on Trigonometry.

Phasor Transforms of Sinusoids

$$P[V_{max}\cos(\omega t + \phi)] = V_{rms}\angle\phi = \mathbf{V}$$

$$P[I_{max}\cos(\omega t + \theta)] = I_{rms}\angle\theta = \mathbf{I}$$

For a circuit element, the impedance is defined as the ratio of phasor voltage to phasor current.

$$\mathbf{Z} = \mathbf{V}/\mathbf{I}$$

For a resistor, $\mathbf{Z}_R = R$

For a capacitor, $\mathbf{Z}_C = \dfrac{1}{j\omega C} = jX_C$

For an inductor,

$$\mathbf{Z}_L = j\omega L = jX_L, \text{ where}$$

X_C and X_L are the capacitive and inductive reactances respectively defined as

$$X_C = -\frac{1}{\omega C} \text{ and } X_L = \omega L$$

Impedances in series combine additively while those in parallel combine according to the reciprocal rule just as in the case of resistors.

Maximum Power-Transfer Theorem

DC Circuits

Maximum power transfer to the load R_L occurs when $R_L = R_{Th}$.

$$P_{max} = \frac{V_{Th}^2}{4\,R_{Th}}$$

$$\text{Efficiency: } \eta = \frac{P_L}{P_S} = \frac{R_L}{R_L + R_{Th}}$$

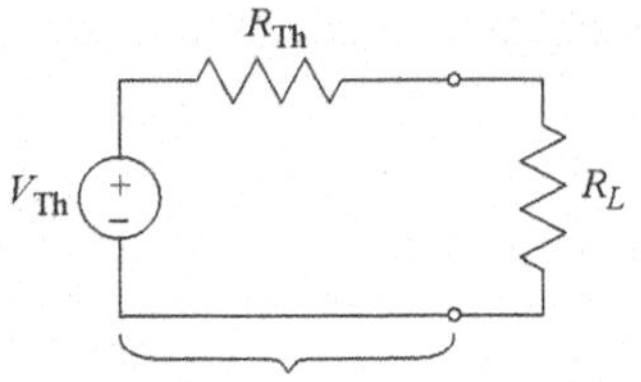

Thevenin Equivalent Circuit

AC Circuits

In an ac circuit maximum power transfer to the load impedance Z_L occurs when the load impedance equals the complex conjugate of the Thevenin equivalent impedance:

$$Z_L = Z_{Th}^*$$

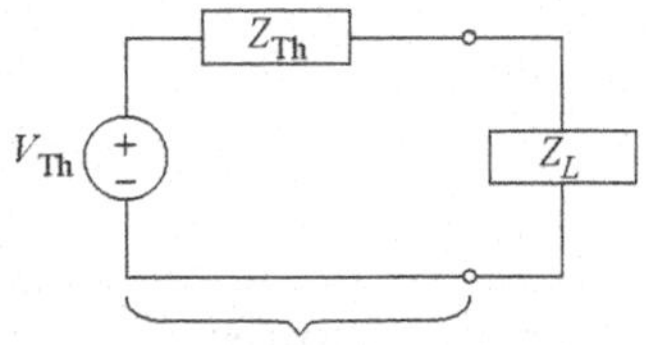

Thevenin Equivalent Circuit

*If the load is purely resistive (R_L) then for maximum power transfer $R_L = |Z_{Th}|$

RC AND RL TRANSIENTS

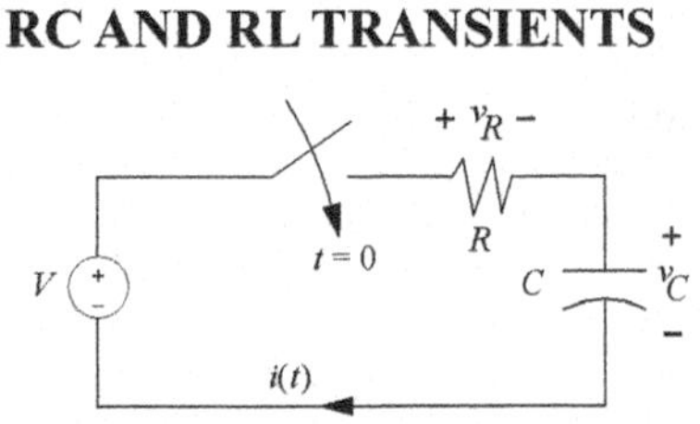

$$t \geq 0;\ v_C(t) = v_C(0)e^{-t/RC} + V\left(1 - e^{-t/RC}\right)$$

$$i(t) = \left\{\left[V - v_C(0)\right]/R\right\}e^{-t/RC}$$

$$v_R(t) = i(t)R = \left[V - v_C(0)\right]e^{-t/RC}$$

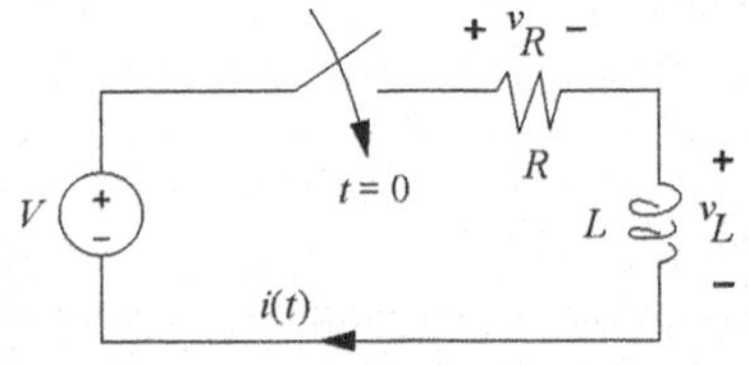

$$t \geq 0;\ i(t) = i(0)e^{-Rt/L} + \frac{V}{R}\left(1 - e^{-Rt/L}\right)$$

$$v_R(t) = i(t)R = i(0)Re^{-Rt/L} + V\left(1 - e^{-Rt/L}\right)$$

$$v_L(t) = L(di/dt) = -\,i(0)Re^{-Rt/L} + Ve^{-Rt/L}$$

where $v(0)$ and $i(0)$ denote the initial conditions and the parameters RC and L/R are termed the respective circuit time constants.

RESONANCE

The radian resonant frequency for both parallel and series resonance situations is

$$\omega_0 = \frac{1}{\sqrt{LC}} = 2\pi f_0 \text{ rad/s}$$

Series Resonance

$$\omega_0 L = \frac{1}{\omega_0 C}$$

$$Z = R \text{ at resonance}$$

$$Q = \frac{\omega_0 L}{R} = \frac{1}{\omega_0 CR}$$

$$BW = \frac{\omega_0}{Q} \text{ rad/s}$$

Parallel Resonance

$$\omega_0 L = \frac{1}{\omega_0 C}$$

$$Z = R \text{ at resonance}$$

$$Q = \omega_0 RC = \frac{R}{\omega_0 L}$$

$$BW = \frac{\omega_0}{Q} \text{ rad/s}$$

AC POWER

Complex Power

Real power P (watts) is defined by

$$P = (\tfrac{1}{2})V_{max} I_{max} \cos\theta$$

$$= V_{rms} I_{rms} \cos\theta$$

where θ is the angle measured from **V** to **I**. If **I** leads (lags) **V**, then the power factor (pf),

$$pf = \cos\theta$$

is said to be a leading (lagging) pf.

Reactive power Q (vars) is defined by

$$Q = (\tfrac{1}{2})V_{max} I_{max} \sin\theta$$

$$= V_{rms} I_{rms} \sin\theta$$

Complex power **S** (volt-amperes) is defined by

$$\mathbf{S} = \mathbf{VI}^* = P + jQ,$$

where **I*** is the complex conjugate of the phasor current.

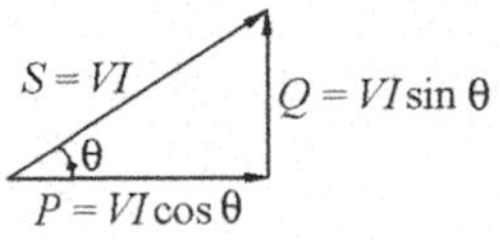

Complex Power Triangle (Inductive Load)

For resistors, $\theta = 0$, so the real power is

$$P = V_{rms}\, I_{rms} = V_{rms}^2/R = I_{rms}^2 R$$

Balanced Three-Phase (3-ϕ) Systems

The 3-phase line-phase relations are

for a delta

$V_L = V_P$

$I_L = \sqrt{3}\, I_P$

for a wye

$V_L = \sqrt{3}\, V_P = \sqrt{3}\, V_{LN}$

$I_L = I_P$

where subscripts L and P denote line and phase respectively.

A balanced 3-ϕ, delta-connected load impedance can be converted to an equivalent wye-connected load impedance using the following relationship

$$\mathbf{Z}_\Delta = 3\mathbf{Z}_Y$$

The following formulas can be used to determine 3-ϕ power for balanced systems.

$$\mathbf{S} = P + jQ$$

$$|\mathbf{S}| = 3V_P I_P = \sqrt{3}\, V_L I_L$$

$$\mathbf{S} = 3V_P I_P^{*} = \sqrt{3}\, V_L I_L \left(\cos\theta_P + j\sin\theta_P\right)$$

For balanced 3-ϕ, wye- and delta-connected loads

$$\mathbf{S} = \frac{V_L^2}{Z_Y^{*}} \qquad \mathbf{S} = 3\frac{V_L^2}{Z_\Delta^{*}}$$

where

$\mathbf{S}$ = total 3-ϕ complex power (VA)

$|\mathbf{S}|$ = total 3-ϕ apparent power (VA)

P = total 3-ϕ real power (W)

Q = total 3-ϕ reactive power (var)

θ_P = power factor angle of each phase

V_L = rms value of the line-to-line voltage

V_{LN} = rms value of the line-to-neutral voltage

I_L = rms value of the line current

I_P = rms value of the phase current

For a 3-ϕ, wye-connected source or load with line-to-neutral voltages and a positive phase sequence

$$\mathbf{V}_{an} = V_P \angle 0°$$

$$\mathbf{V}_{bn} = V_P \angle -120°$$

$$\mathbf{V}_{cn} = V_P \angle 120°$$

The corresponding line-to-line voltages are

$$\mathbf{V}_{ab} = \sqrt{3}\, V_P \angle 30°$$

$$\mathbf{V}_{bc} = \sqrt{3}\, V_P \angle -90°$$

$$\mathbf{V}_{ca} = \sqrt{3}\, V_P \angle 150°$$

Transformers (Ideal)

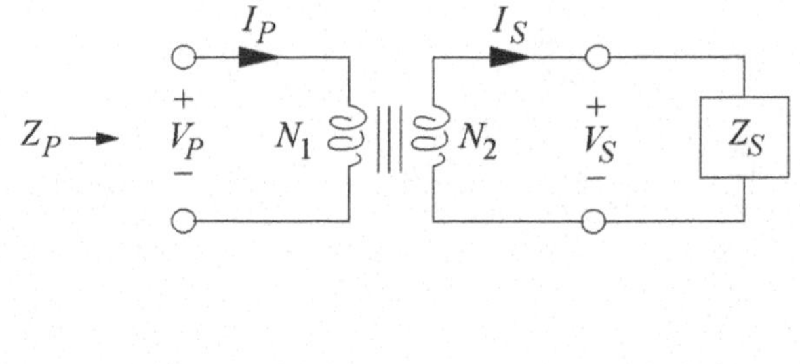

Turns Ratio

$$a = N_1/N_2$$

$$a = \left|\frac{\mathbf{V}_P}{\mathbf{V}_S}\right| = \left|\frac{\mathbf{I}_S}{\mathbf{I}_P}\right|$$

The impedance seen at the input is

$$\mathbf{Z}_P = a^2\mathbf{Z}_S$$

AC Machines

The synchronous speed n_s for ac motors is given by

$n_s = 120f/p$, where

f = the line voltage frequency (Hz)

p = the number of poles

The slip for an induction motor is

slip $= (n_s - n)/n_s$, where

n = the rotational speed (rpm)

DC Machines

The armature circuit of a dc machine is approximated by a series connection of the armature resistance R_a, the armature inductance L_a, and a dependent voltage source of value

$V_a = K_a n\phi$ volts, where

K_a = constant depending on the design

n = armature speed (rpm)

ϕ = the magnetic flux generated by the field

The field circuit is approximated by the field resistance R_f in series with the field inductance L_f. Neglecting saturation, the magnetic flux generated by the field current I_f is

$\phi = K_f I_f$ webers

The mechanical power generated by the armature is

$P_m = V_a I_a$ watts

where I_a is the armature current. The mechanical torque produced is

$T_m = (60/2\pi)K_a\phi I_a$ newton-meters.

Servomotors and Generators

Servomotors are electrical motors tied to a feedback system to obtain precise control. Smaller servomotors typically are dc motors.

A permanent magnet dc generator can be used to convert mechanical energy to electrical energy, as in a tachometer.

DC motor suppliers may provide data sheets with speed torque curves, motor torque constants (K_T), and motor voltage constants (K_E). An idealized dc motor at steady state exhibits the following relationships:

$$V = IR + K_E\,\omega$$

$$T = K_T I$$

where

V = voltage at the motor terminals

I = current through the motor

T = torque applied by the motor

R = resistance of the windings

ω = rotational speed

When using consistent SI units [N•m/A and V/(rad/s)], $K_T = K_E$.

An ideal speed-torque curve for a servomotor, with constant V, would look like this:

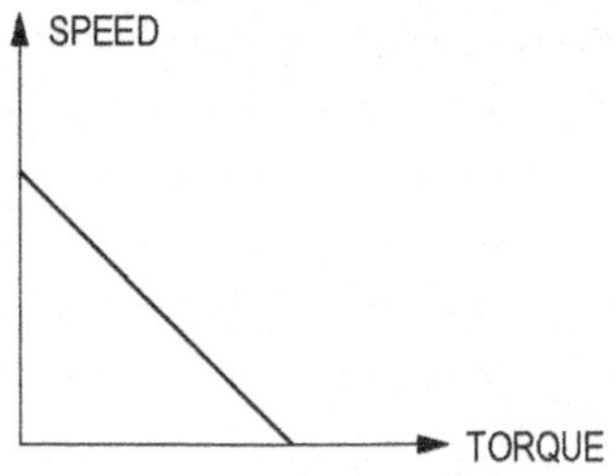

Voltage Regulation

The percent voltage regulation of a power supply is defined as

$$\% \text{ Regulation} = \frac{|V_{NL}| - |V_{FL}|}{|V_{FL}|} \times 100\%$$

where

V_{NL} = voltage under no load conditions

V_{FL} = voltage under full load conditions

TWO-PORT PARAMETERS

A two-port network consists of two input and two output terminals as shown.

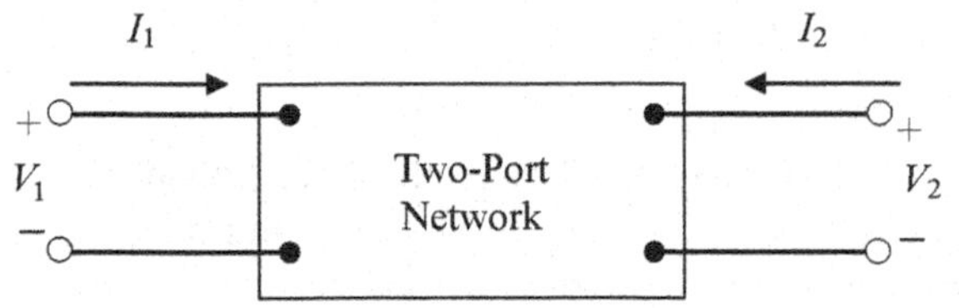

A two-port network may be represented by an equivalent circuit using a set of two-port parameters. Three commonly used sets of parameters are impedance, admittance, and hybrid parameters. The following table describes the equations used for each of these sets of parameters.

Parameter Type	Equations	Definitions
Impedance (z)	$V_1 = z_{11}I_1 + z_{12}I_2$ $V_2 = z_{21}I_1 + z_{22}I_2$	$z_{11} = \frac{V_1}{I_1}\Big\|_{I_2=0}$ $z_{12} = \frac{V_1}{I_2}\Big\|_{I_1=0}$ $z_{21} = \frac{V_2}{I_1}\Big\|_{I_2=0}$ $z_{22} = \frac{V_2}{I_2}\Big\|_{I_1=0}$
Admittance (y)	$I_1 = y_{11}V_1 + y_{12}V_2$ $I_2 = y_{21}V_1 + y_{22}V_2$	$y_{11} = \frac{I_1}{V_1}\Big\|_{V_2=0}$ $y_{12} = \frac{I_1}{V_2}\Big\|_{V_1=0}$ $y_{21} = \frac{I_2}{V_1}\Big\|_{V_2=0}$ $y_{22} = \frac{I_2}{V_2}\Big\|_{V_1=0}$
Hybrid (h)	$V_1 = h_{11}I_1 + h_{12}V_2$ $I_2 = h_{21}I_1 + h_{22}V_2$	$h_{11} = \frac{V_1}{I_1}\Big\|_{V_2=0}$ $h_{12} = \frac{V_1}{V_2}\Big\|_{I_1=0}$ $h_{21} = \frac{I_2}{I_1}\Big\|_{V_2=0}$ $h_{22} = \frac{I_2}{V_2}\Big\|_{I_1=0}$

ELECTROMAGNETIC DYNAMIC FIELDS

The integral and point form of Maxwell's equations are

$$\oint \mathbf{E} \cdot d\ell = -\iint_S (\partial \mathbf{B}/\partial t) \cdot d\mathbf{S}$$
$$\oint \mathbf{H} \cdot d\ell = I_{\text{enc}} + \iint_S (\partial \mathbf{D}/\partial t) \cdot d\mathbf{S}$$
$$\oiint_{S_V} \mathbf{D} \cdot d\mathbf{S} = \iiint_V \rho dv$$
$$\oiint_{S_V} \mathbf{B} \cdot d\mathbf{S} = 0$$
$$\nabla \times \mathbf{E} = -\partial \mathbf{B}/\partial t$$
$$\nabla \times \mathbf{H} = \mathbf{J} + \partial \mathbf{D}/\partial t$$
$$\nabla \cdot \mathbf{D} = \rho$$
$$\nabla \cdot \mathbf{B} = 0$$

LOSSLESS TRANSMISSION LINES

The wavelength, λ, of a sinusoidal signal is defined as the distance the signal will travel in one period.

$$\lambda = \frac{U}{f}$$

where U is the velocity of propagation and f is the frequency of the sinusoid.

The characteristic impedance, $\mathbf{Z}_0$, of a transmission line is the input impedance of an infinite length of the line and is given by

$$\mathbf{Z}_0 = \sqrt{L/C}$$

where L and C are the per unit length inductance and capacitance of the line.

The reflection coefficient at the load is defined as

$$\Gamma = \frac{V^-}{V^+} = \frac{\mathbf{Z}_L - \mathbf{Z}_0}{\mathbf{Z}_L + \mathbf{Z}_0}$$

and the standing wave ratio SWR is

$$\text{SWR} = \frac{1 + |\Gamma|}{1 - |\Gamma|}$$

$$\beta = \text{Propagation constant} = \frac{2\pi}{\lambda}$$

For sinusoidal voltages and currents:

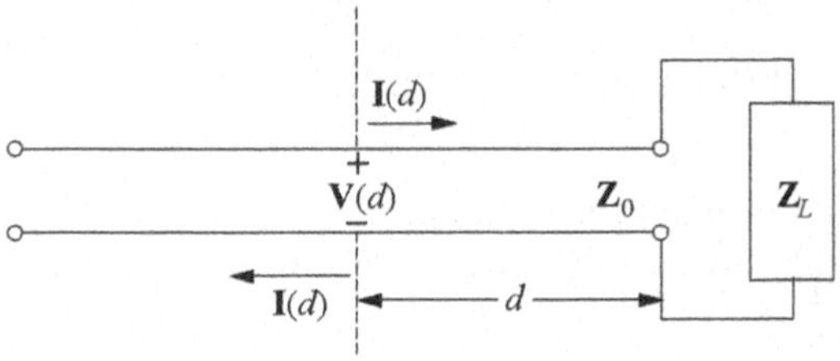

Voltage across the transmission line:

$$\mathbf{V}(d) = \mathbf{V}^+ e^{j\beta d} + \mathbf{V}^- e^{-j\beta d}$$

Current along the transmission line:

$$\mathbf{I}(d) = \mathbf{I}^+ e^{j\beta d} + \mathbf{I}^- e^{-j\beta d}$$

where $\mathbf{I}^+ = \mathbf{V}^+/\mathbf{Z}_0$ and $\mathbf{I}^- = -\mathbf{V}^-/\mathbf{Z}_0$

Input impedance at d

$$\mathbf{Z}_{\text{in}}(d) = \mathbf{Z}_0 \frac{\mathbf{Z}_L + j\mathbf{Z}_0 \tan(\beta d)}{\mathbf{Z}_0 + j\mathbf{Z}_L \tan(\beta d)}$$

DIFFERENCE EQUATIONS

Difference equations are used to model discrete systems. Systems which can be described by difference equations include computer program variables iteratively evaluated in a loop, sequential circuits, cash flows, recursive processes, systems with time-delay components, etc. Any system whose input $x(t)$ and output $y(t)$ are defined only at the equally spaced intervals $t = kT$ can be described by a difference equation.

First-Order Linear Difference Equation

A first-order difference equation is

$$y[k] + a_1 y[k-1] = x[k]$$

Second-Order Linear Difference Equation

A second-order difference equation is

$$y[k] + a_1 y[k-1] + a_2 y[k-2] = x[k]$$

z-Transforms

The transform definition is

$$F(z) = \sum_{k=0}^{\infty} f[k] z^{-k}$$

The inverse transform is given by the contour integral

$$f[k] = \frac{1}{2\pi j} \oint_\Gamma F(z) z^{k-1} dz$$

and it represents a powerful tool for solving linear shift-invariant difference equations. A limited unilateral list of z-transform pairs assuming zero initial conditions follows:

$f[k]$	$F(z)$
$\delta[k]$, Impulse at $k = 0$	1
$u[k]$, Step at $k = 0$	$1/(1 - z^{-1})$
β^k	$1/(1 - \beta z^{-1})$
$y[k-1]$	$z^{-1}Y(z)$
$y[k-2]$	$z^{-2}Y(z)$
$y[k+1]$	$zY(z) - zy[0]$
$y[k+2]$	$z^2Y(z) - z^2y[0] - zy[1]$
$\sum_{m=0}^{\infty} x[k-m]h[m]$	$H(z)X(z)$
$\lim_{k\to 0} f[k]$	$\lim_{z\to\infty} F(z)$
$\lim_{k\to\infty} f[k]$	$\lim_{z\to 1}(1 - z^{-1})F(z)$

[Note: The last two transform pairs represent the Initial Value Theorem (I.V.T.) and the Final Value Theorem (F.V.T.) respectively.]

CONVOLUTION

Continuous-time convolution:

$$v(t) = x(t) * y(t) = \int_{-\infty}^{\infty} x(\tau)y(t - \tau)d\tau$$

Discrete-time convolution:

$$v[n] = x[n] * y[n] = \sum_{k=-\infty}^{\infty} x[k]y[n - k]$$

DIGITAL SIGNAL PROCESSING

A discrete-time, linear, time-invariant (DTLTI) system with a single input $x[n]$ and a single output $y[n]$ can be described by a linear difference equation with constant coefficients of the form

$$y[n] + \sum_{i=1}^{k} b_i y[n - i] = \sum_{i=0}^{l} a_i x[n - i]$$

If all initial conditions are zero, taking a z-transform yields a transfer function

$$H(z) = \frac{Y(z)}{X(z)} = \frac{\sum_{i=0}^{l} a_i z^{-i}}{1 + \sum_{i=1}^{k} b_i z^{-i}}$$

Two common discrete inputs are the unit-step function $u[n]$ and the unit impulse function $\delta[n]$, where

$$u[n] = \begin{Bmatrix} 0 & n < 0 \\ 1 & n \geq 0 \end{Bmatrix} \text{ and } \delta[n] = \begin{Bmatrix} 1 & n = 0 \\ 0 & n \neq 0 \end{Bmatrix}$$

The impulse response $h[n]$ is the response of a discrete-time system to $x[n] = \delta[n]$.

A finite impulse response (FIR) filter is one in which the impulse response $h[n]$ is limited to a finite number of points:

$$h[n] = \sum_{i=0}^{k} a_i \delta[n - i]$$

The corresponding transfer function is given by

$$H(z) = \sum_{i=0}^{k} a_i z^{-i}$$

where k is the order of the filter.

An infinite impulse response (IIR) filter is one in which the impulse response $h[n]$ has an infinite number of points:

$$h[n] = \sum_{i=0}^{\infty} a_i \delta[n - i]$$

COMMUNICATION THEORY AND CONCEPTS

The following concepts and definitions are useful for communications systems analysis.

Functions

Unit step, $u(t)$	$u(t) = \begin{cases} 0 & t < 0 \\ 1 & t > 0 \end{cases}$
Rectangular pulse, $\Pi(t/\tau)$	$\Pi(t/\tau) = \begin{cases} 1 & \lvert t/\tau \rvert < \frac{1}{2} \\ 0 & \lvert t/\tau \rvert > \frac{1}{2} \end{cases}$
Triangular pulse, $\Lambda(t/\tau)$	$\Lambda(t/\tau) = \begin{cases} 1 - \lvert t/\tau \rvert & \lvert t/\tau \rvert < 1 \\ 0 & \lvert t/\tau \rvert > 1 \end{cases}$
Sinc, $\text{sinc}(at)$	$\text{sinc}(at) = \frac{\sin(a\pi t)}{a\pi t}$
Unit impulse, $\delta(t)$	$\int_{-\infty}^{+\infty} x(t + t_0)\delta(t)dt = x(t_0)$ for every $x(t)$ defined and continuous at $t = t_0$. This is equivalent to $\int_{-\infty}^{+\infty} x(t)\delta(t - t_0)dt = x(t_0)$

$$x(t) * h(t) = \int_{-\infty}^{+\infty} x(\lambda)h(t - \lambda)d\lambda$$
$$= h(t) * x(t) = \int_{-\infty}^{+\infty} h(\lambda)x(t - \lambda)d\lambda$$

In particular,

$$x(t) * \delta(t - t_0) = x(t - t_0)$$

The Fourier Transform and its Inverse

$$X(f) = \int_{-\infty}^{+\infty} x(t)e^{-j2\pi ft}dt$$

$$x(t) = \int_{-\infty}^{+\infty} X(f)e^{j2\pi ft}df$$

$x(t)$ and $X(f)$ form a *Fourier transform pair*:

$$x(t) \leftrightarrow X(f)$$

Frequency Response and Impulse Response

The *frequency response* $H(f)$ of a system with input $x(t)$ and output $y(t)$ is given by

$$H(f) = \frac{Y(f)}{X(f)}$$

This gives

$$Y(f) = H(f)X(f)$$

The response $h(t)$ of a linear time-invariant system to a unit-impulse input $\delta(t)$ is called the *impulse response* of the system. The response $y(t)$ of the system to any input $x(t)$ is the convolution of the input $x(t)$ with the impulse response $h(t)$:

$$y(t) = x(t) * h(t) = \int_{-\infty}^{+\infty} x(\lambda)h(t-\lambda)\,d\lambda$$

$$= h(t) * x(t) = \int_{-\infty}^{+\infty} h(\lambda)x(t-\lambda)\,d\lambda$$

Therefore, the impulse response $h(t)$ and frequency response $H(f)$ form a Fourier transform pair:

$$h(t) \leftrightarrow H(f)$$

Parseval's Theorem

The total energy in an energy signal (finite energy) $x(t)$ is given by

$$E = \int_{-\infty}^{+\infty} |x(t)|^2\,dt = \int_{-\infty}^{+\infty} |X(f)|^2\,df$$

Parseval's Theorem for Fourier Series

A periodic signal $x(t)$ with period T_0 and fundamental frequency $f_0 = 1/T_0 = \omega_0/2\pi$ can be represented by a complex-exponential Fourier series

$$x(t) = \sum_{n=-\infty}^{n=+\infty} X_n e^{jn2\pi f_0 t}$$

The average power in the dc component and the first N harmonics is

$$P = \sum_{n=-N}^{n=+N} |X_n|^2 = X_0^2 + 2\sum_{n=0}^{n=N} |X_n|^2$$

The total average power in the periodic signal $x(t)$ is given by *Parseval's theorem*:

$$P = \frac{1}{T_0}\int_{t_0}^{t_0+T_0} |x(t)|^2\,dt = \sum_{n=-\infty}^{n=+\infty} |X_n|^2$$

Decibels and Bode Plots

Decibels is a technique to measure the ratio of two powers:

$$\text{dB} = 10\log_{10}(P_2/P_1)$$

The definition can be modified to measure the ratio of two voltages:

$$\text{dB} = 20\log_{10}(V_2/V_1)$$

Bode plots use a logarithmic scale for the frequency when plotting magnitude and phase response, where the magnitude is plotted in dB using a straight-line (asymptotic) approximation.

The information below summarizes Bode plots for several terms commonly encountered when determining voltage gain, $G_v(j\omega)$. Since logarithms are used to convert gain to decibels, the decibel response when these various terms are multiplied together can be added to determine the overall response.

Term	**Magnitude Response** $\lvert G_v(j\omega)\rvert_{dB}$	**Phase Response** $< G_v(j\omega)$	**Plot**
K_0	$20\log_{10}(K_0)$	0°	a
$(j\omega)^{\pm1}$	$\pm20\log_{10}(\omega)$	±90°	b & c
$(1 + j\omega/\omega_c)^{\pm1}$	0 for $\omega \ll \omega_c$ ±3 dB for $\omega = \omega_c$ $\pm20\log_{10}(\omega)$ for $\omega \gg \omega_c$	0° for $\omega \ll \omega_c$ ±45° for $\omega = \omega_c$ ±90° for $\omega \gg \omega_c$	d & e

(a) K_o

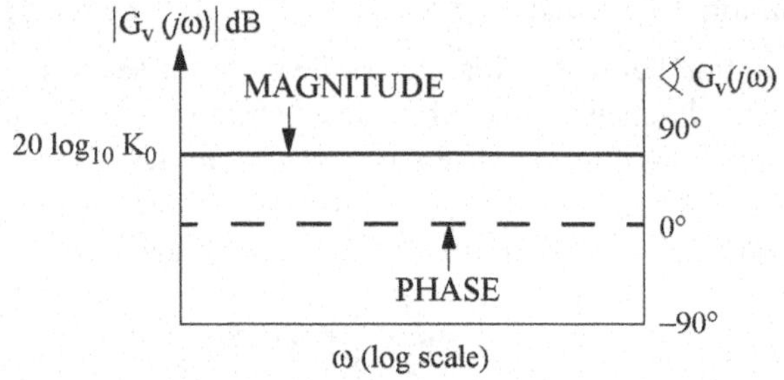

(b) $(j\omega)$

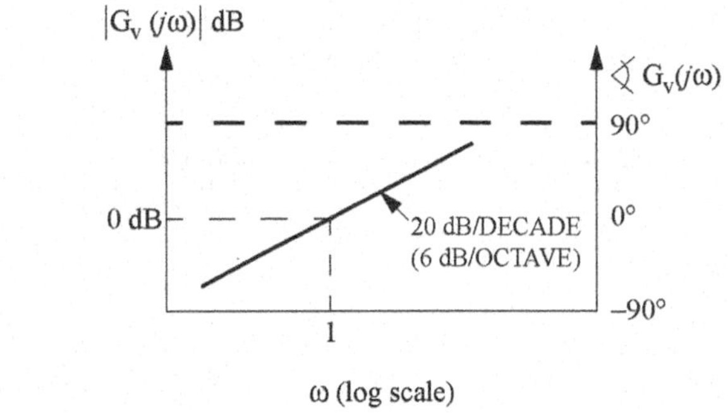

(c) $(j\omega)^{-1}$

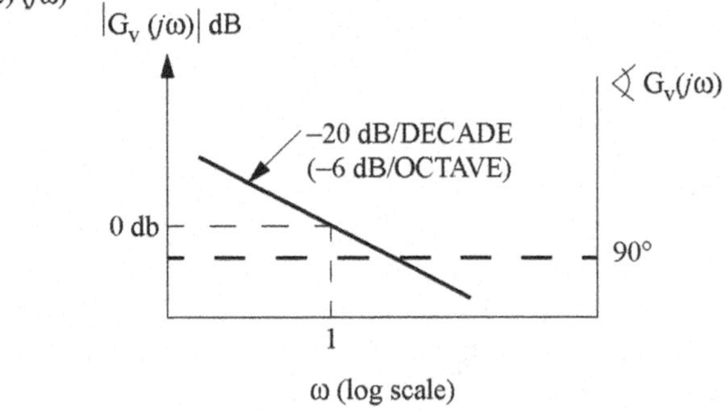

(d) $(1 + j\omega/\omega_c)$

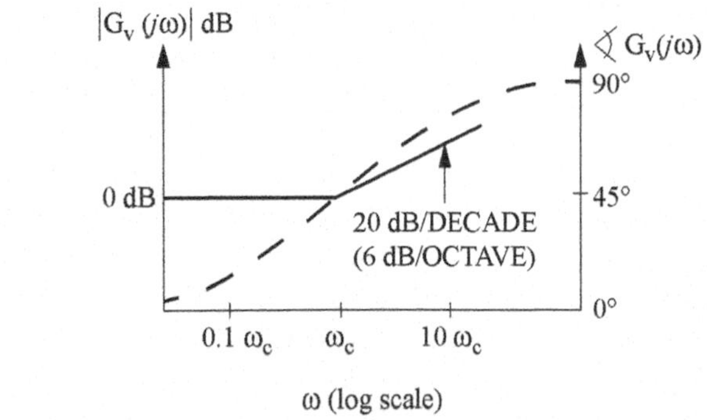

(e) $(1 + j\omega/\omega_c)^{-1}$

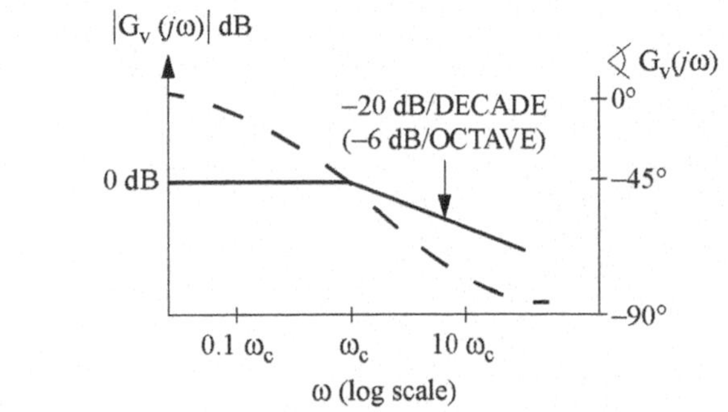

Amplitude Modulation (AM)

$$x_{AM}(t) = A_c\left[A + m(t)\right]\cos(2\pi f_c t) = A'_c\left[1 + a m_n(t)\right]\cos(2\pi f_c t)$$

The *modulation index* is a, and the normalized message is

$$m_n(t) = \frac{m(t)}{\max|m(t)|}$$

The *efficiency* η is the percent of the total transmitted power that contains the message.

$$\eta = \frac{a^2 < m_n^2(t) >}{1 + a^2 < m_n^2(t) >} 100 \text{ percent}$$

where the mean-squared value or normalized average power in $m_n(t)$ is

$$<m_n^2(t)> = \lim_{T \to \infty} \frac{1}{2T} \int_{-T}^{+T} |m_n(t)|^2 dt$$

If $M(f) = 0$ for $|f| > W$, then the *bandwidth* of $x_{AM}(t)$ is $2W$. AM signals can be demodulated with an envelope detector or a synchronous demodulator.

Double-Sideband Modulation (DSB)

$$x_{DSB}(t) = A_c m(t)\cos(2\pi f_c t)$$

If $M(f) = 0$ for $|f| > W$, then the bandwidth of $m(t)$ is W and the bandwidth of $x_{DSB}(t)$ is $2W$. DSB signals must be demodulated with a synchronous demodulator. A Costas loop is often used.

Single-Sideband Modulation (SSB)

Lower Sideband:

$$x_{LSB}(t) \longleftrightarrow X_{LSB}(f) = X_{DSB}(f)\Pi\left(\frac{f}{2f_c}\right)$$

Upper Sideband:

$$x_{USB}(t) \longleftrightarrow X_{USB}(f) = X_{DSB}(f)\left[1 - \Pi\left(\frac{f}{2f_c}\right)\right]$$

In either case, if $M(f) = 0$ for $|f| > W$, then the bandwidth of $x_{LSB}(t)$ or of $x_{USB}(t)$ is W. SSB signals can be demodulated with a synchronous demodulator or by carrier reinsertion and envelope detection.

Angle Modulation

$$x_{Ang}(t) = A_c\cos\left[2\pi f_c t + \phi(t)\right]$$

The *phase deviation* $\phi(t)$ is a function of the message $m(t)$. The *instantaneous phase* is

$$\phi_i(t) = 2\pi f_c t + \phi(t) \text{ rad}$$

The *instantaneous frequency* is

$$\omega_i(t) = \frac{d}{dt}\phi_i(t) = 2\pi f_c + \frac{d}{dt}\phi(t) \text{ rad/s}$$

The *frequency deviation* is

$$\Delta\omega(t) = \frac{d}{dt}\phi(t) \text{ rad/s}$$

The *phase deviation* is

$$\phi(t) = k_P m(t) \text{ rad}$$

The *complete* bandwidth of an angle-modulated signal is infinite.

A discriminator or a phase-lock loop can demodulate angle-modulated signals.

Frequency Modulation (FM)

The *phase deviation* is

$$\phi(t) = k_F \int_{-\infty}^{t} m(\lambda)d\lambda \text{ rad}$$

The *frequency-deviation ratio* is

$$D = \frac{k_F \max|m(t)|}{2\pi W}$$

where W is the message bandwidth. If $D << 1$ (narrowband FM), the 98% power bandwidth B is

$$B \cong 2W$$

If $D > 1$, (wideband FM) the 98% power bandwidth B is given by *Carson's rule*:

$$B \cong 2(D + 1)W$$

Sampled Messages

A low-pass message $m(t)$ can be exactly reconstructed from uniformly spaced samples taken at a sampling frequency of $f_s = 1/T_s$

$$f_s > 2W \text{ where } M(f) = 0 \text{ for } f > W$$

The frequency $2W$ is called the *Nyquist frequency*. Sampled messages are typically transmitted by some form of pulse modulation. The minimum bandwidth B required for transmission of the modulated message is inversely proportional to the pulse length τ.

$$B \propto \frac{1}{\tau}$$

Frequently, for approximate analysis

$$B \cong \frac{1}{2\tau}$$

is used as the *minimum* bandwidth of a pulse of length τ.

Ideal-Impulse Sampling

$$x_\delta(t) = m(t)\sum_{n=-\infty}^{n=+\infty}\delta(t - nT_s) = \sum_{n=-\infty}^{n=+\infty} m(nT_s)\delta(t - nT_s)$$

$$X_\delta(f) = M(f) * \left[f_s \sum_{k=-\infty}^{k=+\infty}\delta(f - kf_s)\right]$$

$$= f_s \sum_{k=-\infty}^{k=+\infty} M(f - kf_s)$$

The message $m(t)$ can be recovered from $x_\delta(t)$ with an ideal low-pass filter of bandwidth W if $f_s > 2W$.

(PAM) Pulse-Amplitude Modulation—Natural Sampling

A PAM signal can be generated by multiplying a message by a pulse train with pulses having duration τ and period $T_s = 1/f_s$

$$x_N(t) = m(t)\sum_{n=-\infty}^{n=+\infty}\Pi\left[\frac{t - nT_s}{\tau}\right] = \sum_{n=-\infty}^{n=+\infty} m(t)\Pi\left[\frac{t - nT_s}{\tau}\right]$$

$$X_N(f) = \tau f_s \sum_{k=-\infty}^{k=+\infty} \text{sinc}(k\tau f_s)M(f - kf_s)$$

The message $m(t)$ can be recovered from $x_N(t)$ with an ideal low-pass filter of bandwidth W.

Pulse-Code Modulation (PCM)

PCM is formed by sampling a message $m(t)$ and digitizing

the sample values with an A/D converter. For an n-bit binary word length, transmission of a pulse-code-modulated low-pass message $m(t)$, with $M(f) = 0$ for $f \geq W$, requires the transmission of at least $2nW$ binary pulses per second. A binary word of length n bits can represent q quantization levels:

$$q = 2^n$$

The minimum bandwidth required to transmit the PCM message will be

$$B \propto 2nW = 2W \log_2 q$$

ANALOG FILTER CIRCUITS

Analog filters are used to separate signals with different frequency content. The following circuits represent simple analog filters used in communications and signal processing.

First-Order Low-Pass Filters

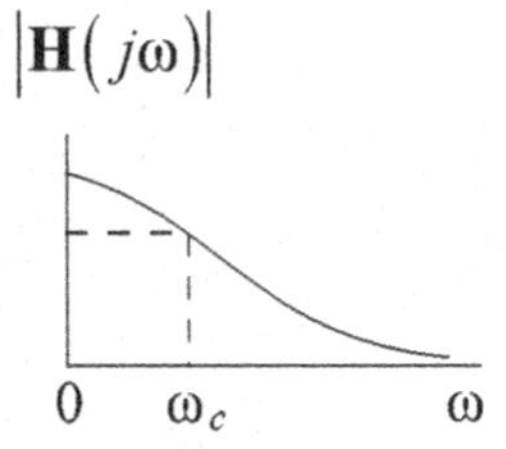

$$|\mathbf{H}(j\omega_c)| = \frac{1}{\sqrt{2}}|\mathbf{H}(0)|$$

Frequency Response

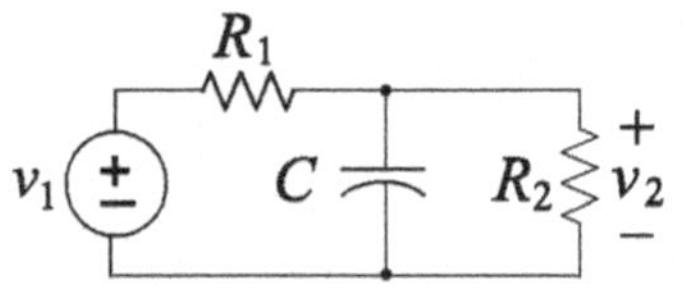

$$\mathbf{H}(s) = \frac{\mathbf{V}_2}{\mathbf{V}_1} = \frac{R_P}{R_1} \bullet \frac{1}{1 + sR_PC}$$

$$R_P = \frac{R_1 R_2}{R_1 + R_2} \qquad \omega_c = \frac{1}{R_PC}$$

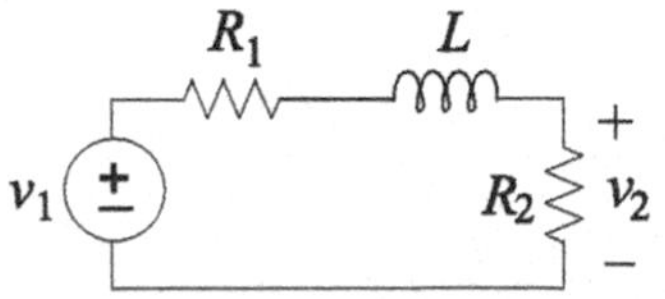

$$\mathbf{H}(s) = \frac{\mathbf{V}_2}{\mathbf{V}_1} = \frac{R_2}{R_S} \bullet \frac{1}{1 + s\,L/R_S}$$

$$R_S = R_1 + R_2 \qquad \omega_c = \frac{R_S}{L}$$

First-Order High-Pass Filters

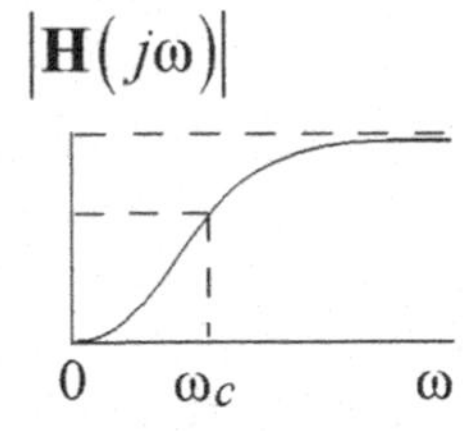

$$|\mathbf{H}(j\omega_c)| = \frac{1}{\sqrt{2}}|\mathbf{H}(j\infty)|$$

Frequency Response

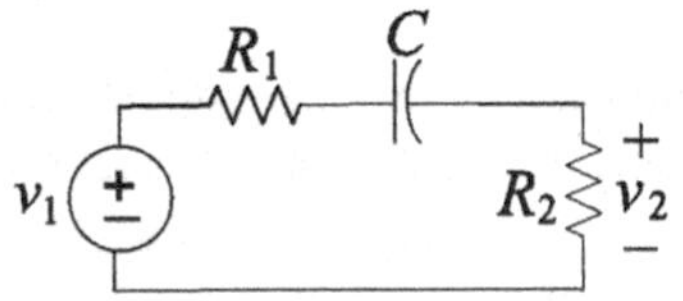

$$\mathbf{H}(s) = \frac{\mathbf{V}_2}{\mathbf{V}_1} = \frac{R_2}{R_S} \bullet \frac{sR_SC}{1 + sR_SC}$$

$$R_S = R_1 + R_2 \qquad \omega_c = \frac{1}{R_SC}$$

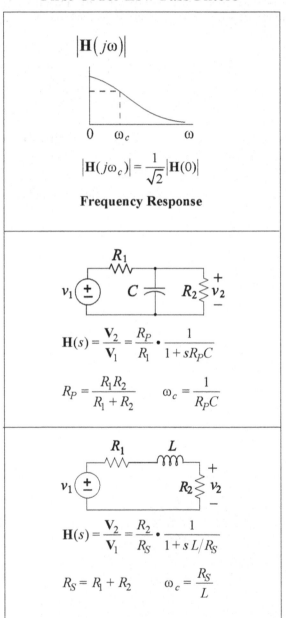

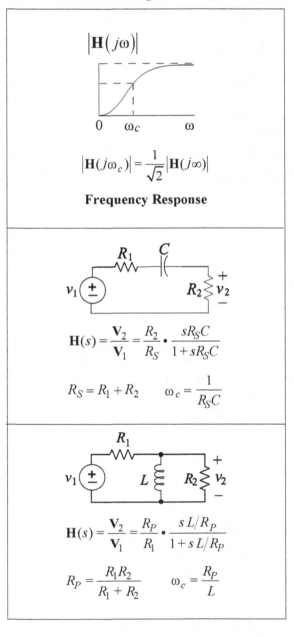

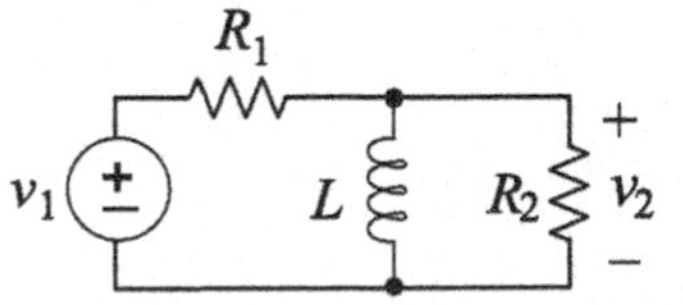

$$\mathbf{H}(s) = \frac{\mathbf{V}_2}{\mathbf{V}_1} = \frac{R_P}{R_1} \bullet \frac{s\,L/R_P}{1 + s\,L/R_P}$$

$$R_P = \frac{R_1 R_2}{R_1 + R_2} \qquad \omega_c = \frac{R_P}{L}$$

Band-Pass Filters

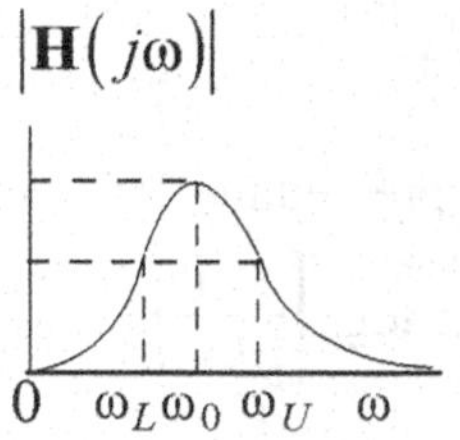

$$|\mathbf{H}(j\omega_L)| = |\mathbf{H}(j\omega_U)| = \frac{1}{\sqrt{2}}|\mathbf{H}(j\omega_0)|$$

3-dB Bandwidth = $BW = \omega_U - \omega_L$

Frequency Response

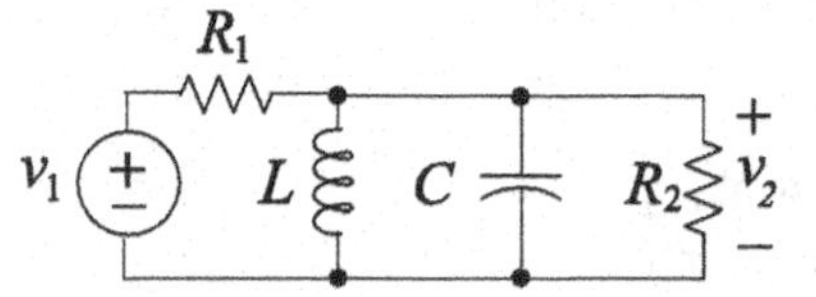

$$\mathbf{H}(s) = \frac{\mathbf{V}_2}{\mathbf{V}_1} = \frac{1}{R_1 C} \bullet \frac{s}{s^2 + s/R_P C + 1/LC}$$

$$R_P = \frac{R_1 R_2}{R_1 + R_2} \qquad \omega_0 = \frac{1}{\sqrt{LC}}$$

$$|\mathbf{H}(j\omega_0)| = \frac{R_2}{R_1 + R_2} = \frac{R_P}{R_1} \qquad BW = \frac{1}{R_P C}$$

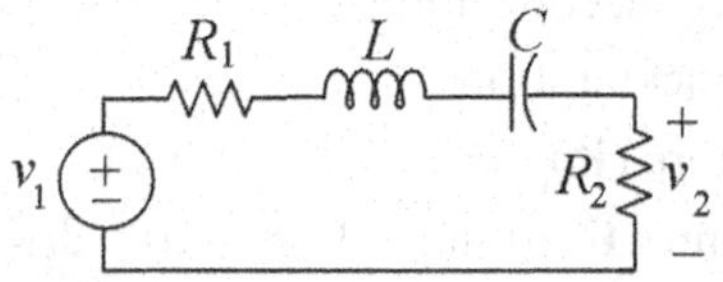

$$\mathbf{H}(s) = \frac{\mathbf{V}_2}{\mathbf{V}_1} = \frac{R_2}{L} \bullet \frac{s}{s^2 + sR_S/L + 1/LC}$$

$$R_S = R_1 + R_2 \qquad \omega_0 = \frac{1}{\sqrt{LC}}$$

$$|\mathbf{H}(j\omega_0)| = \frac{R_2}{R_1 + R_2} = \frac{R_2}{R_S} \qquad BW = \frac{R_S}{L}$$

Band-Reject Filters

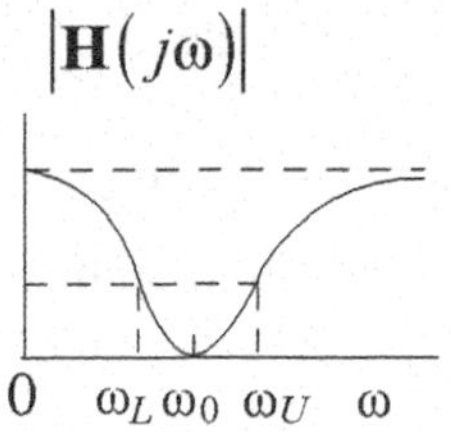

$$|\mathbf{H}(j\omega_L)| = |\mathbf{H}(j\omega_U)| = \left[1 - \frac{1}{\sqrt{2}}\right]|\mathbf{H}(0)|$$

3-dB Bandwidth = $BW = \omega_U - \omega_L$

Frequency Response

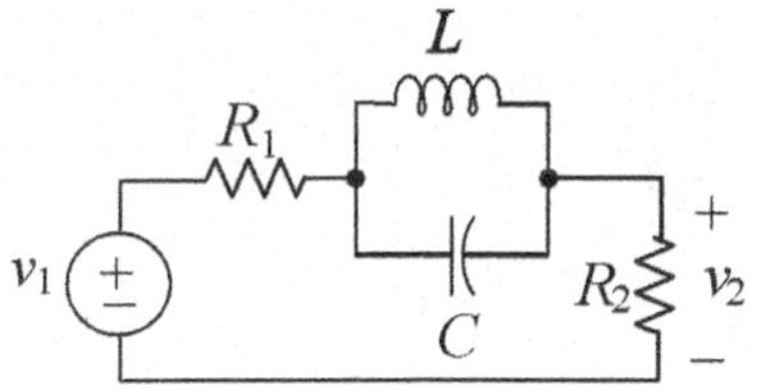

$$\mathbf{H}(s) = \frac{\mathbf{V}_2}{\mathbf{V}_1} = \frac{R_2}{R_S} \bullet \frac{s^2 + 1/LC}{s^2 + s/R_S C + 1/LC}$$

$$R_S = R_1 + R_2 \qquad \omega_0 = \frac{1}{\sqrt{LC}}$$

$$|\mathbf{H}(0)| = \frac{R_2}{R_1 + R_2} = \frac{R_2}{R_S} \qquad BW = \frac{1}{R_S C}$$

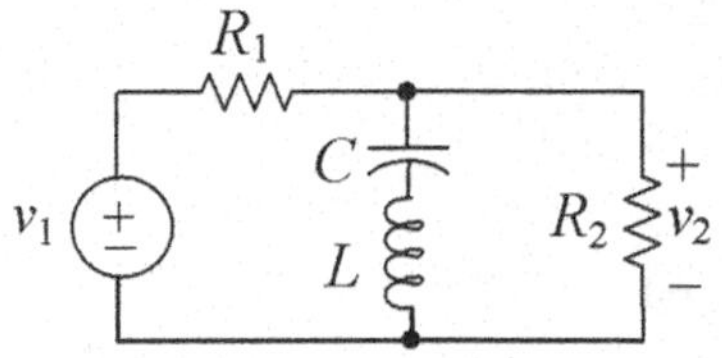

$$\mathbf{H}(s) = \frac{\mathbf{V}_2}{\mathbf{V}_1} = \frac{R_P}{R_1} \bullet \frac{s^2 + 1/LC}{s^2 + sR_P/L + 1/LC}$$

$$R_P = \frac{R_1 R_2}{R_1 + R_2} \qquad \omega_0 = \frac{1}{\sqrt{LC}}$$

$$|\mathbf{H}(0)| = \frac{R_2}{R_1 + R_2} = \frac{R_P}{R_1} \qquad BW = \frac{R_P}{L}$$

OPERATIONAL AMPLIFIERS

Ideal

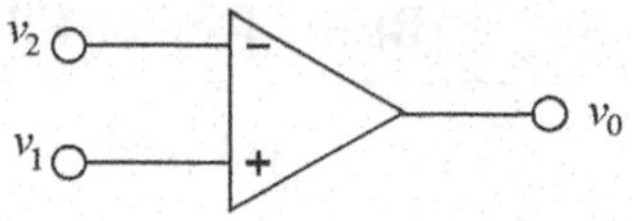

$v_0 = A(v_1 - v_2)$

where

A is large ($> 10^4$), and

$v_1 - v_2$ is small enough so as not to saturate the amplifier.

For the ideal operational amplifier, assume that the input currents are zero and that the gain A is infinite so when operating linearly $v_2 - v_1 = 0$.

For the two-source configuration with an ideal operational amplifier,

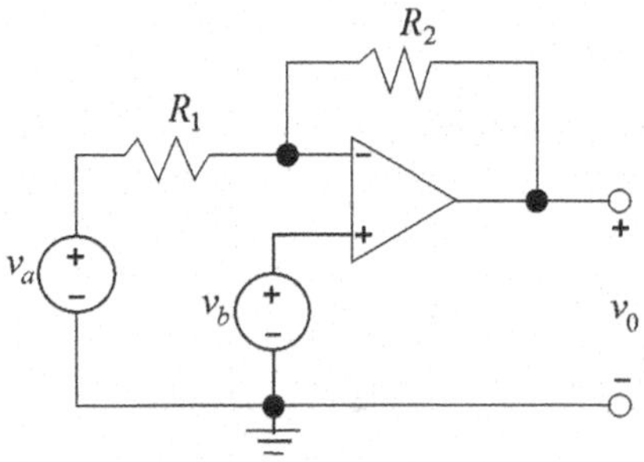

$$v_0 = -\frac{R_2}{R_1} v_a + \left(1 + \frac{R_2}{R_1}\right) v_b$$

If $v_a = 0$, we have a non-inverting amplifier with

$$v_0 = \left(1 + \frac{R_2}{R_1}\right) v_b$$

If $v_b = 0$, we have an inverting amplifier with

$$v_0 = -\frac{R_2}{R_1} v_a$$

Common Mode Rejection Ratio (*CMRR*)

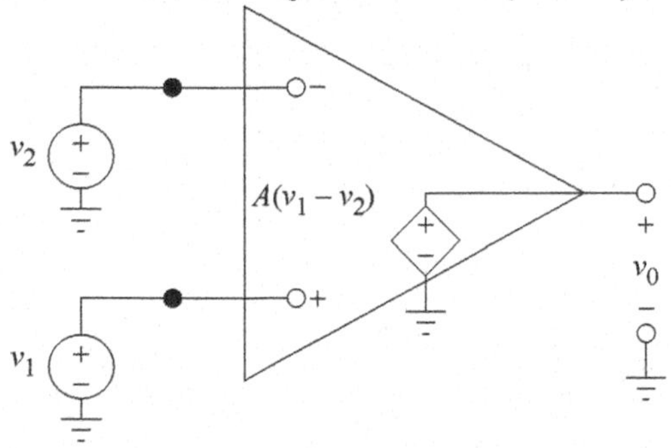

Equivalent Circuit of an Ideal Op Amp

In the op amp circuit shown the differential input is defined as:

$$v_{id} = v_1 - v_2$$

The common-mode input voltage is defined as:

$$v_{icm} = (v_1 + v_2)/2$$

The output voltage is given by:

$$v_O = A v_{id} + A_{cm} v_{icm}$$

In an ideal op amp $A_{cm} = 0$. In a non-ideal op amp the *CMRR* is used to measure the relative degree of rejection between the differential gain and common-mode gain.

$$CMRR = \frac{|A|}{|A_{cm}|}$$

CMRR is usually expressed in decibels as:

$$CMRR = 20 \log_{10}\left[\frac{|A|}{|A_{cm}|}\right]$$

SOLID-STATE ELECTRONICS AND DEVICES

Conductivity of a semiconductor material:

$\sigma = q\,(n\mu_n + p\mu_p)$, where

$\mu_n \equiv$ electron mobility

$\mu_p \equiv$ hole mobility

$n \equiv$ electron concentration

$p \equiv$ hole concentration

$q \equiv$ charge on an electron (1.6×10^{-19}C)

Doped material:

p-type material; $p_p \approx N_a$

n-type material; $n_n \approx N_d$

Carrier concentrations at equilibrium

$(p)(n) = n_i^2$, where

$n_i \equiv$ intrinsic concentration.

Built-in potential (contact potential) of a p-n junction:

$$V_0 = \frac{kT}{q} \ln \frac{N_a N_d}{n_i^2}$$

Thermal voltage

$$V_T = \frac{kT}{q} \approx 0.026 \text{ V at 300K}$$

N_a = acceptor concentration

N_d = donor concentration

T = temperature (K)

k = Boltzmann's Constant = 1.38×10^{-23} J/K

Capacitance of abrupt p–n junction diode

$C(V) = C_0/\sqrt{1 - V/V_{bi}}$, where:

C_0 = junction capacitance at $V = 0$

V = potential of anode with respect to cathode

V_{bi} = junction contact potential

Resistance of a diffused layer is

$R = R_s(L/W)$, where:

R_s = sheet resistance = ρ/d in ohms per square

ρ = resistivity

d = thickness

L = length of diffusion

W = width of diffusion

♦ Differential Amplifier

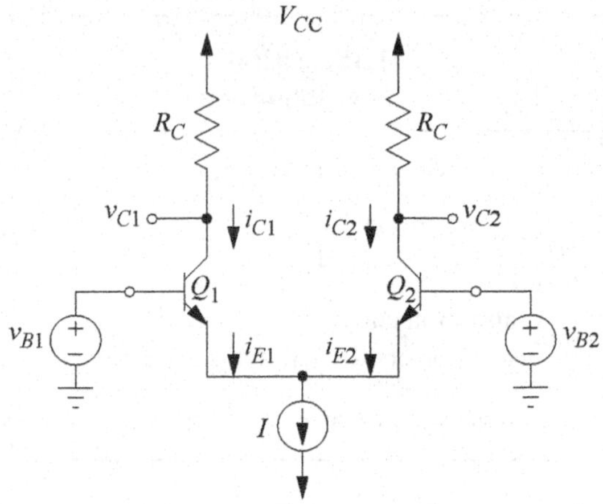

A Basic BJT Differential Amplifier

A basic BJT differential amplifier consists of two matched transistors whose emitters are connected and that are biased by a constant-current source. The following equations govern the operation of the circuit given that neither transistor is operating in the saturation region.

$$\frac{i_{E1}}{i_{E2}} = e^{(v_{B1} - v_{B2})/V_T}$$

$$i_{E1} + i_{E2} = I$$

$$i_{E1} = \frac{I}{1 + e^{(v_{B2} - v_{B1})/V_T}} \qquad i_{E2} = \frac{I}{1 + e^{(v_{B1} - v_{B2})/V_T}}$$

$$i_{C1} = \alpha I_{E1} \qquad i_{C2} = \alpha I_{E2}$$

The following figure shows a plot of two normalized collector currents versus normalized differential input voltage for a circuit using transistors with $\alpha \cong 1$.

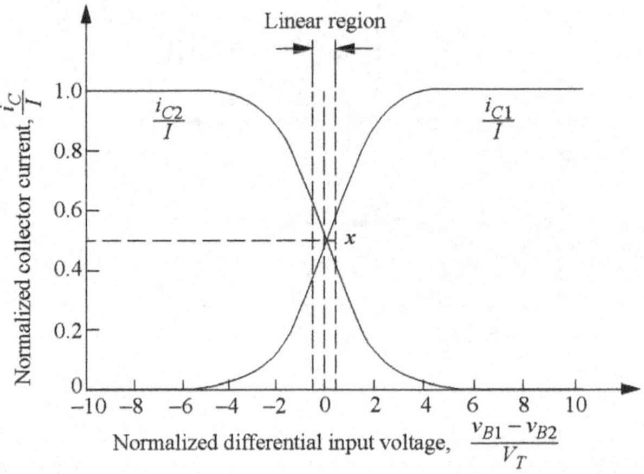

Transfer characteristics of the BJT differential amplifier with $\alpha \cong 1$

♦ Sedra, Adel, and Kenneth Smith, *Microelectronic Circuits*, 3rd ed., ©1991, pp. 408 and 412, by permission of Oxford University Press.

DIODES			
Device and Schematic Symbol	**Ideal $I-V$ Relationship**	**Piecewise-Linear Approximation of the $I-V$ Relationship**	**Mathematical $I-V$ Relationship**
(Junction Diode) 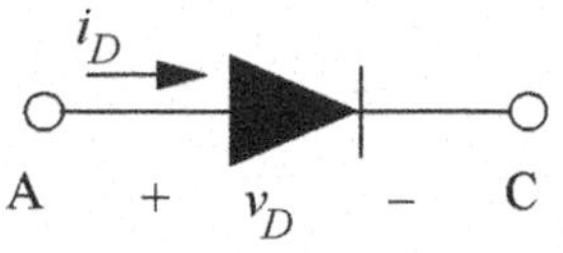	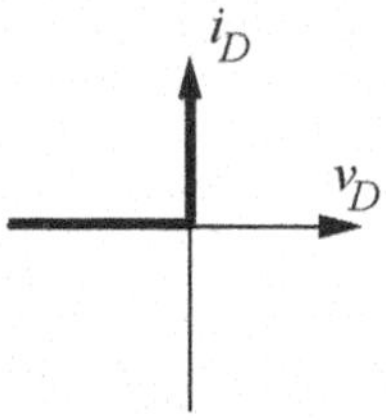	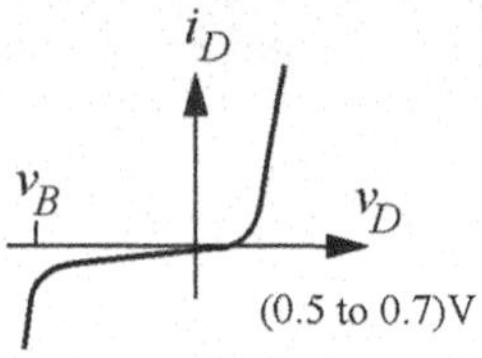 v_B = breakdown voltage	Shockley Equation $i_D \approx I_s\left[e^{(v_D/\eta V_T)}-1\right]$ where I_s = saturation current η = emission coefficient, typically 1 for Si V_T = thermal voltage $= \dfrac{kT}{q}$
(Zener Diode) 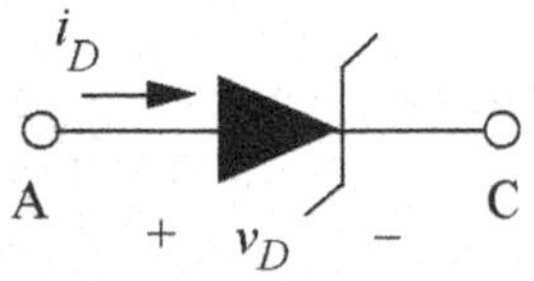	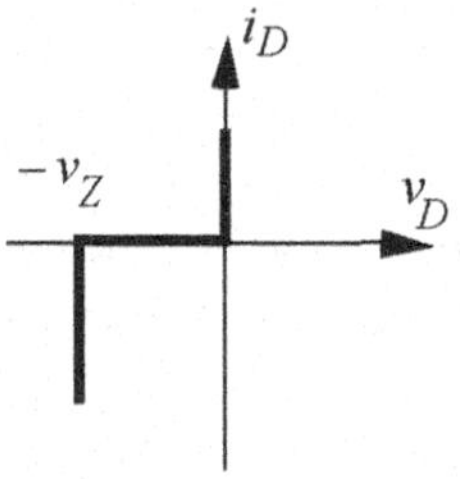	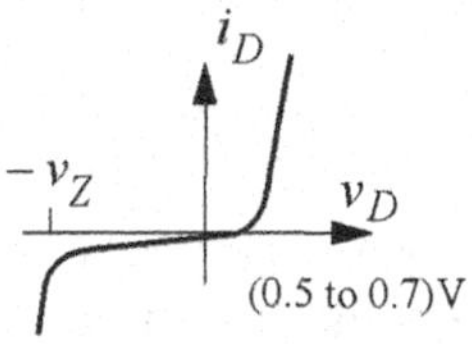 v_Z = Zener voltage	Same as above.

Thyristor or Silicon Controlled Rectifier (SCR)		
Schematic Symbol	**Ideal *I-V* Relationship**	**Piecewise-Linear Approximation of the *I-V* Relationship**
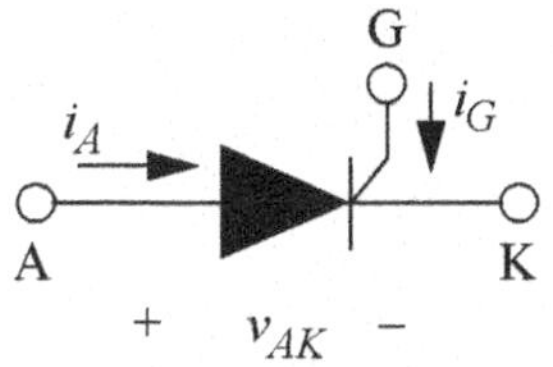	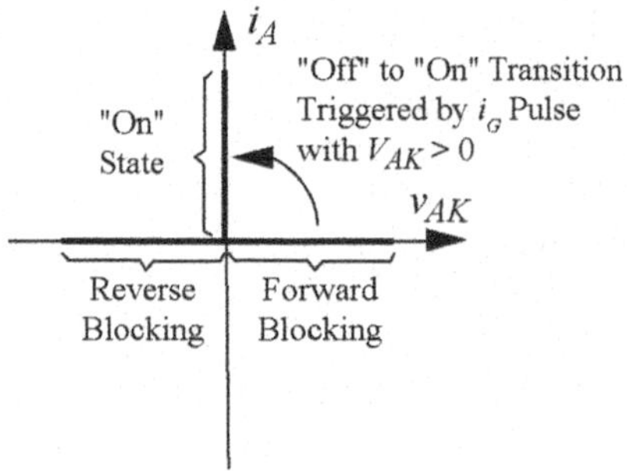	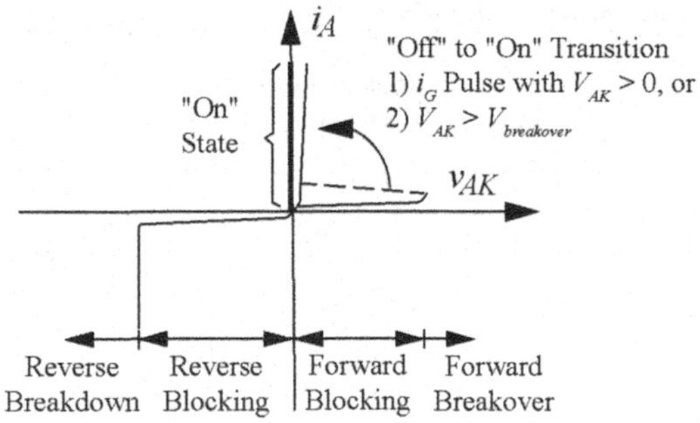

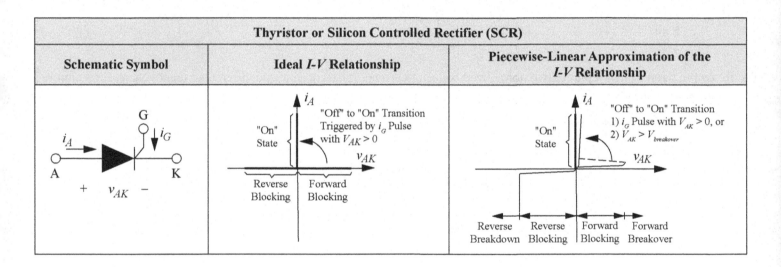

<table>
<tr><th colspan="4">NPN Bipolar Junction Transistor (BJT)</th></tr>
<tr><th>Schematic Symbol</th><th>Mathematical Relationships</th><th>Large-Signal (DC) Equivalent Circuit</th><th>Low-Frequency Small-Signal (AC) Equivalent Circuit</th></tr>
<tr>
<td>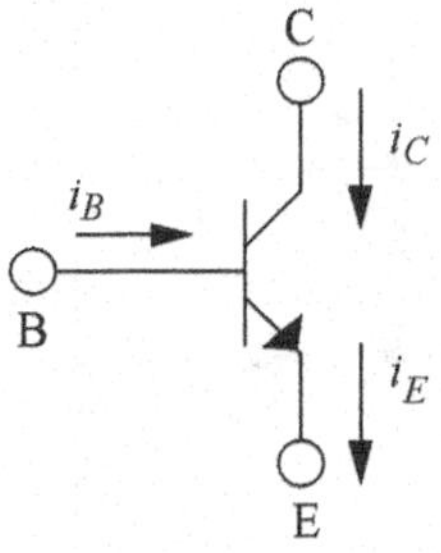

NPN – Transistor</td>
<td>$i_E = i_B + i_C$
$i_C = \beta i_B$
$i_C = \alpha i_E$
$\alpha = \beta/(\beta + 1)$
$i_C \approx I_S e^{(V_{BE}/V_T)}$
I_S = emitter saturation current
V_T = thermal voltage

<u>Note</u>: These relationships are valid in the active mode of operation.</td>
<td><u>Active Region</u>:
base emitter junction forward biased; base collector juction reverse biased
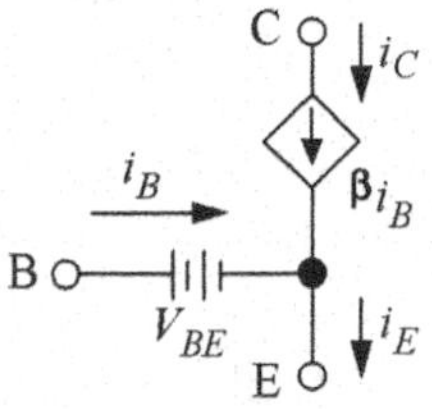

<hr>
<u>Saturation Region</u>:
both junctions forward biased
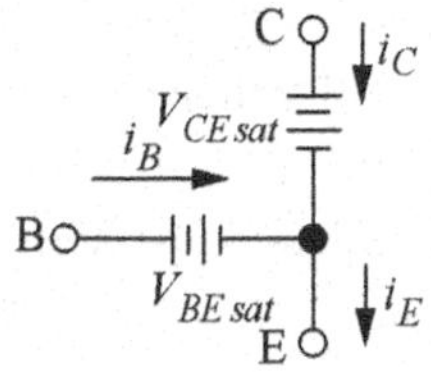
</td>
<td><u>Low Frequency</u>:
$g_m \approx I_{CQ}/V_T$
$r_\pi \approx \beta/g_m,$
$$r_o = \left[\frac{\partial v_{CE}}{\partial i_c}\right]_{Q_{point}} \approx \frac{V_A}{I_{CQ}}$$
where
I_{CQ} = dc collector current at the Q_{point}
V_A = Early voltage
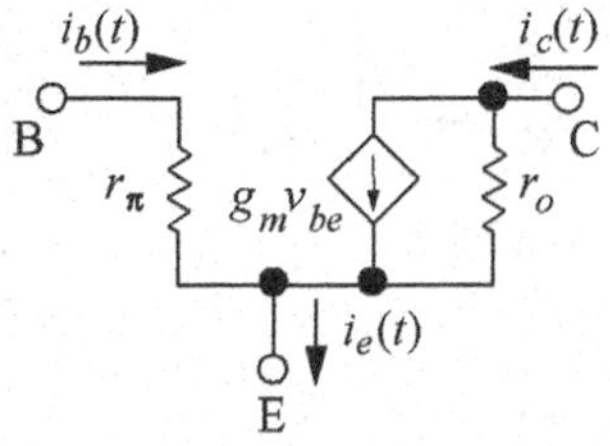
</td>
</tr>
<tr>
<td>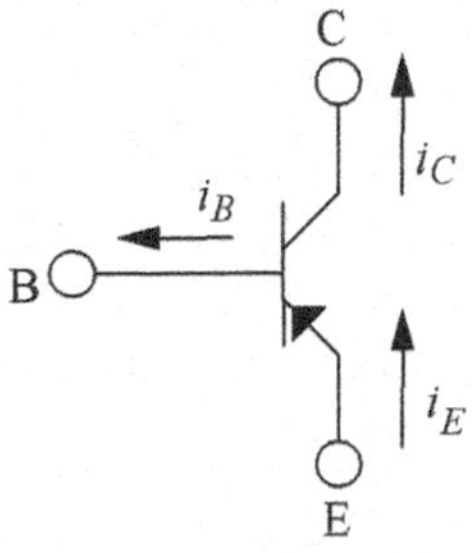

PNP – Transistor</td>
<td>Same as for NPN with current directions and voltage polarities reversed.</td>
<td><u>Cutoff Region</u>:
both junctions reverse biased
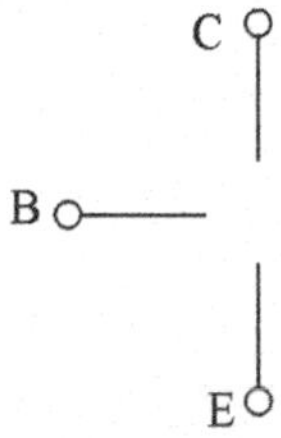

<hr>
Same as NPN with current directions and voltage polarities reversed</td>
<td>Same as for NPN.</td>
</tr>
</table>

N-Channel Junction Field Effect Transistors (JFETs) and Depletion MOSFETs (Low and Medium Frequency)		
Schematic Symbol	**Mathematical Relationships**	**Small-Signal (AC) Equivalent Circuit**
N-CHANNEL JFET 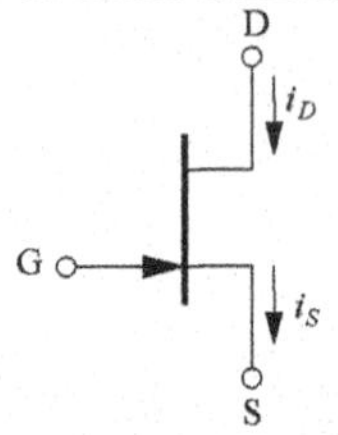 P-CHANNEL JFET 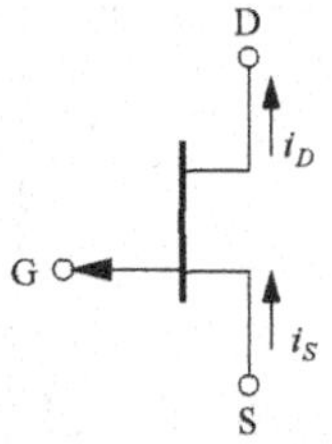 N-CHANNEL DEPLETION MOSFET (NMOS) 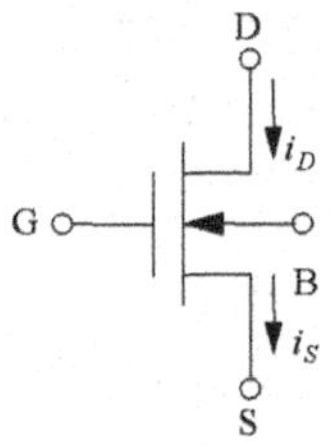 SIMPLIFIED SYMBOL 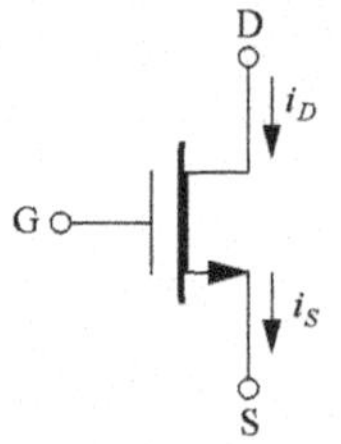 	<u>Cutoff Region:</u> $v_{GS} < V_p$ $i_D = 0$ <u>Triode Region:</u> $v_{GS} > V_p$ and $v_{GD} > V_p$ $i_D = (I_{DSS}/V_p^2)[2v_{DS}(v_{GS} - V_p) - v_{DS}^2]$ <u>Saturation Region:</u> $v_{GS} > V_p$ and $v_{GD} < V_p$ $i_D = I_{DSS}(1 - v_{GS}/V_p)^2$ where I_{DSS} = drain current with $v_{GS} = 0$ (in the saturation region) $= KV_p^2$, K = conductivity factor For JFETs, V_p = pinch-off voltage For MOSFETs, $V_p = V_T$ = threshold voltage	$g_m = \dfrac{2\sqrt{I_{DSS} I_D}}{\lvert V_p \rvert}$ in saturation region 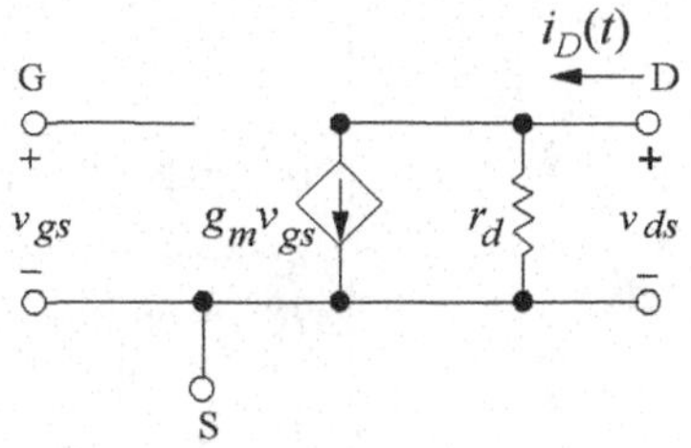 where $r_d = \left\lvert \dfrac{\partial v_{ds}}{\partial i_d} \right\rvert_{Q_{point}}$
P-CHANNEL DEPLETION MOSFET (PMOS) 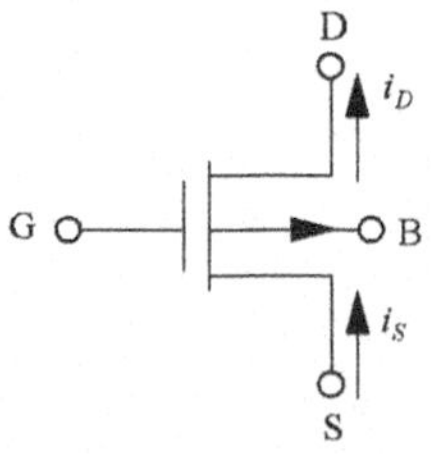 SIMPLIFIED SYMBOL 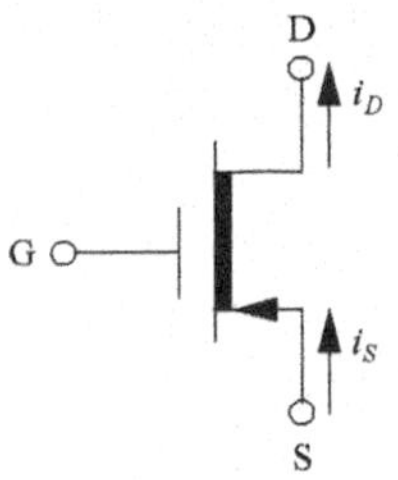 	Same as for N-Channel with current directions and voltage polarities reversed.	Same as for N-Channel.

Enhancement MOSFET (Low and Medium Frequency)		
Schematic Symbol	**Mathematical Relationships**	**Small-Signal (AC) Equivalent Circuit**
N-CHANNEL ENHANCEMENT MOSFET (NMOS) 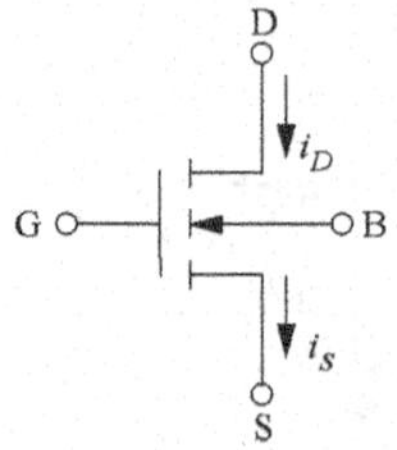 SIMPLIFIED SYMBOL 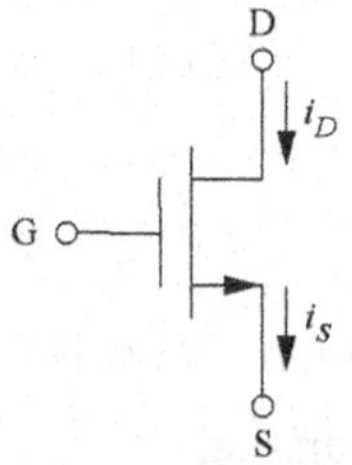	<u>Cutoff Region</u>: $v_{GS} < V_t$ $i_D = 0$ <u>Triode Region</u>: $v_{GS} > V_t$ and $v_{GD} > V_t$ $i_D = K\,[2v_{DS}\,(v_{GS} - V_t) - v_{DS}^{\,2}\,]$ <u>Saturation Region</u>: $v_{GS} > V_t$ and $v_{GD} < V_t$ $i_D = K\,(v_{GS} - V_t)^2$ where $K =$ conductivity factor $V_t =$ threshold voltage	$g_m = 2K(v_{GS} - V_t)$ in saturation region 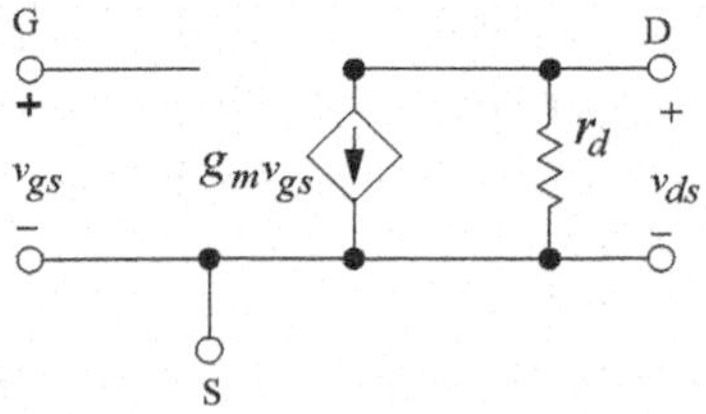 where $r_d = \left\|\dfrac{\partial v_{ds}}{\partial i_d}\right\|_{Q_{\text{point}}}$
P-CHANNEL ENHANCEMENT MOSFET (PMOS) 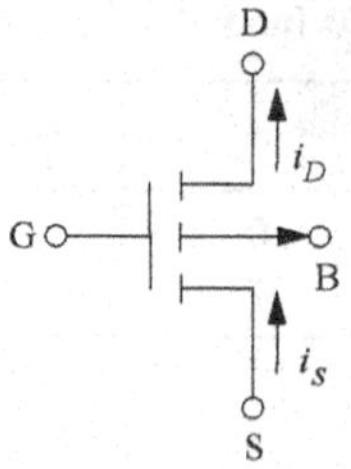 SIMPLIFIED SYMBOL 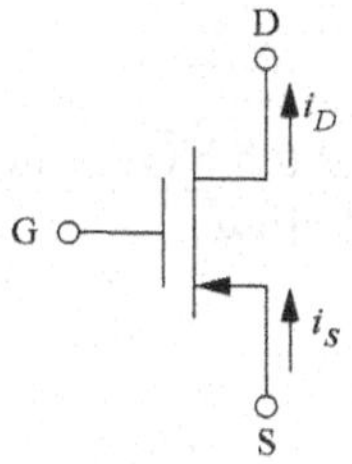	Same as for N-channel with current directions and voltage polarities reversed.	Same as for N-channel.

NUMBER SYSTEMS AND CODES

An unsigned number of base-r has a decimal equivalent D defined by

$$D = \sum_{k=0}^{n} a_k r^k + \sum_{i=1}^{m} a_i r^{-i}, \text{ where}$$

a_k = the $(k + 1)$ digit to the left of the radix point and

a_i = the ith digit to the right of the radix point.

Binary Number System

In digital computers, the base-2, or binary, number system is normally used. Thus the decimal equivalent, D, of a binary number is given by

$$D = a_k 2^k + a_{k-1} 2^{k-1} + \ldots + a_0 + a_{-1} 2^{-1} + \ldots$$

Since this number system is so widely used in the design of digital systems, we use a short-hand notation for some powers of two:

$2^{10} = 1{,}024$ is abbreviated "k" or "kilo"

$2^{20} = 1{,}048{,}576$ is abbreviated "M" or "mega"

Signed numbers of base-r are often represented by the radix complement operation. If M is an N-digit value of base-r, the radix complement $R(M)$ is defined by

$$R(M) = r^N - M$$

The 2's complement of an N-bit binary integer can be written

$$\text{2's Complement } (M) = 2^N - M$$

This operation is equivalent to taking the 1's complement (inverting each bit of M) and adding one.

The following table contains equivalent codes for a four-bit binary value.

Binary Base-2	Decimal Base-10	Hexa-decimal Base-16	Octal Base-8	BCD Code	Gray Code
0000	0	0	0	0	0000
0001	1	1	1	1	0001
0010	2	2	2	2	0011
0011	3	3	3	3	0010
0100	4	4	4	4	0110
0101	5	5	5	5	0111
0110	6	6	6	6	0101
0111	7	7	7	7	0100
1000	8	8	10	8	1100
1001	9	9	11	9	1101
1010	10	A	12	---	1111
1011	11	B	13	---	1110
1100	12	C	14	---	1010
1101	13	D	15	---	1011
1110	14	E	16	---	1001
1111	15	F	17	---	1000

LOGIC OPERATIONS AND BOOLEAN ALGEBRA

Three basic logic operations are the "AND (•)," "OR (+)," and "Exclusive-OR ⊕" functions. The definition of each function, its logic symbol, and its Boolean expression are given in the following table.

Function / Inputs	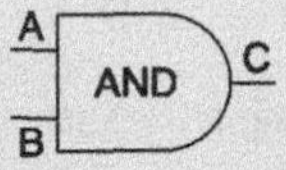	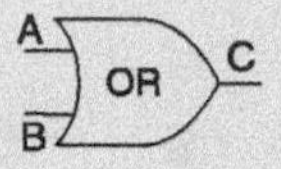	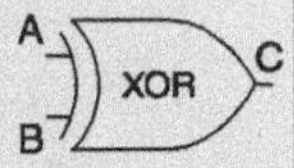
$A\ B$	$C = A \cdot B$	$C = A + B$	$C = A \oplus B$
0 0	0	0	0
0 1	0	1	1
1 0	0	1	1
1 1	1	1	0

As commonly used, A AND B is often written AB or $A \bullet B$.

The not operator inverts the sense of a binary value $(0 \rightarrow 1, 1 \rightarrow 0)$

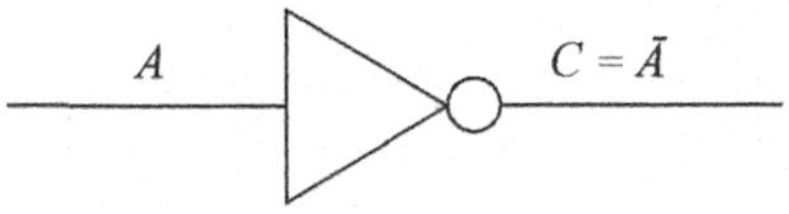

LOGIC SYMBOL

NOT OPERATOR

Input	Output
A	$C = \bar{A}$
0	1
1	0

De Morgan's Theorems

first theorem: $\overline{A + B} = \bar{A} \cdot \bar{B}$

second theorem: $\overline{A \cdot B} = \bar{A} + \bar{B}$

These theorems define the NAND gate and the NOR gate. Logic symbols for these gates are shown below.

NAND Gates: $\overline{A \bullet B} = \bar{A} + \bar{B}$

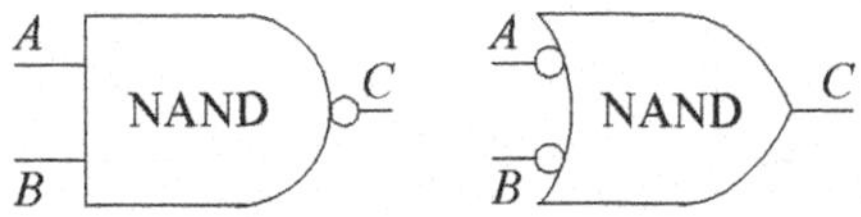

NOR Gates: $\overline{A + B} = \bar{A} \bullet \bar{B}$

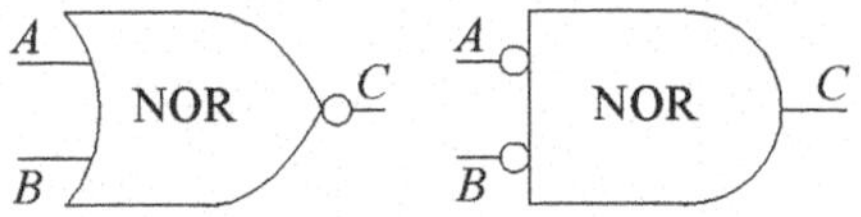

FLIP-FLOPS

A flip-flop is a device whose output can be placed in one of two states, 0 or 1. The flip-flop output is synchronized with a clock (CLK) signal. Q_n represents the value of the flip-flop output before CLK is applied, and Q_{n+1} represents the output after CLK has been applied. Three basic flip-flops are described below.

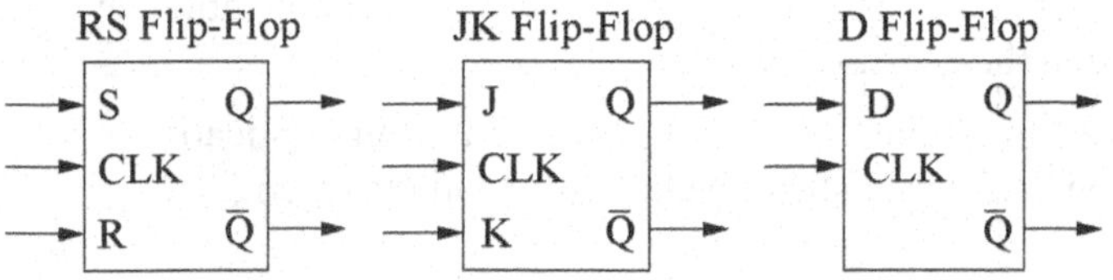

SR	Q_{n+1}
00	Q_n no change
01	0
10	1
11	x invalid

JK	Q_{n+1}
00	Q_n no change
01	0
10	1
11	$\overline{Q}_n$ toggle

D	Q_{n+1}
0	0
1	1

Composite Flip-Flop State Transition						
Q_n	Q_{n+1}	*S*	*R*	*J*	*K*	*D*
0	0	0	x	0	x	0
0	1	1	0	1	x	1
1	0	0	1	x	1	0
1	1	x	0	x	0	1

Switching Function Terminology

Minterm, m_i – A product term which contains an occurrence of every variable in the function.

Maxterm, M_i – A sum term which contains an occurrence of every variable in the function.

Implicant – A Boolean algebra term, either in sum or product form, which contains one or more minterms or maxterms of a function.

Prime Implicant – An implicant which is not entirely contained in any other implicant.

Essential Prime Implicant – A prime implicant which contains a minterm or maxterm which is not contained in any other prime implicant.

A function can be described as a sum of minterms using the notation

$$F(ABCD) = \Sigma m(h, i, j, \ldots) = m_h + m_i + m_j + \ldots$$

A function can be described as a product of maxterms using the notation

$$G(ABCD) = \Pi M(h, i, j, \ldots) = M_h \bullet M_i \bullet M_j \bullet \ldots$$

A function represented as a sum of minterms only is said to be in *canonical sum of products* (SOP) form. A function represented as a product of maxterms only is said to be in *canonical product of sums* (POS) form. A function in canonical SOP form is often represented as a *minterm list*, while a function in canonical POS form is often represented as a *maxterm list*.

A *Karnaugh Map* (K-Map) is a graphical technique used to represent a truth table. Each square in the K-Map represents one minterm, and the squares of the K-Map are arranged so that the adjacent squares differ by a change in exactly one variable. A four-variable K-Map with its corresponding minterms is shown below. K-Maps are used to simplify switching functions by visually identifying all essential prime implicants.

AB \ CD	00	01	11	10
00	m_0	m_1	m_3	m_2
01	m_4	m_5	m_7	m_6
11	m_{12}	m_{13}	m_{15}	m_{14}
10	m_8	m_9	m_{11}	m_{10}

COMPUTER NETWORKING

Modern computer networks are primarily packet switching networks. This means that the messages in the system are broken down, or segmented into packets, and the packets are transmitted separately into the network. The primary purpose of the network is to exchange messages between endpoints of the network called hosts or nodes, typically computers, servers, or handheld devices. At the host, the packets are reassembled into the message and delivered to a software application, e.g., a browser, email, or video player.

Two widely used abstract models for modern computer networks are the open systems interconnect (OSI) model and the TCP/IP model shown in the figure below.

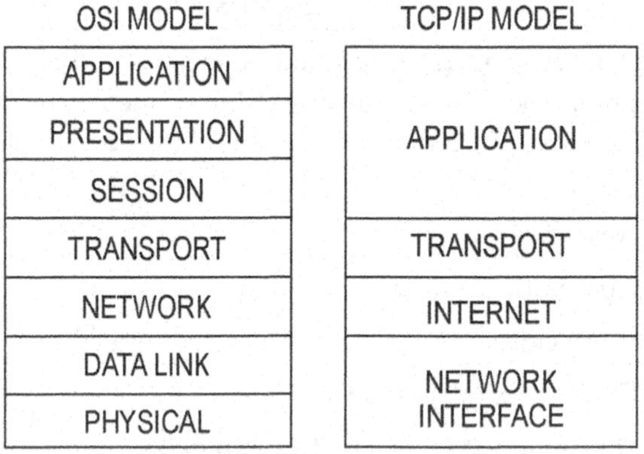

Tanenbaum, Andrew S., *Computer Networks*, 3 ed., Prentice Hall, 1996, p. 36.

The application layer on the TCP/IP model corresponds to the three upper layers (application, presentation, and session) of the OSI model. The network interface layer of the TCP/IP model corresponds to the bottom two layers (data link and physical) of the OSI model.

The application layer is the network layer closest to the end user, which means both the application layer and the user interact directly with the software application. This layer interacts with software applications that implement a communicating component.

In the OSI model, the application layer interacts with the presentation layer. The presentation layer is responsible for the delivery and formatting of information to the application layer for further processing or display. It relieves the application layer of concern regarding syntactical differences in data representation within the end-user systems.

The OSI session layer provides the mechanism for opening, closing, and managing a session between end-user application processes. It provides for full-duplex, half-duplex, or simplex operation, and establishes checkpointing, adjournment, termination, and restart procedures.

The transport layer adds a transport header normally containing TCP and UDP protocol information. The transport layer provides logical process-to-process communication primitives. Optionally, it may provide other services, such as reliability, in-order delivery, flow control, and congestion control.

The network layer or Internet layer adds another header normally containing the IP protocol; the main role of the networking layer is finding appropriate routes between end hosts, and forwarding the packets along these routes.

The link layer or data link layer contains protocols for transmissions between devices on the same link and usually handles error detection and correction and medium-access control.

The physical layer specifies physical transmission parameters (e.g., modulation, coding, channels, data rates) and governs the transmission of frames from one network element to another sharing a common link.

Hosts, routers, and link-layer switches showing the four-layer protocol stack with different sets of layers for hosts, a switch, and a router are shown in the figure below.

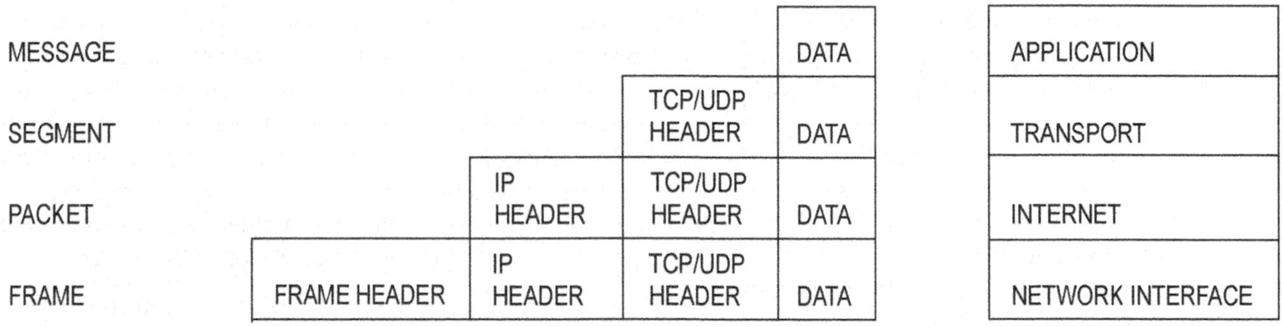

ENCAPSULATION OF APPLICATION DATA THROUGH EACH LAYER

In computer networking, encapsulation is a method of designing modular communication protocols in which logically separate functions in the network are abstracted from their underlying structures by inclusion or information hiding within higher-level objects. For example, a network layer packet is encapsulated in a data link layer frame.

Acronyms

ACK Acknowledge

ARQ Automatic request

BW Bandwidth

CRC Cyclic redundancy code

DHCP Dynamic host configuration protocol

IP Internet protocol

LAN Local area network

NAK Negative acknowledgement

OSI Open systems interconnect

TCP Transmission control protocol

Protocol Definitions

- TCP/IP is the basic communication protocol suite for communication over the Internet.
- Internet Protocol (IP) provides end-to-end addressing and is used to encapsulate TCP or UDP datagrams. Both version 4 (IPv4) and version 6 (IPv6) are used and can co-exist on the same network.
- Transmission Control Protocol (TCP) is a connection-oriented protocol that detects lost packets, duplicated packets, or packets that are received out of order and has mechanisms to correct these problems.
- User Datagram Protocol (UDP), is a connectionless-oriented protocol that has less network overhead than TCP but provides no guarantee of delivery, ordering, or duplicate protection.
- Internet Control Message Protocol (ICMP) is a supporting protocol used to send error messages and operational information.

Local Area Network (LAN)

There are different methods for assigning IP addresses for devices entering a network.

- Dynamic host configuration protocol (DHCP) is a networking protocol that allows a router to assign the IP address and other configuration information for all stations joining a network.
- Static IP addressing implies each station joining a network is manually configured with its own IP address.
- Stateless address autoconfiguration (SLAAC) allows for hosts to automatically configure themselves when connecting to an IPv6 network.

Error Coding

Error coding is a method of detecting and correcting errors that may have been introduced into a frame during data transmission. A system that is capable of detecting errors may be able to detect single or multiple errors at the receiver based on the error coding method. Below are a few examples of error detecting error coding methods.

Parity – For parity bit coding, a parity bit value is added to the transmitted frame to make the total number of ones odd (odd parity) or even (even parity). Parity bit coding can detect single bit errors.

Cyclical Redundancy Code (*CRC*) – CRC can detect multiple errors. To generate the transmitted frame from the receiver, the following equation is used:

$$T(x)/G(x) = E(x)$$

where T(x) is the frame, G(x) is the generator, and E(x) is the remainder. The transmitted code is T(x) + E(x).

On the receiver side, if

$$[T(x) + E(x)]/G(x) = 0$$

then no errors were detected.

To detect and correct errors, redundant bits need to be added to the transmitted data. Some error detecting and correcting algorithms include block code, Hamming code, and Reed Solomon.

Delays in Computer Networks

Transmission Delay – the time it takes to transmit the bits in the packet on the transmission link:

$$d_{trans} = L/R$$

where L is the packet size in bits/packet and

R is the rate of transmission in bits/sec.

Propagation Delay – the time taken for a bit to travel from one end of the link to the other:

$$d_{prop} = d/s$$

where d is the distance or length of the link and s is the propagation speed

The propagation speed is usually somewhere between the speed of light c and 2/3 c.

Nodal Processing Delay – It takes time to examine the packet's header and determine where to direct the packet to its destination.

Queueing Delay – The packet may experience delay as it waits to be transmitted onto the link.

Ignoring nodal and queueing delays, the round-trip delay of delivering a packet from one node to another in the stop-and-wait system is

$$D = 2\,d_{prop} + d_{transAck} + d_{transData}$$

Because the sending host must wait until the ACK packet is received before sending another packet, this leads to a very poor utilization, U, of resources for stop-and-wait links with relatively large propagation delays:

$$U = d_{trans}/D$$

For this reason, for paths with large propagation delays, most computer networking systems use a pipelining system called go-back-N, in which N packets are transmitted in sequence before the transmitter receives an ACK for the first packet.

Automatic Request for Retransmission (ARQ)

Links in the network are most often twisted pair, optical fiber, coaxial cable, or wireless channels. These are all subject to errors and are often unreliable. The ARQ system is designed to provide reliable communications over these unreliable links. In ARQ, each packet contains an error detection process (at the link layer). If no errors are detected in the packet, the host (or intermediate switch) transmits a positive acknowledgement (ACK) packet back to the transmitting element indicating that the packet was received correctly. If any error is detected, the receiving host (or switch) automatically discards the packet and sends a negative acknowledgement (NAK) packet back to the originating element (or stays silent, allowing the transmitter to timeout). Upon receiving the NAK packet or by the trigger of a timeout, the transmitting host (or switch) retransmits the message packet that was in error. A diagram of a simple stop and wait ARQ system with a positive acknowledgement is shown below.

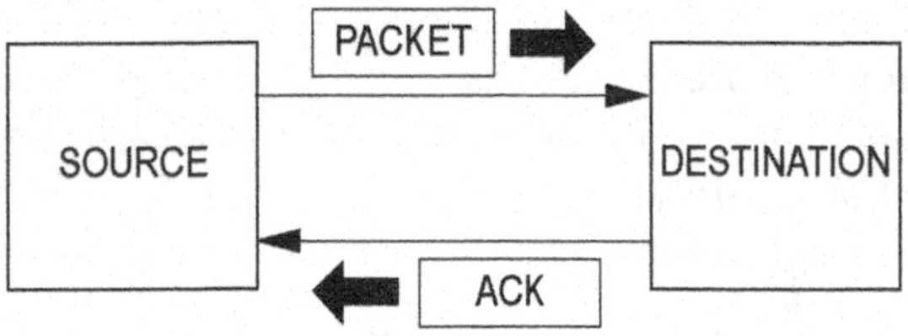

Transmission Algorithms
Sliding window protocol is used where delivery of data is required while maximizing channel capacity. In the sliding window protocol, each outbound frame contains a sequence number. When the transmitted frame is received, the receiver is required to transmit an ACK for each received frame before an additional frame can be transmitted. If the frame is not received, the receiver will transmit a NAK message indicating the frame was not received after an appropriate time has expired. Sliding window protocols automatically adjust the transmission speed to both the speed of the network and the rate at which the receiver sends new acknowledgements.

Shannon Channel Capacity Formula

$$C = BW \log_2 (1+S/N)$$

where BW is bandwidth, S is the power of the signal at the transmitting device, and N is the noise received at the destination

Network Topologies

Point-to-Point

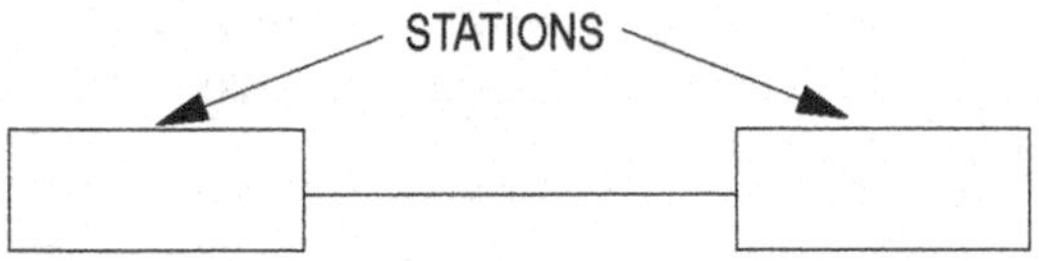

Token Ring

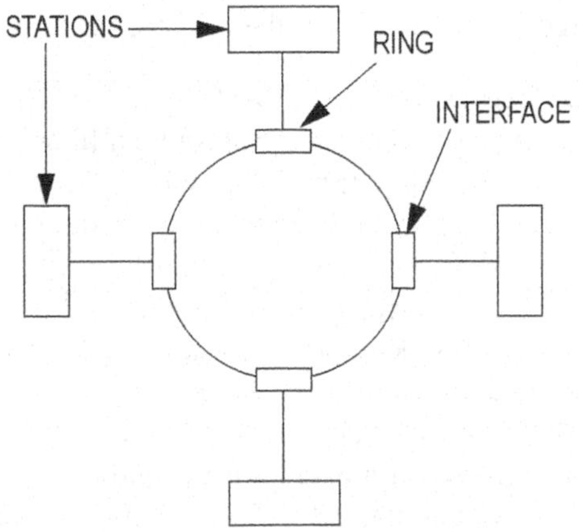

Bus

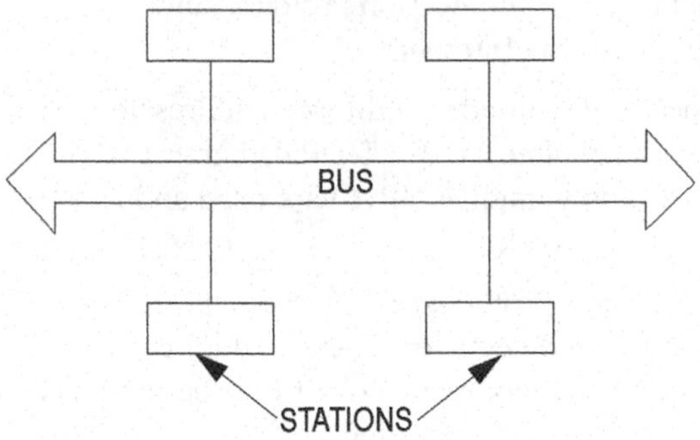

Star

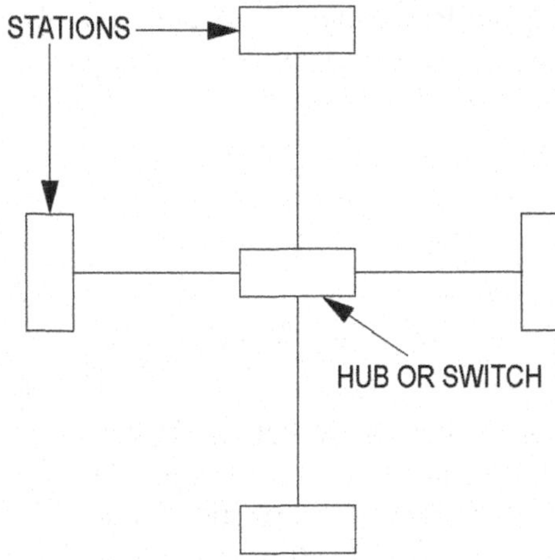

Mesh

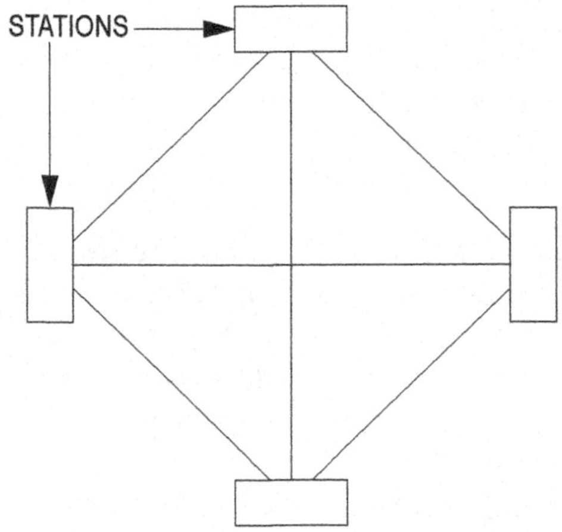

Tree

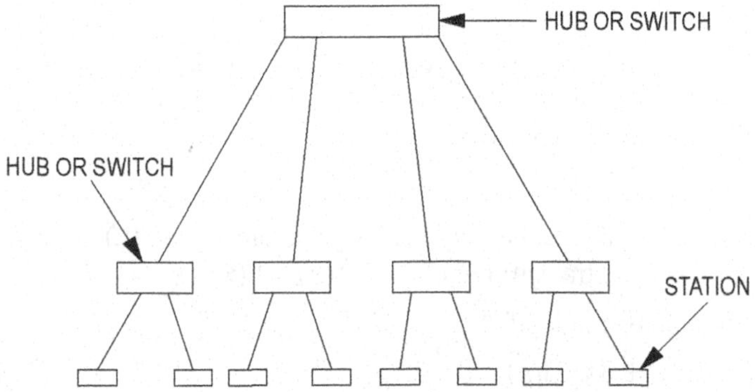

COMPUTER SYSTEMS

Memory/Storage Types

RAM – Primary memory system in computing systems, volatile

Cache – faster, but smaller segment of memory used for buffering immediate data from slower memories

- L1: Level 1 cache, fastest memory available
- L2: Level 2 cache, next level away from CPU. May or may not be exclusive of L1 depending on architecture

ROM – nonvolatile. Contains system instructions or constant data for the system

Replacement Policy – For set associative and fully associative caches, if there is a miss and the set or cache (respectively) is full, then a block must be selected for replacement. The replacement policy determines which block is replaced. Common replacement policies are:

- Least recently used (LRU): Replace the least recently used block.
- Most recently used (MRU): Replace the most recently used block.
- First-in, first-out (FIFO): Replace blocks based on the order in which they were fetched into the set or cache.
- Random: Choose a block at random for replacement.
- Least frequently used (LFU): Replace the block that had the fewest references among the candidate blocks.

Write Policy – With caches, multiple copies of a memory block may exist in the system (e.g., a copy in the cache and a copy in main memory). There are two possible write policies.

- Write-through: Write to both the cache's copy and the main memory's copy.
- Write-back: Write only to the cache's copy. This requires adding a "dirty bit" for each block in the cache. When a block in the cache is written to, its dirty bit is set to indicate that the main memory's copy is stale. When a dirty block is evicted from the cache (due to a replacement), the entire block must be written back to main memory. Clean blocks need not be written back when they are evicted.

Cache Size – C (bytes) = S*A*B

where S = Number of sets

A = Set associativity

B = Block size (bytes)

To search for the requested block in the cache, the CPU will generally divide the address into three fields: the tag, index, and block offset.

TAG	INDEX	BLOCK OFFSET

- *Tag* – These are the most significant bits of the address, which are checked against the current row (the row that has been retrieved by index) to see if it is the one needed or another, irrelevant memory location that happened to have the same index bits as the one wanted.

  # tag bits = # address bits – # index bits – # block offset bits

- *Index* – These bits specify which cache row (set) that the data has been put in.

  # index bits = $\log_2$(# sets) = $\log_2(S)$

- *Block Offset* – These are the lower bits of the address that select a byte within the block.

  # block offset bits = $\log_2$(block size) = $\log_2(B)$

Pipeline Type

The following is a basic five-stage pipeline in a RISC machine:

INSTR. NO.	PIPELINE STAGE						
1	IF	ID	EX	MEM	WB		
2		IF	ID	EX	MEM	WB	
3			IF	ID	EX	MEM	WB
4				IF	ID	EX	MEM
5					IF	ID	EX
CLOCK CYCLE	1	2	3	4	5	6	7

IF = Instruction Fetch

ID = Instruction Decode

EX = Execute

MEM = Memory Access

WB = Register Write Back

In the fourth clock cycle (the highlighted column), the earliest instruction is in MEM stage, and the latest instruction has not yet entered the pipeline.

Microprocessor Architecture – Harvard

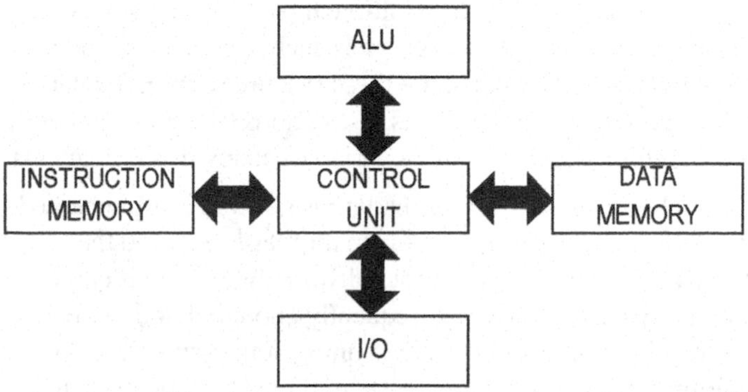

Multicore

A multicore processor is a single computing component with two or more independent actual processing units (called cores), which are the units that read and execute program instructions. The instructions are ordinary CPU instructions such as Add, Move Data, and Branch, but the multiple cores can run multiple instructions at the same time, increasing overall speed for programs amenable to parallel computing.

A multicore processor implements multiprocessing in a single physical package. Designers may couple cores in a multicore device tightly or loosely. For example, cores may or may not share caches, and they may implement message passing or shared memory intercore communication methods. Common network topologies to interconnect cores include bus, ring, two-dimensional mesh, and crossbar. Homogeneous multicore systems include only identical cores; heterogeneous multicore systems have cores that are not identical. Just as with single-processor systems, cores in multicore systems may implement architectures such as superscalar, VLIW, vector processing, SIMD, or multithreading.

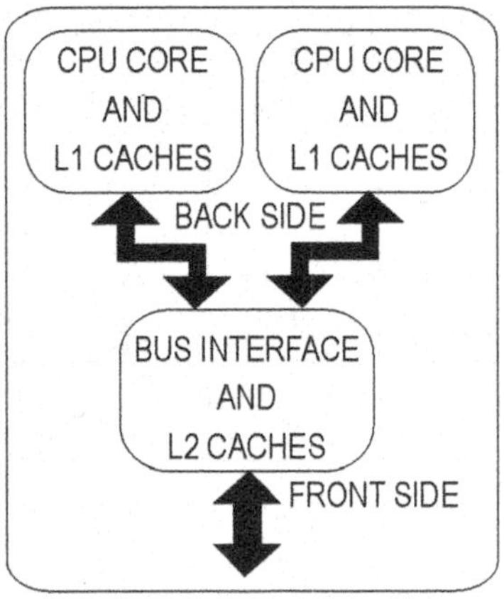

Generic dual-core processor, with CPU-local level 1 caches, and a shared, on-die level 2 cache

Threading

In computer science, a thread of execution is the smallest sequence of programmed instructions that can be managed independently by a scheduler, which is typically a part of the operating system. The implementation of threads and processes differs between operating systems, but in most cases a thread is a component of a process. Multiple threads can exist within the same process and share resources such as memory, while different processes do not share these resources. In particular, the threads of a process share its instructions (executable code) and its context (the values of its variables at any given moment).

On a single processor, multithreading is generally implemented by time-division multiplexing (as in multitasking), and the CPU switches between different software threads. This context switching generally happens frequently enough that the user perceives the threads or tasks as running at the same time. On a multiprocessor or multicore system, threads can be executed in a true concurrent manner, with every processor or core executing a separate thread simultaneously. To implement multiprocessing, the operating system may use hardware threads that exist as a hardware-supported method for better utilization of a particular CPU. These are different from the software threads that are a pure software construct with no CPU-level representation.

Acronyms

CISC	Complex instruction set computing
CPU	Central processing unit
FIFO	First-in, first out
I/O	Input/output
LFU	Least frequently used
LRU	Least recently used
MRU	Most recently used
RISC	Reduced instruction set computing
RAM	Random access memory
ROM	Read only memory

SOFTWARE DEVELOPMENT

Algorithm Efficiency (Big-O)

The concept of Big O Notation is used in software engineering to determine the efficiency of an algorithm. Big O equations are written as:

$$O(n) = f(n)$$

When comparing the efficiency of two algorithms, compare two O(n) values as n approaches infinity.

Endian-ness

MSB – most significant bit first. Also known as Big-endian.

LSB – least significant bit first. Also known as Little-endian.

Pointers

A pointer is a reference to an object. The literal value of a pointer is the object's location in memory. Extracting the object referenced by a pointer is defined as dereferencing.

Object Oriented

Object-oriented programming constructs systems in terms of objects that represent things in the real world naturally and effectively. This makes the software system easier to understand for designers, programmers, and users.

- *Class* – a collection of members, which can be defined as variables and methods. Members can be public, protected, or private.
- *Derived Class* – a class with members that have been inherited from another class
- *Base Class* – a class with one or more derived classes associated with it
- *Abstract Class* – a base class that cannot be implemented
- *Interface* – a collection of public methods that are not defined. A class that implements an interface must define these methods.

Algorithms

An algorithm is a specific sequence of steps that describe a process.

Sorting Algorithm – an algorithm that transforms a random collection of elements into a sorted collection of elements. Examples include:

- Bubble Sort
- Insertion Sort
- Merge Sort
- Heap Sort
- Quick Sort

Searching Algorithm – an algorithm that determines if an element exists in a collection of elements. If the element does exist, its location is also returned. Examples include:

- Binary search
- Hashing

Data Structures

Collection – a grouping of elements that are stored and accessed using algorithms. Examples include:

- Array
- Simple list
- Linked list
- Queue
- Stack
- Map
- Set
- Tree
- Graph

Tree Traversal

There are three primary algorithms that are used to traverse a binary tree data structure.

Pre-Order Traversal

1. Traverse the left sub-tree.
2. Visit the root node.
3. Traverse the right sub-tree.

In-Order Traversal

1. Visit the root node.
2. Traverse the left sub-tree.
3. Traverse the right sub-tree.

Post-Order Traversal

1. Traverse the left sub-tree.
2. Traverse the right sub-tree.
3. Visit the root node.

Graph Traversal

A graph is a collection of nodes (also called vertices) that are connected by edges (also known as arcs or lines). It can be useful to parse through each node in the graph. Primarily two algorithms will perform this task.

Breadth First Search–Beginning at a given node, the algorithm visits all connected nodes that have not been visited. The algorithm repeats for each visited node. The output of the algorithm is a list of nodes in the order that they have been visited. A queue data structure can be used to facilitate this algorithm.

Depth First Search–Beginning at a given node, the algorithm visits one connected node that has not been visited. This is repeated until a node does not have any connected nodes that have not been visited. At this point the algorithm backtracks to the last visited node and repeats the algorithm. The output of the algorithm is a list of nodes in the order that they have been visited. A stack can be used to facilitate this algorithm.

Software Process Flows

Waterfall

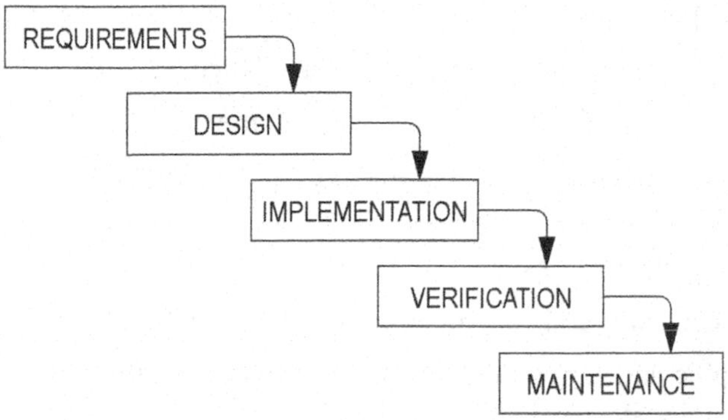

Westfall, Linda, *The Certified Software Quality Engineer Handbook*, Milwaukee, Quality Press, 2010, p. 131.

Iterative

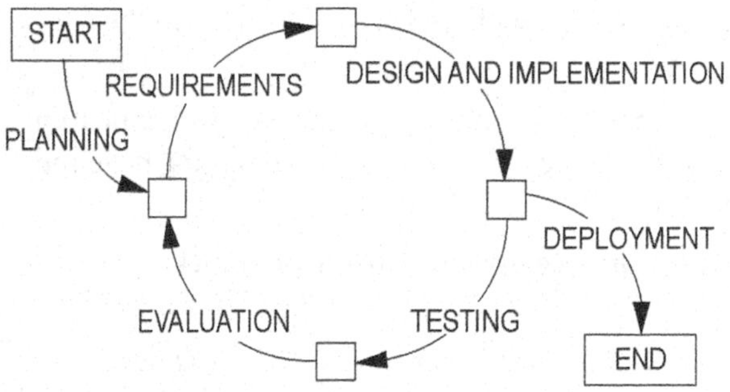

V-Model

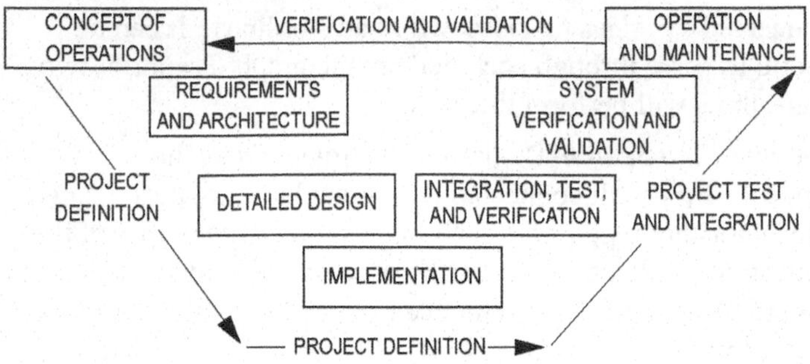

Clarus Concept of Operations, Publication No. FHWA-JPO-05-072, Federal Highway Administration, 2005.

Software Syntax Guidelines

- Code is pseudocode, no specific language
- No end-of-line punctuation (e.g., semicolon) is used
- Comments are indicated with "--" double hyphen
- Loop structures end with "end" followed by structure name, e.g., "end while"
- "do-while" begins with "do" and ends with "while"—no "end" per se
- "if-then" statements have both "if" and "then"
- "else if" is a substitute for the "end" on the preceding "if"
- "=" is used to designate assignment. "==" refers to comparison in a conditional statement.
- Not equals is represented by <>
- Logical "and" and "or" are spelled out as "and" and "or"
- Variable and argument declarations are Pascal style—"name: type"
- Numeric data types are "integer" and "float"
- Text is a procedural variable, unless specified to be an object of type String
- Variables can be constant, and are declared with the "const" modifier
- Variables whose type is object and the exact specification of that object is not critical to the problem must have the data type obj
- Array indices are designated with square brackets [], not parentheses
- Unless otherwise specified, arrays begin at 1 (one)
- Compilation units are "procedure" and "function". "Module" is not a compilation unit
- Function parameters are designated with parentheses ()
- Unless specified, procedures and functions must have the return type "void"
- Arguments in a function/procedure call are separated by semicolons
- Class definitions start with "cls" (e.g., clsClassName)
- Classes, properties, and procedures are by default public and may be optionally modified by "private" or "protected"
- To instantiate an object, the follow syntax must be used: new clsName objName
- For input, read ("filename.ext", <variable list>)—if reading from console, do not use the first argument
- For output, write ("filename.ext", <expression list>)—if writing to console, do not use the first argument
- The Boolean data type is "boolean"; the return result of all comparison operators is a boolean type
- The operator "*" in front of a variable is used to return the data at the address location within that variable
- The operator "&" in front of a variable is used to return the address of a given variable. The declaration of "pointer_to" is used to define a variable of a pointer type

Flow Chart Definition

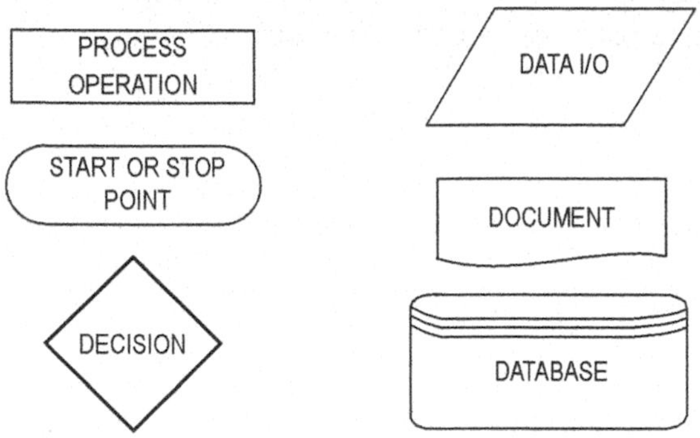

UML Definition

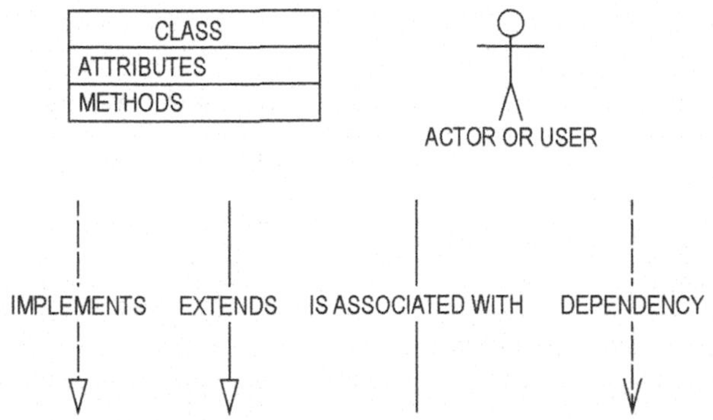

Key Equations

Assume that "*" implies multiplication.

McCabe's Cyclomatic Complexity

$c = e - n + 2$

where for a single program graph, n is the number of nodes, e is the number of edges, and c is the cyclomatic complexity.

The RSA Public-Key Cryptosystem

$n = p * q$ where p and q are both primes.

$e * d = 1 \pmod{t}$, where t = least common multiple $(p - 1, q - 1)$

- The encrypted cyphertext c of a message m is $c = m^e \pmod{n}$
- The decrypted message is $m = c^d \pmod{n}$
- The signature s of a message m is $s = m^d \pmod{n}$

Diffie-Hellman Key-Exchange Protocol

A sender and receiver separately select private keys x and y. Generator value g and prime number p is shared between the two. Their shared secret key k is:

$k = (g^x)^y \pmod{p} = (g^y)^x \pmod{p}$.

INDUSTRIAL AND SYSTEMS ENGINEERING

LINEAR PROGRAMMING

The general linear programming (LP) problem is:

Maximize $Z = c_1x_1 + c_2x_2 + \ldots + c_nx_n$

Subject to:

$$a_{11}x_1 + a_{12}x_2 + \ldots + a_{1n}x_n \le b_1$$
$$a_{21}x_1 + a_{22}x_2 + \ldots + a_{2n}x_n \le b_2$$
$$\vdots$$
$$a_{m1}x_1 + a_{m2}x_2 + \ldots + a_{mn}x_n \le b_m$$
$$x_1, \ldots, x_n \ge 0$$

An LP problem is frequently reformulated by inserting non-negative slack and surplus variables. Although these variables usually have zero costs (depending on the application), they can have non-zero cost coefficients in the objective function. A slack variable is used with a "less than" inequality and transforms it into an equality. For example, the inequality $5x_1 + 3x_2 + 2x_3 \le 5$ could be changed to $5x_1 + 3x_2 + 2x_3 + s_1 = 5$ if s_1 were chosen as a slack variable. The inequality $3x_1 + x_2 - 4x_3 \ge 10$ might be transformed into $3x_1 + x_2 - 4x_3 - s_2 = 10$ by the addition of the surplus variable s_2. Computer printouts of the results of processing an LP usually include values for all slack and surplus variables, the dual prices, and the reduced costs for each variable.

Dual Linear Program

Associated with the above linear programming problem is another problem called the dual linear programming problem. If we take the previous problem and call it the primal problem, then in matrix form the primal and dual problems are respectively:

Primal	Dual
Maximize $Z = \boldsymbol{cx}$	Minimize $W = \boldsymbol{yb}$
Subject to: $\boldsymbol{Ax} \le \boldsymbol{b}$	Subject to: $\boldsymbol{yA} \ge \boldsymbol{c}$
$\boldsymbol{x} \ge 0$	$\boldsymbol{y} \ge 0$

It is assumed that if $\boldsymbol{A}$ is a matrix of size $[m \times n]$, then $\boldsymbol{y}$ is a $[1 \times m]$ vector, $\boldsymbol{c}$ is a $[1 \times n]$ vector, $\boldsymbol{b}$ is an $[m \times 1]$ vector, and $\boldsymbol{x}$ is an $[n \times 1]$ vector.

Network Optimization

Assume we have a graph $G(N, A)$ with a finite set of nodes N and a finite set of arcs A. Furthermore, let

$N = \{1, 2, \ldots, n\}$

x_{ij} = flow from node i to node j

c_{ij} = cost per unit flow from i to j

u_{ij} = capacity of arc (i, j)

b_i = net flow generated at node i

We wish to minimize the total cost of sending the available supply through the network to satisfy the given demand. The minimal cost flow model is formulated as follows:

$$\text{Minimize } Z = \sum_{i=1}^{n} \sum_{j=1}^{n} c_{ij}x_{ij}$$

subject to

$$\sum_{j=1}^{n} x_{ij} - \sum_{j=1}^{n} x_{ji} = b_i \text{ for each node } i \in N$$

and

$$0 \le x_{ij} \le u_{ij} \text{ for each arc } (i, j) \in A$$

The constraints on the nodes represent a conservation of flow relationship. The first summation represents total flow out of node i, and the second summation represents total flow into node i. The net difference generated at node i is equal to b_i.

Many models, such as shortest-path, maximal-flow, assignment and transportation models, can be reformulated as minimal-cost network flow models.

PROCESS CAPABILITY

Actual Capability

$$PCR_k = C_{pk} = \min\left(\frac{\mu - LSL}{3\sigma}, \frac{USL - \mu}{3\sigma}\right)$$

Potential Capability (i.e., Centered Process)

$$PCR = C_p = \frac{USL - LSL}{6\sigma}, \text{ where}$$

μ and σ are the process mean and standard deviation, respectively, and LSL and USL are the lower and upper specification limits, respectively.

QUEUEING MODELS

Definitions

P_n = probability of n units in system

L = expected number of units in the system

L_q = expected number of units in the queue

W = expected waiting time in system

W_q = expected waiting time in queue

λ = mean arrival rate (constant)

$\tilde{\lambda}$ = effective arrival rate

μ = mean service rate (constant)

ρ = server utilization factor

s = number of servers

Kendall notation for describing a queueing system:

$A / B / s / M$

A = the arrival process

B = the service time distribution

s = the number of servers

M = the total number of customers including those in service

Fundamental Relationships

$L = \lambda W$

$L_q = \lambda W_q$

$W = W_q + 1/\mu$

$\rho = \lambda/(s\mu)$

Single Server Models (s = 1)

Poisson Input—Exponential Service Time: $M = \infty$

$P_0 = 1 - \lambda/\mu = 1 - \rho$

$P_n = (1 - \rho)\rho^n = P_0\rho^n$

$L = \rho/(1 - \rho) = \lambda/(\mu - \lambda)$

$L_q = \lambda^2/[\mu\,(\mu - \lambda)]$

$W = 1/[\mu\,(1 - \rho)] = 1/(\mu - \lambda)$

$W_q = W - 1/\mu = \lambda/[\mu\,(\mu - \lambda)]$

Finite queue: $M < \infty$

$\tilde{\lambda} = \lambda(1 - P_m)$

$P_0 = (1 - \rho)/(1 - \rho^{M+1})$

$P_n = [(1 - \rho)/(1 - \rho^{M+1})]\rho^n$

$L = \rho/(1 - \rho) - (M + 1)\rho^{M+1}/(1 - \rho^{M+1})$

$L_q = L - (1 - P_0)$

$W = L/\tilde{\lambda}$

$W = W_q + 1/\mu$

Poisson Input—Arbitrary Service Time

Variance σ^2 is known. For constant service time, $\sigma^2 = 0$.

$P_0 = 1 - \rho$

$L_q = (\lambda^2\sigma^2 + \rho^2)/[2\,(1 - \rho)]$

$L = \rho + L_q$

$W_q = L_q / \lambda$

$W = W_q + 1/\mu$

Poisson Input—Erlang Service Times, $\sigma^2 = 1/(k\mu^2)$

$L_q = [(1 + k)/(2k)][(\lambda^2)/(\mu\,(\mu - \lambda))]$

$= [\lambda^2/(k\mu^2) + \rho^2]/[2(1 - \rho)]$

$W_q = [(1 + k)/(2k)]\{\lambda/[\mu\,(\mu - \lambda)]\}$

$W = W_q + 1/\mu$

Multiple Server Model (s > 1)

Poisson Input—Exponential Service Times

$$P_0 = \left[\sum_{n=0}^{s-1}\frac{\left(\frac{\lambda}{\mu}\right)^n}{n!} + \frac{\left(\frac{\lambda}{\mu}\right)^s}{s!}\left(\frac{1}{1 - \frac{\lambda}{s\mu}}\right)\right]^{-1}$$

$$= 1\Bigg/\left[\sum_{n=0}^{s-1}\frac{(s\rho)^n}{n!} + \frac{(s\rho)^s}{s!(1 - \rho)}\right]$$

$$L_q = \frac{P_0\left(\frac{\lambda}{\mu}\right)^s \rho}{s!(1 - \rho)^2}$$

$$= \frac{P_0 s^s \rho^{s+1}}{s!(1 - \rho)^2}$$

$P_n = P_0(\lambda/\mu)^n/n! \qquad 0 \le n \le s$

$P_n = P_0(\lambda/\mu)^n/(s!s^{n-s}) \qquad n \ge s$

$W_q = L_q/\lambda$

$W = W_q + 1/\mu$

$L = L_q + \lambda/\mu$

Calculations for P_0 and L_q can be time consuming; however, the following table gives formulas for 1, 2, and 3 servers.

s	P_0	L_q
1	$1 - \rho$	$\rho^2/(1 - \rho)$
2	$(1 - \rho)/(1 + \rho)$	$2\rho^3/(1 - \rho^2)$
3	$\dfrac{2(1 - \rho)}{2 + 4\rho + 3\rho^2}$	$\dfrac{9\rho^4}{2 + 2\rho - \rho^2 - 3\rho^3}$

SIMULATION

1. Random Variate Generation

The linear congruential method of generating pseudo-random numbers U_i between 0 and 1 is obtained using $Z_n = (aZ_{n-1}+C)$ (mod m) where a, C, m, and Z_0 are given nonnegative integers and where $U_i = Z_i/m$. Two integers are equal (mod m) if their remainders are the same when divided by m.

2. Inverse Transform Method

If X is a continuous random variable with cumulative distribution function $F(x)$, and U_i is a random number between 0 and 1, then the value of X_i corresponding to U_i can be calculated by solving $U_i = F(x_i)$ for x_i. The solution obtained is $x_i = F^{-1}(U_i)$, where F^{-1} is the inverse function of $F(x)$.

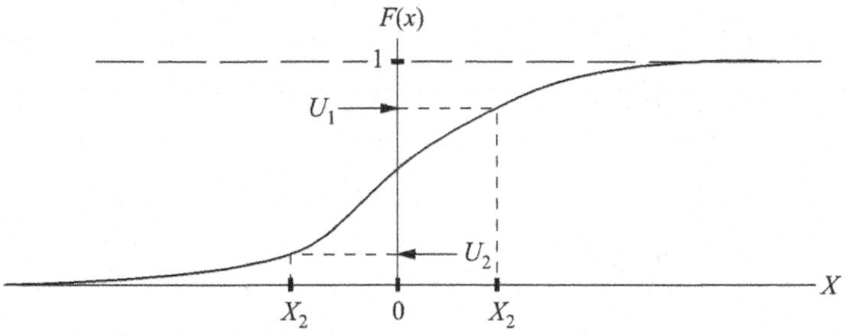

Inverse Transform Method for Continuous Random Variables

FORECASTING

Moving Average

$$\hat{d}_t = \frac{\sum_{i=1}^{n} d_{t-i}}{n}, \text{ where}$$

$\hat{d}_t$ = forecasted demand for period t

d_{t-i} = actual demand for ith period preceding t

n = number of time periods to include in the moving average

Exponentially Weighted Moving Average

$$\hat{d}_t = \alpha d_{t-1} + (1-\alpha)\hat{d}_{t-1}, \text{ where}$$

$\hat{d}_t$ = forecasted demand for t

α = smoothing constant, $0 \le \alpha \le 1$

2^n FACTORIAL EXPERIMENTS

Factors: $X_1, X_2, \ldots, X_n$

Levels of each factor: 1, 2 (sometimes these levels are represented by the symbols – and +, respectively)

r = number of observations for each experimental condition (treatment)

E_i = estimate of the effect of factor X_i, $i = 1, 2, \ldots, n$

E_{ij} = estimate of the effect of the interaction between factors X_i and X_j

$\overline{Y}_{ik}$ = average response value for all $r2^{n-1}$ observations having X_i set at level k, $k = 1, 2$

$\overline{Y}_{ij}^{km}$ = average response value for all $r2^{n-2}$ observations having X_i set at level k, $k = 1, 2$, and X_j set at level m, $m = 1, 2$.

$$E_i = \overline{Y}_{i2} - \overline{Y}_{i1}$$

$$E_{ij} = \frac{\left(\overline{Y}_{ij}^{22} - \overline{Y}_{ij}^{21}\right) - \left(\overline{Y}_{ij}^{12} - \overline{Y}_{ij}^{11}\right)}{2}$$

ANALYSIS OF VARIANCE FOR 2^n FACTORIAL DESIGNS

Main Effects

Let E be the estimate of the effect of a given factor, let L be the orthogonal contrast belonging to this effect. It can be proved that

$$E = \frac{L}{2^{n-1}}$$

$$L = \sum_{c=1}^{m} a_{(c)}\overline{Y}_{(c)}$$

$$SS_L = \frac{rL^2}{2^n}, \text{ where}$$

m = number of experimental conditions ($m = 2^n$ for n factors)

$a_{(c)}$ = –1 if the factor is set at its low level (level 1) in experimental condition c

$a_{(c)}$ = +1 if the factor is set at its high level (level 2) in experimental condition c

r = number of replications for each experimental condition

$\overline{Y}_{(c)}$ = average response value for experimental condition c

SS_L = sum of squares associated with the factor

Interaction Effects

Consider any group of two or more factors.

$a_{(c)}$ = +1 if there is an even number (or zero) of factors in the group set at the low level (level 1) in experimental condition $c = 1, 2, \ldots, m$

$a_{(c)}$ = –1 if there is an odd number of factors in the group set at the low level (level 1) in experimental condition $c = 1, 2, \ldots, m$

It can be proved that the interaction effect E for the factors in the group and the corresponding sum of squares SS_L can be determined as follows:

$$E = \frac{L}{2^{n-1}}$$

$$L = \sum_{c=1}^{m} a_{(c)}\overline{Y}_{(c)}$$

$$SS_L = \frac{rL^2}{2^n}$$

Sum of Squares of Random Error

The sum of the squares due to the random error can be computed as

$$SS_{\text{error}} = SS_{\text{total}} - \Sigma_i SS_i - \Sigma_i \Sigma_j SS_{ij} - \ldots - SS_{12\ldots n}$$

where SS_i is the sum of squares due to factor X_i, SS_{ij} is the sum of squares due to the interaction of factors X_i and X_j, and so on. The total sum of squares is equal to

$$SS_{\text{total}} = \sum_{c=1}^{m} \sum_{k=1}^{r} Y_{ck}^2 - \frac{T^2}{N}$$

where Y_{ck} is the kth observation taken for the cth experimental condition, $m = 2^n$, T is the grand total of all observations, and $N = r2^n$.

RELIABILITY

If P_i is the probability that component i is functioning, a reliability function $R(P_1, P_2, \ldots, P_n)$ represents the probability that a system consisting of n components will work.

For n independent components connected in series,

$$R(P_1, P_2, \ldots P_n) = \prod_{i=1}^{n} P_i$$

For n independent components connected in parallel,

$$R(P_1, P_2, \ldots P_n) = 1 - \prod_{i=1}^{n} (1 - P_i)$$

LEARNING CURVES

The time to do the repetition N of a task is given by

$T_N = KN^s$, where

K = constant

s = ln (learning rate, as a decimal)/ln 2; or, learning rate = 2^s

If N units are to be produced, the average time per unit is given by

$$T_{\text{avg}} = \frac{K}{N(1+s)}\left[(N+0.5)^{(1+s)} - 0.5^{(1+s)}\right]$$

INVENTORY MODELS

For instantaneous replenishment (with constant demand rate, known holding and ordering costs, and an infinite stockout cost), the economic order quantity is given by

$$EOQ = \sqrt{\frac{2AD}{h}}, \text{ where}$$

A = cost to place one order

D = number of units used per year

h = holding cost per unit per year

Under the same conditions as above with a finite replenishment rate, the economic manufacturing quantity is given by

$$EMQ = \sqrt{\frac{2AD}{h(1 - D/R)}}, \text{ where}$$

R = the replenishment rate

FACILITY PLANNING

Equipment Requirements

$$M_j = \sum_{i=1}^{n} \frac{P_{ij} T_{ij}}{C_{ij}} \text{ where}$$

M_j = number of machines of type j required per production period

P_{ij} = desired production rate for product i on machine j, measured in pieces per production period

T_{ij} = production time for product i on machine j, measured in hours per piece

C_{ij} = number of hours in the production period available for the production of product i on machine j

n = number of products

People Requirements

$$A_j = \sum_{i=1}^{n} \frac{P_{ij} T_{ij}}{C_{ij}}, \text{ where}$$

A_j = number of crews required for assembly operation j

P_{ij} = desired production rate for product i and assembly operation j (pieces per day)

T_{ij} = standard time to perform operation j on product i (minutes per piece)

C_{ij} = number of minutes available per day for assembly operation j on product i

n = number of products

Standard Time Determination

$$ST = NT \times AF$$

where

NT = normal time

AF = allowance factor

Case 1: Allowances are based on the *job time*.

$AF_{\text{job}} = 1 + A_{\text{job}}$

A_{job} = allowance fraction (percentage/100) based on *job time*.

Case 2: Allowances are based on *workday*.

$AF_{\text{time}} = 1/(1 - A_{\text{day}})$

A_{day} = allowance fraction (percentage/100) based on *workday*.

- ♦ Predetermined time systems are useful in cases where either (1) the task does not yet exist or (2) changes to a task are being designed and normal times have not yet been established for all elements of the new task or changed task. In such cases no opportunity exists to measure the element time. Unfortunately, there is no scientific basis for predicting element times without breaking them down into motion-level parts. A task consists of elements. An organization may develop its own database of normal element durations, and normal times for new or changed tasks may be predicted if the tasks consist entirely of elements whose normal times are already in the database. But new elements can be decomposed into motions, for which scientifically predetermined times exist in databases called MTM-1, MTM-2, and MTM-3. These databases and software to manipulate them are commercially available. To use one of them effectively requires about 50 hours of training.

Plant Location

The following is one formulation of a discrete plant location problem.

Minimize

$$z = \sum_{i=1}^{m} \sum_{j=1}^{n} c_{ij} y_{ij} + \sum_{j=1}^{n} f_j x_j$$

subject to

$$\sum_{i=1}^{m} y_{ij} \le m x_j, \quad j = 1, \ldots, n$$

$$\sum_{j=1}^{n} y_{ij} = 1, \quad i = 1, \ldots, m$$

$$y_{ij} \ge 0, \text{ for all } i, j$$

$$x_j = (0, 1), \text{ for all } j, \text{ where}$$

m = number of customers

n = number of possible plant sites

y_{ij} = fraction or proportion of the demand of customer i which is satisfied by a plant located at site j; $i = 1, \ldots, m$; $j = 1, \ldots, n$

x_j = 1, if a plant is located at site j

x_j = 0, otherwise

c_{ij} = cost of supplying the entire demand of customer i from a plant located at site j

f_j = fixed cost resulting from locating a plant at site j

Material Handling

Distances between two points (x_1, y_1) and (x_2, y_2) under different metrics:

Euclidean:

$$D = \sqrt{(x_1 - x_2)^2 + (y_1 - y_2)^2}$$

Rectilinear (or Manhattan):

$$D = |x_1 - x_2| + |y_1 - y_2|$$

Chebyshev (simultaneous x and y movement):

$$D = \max(|x_1 - x_2|, |y_1 - y_2|)$$

Line Balancing

$$N_{\min} = \left(OR \times \sum_i t_i / OT\right)$$

= theoretical minimum number of stations

Idle Time/Station = $CT - ST$

Idle Time/Cycle = $\Sigma(CT - ST)$

$$\text{Percent Idle Time} = \frac{\text{Idle Time/Cycle}}{N_{\text{actual}} \times CT} \times 100, \text{ where}$$

CT = cycle time (time between units)

OT = operating time/period

OR = output rate/period

ST = station time (time to complete task at each station)

t_i = individual task times

N = number of stations

Job Sequencing

Two Work Centers—Johnson's Rule

1. Select the job with the shortest time, from the list of jobs, and its time at each work center.
2. If the shortest job time is the time at the first work center, schedule it first, otherwise schedule it last. Break ties arbitrarily.
3. Eliminate that job from consideration.
4. Repeat 1, 2, and 3 until all jobs have been scheduled.

CRITICAL PATH METHOD (CPM)

d_{ij} = duration of activity (i, j)

CP = critical path (longest path)

T = duration of project

$$T = \sum_{(i,j) \in CP} d_{ij}$$

♦ Kennedy, W.J., and Daniel P. Rogers, *Review for the Professional Engineers' Examination in Industrial Engineering*, 2012.

PERT

(a_{ij}, b_{ij}, c_{ij}) = (optimistic, most likely, pessimistic) durations for activity (i, j)

μ_{ij} = mean duration of activity (i, j)

σ_{ij} = standard deviation of the duration of activity (i, j)

μ = project mean duration

σ = standard deviation of project duration

$$\mu_{ij} = \frac{a_{ij} + 4b_{ij} + c_{ij}}{6}$$

$$\sigma_{ij} = \frac{c_{ij} - a_{ij}}{6}$$

$$\mu = \sum_{(i,j)\in CP} \mu_{ij}$$

$$\sigma^2 = \sum_{(i,j)\in CP} \sigma_{ij}^2$$

TAYLOR TOOL LIFE FORMULA

$VT^n = C$, where

V = speed in surface feet per minute

T = tool life in minutes

C, n = constants that depend on the material and on the tool

WORK SAMPLING FORMULAS

$$D = Z_{\alpha/2}\sqrt{\frac{p(1-p)}{n}} \text{ and } R = Z_{\alpha/2}\sqrt{\frac{1-p}{pn}}, \text{ where}$$

p = proportion of observed time in an activity

D = absolute error

R = relative error ($R = D/p$)

n = sample size

ANTHROPOMETRIC MEASUREMENTS

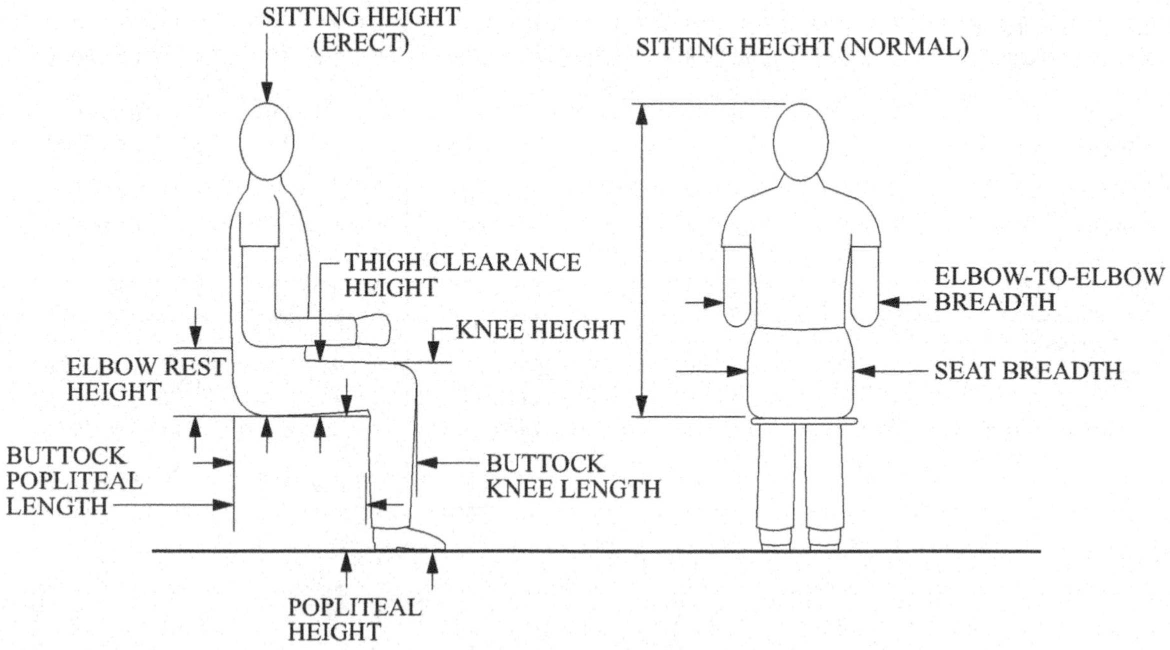

After Sanders and Mccormick, *Human Factors In Design*, McGraw-Hill, 1987.

U.S. Civilian Body Dimensions, Female/Male, for Ages 20 to 60 Years (Centimeters)				
(See Anthropometric Measurements Figure)	Percentiles			
	5th	50th	95th	Std. Dev.
HEIGHTS				
Stature (height)	149.5 / 161.8	160.5 / 173.6	171.3 / 184.4	6.6 / 6.9
Eye height	138.3 / 151.1	148.9 / 162.4	159.3 / 172.7	6.4 / 6.6
Shoulder (acromion) height	121.1 / 132.3	131.1 / 142.8	141.9 / 152.4	6.1 / 6.1
Elbow height	93.6 / 100.0	101.2 / 109.9	108.8 / 119.0	4.6 / 5.8
Knuckle height	**64.3 / 69.8**	**70.2 / 75.4**	**75.9 / 80.4**	**3.5 / 3.2**
Height, sitting (erect)	78.6 / 84.2	85.0 / 90.6	90.7 / 96.7	3.5 / 3.7
Eye height, sitting	67.5 / 72.6	73.3 / 78.6	78.5 / 84.4	3.3 / 3.6
Shoulder height, sitting	49.2 / 52.7	55.7 / 59.4	61.7 / 65.8	3.8 / 4.0
Elbow rest height, sitting	18.1 / 19.0	23.3 / 24.3	28.1 / 29.4	2.9 / 3.0
Knee height, sitting	**45.2 / 49.3**	**49.8 / 54.3**	**54.5 / 59.3**	**2.7 / 2.9**
Popliteal height, sitting	35.5 / 39.2	39.8 / 44.2	44.3 / 48.8	2.6 / 2.8
Thigh clearance height	10.6 / 11.4	13.7 / 14.4	17.5 / 17.7	1.8 / 1.7
DEPTHS				
Chest depth	21.4 / 21.4	24.2 / 24.2	29.7 / 27.6	2.5 / 1.9
Elbow-fingertip distance	38.5 / 44.1	42.1 / 47.9	46.0 / 51.4	2.2 / 2.2
Buttock-knee length, sitting	51.8 / 54.0	56.9 / 59.4	62.5 / 64.2	3.1 / 3.0
Buttock-popliteal length, sitting	43.0 / 44.2	48.1 / 49.5	53.5 / 54.8	3.1 / 3.0
Forward reach, functional	64.0 / 76.3	71.0 / 82.5	79.0 / 88.3	4.5 / 5.0
BREADTHS				
Elbow-to-elbow breadth	31.5 / 35.0	38.4 / 41.7	49.1 / 50.6	5.4 / 4.6
Seat (hip) breadth, sitting	31.2 / 30.8	36.4 / 35.4	43.7 / 40.6	3.7 / 2.8
HEAD DIMENSIONS				
Head breadth	13.6 / 14.4	14.54 / 15.42	15.5 / 16.4	0.57 / 0.59
Head circumference	52.3 / 53.8	54.9 / 56.8	57.7 / 59.3	1.63 / 1.68
Interpupillary distance	5.1 / 5.5	5.83 / 6.20	6.5 / 6.8	0.4 / 0.39
HAND DIMENSIONS				
Hand length	16.4 / 17.6	17.95 / 19.05	19.8 / 20.6	1.04 / 0.93
Breadth, metacarpal	7.0 / 8.2	7.66 / 8.88	8.4 / 9.8	0.41 / 0.47
Circumference, metacarpal	16.9 / 19.9	18.36 / 21.55	19.9 / 23.5	0.89 / 1.09
Thickness, metacarpal III	2.5 / 2.4	2.77 / 2.76	3.1 / 3.1	0.18 / 0.21
Digit 1				
Breadth, interphalangeal	1.7 / 2.1	1.98 / 2.29	2.1 / 2.5	0.12 / 0.13
Crotch-tip length	4.7 / 5.1	5.36 / 5.88	6.1 / 6.6	0.44 / 0.45
Digit 2				
Breadth, distal joint	1.4 / 1.7	1.55 / 1.85	1.7 / 2.0	0.10 / 0.12
Crotch-tip length	6.1 / 6.8	6.88 / 7.52	7.8 / 8.2	0.52 / 0.46
Digit 3				
Breadth, distal joint	1.4 / 1.7	1.53 / 1.85	1.7 / 2.0	0.09 / 0.12
Crotch-tip length	7.0 / 7.8	7.77 / 8.53	8.7 / 9.5	0.51 / 0.51
Digit 4				
Breadth, distal joint	1.3 / 1.6	1.42 / 1.70	1.6 / 1.9	0.09 / 0.11
Crotch-tip length	6.5 / 7.4	7.29 / 7.99	8.2 / 8.9	0.53 / 0.47
Digit 5				
Breadth, distal joint	1.2 / 1.4	1.32 / 1.57	1.5 / 1.8	0.09/0.12
Crotch-tip length	4.8 / 5.4	5.44 / 6.08	6.2 / 6.99	0.44/0.47
FOOT DIMENSIONS				
Foot length	22.3 / 24.8	24.1 / 26.9	26.2 / 29.0	1.19 / 1.28
Foot breadth	8.1 / 9.0	8.84 / 9.79	9.7 / 10.7	0.50 / 0.53
Lateral malleolus height	5.8 / 6.2	6.78 / 7.03	7.8 / 8.0	0.59 / 0.54
Weight (kg)	46.2 / 56.2	61.1 / 74.0	89.9 / 97.1	13.8 / 12.6

Kroemer, Karl H. E., "Engineering Anthropometry," *Ergonomics*, Vol. 32, No. 7, pp. 779-780, 1989.

ERGONOMICS—HEARING

The average shifts with age of the threshold of hearing for pure tones of persons with "normal" hearing, using a 25-year-old group as a reference group.

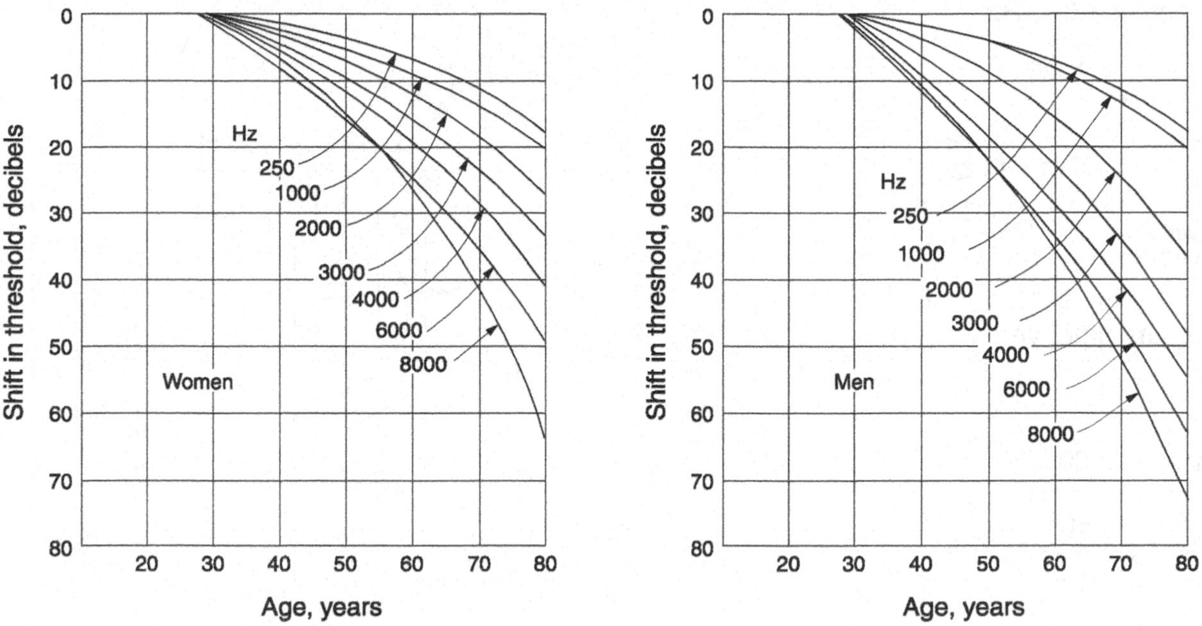

Equivalent sound-level contours used in determining the A-weighted sound level on the basis of an octave-band analysis. The curve at the point of the highest penetration of the noise spectrum reflects the A-weighted sound level.

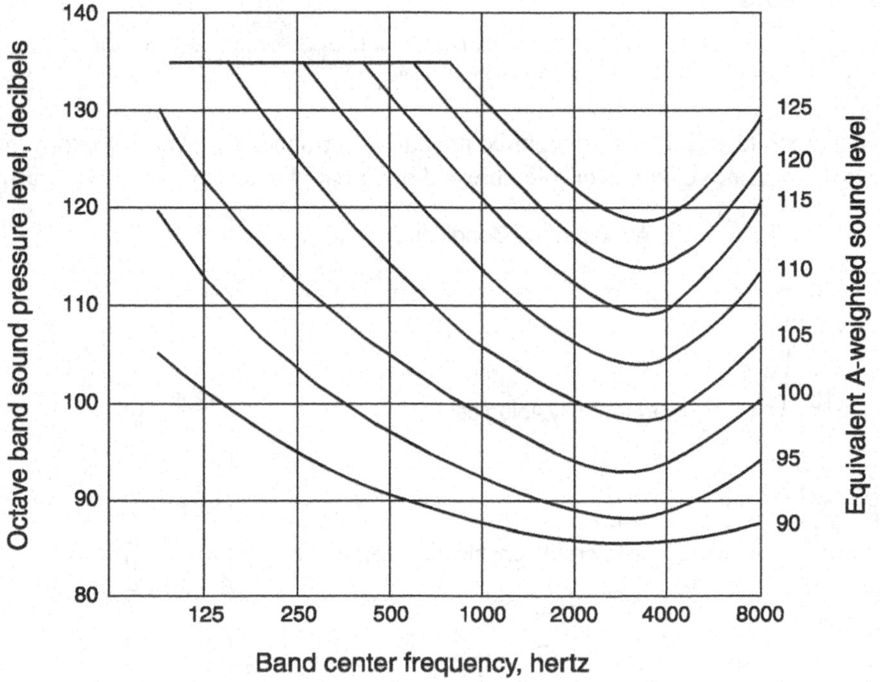

Estimated average trend curves for net hearing loss at 1,000, 2,000, and 4,000 Hz after continuous exposure to steady noise. Data are corrected for age, but not for temporary threshold shift. Dotted portions of curves represent extrapolation from available data.

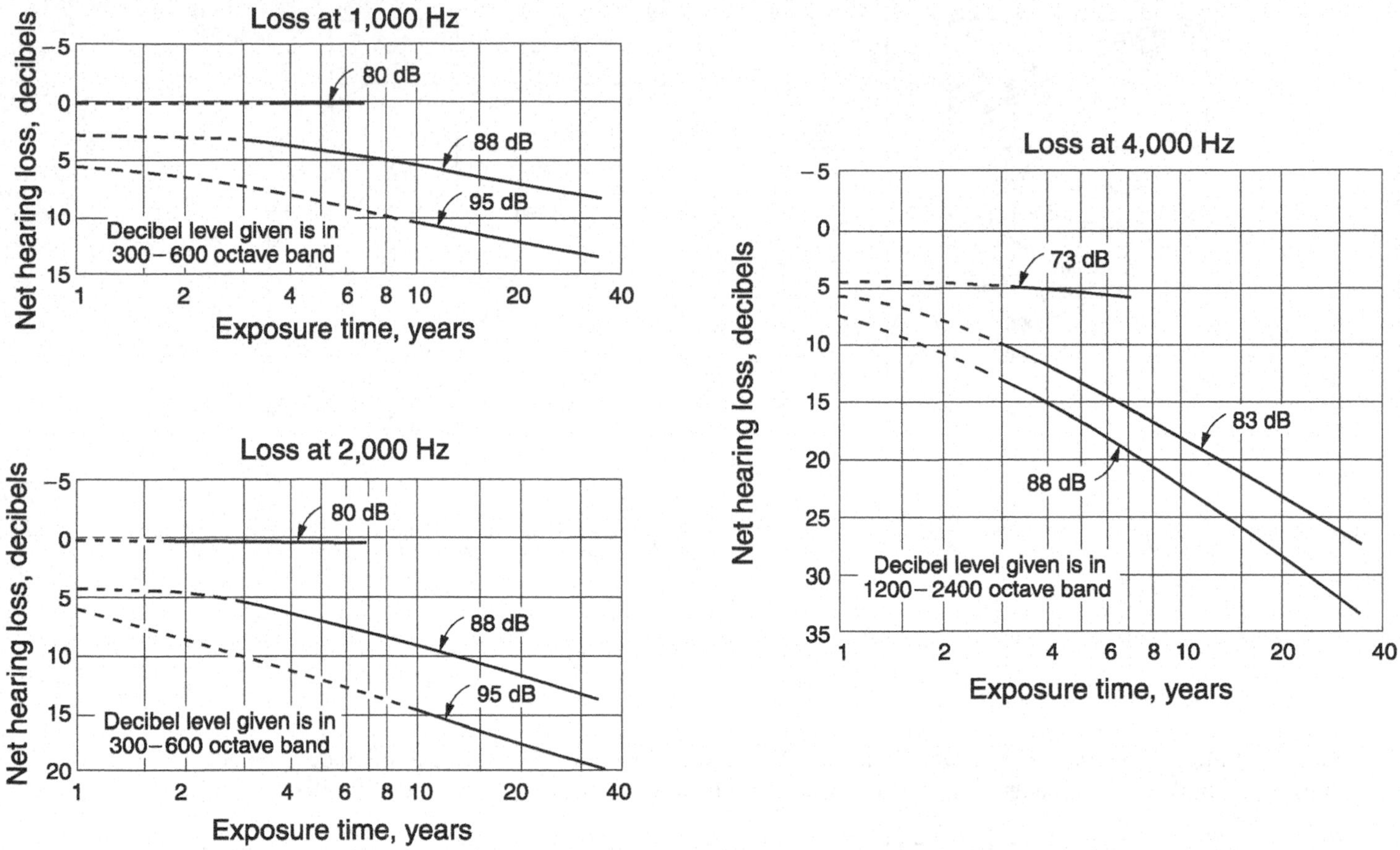

"The Relations of Hearing Loss to Noise Exposure," Exploratory Subcommittee Z24-X-2 of the American Standards Association Z24 Special Committee on Acoustics, Vibration, and Mechanical Shock, sponsored by the Acoustical Society of America, American Standards Association, 1954, pp. 31–33.

Tentative upper limit of effective temperature (ET) for unimpaired mental performance as related to exposure time; data are based on an analysis of 15 studies. Comparative curves of tolerable and marginal physiological limits are also given.

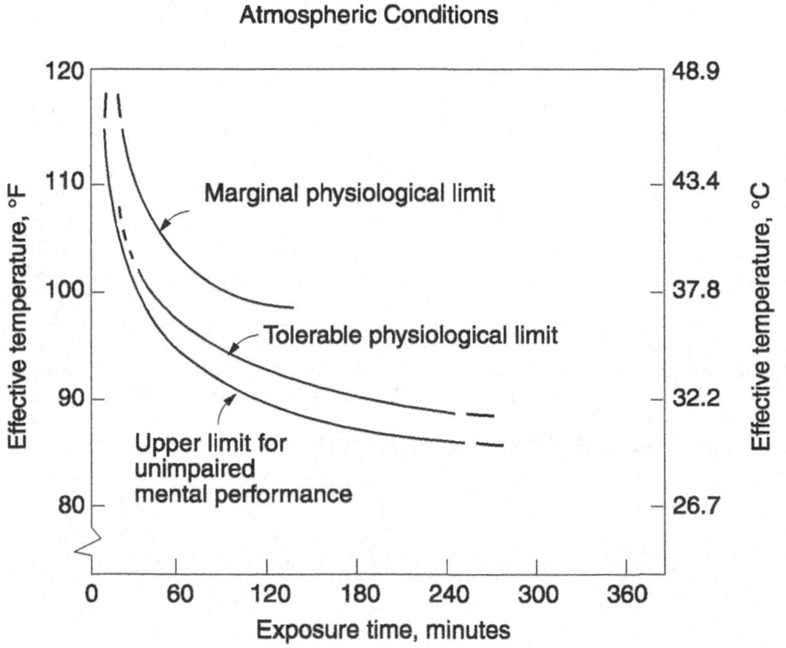

Effective temperature (ET) is the dry bulb temperature at 50% relative humidity, which results in the same physiological effect as the present conditions.

MECHANICAL ENGINEERING

MECHANICAL DESIGN AND ANALYSIS

Springs

<u>Mechanical Springs</u>

Helical Linear Springs: The shear stress in a helical linear spring is

$$\tau = K_s \frac{8FD}{\pi d^3}, \text{ where}$$

d = wire diameter
F = applied force
D = mean spring diameter
$K_s = (2C + 1)/(2C)$
$C = D/d$

The deflection and force are related by $F = kx$ where the spring rate (spring constant) k is given by

$$k = \frac{d^4 G}{8D^3 N}$$

where G is the shear modulus of elasticity and N is the number of active coils.

Spring Material: The minimum tensile strength of common spring steels may be determined from

$$S_{ut} = A/d^m$$

where S_{ut} is the tensile strength in MPa, d is the wire diameter in millimeters, and A and m are listed in the following table:

Material	ASTM	m	A
Music wire	A228	0.163	2060
Oil-tempered wire	A229	0.193	1610
Hard-drawn wire	A227	0.201	1510
Chrome vanadium	A232	0.155	1790
Chrome silicon	A401	0.091	1960

Maximum allowable torsional stress for static applications may be approximated as

$S_{sy} = \tau = 0.45 S_{ut}$ cold-drawn carbon steel (A227, A228, A229)

$S_{sy} = \tau = 0.50 S_{ut}$ hardened and tempered carbon and low-alloy steels (A232, A401)

Compression Spring Dimensions

Type of Spring Ends		
Term	**Plain**	**Plain and Ground**
End coils, N_e	0	1
Total coils, N_t	N	$N + 1$
Free length, L_0	$pN + d$	$p(N + 1)$
Solid length, L_s	$d(N_t + 1)$	dN_t
Pitch, p	$(L_0 - d)/N$	$L_0/(N + 1)$

Term	Squared or Closed	Squared and Ground
End coils, N_e	2	2
Total coils, N_t	$N + 2$	$N + 2$
Free length, L_0	$pN + 3d$	$pN + 2d$
Solid length, L_s	$d(N_t + 1)$	dN_t
Pitch, p	$(L_0 - 3d)/N$	$(L_0 - 2d)/N$

Helical Torsion Springs: The bending stress is given as

$$\sigma = K_i [32Fr/(\pi d^3)]$$

where F is the applied load and r is the radius from the center of the coil to the load.

K_i = correction factor
$= (4C^2 - C - 1) / [4C(C - 1)]$
$C = D/d$

The deflection θ and moment Fr are related by

$$Fr = k\theta$$

where the spring rate k is given by

$$k = \frac{d^4 E}{64DN}$$

where k has units of N•m/rad and θ is in radians.

Spring Material: The strength of the spring wire may be found as shown in the section on linear springs. The allowable stress σ is then given by

$S_y = \sigma = 0.78 S_{ut}$ cold-drawn carbon steel (A227, A228, A229)

$S_y = \sigma = 0.87 S_{ut}$ hardened and tempered carbon and low-alloy steel (A232, A401)

Bearings

Ball/Roller Bearing Selection

The minimum required *basic load rating* (load for which 90% of the bearings from a given population will survive 1 million revolutions) is given by

$$C = PL^{1/a}, \text{ where}$$

C = minimum required basic load rating

P = design radial load

L = design life (in millions of revolutions)

a = 3 for ball bearings, 10/3 for roller bearings

When a ball bearing is subjected to both radial and axial loads, an equivalent radial load must be used in the equation above. The equivalent radial load is

$$P_{eq} = XVF_r + YF_a, \text{ where}$$

P_{eq} = equivalent radial load

F_r = applied constant radial load

F_a = applied constant axial (thrust) load

For radial contact, deep-groove ball bearings:

V = 1 if inner ring rotating, 1.2 if outer ring rotating,

If $F_a/(VF_r) > e$,

$$X = 0.56, \text{ and } Y = 0.840\left(\frac{F_a}{C_0}\right)^{-0.247}$$

$$\text{where } e = 0.513\left(\frac{F_a}{C_0}\right)^{0.236}$$

C_0 = basic static load rating from bearing catalog

If $F_a/(VF_r) \le e$, X = 1 and Y = 0.

Power Screws

Square Thread Power Screws: The torque required to raise, T_R, or to lower, T_L, a load is given by

$$T_R = \frac{Fd_m}{2}\left(\frac{l + \pi\mu d_m}{\pi d_m - \mu l}\right) + \frac{F\mu_c d_c}{2}$$

$$T_L = \frac{Fd_m}{2}\left(\frac{\pi\mu d_m - l}{\pi d_m + \mu l}\right) + \frac{F\mu_c d_c}{2}, \text{ where}$$

d_c = mean collar diameter

d_m = mean thread diameter

l = lead

F = load

μ = coefficient of friction for thread

μ_c = coefficient of friction for collar

The efficiency of a power screw may be expressed as

$$\eta = Fl/(2\pi T)$$

Power Transmission

Shafts and Axles

Static Loading: The maximum shear stress and the von Mises stress may be calculated in terms of the loads from

$$\tau_{max} = \frac{2}{\pi d^3}\left[(8M + Fd)^2 + (8T)^2\right]^{1/2},$$

$$\sigma' = \frac{4}{\pi d^3}\left[(8M + Fd)^2 + 48T^2\right]^{1/2}, \text{ where}$$

M = the bending moment

F = the axial load

T = the torque

d = the diameter

Fatigue Loading: Using the maximum-shear-stress theory combined with the Soderberg line for fatigue, the diameter and safety factor are related by

$$\frac{\pi d^3}{32} = n\left[\left(\frac{M_m}{S_y} + \frac{K_f M_a}{S_e}\right)^2 + \left(\frac{T_m}{S_y} + \frac{K_{fs} T_a}{S_e}\right)^2\right]^{1/2}$$

where

d = diameter

n = safety factor

M_a = alternating moment

M_m = mean moment

T_a = alternating torque

T_m = mean torque

S_e = fatigue limit

S_y = yield strength

K_f = fatigue strength reduction factor

K_{fs} = fatigue strength reduction factor for shear

Gearing

Involute Gear Tooth Nomenclature

Circular pitch	$p_c = \pi d/N$
Base pitch	$p_b = p_c \cos \phi$
Module	$m = d/N$
Center distance	$C = (d_1 + d_2)/2$

where

N = number of teeth on pinion or gear

d = pitch circle diameter

ϕ = pressure angle

Gear Trains: *Velocity ratio*, m_v, is the ratio of the output velocity to the input velocity. Thus, $m_v = \omega_{out}/\omega_{in}$. For a two-gear train, $m_v = -N_{in}/N_{out}$ where N_{in} is the number of teeth on the input gear and N_{out} is the number of teeth on the output gear. The negative sign indicates that the output gear rotates in the opposite sense with respect to the input gear. In a *compound gear train*, at least one shaft carries more than one gear (rotating at the same speed). The velocity ratio for a compound train is:

$$m_v = \pm\frac{\text{product of number of teeth on driver gears}}{\text{product of number of teeth on driven gears}}$$

A *simple planetary gearset* has a sun gear, an arm that rotates about the sun gear axis, one or more gears (planets) that rotate about a point on the arm, and a ring (internal) gear that is concentric with the sun gear. The planet gear(s) mesh with the sun gear on one side and with the ring gear on the other. A planetary gearset has two independent inputs and one output (or two outputs and one input, as in a differential gearset).

Often one of the inputs is zero, which is achieved by grounding either the sun or the ring gear. The velocities in a planetary set are related by

$$\frac{\omega_L - \omega_{arm}}{\omega_f - \omega_{arm}} = \pm\, m_v, \text{ where}$$

ω_f = speed of the first gear in the train
ω_L = speed of the last gear in the train
ω_{arm} = speed of the arm

Neither the first nor the last gear can be one that has planetary motion. In determining m_v, it is helpful to invert the mechanism by grounding the arm and releasing any gears that are grounded.

Dynamics of Mechanisms

Gearing

Loading on Straight Spur Gears: The load, W, on straight spur gears is transmitted along a plane that, in edge view, is called the *line of action*. This line makes an angle with a tangent line to the pitch circle that is called the *pressure angle* ϕ. Thus, the contact force has two components: one in the tangential direction, W_t, and one in the radial direction, W_r. These components are related to the pressure angle by

$$W_r = W_t \tan(\phi).$$

Only the tangential component W_t transmits torque from one gear to another. Neglecting friction, the transmitted force may be found if either the transmitted torque or power is known:

$$W_t = \frac{2T}{d} = \frac{2T}{mN},$$

$$W_t = \frac{2H}{d\omega} = \frac{2H}{mN\omega}, \text{ where}$$

W_t = transmitted force (newtons)
T = torque on the gear (newton-mm)
d = pitch diameter of the gear (mm)
N = number of teeth on the gear
m = gear module (mm) (same for both gears in mesh)
H = power (kW)
ω = speed of gear (rad/s)

Joining Methods

Threaded Fasteners: The load carried by a bolt in a threaded connection is given by

$$F_b = CP + F_i \qquad F_m < 0$$

while the load carried by the members is

$$F_m = (1 - C)\,P - F_i \qquad F_m < 0, \text{ where}$$

C = joint coefficient
$= k_b/(k_b + k_m)$
F_b = total bolt load
F_i = bolt preload
F_m = total material load
P = externally applied load
k_b = effective stiffness of the bolt or fastener in the grip
k_m = effective stiffness of the members in the grip

Bolt stiffness may be calculated from

$$k_b = \frac{A_d A_t E}{A_d l_t + A_t l_d}, \text{ where}$$

A_d = major-diameter area
A_t = tensile-stress area
E = modulus of elasticity
l_d = length of unthreaded shank
l_t = length of threaded shank contained within the grip

If all members within the grip are of the same material, *member stiffness* may be obtained from

$$k_m = dEAe^{b(d/l)}, \text{ where}$$

d = bolt diameter
E = modulus of elasticity of members
l = grip length

Coefficients A and b are given in the table below for various joint member materials.

Material	*A*	*b*
Steel	0.78715	0.62873
Aluminum	0.79670	0.63816
Copper	0.79568	0.63553
Gray cast iron	0.77871	0.61616

The approximate tightening torque required for a given preload F_i and for a steel bolt in a steel member is given by $T = 0.2\, F_i d$.

Threaded Fasteners – Design Factors: The bolt load factor is

$$n_b = (S_p A_t - F_i)/CP, \text{ where}$$

S_p = proof strength

The factor of safety guarding against joint separation is

$$n_s = F_i / [P\,(1 - C)]$$

Threaded Fasteners – Fatigue Loading: If the externally applied load varies between zero and P, the alternating stress is

$$\sigma_a = CP/(2A_t)$$

and the mean stress is

$$\sigma_m = \sigma_a + F_i/A_t$$

Bolted and Riveted Joints Loaded in Shear:

Failure by Pure Shear

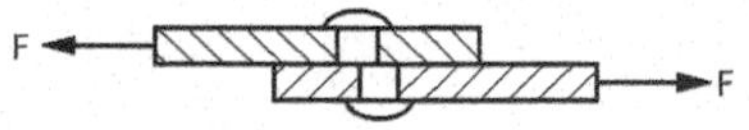

FASTENER IN SHEAR

$$\tau = F/A, \text{ where}$$

F = shear load

A = cross-sectional area of bolt or rivet

Failure by Rupture

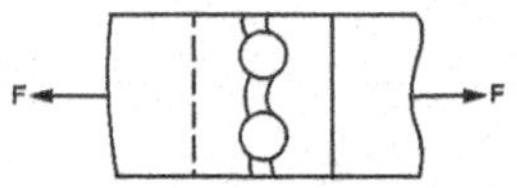

MEMBER RUPTURE

$$\sigma = F/A, \text{ where}$$

F = load

A = net cross-sectional area of thinnest member

Failure by Crushing of Rivet or Member

MEMBER OR FASTENER CRUSHING

$$\sigma = F/A, \text{ where}$$

F = load

A = projected area of a single rivet

Fastener Groups in Shear

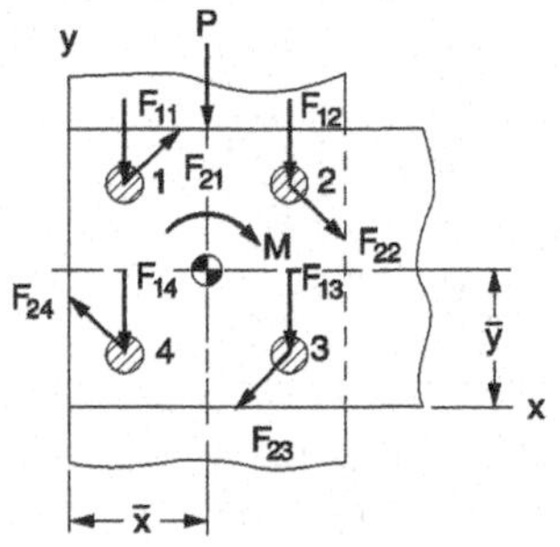

FASTENER GROUPS

The location of the centroid of a fastener group with respect to any convenient coordinate frame is:

$$\bar{x} = \frac{\sum_{i=1}^{n} A_i x_i}{\sum_{i=1}^{n} A_i}, \quad \bar{y} = \frac{\sum_{i=1}^{n} A_i y_i}{\sum_{i=1}^{n} A_i}, \text{ where}$$

n = total number of fasteners

i = the index number of a particular fastener

A_i = cross-sectional area of the ith fastener

x_i = x-coordinate of the center of the ith fastener

y_i = y-coordinate of the center of the ith fastener

The total shear force on a fastener is the **vector** sum of the force due to direct shear P and the force due to the moment M acting on the group at its centroid.

The magnitude of the direct shear force due to P is

$$|F_{1i}| = \frac{P}{n}.$$

This force acts in the same direction as P.

The magnitude of the shear force due to M is

$$|F_{2i}| = \frac{Mr_i}{\sum_{i=1}^{n} r_i^2}.$$

This force acts perpendicular to a line drawn from the group centroid to the center of a particular fastener. Its sense is such that its moment is in the same direction (CW or CCW) as M.

Press/Shrink Fits

The interface pressure induced by a press/shrink fit is

$$p = \frac{0.5\delta}{\frac{r}{E_o}\left(\frac{r_o^2 + r^2}{r_o^2 - r^2} + \nu_o\right) + \frac{r}{E_i}\left(\frac{r^2 + r_i^2}{r^2 - r_i^2} - \nu_i\right)}$$

where the subscripts i and o stand for the inner and outer member, respectively, and

p = inside pressure on the outer member and outside pressure on the inner member

δ = the diametral interference

r = nominal interference radius

r_i = inside radius of inner member

r_o = outside radius of outer member

E = Young's modulus of respective member

ν = Poisson's ratio of respective member

The *maximum torque* that can be transmitted by a press fit joint is approximately

$$T = 2\pi r^2 \mu p l,$$

where r and p are defined above,

T = torque capacity of the joint

μ = coefficient of friction at the interface

l = length of hub engagement

MANUFACTURABILITY

Limits and Fits

The designer is free to adopt any geometry of fit for shafts and holes that will ensure intended function. Over time, sufficient experience with common situations has resulted in the development of a standard. The metric version of the standard is newer and will be presented. The standard specifies that uppercase letters always refer to the hole, while lowercase letters always refer to the shaft.

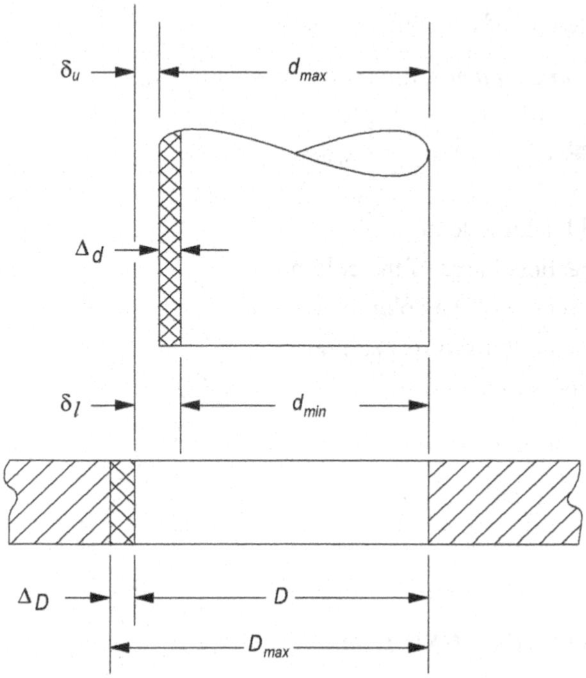

Definitions

Basic Size or *nominal size*, D or d, is the size to which the limits or deviations are applied. It is the same for both components.

Deviation is the algebraic difference between the actual size and the corresponding basic size.

Upper Deviation, δ_u, is the algebraic difference between the maximum limit and the corresponding basic size.

Lower Deviation, δ_l, is the algebraic difference between the minimum limit and the corresponding basic size.

Fundamental Deviation, δ_F, is the upper or lower deviation, depending on which is closer to the basic size.

Tolerance, Δ_D or Δ_d, is the difference between the maximum and minimum size limits of a part.

International tolerance (IT) grade numbers designate groups of tolerances such that the tolerance for a particular IT number will have the same relative accuracy for a basic size.

Hole basis represents a system of fits corresponding to a basic hole size. The fundamental deviation is H.

Some Preferred Fits

Clearance	
Free running fit: not used where accuracy is essential but good for large temperature variations, high running speeds, or heavy journal loads.	H9/d9
Sliding fit: where parts are not intended to run freely but must move and turn freely and locate accurately.	H7/g6
Locational clearance fit: provides snug fit for location of stationary parts, but can be freely assembled and disassembled.	H7/h6
Loose running fit: for wide commercial tolerances or allowances on external members	H11/c11
Close running fit: for running on accurate machines and for accurate location at moderate speeds and journal pressures	H8/f7
Transition	
Locational transition fit: for accurate location, a compromise between clearance and interference	H7/k6
Locational transition fit: for more accurate location where greater interface is permissible	H7/n6
Interference	
Location interference fit: for parts requiring rigidity and alignment with prime accuracy of location but without special bore pressure requirements.	H7/p6
Medium drive fit: for ordinary steel parts or shrink fits on light sections. The tightest fit usable on cast iron.	H7/s6
Force fit: suitable for parts which can be highly stressed or for shrink fits where the heavy pressing forces required are impractical.	H7/u6

For the hole

$$D_{max} = D + \Delta_D$$
$$D_{min} = D$$

For a shaft with clearance fits *d, g, h, c,* or *f*

$$d_{max} = d + \delta_F$$
$$d_{min} = d_{max} - \Delta_d$$

For a shaft with transition or interference fits *k, p, s, u,* or *n*

$$d_{min} = d + \delta_F$$
$$d_{max} = d_{min} + \Delta_d$$

where

D = basic size of hole
d = basic size of shaft
δ_u = upper deviation
δ_l = lower deviation
δ_F = fundamental deviation
Δ_D = tolerance grade for hole
Δ_d = tolerance grade for shaft

International Tolerance (IT) Grades

Lower limit < Basic Size ≤ Upper Limit
All values in mm

Basic Size	Tolerance Grade, (Δ_D or Δ_d)		
	IT6	IT7	IT9
0–3	0.006	0.010	0.025
3–6	0.008	0.012	0.030
6–10	0.009	0.015	0.036
10–18	0.011	0.018	0.043
18–30	0.013	0.021	0.052
30–50	0.016	0.025	0.062

Source: Preferred Metric Limits and Fits, ANSI B4.2-1978

Deviations for shafts

Lower limit < Basic Size ≤ Upper Limit
All values in mm

Basic Size	Upper Deviation Letter, (δ_u)					Lower Deviation Letter, (δ_l)				
	c	d	f	g	h	k	n	p	s	u
0–3	–0.060	–0.020	–0.006	–0.002	0	0	+0.004	+0.006	+0.014	+0.018
3–6	–0.070	–0.030	–0.010	–0.004	0	+0.001	+0.008	+0.012	+0.019	+0.023
6–10	–0.080	–0.040	–0.013	–0.005	0	+0.001	+0.010	+0.015	+0.023	+0.028
10–14	–0.095	–0.050	–0.016	–0.006	0	+0.001	+0.012	+0.018	+0.028	+0.033
14–18	–0.095	–0.050	–0.016	–0.006	0	+0.001	+0.012	+0.018	+0.028	+0.033
18–24	–0.110	–0.065	–0.020	–0.007	0	+0.002	+0.015	+0.022	+0.035	+0.041
24–30	–0.110	–0.065	–0.020	–0.007	0	+0.002	+0.015	+0.022	+0.035	+0.048
30–40	–0.120	–0.080	–0.025	–0.009	0	+0.002	+0.017	+0.026	+0.043	+0.060
40–50	–0.130	–0.080	–0.025	–0.009	0	+0.002	+0.017	+0.026	+0.043	+0.070

Source: ASME B4.2:2009

As an example, 34H7/s6 denotes a basic size of $D = d = 34$ mm, an IT class of 7 for the hole, and an IT class of 6 and an "s" fit class for the shaft.

Maximum Material Condition (MMC)
The maximum material condition defines the dimension of a part such that the part weighs the most. The MMC of a shaft is at the maximum size of the shaft while the MMC of a hole is at the minimum size of the hole.

Least Material Condition (LMC)
The least material condition or minimum material condition defines the dimensions of a part such that the part weighs the least. The LMC of a shaft is the minimum size of the shaft while the LMC of a hole is at the maximum size of the hole.

Intermediate- and Long-Length Columns

For both intermediate and long columns, the effective column length depends on the end conditions. The AISC recommended design values for the effective lengths of columns are, for: rounded-rounded or pinned-pinned ends, $l_{eff} = l$; fixed-free, $l_{eff} = 2.1l$; fixed-pinned, $l_{eff} = 0.80l$; fixed-fixed, $l_{eff} = 0.65l$. The effective column length should be used when calculating the slenderness ratio.

The slenderness ratio of a column is $S_r = l/r$, where l is the length of the column and r is the radius of gyration. The radius of gyration of a column cross-section is, $r = \sqrt{I/A}$ where I is the area moment of inertia and A is the cross-sectional area of the column. A column is considered to be intermediate if its slenderness ratio is less than or equal to $(S_r)_D$, where

$$(S_r)_D = \pi\sqrt{\frac{2E}{S_y}}, \text{ and}$$

E = Young's modulus of respective member
S_y = yield strength of the column material

For intermediate columns, the critical load is

$$P_{cr} = A\left[S_y - \frac{1}{E}\left(\frac{S_y S_r}{2\pi}\right)^2\right], \text{ where}$$

P_{cr} = critical buckling load
A = cross-sectional area of the column
S_y = yield strength of the column material
E = Young's modulus of respective member
S_r = slenderness ratio

For long columns, the critical load is

$$P_{cr} = \frac{\pi^2 EA}{S_r^2}$$

where the variables are as defined above.

STATIC LOADING FAILURE THEORIES

Brittle Materials

Maximum-Normal-Stress Theory
The maximum-normal-stress theory states that failure occurs when one of the three principal stresses equals the strength of the material. If $\sigma_1 \geq \sigma_2 \geq \sigma_3$, then the theory predicts that failure occurs whenever $\sigma_1 \geq S_{ut}$ or $\sigma_3 \leq - S_{uc}$ where S_{ut} and S_{uc} are the tensile and compressive strengths, respectively.

Coulomb-Mohr Theory
The Coulomb-Mohr theory is based upon the results of tensile and compression tests. On the σ, τ coordinate system, one circle is plotted for S_{ut} and one for S_{uc}. As shown in the figure, lines are then drawn tangent to these circles. The Coulomb-Mohr theory then states that fracture will occur for any stress situation that produces a circle that is either tangent to or crosses the envelope defined by the lines tangent to the S_{ut} and S_{uc} circles.

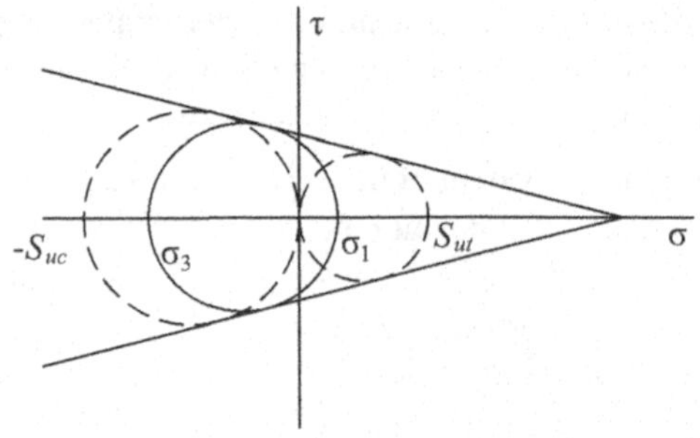

If $\sigma_1 \geq \sigma_2 \geq \sigma_3$ and $\sigma_3 < 0$, then the theory predicts that yielding will occur whenever

$$\frac{\sigma_1}{S_{ut}} - \frac{\sigma_3}{S_{uc}} \geq 1$$

Ductile Materials

Maximum-Shear-Stress Theory

The maximum-shear-stress theory states that yielding begins when the maximum shear stress equals the maximum shear stress in a tension-test specimen of the same material when that specimen begins to yield. If $\sigma_1 \geq \sigma_2 \geq \sigma_3$, then the theory predicts that yielding will occur whenever $\tau_{max} \geq S_y/2$ where S_y is the yield strength.

$$\tau_{max} = \frac{\sigma_1 - \sigma_3}{2}.$$

Distortion-Energy Theory

The distortion-energy theory states that yielding begins whenever the distortion energy in a unit volume equals the distortion energy in the same volume when uniaxially stressed to the yield strength. The theory predicts that yielding will occur whenever

$$\left[\frac{(\sigma_1 - \sigma_2)^2 + (\sigma_2 - \sigma_3)^2 + (\sigma_1 - \sigma_3)^2}{2}\right]^{1/2} \geq S_y$$

The term on the left side of the inequality is known as the effective or Von Mises stress. For a biaxial stress state the effective stress becomes

$$\sigma' = \left(\sigma_A^2 - \sigma_A\sigma_B + \sigma_B^2\right)^{1/2}$$

or

$$\sigma' = \left(\sigma_x^2 - \sigma_x\sigma_y + \sigma_y^2 + 3\tau_{xy}^2\right)^{1/2}$$

where σ_A and σ_B are the two nonzero principal stresses and σ_x, σ_y, and τ_{xy} are the stresses in orthogonal directions.

VARIABLE LOADING FAILURE THEORIES

Modified Goodman Theory: The modified Goodman criterion states that a fatigue failure will occur whenever

$$\frac{\sigma_a}{S_e} + \frac{\sigma_m}{S_{ut}} \geq 1 \quad \text{or} \quad \frac{\sigma_{max}}{S_y} \geq 1, \quad \sigma_m \geq 0,$$

where

S_e = fatigue strength

S_{ut} = ultimate strength

S_y = yield strength

σ_a = alternating stress

σ_m = mean stress

$\sigma_{max} = \sigma_m + \sigma_a$

Soderberg Theory: The Soderberg theory states that a fatigue failure will occur whenever

$$\frac{\sigma_a}{S_e} + \frac{\sigma_m}{S_y} \geq 1 \qquad \sigma_m \geq 0$$

Endurance Limit for Steels: When test data is unavailable, the endurance limit for steels may be estimated as

$$S'_e = \begin{Bmatrix} 0.5\,S_{ut}, S_{ut} \leq 1{,}400\text{ MPa} \\ 700\text{ MPa}, S_{ut} > 1{,}400\text{ MPa} \end{Bmatrix}$$

Endurance Limit Modifying Factors: Endurance limit modifying factors are used to account for the differences between the endurance limit as determined from a rotating beam test, S'_e, and that which would result in the real part, S_e.

$$S_e = k_a k_b k_c k_d k_e S'_e$$

where

Surface Factor, $k_a = aS_{ut}^b$

Surface Finish	Factor *a*		Exponent *b*
	kpsi	MPa	
Ground	1.34	1.58	–0.085
Machined or CD	2.70	4.51	–0.265
Hot rolled	14.4	57.7	–0.718
As forged	39.9	272.0	–0.995

Size Factor, k_b:

For bending and torsion:

$d \leq 8$ mm; $k_b = 1$

8 mm $\leq d \leq$ 250 mm; $k_b = 1.189 d_{eff}^{-0.097}$

$d > 250$ mm; $0.6 \leq k_b \leq 0.75$

For axial loading: $k_b = 1$

Load Factor, k_c:

$k_c = 0.923$ axial loading, $S_{ut} \leq 1{,}520$ MPa

$k_c = 1$ axial loading, $S_{ut} > 1{,}520$ MPa

$k_c = 1$ bending

$k_c = 0.577$ torsion

Temperature Factor, k_d:

for T $\leq$ 450°C, $k_d = 1$

Miscellaneous Effects Factor, k_e: Used to account for strength reduction effects such as corrosion, plating, and residual stresses. In the absence of known effects, use $k_e = 1$.

KINEMATICS, DYNAMICS, AND VIBRATIONS

Kinematics of Mechanisms

Four-Bar Linkage

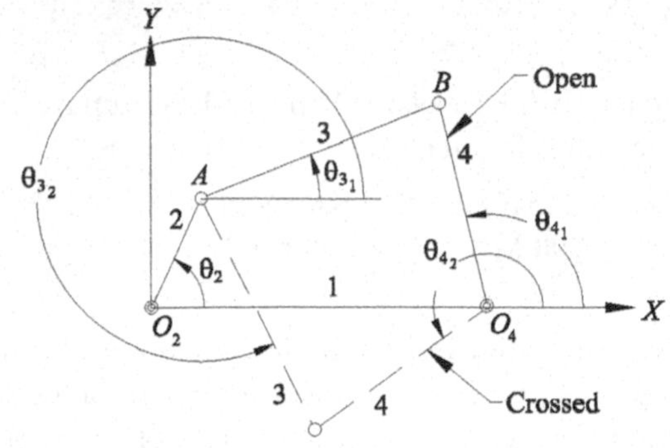

The four-bar linkage shown above consists of a reference (usually grounded) link (1), a crank (input) link (2), a coupler link (3), and an output link (4). Links 2 and 4 rotate about the fixed pivots O_2 and O_4, respectively. Link 3 is joined to link 2 at the moving pivot A and to link 4 at the moving pivot B. The lengths of links 2, 3, 4, and 1 are a, b, c, and d, respectively.

Taking link 1 (ground) as the reference (X-axis), the angles that links 2, 3, and 4 make with the axis are θ_2, θ_3, and θ_4, respectively. It is possible to assemble a four-bar in two different configurations for a given position of the input link (2). These are known as the "open" and "crossed" positions or circuits.

Position Analysis. Given a, b, c, and d, and θ_2

$$\theta_{4_{1,2}} = 2\arctan\left(\frac{-B \pm \sqrt{B^2 - 4AC}}{2A}\right)$$

where $A = \cos\theta_2 - K_1 - K_2\cos\theta_2 + K_3$

$B = -2\sin\theta_2$

$C = K_1 - (K_2 + 1)\cos\theta_2 + K_3$

$$K_1 = \frac{d}{a}, K_2 = \frac{d}{c}, K_3 = \frac{a^2 - b^2 + c^2 + d^2}{2ac}$$

In the equation for θ_4, using the minus sign in front of the radical yields the open solution. Using the plus sign yields the crossed solution.

$$\theta_{3_{1,2}} = 2\arctan\left(\frac{-E \pm \sqrt{E^2 - 4DF}}{2D}\right)$$

where $D = \cos\theta_2 - K_1 + K_4\cos\theta_2 + K_5$

$E = -2\sin\theta_2$

$F = K_1 + (K_4 - 1)\cos\theta_2 + K_5$

$$K_4 = \frac{d}{b}, K_5 = \frac{c^2 - d^2 - a^2 - b^2}{2ab}$$

In the equation for θ_3, using the minus sign in front of the radical yields the open solution. Using the plus sign yields the crossed solution.

Velocity Analysis. Given a, b, c, and d, θ_2, θ_3, θ_4, and ω_2

$$\omega_3 = \frac{a\omega_2}{b}\frac{\sin(\theta_4 - \theta_2)}{\sin(\theta_3 - \theta_4)}$$

$$\omega_4 = \frac{a\omega_2}{c}\frac{\sin(\theta_2 - \theta_3)}{\sin(\theta_4 - \theta_3)}$$

$$V_{Ax} = -a\omega_2\sin\theta_2, \qquad V_{Ay} = a\omega_2\cos\theta_2$$

$$V_{BAx} = -b\omega_3\sin\theta_3, \qquad V_{BAy} = b\omega_3\cos\theta_3$$

$$V_{Bx} = -c\omega_4\sin\theta_4, \qquad V_{By} = c\omega_4\cos\theta_4$$

Acceleration Analysis. Given a, b, c, and d, θ_2, θ_3, θ_4, and ω_2, ω_3, ω_4, and α_2

$$\alpha_3 = \frac{CD - AF}{AE - BD}, \quad \alpha_4 = \frac{CE - BF}{AE - BD}, \text{ where}$$

$$A = c\sin\theta_4, B = b\sin\theta_3$$

$$C = a\alpha_2\sin\theta_2 + a\omega_2^2\cos\theta_2 + b\omega_3^2\cos\theta_3 - c\omega_4^2\cos\theta_4$$

$$D = c\cos\theta_4, E = b\cos\theta_3$$

$$F = a\alpha_2\cos\theta_2 - a\omega_2^2\sin\theta_2 - b\omega_3^2\sin\theta_3 + c\omega_4^2\sin\theta_4$$

HVAC

Nomenclature

h = specific enthalpy

h_f = specific enthalpy of saturated liquid

$\dot{m}_a$ = mass flowrate of dry air

$\dot{m}_w$ = mass flowrate of water

$\dot{Q}$ = rate of heat transfer

T_{wb} = wet bulb temperature

ω = specific humidity (absolute humidity, humidity ratio)

HVAC—Pure Heating and Cooling

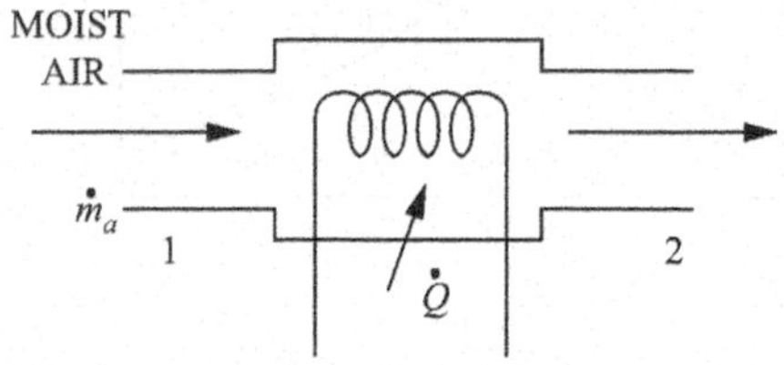

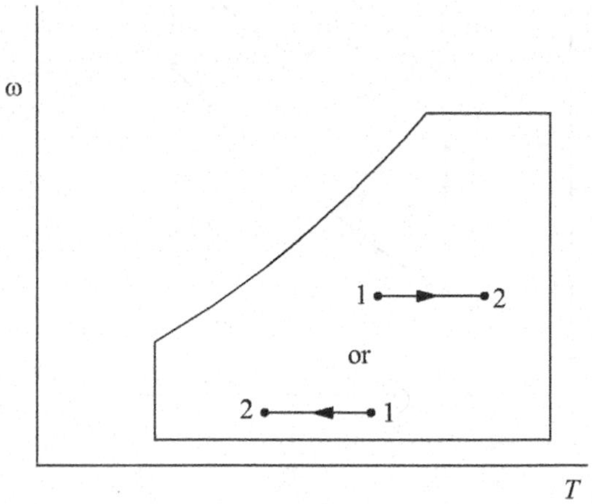

$$\dot{Q} = \dot{m}_a(h_2 - h_1)$$

Cooling and Dehumidification

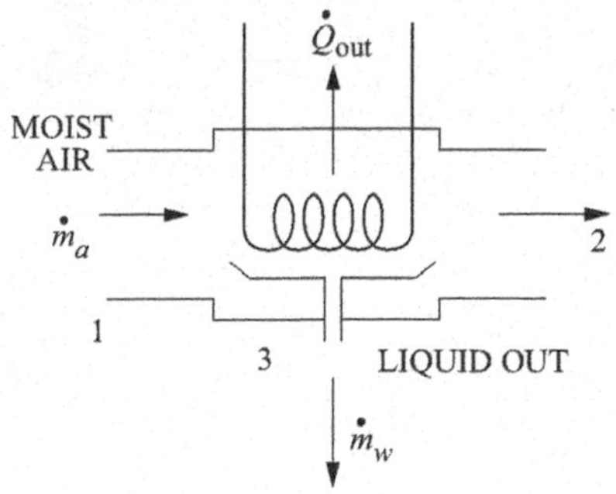

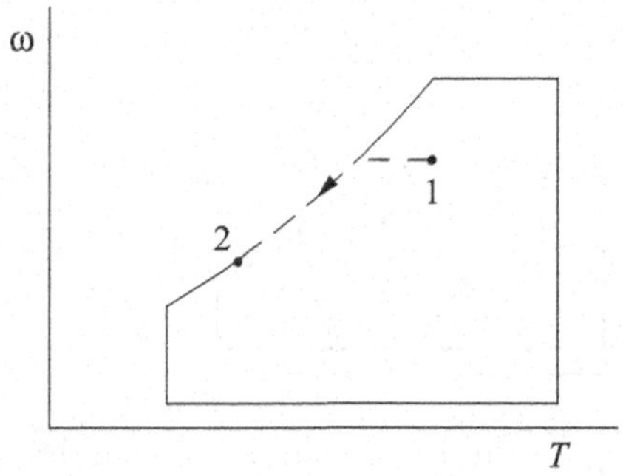

$$\dot{Q}_{out} = \dot{m}_a\left[(h_1 - h_2) - h_{f3}(\omega_1 - \omega_2)\right]$$

$$\dot{m}_w = \dot{m}_a(\omega_1 - \omega_2)$$

Heating and Humidification

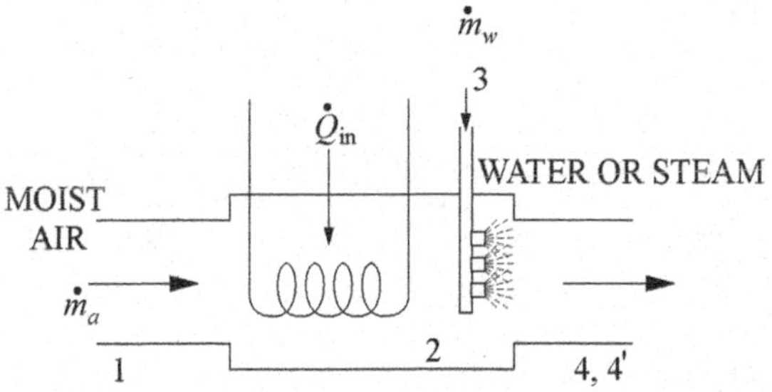

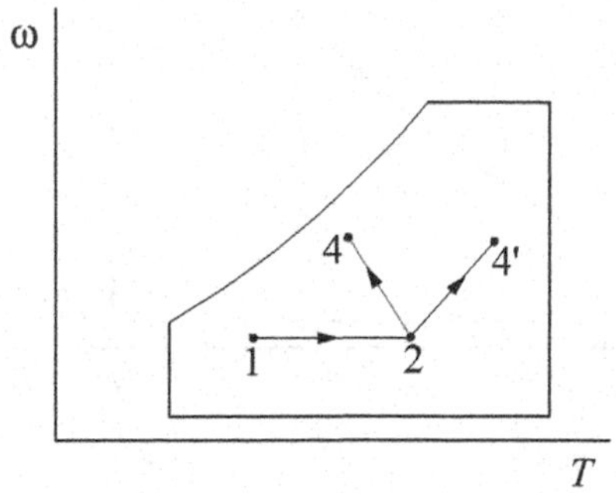

$$\dot{Q}_{in} = \dot{m}_a(h_2 - h_1)$$

$$\dot{m}_w = \dot{m}_a(\omega_{4'} - \omega_2) \text{ or } \dot{m}_w = \dot{m}_a(\omega_4 - \omega_2)$$

Adiabatic Humidification (evaporative cooling)

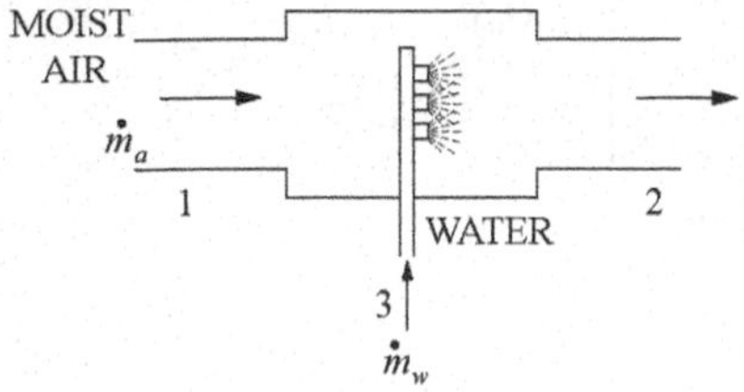

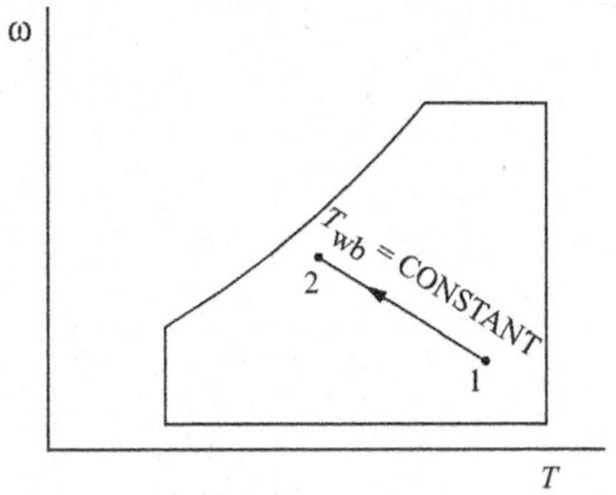

$$h_2 = h_1 + h_3(\omega_2 - \omega_1)$$
$$\dot{m}_w = \dot{m}_a(\omega_2 - \omega_1)$$
$$h_3 = h_f \text{ at } T_{wb}$$

Adiabatic Mixing

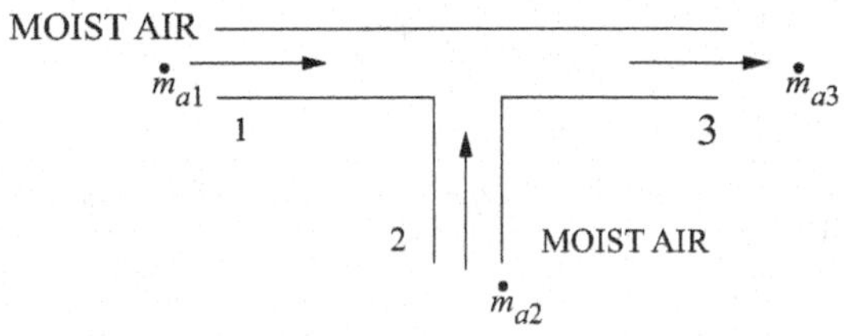

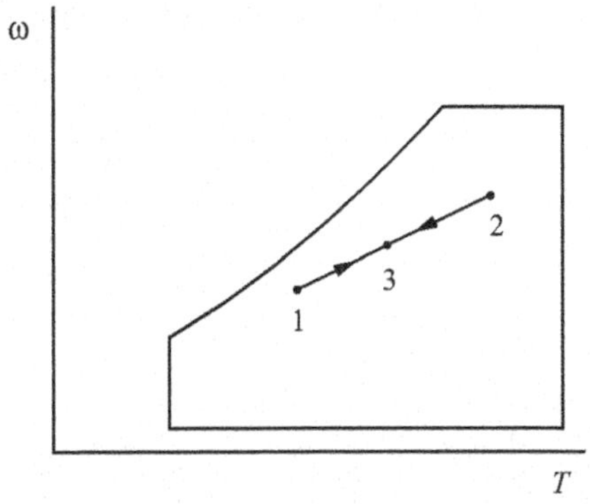

$$\dot{m}_{a3} = \dot{m}_{a1} + \dot{m}_{a2}$$
$$h_3 = \frac{\dot{m}_{a1}h_1 + \dot{m}_{a2}h_2}{\dot{m}_{a3}}$$
$$\omega_3 = \frac{\dot{m}_{a1}\omega_1 + \dot{m}_{a2}\omega_2}{\dot{m}_{a3}}$$

distance $\overline{13} = \dfrac{\dot{m}_{a2}}{\dot{m}_{a3}} \times$ distance $\overline{12}$ measured on psychrometric chart

Cycles and Processes

Internal Combustion Engines

Diesel Cycle

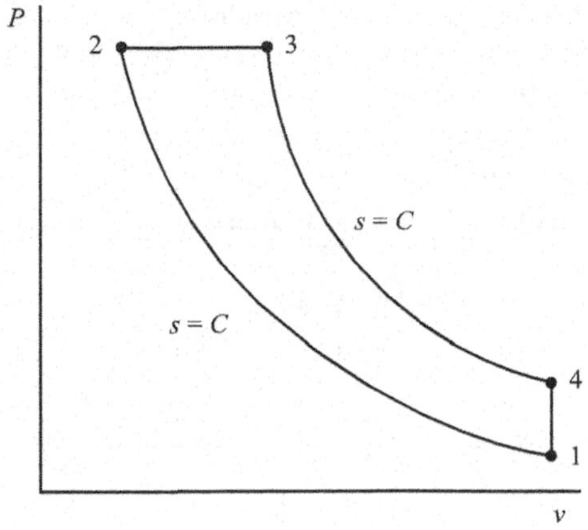

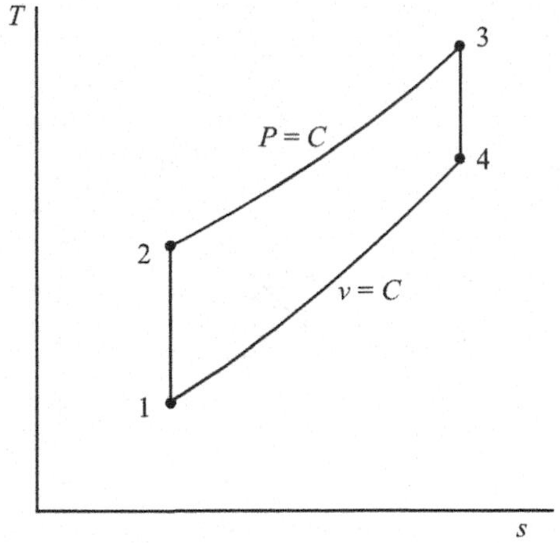

$$r = V_1/V_2$$
$$r_c = V_3/V_2$$
$$\eta = 1 - \frac{1}{r^{k-1}}\left[\frac{r_c^k - 1}{k(r_c - 1)}\right]$$
$$k = c_p/c_v$$

Brake Power

$$\dot{W}_b = 2\pi TN = 2\pi FRN, \text{ where}$$

$\dot{W}_b$ = brake power (W)
T = torque (N•m)
N = rotation speed (rev/s)
F = force at end of brake arm (N)
R = length of brake arm (m)

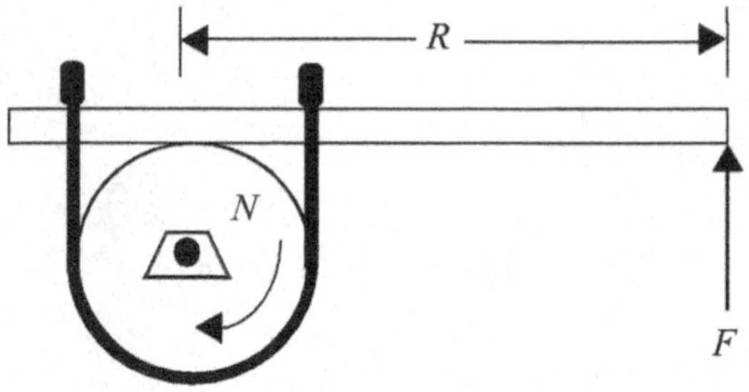

Indicated Power

$$\dot{W}_i = \dot{W}_b + \dot{W}_f, \text{ where}$$

$\dot{W}_i$ = *indicated power* (W)
$\dot{W}_f$ = friction power (W)

Brake Thermal Efficiency

$$\eta_b = \frac{\dot{W}_b}{\dot{m}_f (HV)}, \text{ where}$$

η_b = brake thermal efficiency
$\dot{m}_f$ = fuel consumption rate (kg/s)
HV = heating value of fuel (J/kg)

Indicated Thermal Efficiency

$$\eta_i = \frac{\dot{W}_i}{\dot{m}_f (HV)}$$

Mechanical Efficiency

$$\eta_i = \frac{\dot{W}_b}{\dot{W}_i} = \frac{\eta_b}{\eta_i}$$

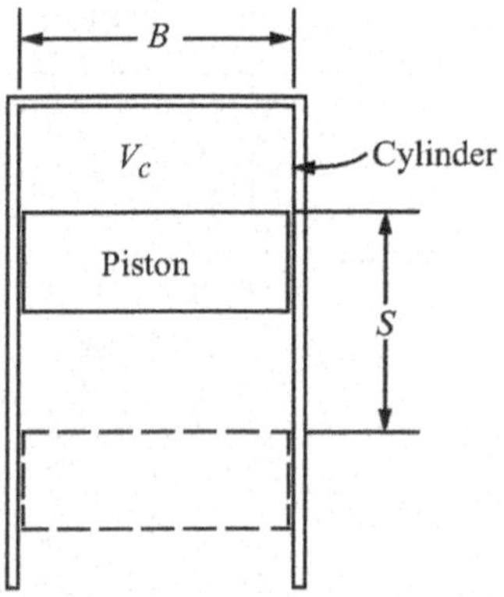

Displacement Volume

$$V_d = \frac{\pi B^2 S}{4}, \text{ m}^3 \text{ for each cylinder}$$

Total volume = $V_t = V_d + V_c$ (m^3)

V_c = clearance volume (m^3)

Compression Ratio

$$r_c = V_t / V_c$$

Mean Effective Pressure (mep)

$$mep = \frac{\dot{W} n_s}{V_d n_c N}, \text{ where}$$

n_s = number of crank revolutions per power stroke
n_c = number of cylinders
V_d = displacement volume per cylinder

mep can be based on brake power (*bmep*), indicated power (*imep*), or friction power (*fmep*).

Volumetric Efficiency

$$\eta_v = \frac{2\dot{m}_a}{\rho_a V_d n_c N} \qquad \text{(four-stroke cycles only)}$$

where
$\dot{m}_a$ = mass flow rate of air into engine (kg/s)
ρ_a = density of air (kg/m^3)

Specific Fuel Consumption (SFC)

$$sfc = \frac{\dot{m}_f}{\dot{W}} = \frac{1}{\eta HV}, \text{ kg/J}$$

Use η_b and $\dot{W}_b$ for *bsfc* and η_i and $\dot{W}_i$ for *isfc*.

Gas Turbines

Brayton Cycle (Steady-Flow Cycle)

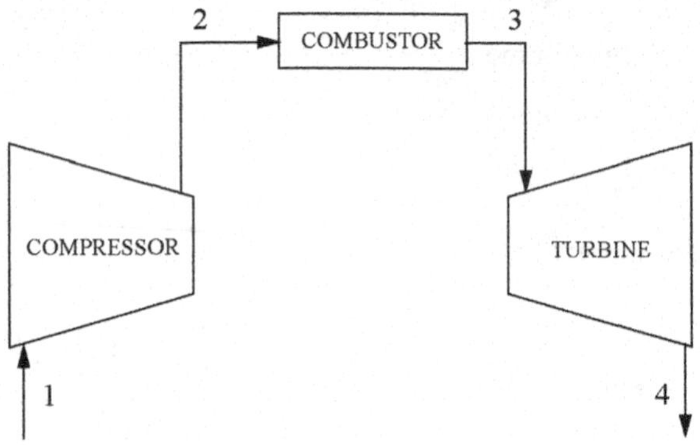

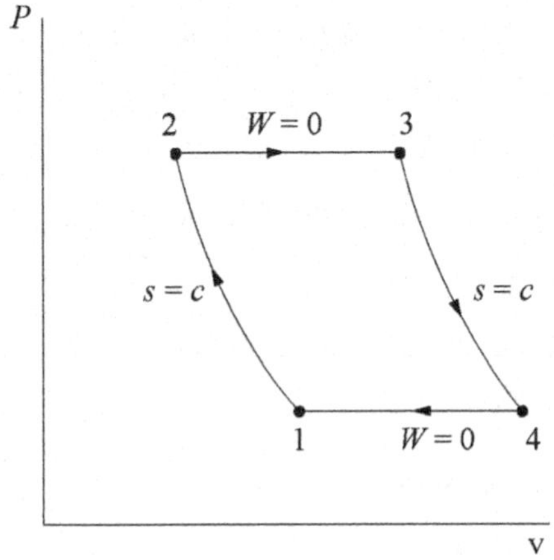

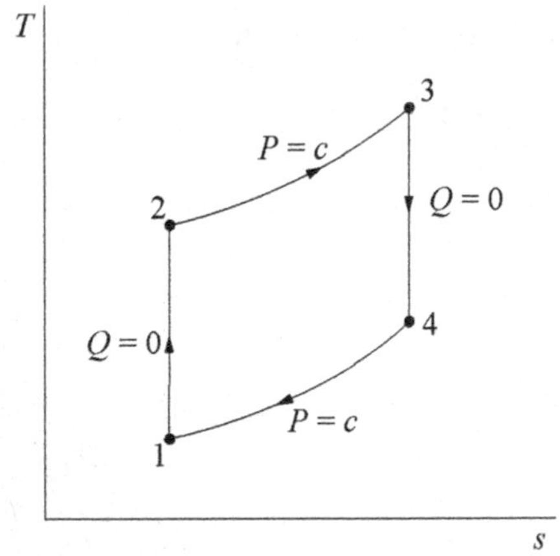

$w_{12} = h_1 - h_2 = c_p(T_1 - T_2)$

$w_{34} = h_3 - h_4 = c_p(T_3 - T_4)$

$w_{\text{net}} = w_{12} + w_{34}$

$q_{23} = h_3 - h_2 = c_p(T_3 - T_2)$

$q_{41} = h_1 - h_4 = c_p(T_1 - T_4)$

$q_{\text{net}} = q_{23} + q_{41}$

$\eta = w_{\text{net}}/q_{23}$

Steam Power Plants

Feedwater Heaters

- *Open (mixing)*

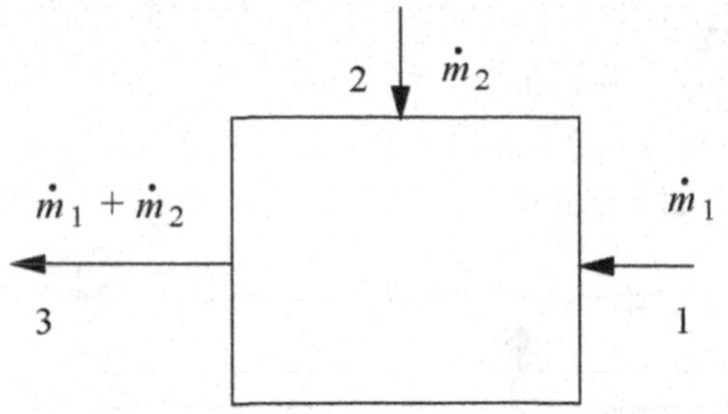

$$\dot{m}_1 h_1 + \dot{m}_2 h_2 = h_3(\dot{m}_1 + \dot{m}_2)$$

- *Closed (no mixing)*

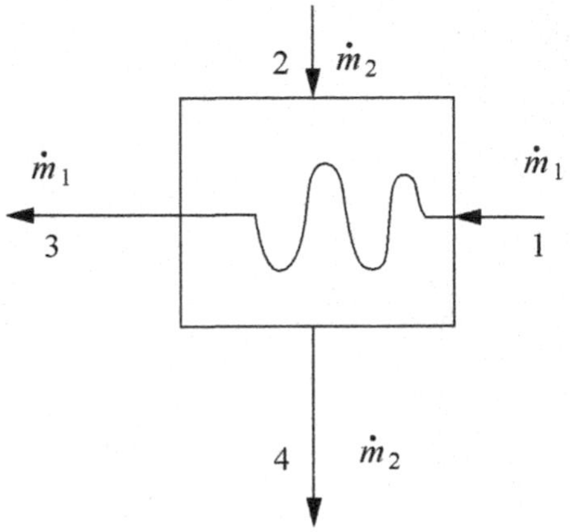

$$\dot{m}_1 h_1 + \dot{m}_2 h_2 = \dot{m}_1 h_3 + \dot{m}_2 h_4$$

Steam Trap

A steam trap removes condensate from steam piping or a heat exchanger.

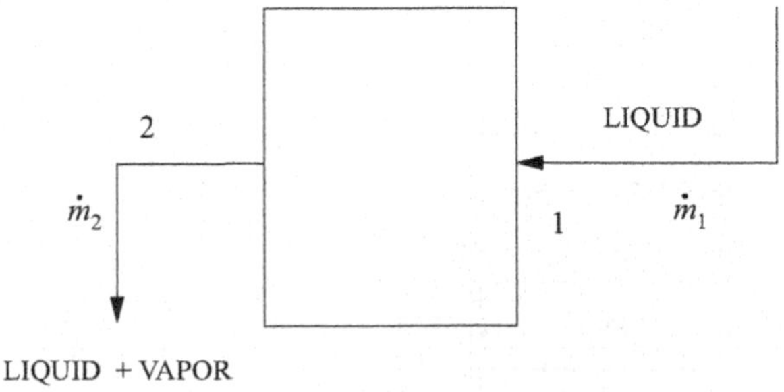

$h_2 = h_1$ (if adiabatic)

Junction

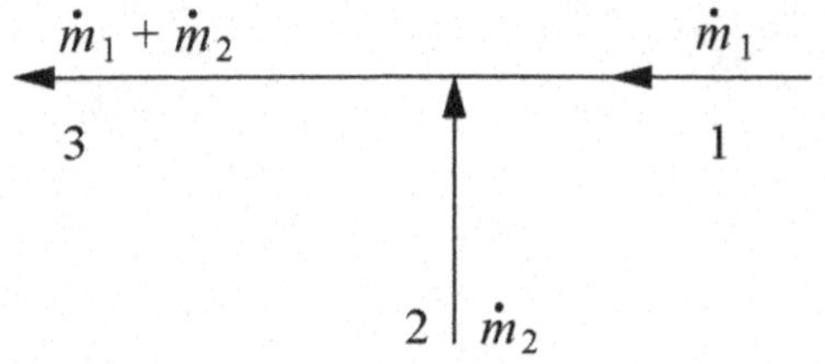

$$\dot{m}_1 h_1 + \dot{m}_2 h_2 = h_3 (\dot{m}_1 + \dot{m}_2)$$

Pump

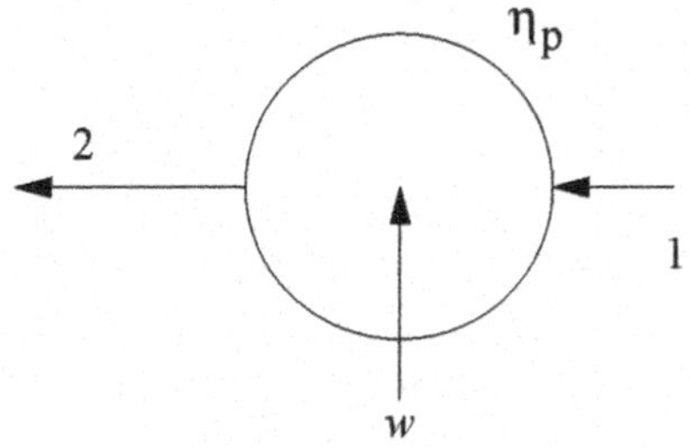

$$w = h_1 - h_2 = (h_1 - h_{2S})/\eta_P$$

$$h_{2S} - h_1 = v(P_2 - P_1)$$

$$w = -\frac{v(P_2 - P_1)}{\eta_P}$$

INDEX

D

E

F

G

H

I

J

K

L

M

N

O

P

Q

R

S

T

U

V

W

X

Y

Z

APPENDIX
FE EXAM SPECIFICATIONS

NCEES
advancing licensure for engineers and surveyors

Fundamentals of Engineering (FE)
CHEMICAL CBT Exam Specifications

Effective Beginning with the January 2014 Examinations

- The FE exam is a computer-based test (CBT). It is closed book with an electronic reference.
- Examinees have 6 hours to complete the exam, which contains 110 multiple-choice questions. The 6-hour time also includes a tutorial and an optional scheduled break.
- The FE exam uses both the International System of Units (SI) and the US Customary System (USCS).

Knowledge	Number of Questions
1. Mathematics A. Analytic geometry B. Roots of equations C. Calculus D. Differential equations	**8–12**
2. Probability and Statistics A. Probability distributions (e.g., discrete, continuous, normal, binomial) B. Expected value (weighted average) in decision making C. Hypothesis testing D. Measures of central tendencies and dispersions (e.g., mean, mode, standard deviation) E. Estimation for a single mean (e.g., point, confidence intervals) F. Regression and curve fitting	**4–6**
3. Engineering Sciences A. Applications of vector analysis (e.g., statics) B. Basic dynamics (e.g., friction, force, mass, acceleration, momentum) C. Work, energy, and power (as applied to particles or rigid bodies) D. Electricity and current and voltage laws (e.g., charge, energy, current, voltage, power, Kirchhoff, Ohm)	**4–6**
4. Computational Tools A. Numerical methods and concepts (e.g., convergence, tolerance) B. Spreadsheets for chemical engineering calculations C. Simulators	**4–6**
5. Materials Science A. Chemical, electrical, mechanical, and physical properties (e.g., effect of temperature, pressure, stress, strain) B. Material types and compatibilities (e.g., engineered materials, ferrous and nonferrous metals) C. Corrosion mechanisms and control	**4–6**

6. Chemistry **8–12**

A. Inorganic chemistry (e.g., molarity, normality, molality, acids, bases, redox reactions, valence, solubility product, pH, pK, electrochemistry, periodic table)
B. Organic chemistry (e.g., nomenclature, structure, qualitative and quantitative analyses, balanced equations, reactions, synthesis, basic biochemistry)

7. Fluid Mechanics/Dynamics **8–12**

A. Fluid properties
B. Dimensionless numbers (e.g., Reynolds number)
C. Mechanical energy balance (e.g., pipes, valves, fittings, pressure losses across packed beds, pipe networks)
D. Bernoulli equation (hydrostatic pressure, velocity head)
E. Laminar and turbulent flow
F. Flow measurement (e.g., orifices, Venturi meters)
G. Pumps, turbines, and compressors
H. Compressible flow and non-Newtonian fluids

8. Thermodynamics **8–12**

A. Thermodynamic properties (e.g. specific volume, internal energy, enthalpy, entropy, free energy)
B. Properties data and phase diagrams (e.g. steam tables, psychrometric charts, T-s, P-h, x-y, T-x-y)
C. Thermodynamic laws (e.g., 1st law, 2nd law)
D. Thermodynamic processes (e.g., isothermal, adiabatic, isentropic)
E. Cyclic processes and efficiency (e.g., power, refrigeration, heat pump)
F. Phase equilibrium (e.g., fugacity, activity coefficient)
G. Chemical equilibrium
H. Heats of reaction and mixing

9. Material/Energy Balances **8–12**

A. Mass balance (steady and unsteady state)
B. Energy balance (steady and unsteady state)
C. Recycle/bypass processes
D. Reactive systems (e.g., combustion)

10. Heat Transfer **8–12**

A. Conductive heat transfer
B. Convective heat transfer (natural and forced)
C. Radiation heat transfer
D. Heat transfer coefficients (e.g., overall, local, fouling)
E. Heat transfer equipment, operation, and design (e.g., double pipe, shell and tube, fouling, number of transfer units, log-mean temperature difference, flow configuration)

11. Mass Transfer and Separation **8–12**

A. Molecular diffusion (e.g., steady and unsteady state, physical property estimation)
B. Convective mass transfer (e.g., mass transfer coefficient, eddy diffusion)
C. Separation systems (e.g., distillation, absorption, extraction, membrane processes)

D. Equilibrium stage methods (e.g., graphical methods, McCabe-Thiele, efficiency)
E. Continuous contact methods (e.g., number of transfer units, height equivalent to a theoretical plate, height of transfer unit, number of theoretical plates)
F. Humidification and drying

12. Chemical Reaction Engineering **8–12**

A. Reaction rates and order
B. Rate constant (e.g., Arrhenius function)
C. Conversion, yield, and selectivity
D. Type of reactions (e.g., series, parallel, forward, reverse, homogeneous, heterogeneous, catalysis, biocatalysis)
E. Reactor types (e.g., batch, semibatch, continuous stirred tank, plug flow, gas phase, liquid phase)

13. Process Design and Economics **8–12**

A. Process flow diagrams and piping and instrumentation diagrams
B. Equipment selection (e.g., sizing and scale-up)
C. Cost estimation
D. Comparison of economic alternatives (e.g., net present value, discounted cash flow, rate of return, expected value and risk)
E. Process design and optimization (e.g., sustainability, efficiency, green engineering, inherently safer design, evaluation of specifications)

14. Process Control **5–8**

A. Dynamics (e.g., time constants and 2nd order, underdamped, and transfer functions)
B. Control strategies (e.g., feedback, feed-forward, cascade, ratio, and PID)
C. Control loop design and hardware (e.g., matching measured and manipulated variables, sensors, control valves, and conceptual process control)

15. Safety, Health, and Environment **5–8**

A. Hazardous properties of materials (e.g., corrosivity, flammability, toxicity, reactivity, handling and storage), including MSDS
B. Industrial hygiene (e.g., noise, PPE, ergonomics)
C. Process safety and hazard analysis [e.g., layer of protection analysis, hazard and operability studies (HazOps), fault-tree analysis or event tree]
D. Overpressure and underpressure protection (e.g., relief, redundant control, intrinsically safe)
E. Waste minimization, waste treatment, and regulation (e.g., air, water, solids, RCRA, CWA, EPA, OSHA)

16. Ethics and Professional Practice **2–3**

A. Codes of ethics (professional and technical societies)
B. Agreements and contracts
C. Ethical and legal considerations
D. Professional liability
E. Public protection issues (e.g., licensing boards)

NCEES

advancing licensure for engineers and surveyors

Fundamentals of Engineering (FE)
CIVIL CBT Exam Specifications

Effective Beginning with the January 2014 Examinations

- The FE exam is a computer-based test (CBT). It is closed book with an electronic reference.
- Examinees have 6 hours to complete the exam, which contains 110 multiple-choice questions. The 6-hour time also includes a tutorial and an optional scheduled break.
- The FE exam uses both the International System of Units (SI) and the US Customary System (USCS).

Knowledge	Number of Questions
1. Mathematics A. Analytic geometry B. Calculus C. Roots of equations D. Vector analysis	**7–11**
2. Probability and Statistics A. Measures of central tendencies and dispersions (e.g., mean, mode, standard deviation) B. Estimation for a single mean (e.g., point, confidence intervals) C. Regression and curve fitting D. Expected value (weighted average) in decision making	**4–6**
3. Computational Tools A. Spreadsheet computations B. Structured programming (e.g., if-then, loops, macros)	**4–6**
4. Ethics and Professional Practice A. Codes of ethics (professional and technical societies) B. Professional liability C. Licensure D. Sustainability and sustainable design E. Professional skills (e.g., public policy, management, and business) F. Contracts and contract law	**4–6**
5. Engineering Economics A. Discounted cash flow (e.g., equivalence, PW, equivalent annual worth, FW, rate of return) B. Cost (e.g., incremental, average, sunk, estimating) C. Analyses (e.g., breakeven, benefit-cost, life cycle) D. Uncertainty (e.g., expected value and risk)	**4–6**
6. Statics A. Resultants of force systems B. Equivalent force systems C. Equilibrium of rigid bodies D. Frames and trusses	**7–11**

E. Centroid of area
F. Area moments of inertia
G. Static friction

7. Dynamics **4–6**

A. Kinematics (e.g., particles and rigid bodies)
B. Mass moments of inertia
C. Force acceleration (e.g., particles and rigid bodies)
D. Impulse momentum (e.g., particles and rigid bodies)
E. Work, energy, and power (e.g., particles and rigid bodies)

8. Mechanics of Materials **7–11**

A. Shear and moment diagrams
B. Stresses and strains (e.g., axial, torsion, bending, shear, thermal)
C. Deformations (e.g., axial, torsion, bending, thermal)
D. Combined stresses
E. Principal stresses
F. Mohr's circle
G. Column analysis (e.g., buckling, boundary conditions)
H. Composite sections
I. Elastic and plastic deformations
J. Stress-strain diagrams

9. Materials **4–6**

A. Mix design (e.g., concrete and asphalt)
B. Test methods and specifications (e.g., steel, concrete, aggregates, asphalt, wood)
C. Physical and mechanical properties of concrete, ferrous and nonferrous metals, masonry, wood, engineered materials (e.g., FRP, laminated lumber, wood/plastic composites), and asphalt

10. Fluid Mechanics **4–6**

A. Flow measurement
B. Fluid properties
C. Fluid statics
D. Energy, impulse, and momentum equations

11. Hydraulics and Hydrologic Systems **8–12**

A. Basic hydrology (e.g., infiltration, rainfall, runoff, detention, flood flows, watersheds)
B. Basic hydraulics (e.g., Manning equation, Bernoulli theorem, open-channel flow, pipe flow)
C. Pumping systems (water and wastewater)
D. Water distribution systems
E. Reservoirs (e.g., dams, routing, spillways)
F. Groundwater (e.g., flow, wells, drawdown)
G. Storm sewer collection systems

12. Structural Analysis **6–9**

A. Analysis of forces in statically determinant beams, trusses, and frames
B. Deflection of statically determinant beams, trusses, and frames
C. Structural determinacy and stability analysis of beams, trusses, and frames

D. Loads and load paths (e.g., dead, live, lateral, influence lines and moving loads, tributary areas)
E. Elementary statically indeterminate structures

13. Structural Design **6–9**

A. Design of steel components (e.g., codes and design philosophies, beams, columns, beam-columns, tension members, connections)
B. Design of reinforced concrete components (e.g., codes and design philosophies, beams, slabs, columns, walls, footings)

14. Geotechnical Engineering **9–14**

A. Geology
B. Index properties and soil classifications
C. Phase relations (air-water-solid)
D. Laboratory and field tests
E. Effective stress (buoyancy)
F. Stability of retaining walls (e.g., active pressure/passive pressure)
G. Shear strength
H. Bearing capacity (cohesive and noncohesive)
I. Foundation types (e.g., spread footings, deep foundations, wall footings, mats)
J. Consolidation and differential settlement
K. Seepage/flow nets
L. Slope stability (e.g., fills, embankments, cuts, dams)
M. Soil stabilization (e.g., chemical additives, geosynthetics)
N. Drainage systems
O. Erosion control

15. Transportation Engineering **8–12**

A. Geometric design of streets and highways
B. Geometric design of intersections
C. Pavement system design (e.g., thickness, subgrade, drainage, rehabilitation)
D. Traffic safety
E. Traffic capacity
F. Traffic flow theory
G. Traffic control devices
H. Transportation planning (e.g., travel forecast modeling)

16. Environmental Engineering **6–9**

A. Water quality (ground and surface)
B. Basic tests (e.g., water, wastewater, air)
C. Environmental regulations
D. Water supply and treatment
E. Wastewater collection and treatment

17. Construction **4–6**

A. Construction documents
B. Procurement methods (e.g., competitive bid, qualifications-based)
C. Project delivery methods (e.g., design-bid-build, design build, construction management, multiple prime)
D. Construction operations and methods (e.g., lifting, rigging, dewatering and pumping, equipment production, productivity analysis and improvement, temporary erosion control)
E. Project scheduling (e.g., CPM, allocation of resources)
F. Project management (e.g., owner/contractor/client relations)
G. Construction safety
H. Construction estimating

18. Surveying **4–6**

A. Angles, distances, and trigonometry
B. Area computations
C. Earthwork and volume computations
D. Closure
E. Coordinate systems (e.g., state plane, latitude/longitude)
F. Leveling (e.g., differential, elevations, percent grades)

NCEES
advancing licensure for engineers and surveyors

Fundamentals of Engineering (FE)
ELECTRICAL AND COMPUTER CBT Exam Specifications

Effective Beginning with the January 2014 Examinations

- The FE exam is a computer-based test (CBT). It is closed book with an electronic reference.
- Examinees have 6 hours to complete the exam, which contains 110 multiple-choice questions. The 6-hour time also includes a tutorial and an optional scheduled break.
- The FE exam uses both the International System of Units (SI) and the US Customary System (USCS).

Knowledge	Number of Questions
1. Mathematics A. Algebra and trigonometry B. Complex numbers C. Discrete mathematics D. Analytic geometry E. Calculus F. Differential equations G. Linear algebra H. Vector analysis	**11–17**
2. Probability and Statistics A. Measures of central tendencies and dispersions (e.g., mean, mode, standard deviation) B. Probability distributions (e.g., discrete, continuous, normal, binomial) C. Expected value (weighted average) in decision making D. Estimation for a single mean (e.g., point, confidence intervals, conditional probability)	**4–6**
3. Ethics and Professional Practice A. Codes of ethics (professional and technical societies) B. NCEES *Model Law* and *Model Rules* C. Intellectual property (e.g., copyright, trade secrets, patents)	**3–5**
4. Engineering Economics A. Time value of money (e.g., present value, future value, annuities) B. Cost estimation C. Risk identification D. Analysis (e.g., cost-benefit, trade-off, breakeven)	**3–5**
5. Properties of Electrical Materials A. Chemical (e.g., corrosion, ions, diffusion) B. Electrical (e.g., conductivity, resistivity, permittivity, magnetic permeability) C. Mechanical (e.g., piezoelectric, strength) D. Thermal (e.g., conductivity, expansion)	**4–6**

6.	**Engineering Sciences**	**6–9**

A. Work, energy, power, heat
B. Charge, energy, current, voltage, power
C. Forces (e.g., between charges, on conductors)
D. Work done in moving a charge in an electric field (relationship between voltage and work)
E. Capacitance
F. Inductance

7.	**Circuit Analysis (DC and AC Steady State)**	**10–15**

A. KCL, KVL
B. Series/parallel equivalent circuits
C. Thevenin and Norton theorems
D. Node and loop analysis
E. Waveform analysis (e.g., RMS, average, frequency, phase, wavelength)
F. Phasors
G. Impedance

8.	**Linear Systems**	**5–8**

A. Frequency/transient response
B. Resonance
C. Laplace transforms
D. Transfer functions
E. 2-port theory

9.	**Signal Processing**	**5–8**

A. Convolution (continuous and discrete)
B. Difference equations
C. Z-transforms
D. Sampling (e.g., aliasing, Nyquist theorem)
E. Analog filters
F. Digital filters

10.	**Electronics**	**7–11**

A. Solid-state fundamentals (e.g., tunneling, diffusion/drift current, energy bands, doping bands, p-n theory)
B. Discrete devices (diodes, transistors, BJT, CMOS) and models and their performance
C. Bias circuits
D. Amplifiers (e.g., single-stage/common emitter, differential)
E. Operational amplifiers (ideal, non-ideal)
F. Instrumentation (e.g., measurements, data acquisition, transducers)
G. Power electronics

11.	**Power**	**8–12**

A. Single phase and three phase
B. Transmission and distribution
C. Voltage regulation
D. Transformers
E. Motors and generators
F. Power factor (pf)

12. Electromagnetics **5–8**

A. Maxwell equations
B. Electrostatics/magnetostatics (e.g., measurement of spatial relationships, vector analysis)
C. Wave propagation
D. Transmission lines (high frequency)
E. Electromagnetic compatibility

13. Control Systems **6–9**

A. Block diagrams (feed-forward, feedback)
B. Bode plots
C. Closed-loop and open-loop response
D. Controller performance (gain, PID), steady-state errors
E. Root locus
F. Stability
G. State variables

14. Communications **5–8**

A. Basic modulation/demodulation concepts (e.g., AM, FM, PCM)
B. Fourier transforms/Fourier series
C. Multiplexing (e.g., time division, frequency division)
D. Digital communications

15. Computer Networks **3–5**

A. Routing and switching
B. Network topologies/frameworks/models
C. Local area networks

16. Digital Systems **7–11**

A. Number systems
B. Boolean logic
C. Logic gates and circuits
D. Logic minimization (e.g., SOP, POS, Karnaugh maps)
E. Flip-flops and counters
F. Programmable logic devices and gate arrays
G. State machine design
H. Data path/controller design
I. Timing (diagrams, asynchronous inputs, races, hazards)

17. Computer Systems **4–6**

A. Architecture (e.g., pipelining, cache memory)
B. Microprocessors
C. Memory technology and systems
D. Interfacing

18. Software Development **4–6**

A. Algorithms
B. Data structures
C. Software design methods (structured, object-oriented)
D. Software implementation (e.g., procedural, scripting languages)
E. Software testing

NCEES
advancing licensure for engineers and surveyors

Fundamentals of Engineering (FE)
ENVIRONMENTAL CBT Exam Specifications

Effective Beginning with the January 2014 Examinations

- The FE exam is a computer-based test (CBT). It is closed book with an electronic reference.
- Examinees have 6 hours to complete the exam, which contains 110 multiple-choice questions. The 6-hour time also includes a tutorial and an optional scheduled break.
- The FE exam uses both the International System of Units (SI) and the US Customary System (USCS).

Knowledge	Number of Questions
1. Mathematics	**4–6**
A. Analytic geometry	
B. Numerical methods	
C. Roots of equations	
D. Calculus	
E. Differential equations	
2. Probability and Statistics	**3–5**
A. Measures of central tendencies and dispersions (e.g., mean, mode, standard deviation)	
B. Probability distributions (e.g., discrete, continuous, normal, binomial)	
C. Estimation (point, confidence intervals) for a single mean	
D. Regression and curve fitting	
E. Expected value (weighted average) in decision making	
F. Hypothesis testing	
3. Ethics and Professional Practice	**5–8**
A. Codes of ethics (professional and technical societies)	
B. Agreements and contracts	
C. Ethical and legal considerations	
D. Professional liability	
E. Public protection issues (e.g., licensing boards)	
F. Regulations (e.g., water, wastewater, air, solid/hazardous waste, groundwater/soils)	
4. Engineering Economics	**4–6**
A. Discounted cash flow (e.g., life cycle, equivalence, PW, equivalent annual worth, FW, rate of return)	
B. Cost (e.g., incremental, average, sunk, estimating)	
C. Analyses (e.g., breakeven, benefit-cost)	
D. Uncertainty (expected value and risk)	
5. Materials Science	**3–5**
A. Properties (e.g., chemical, electrical, mechanical, physical)	
B. Corrosion mechanisms and controls	
C. Material selection and compatibility	

6. Environmental Science and Chemistry **11–17**

A. Reactions (e.g., equilibrium, acid base, oxidation-reduction, precipitation)
B. Stoichiometry
C. Kinetics (chemical, microbiological)
D. Organic chemistry (e.g., nomenclature, functional group reactions)
E. Ecology (e.g., Streeter-Phelps, fluviology, limnology, eutrophication)
F. Multimedia equilibrium partitioning (e.g., Henry's law, octonal partitioning coefficient)

7. Risk Assessment **5–8**

A. Dose-response toxicity (carcinogen, noncarcinogen)
B. Exposure routes

8. Fluid Mechanics **9–14**

A. Fluid statics
B. Closed conduits (e.g., Darcy-Weisbach, Hazen-Williams, Moody)
C. Open channel (Manning)
D. Pumps (e.g., power, operating point, parallel and series)
E. Flow measurement (e.g., weirs, orifices, flowmeters)
F. Blowers (e.g., power, operating point, parallel, and series)

9. Thermodynamics **3–5**

A. Thermodynamic laws (e.g., 1st law, 2nd law)
B. Energy, heat, and work
C. Ideal gases
D. Mixture of nonreacting gases
E. Heat transfer

10. Water Resources **10–15**

A. Demand calculations
B. Population estimations
C. Runoff calculations (e.g., land use, land cover, time of concentration, duration, intensity, frequency)
D. Reservoir sizing
E. Routing (e.g., channel, reservoir)
F. Water quality and modeling (e.g., erosion, channel stability, stormwater quality management)

11. Water and Wastewater **14–21**

A. Water and wastewater characteristics
B. Mass and energy balances
C. Conventional water treatment processes (e.g., clarification, disinfection, filtration, flocculation, softening, rapid mix)
D. Conventional wastewater treatment processes (e.g., activated sludge, decentralized wastewater systems, fixed-film system, disinfection, flow equalization, headworks, lagoons)
E. Alternative treatment process (e.g., conservation and reuse, membranes, nutrient removal, ion exchange, activated carbon, air stripping)
F. Sludge treatment and handling (e.g., land application, sludge digestion, sludge dewatering)

12. Air Quality **10–15**

A. Chemical principles (e.g., ideal gas, mole fractions, stoichiometry, Henry's law)
B. Mass balances
C. Emissions (factors, rates)
D. Atmospheric sciences (e.g., stability classes, dispersion modeling, lapse rates)
E. Gas handling and treatment technologies (e.g., hoods, ducts, coolers, biofiltration, scrubbers, adsorbers, incineration)
F. Particle handling and treatment technologies (e.g., baghouses, cyclones, electrostatic precipitators, settling velocity)

13. Solid and Hazardous Waste **10–15**

A. Composting
B. Mass balances
C. Compatibility
D. Landfilling (e.g., siting, design, leachate, material and energy recovery)
E. Site characterization and remediation
F. Hazardous waste treatment (e.g., physical, chemical, thermal)
G. Radioactive waste treatment and disposal

14. Groundwater and Soils **9–14**

A. Basic hydrogeology (e.g., aquifers, permeability, water table, hydraulic conductivity, saturation, soil characteristics)
B. Drawdown (e.g., Jacob, Theis, Thiem)
C. Groundwater flow (e.g., Darcy's law, specific capacity, velocity, gradient)
D. Soil and groundwater remediation

NCEES
advancing licensure for engineers and surveyors

Fundamentals of Engineering (FE)
INDUSTRIAL AND SYSTEMS CBT Exam Specifications

Effective Beginning with the January 2014 Examinations

- The FE exam is a computer-based test (CBT). It is closed book with an electronic reference.
- Examinees have 6 hours to complete the exam, which contains 110 multiple-choice questions. The 6-hour time also includes a tutorial and an optional scheduled break.
- The FE exam uses both the International System of Units (SI) and the US Customary System (USCS).

Knowledge	Number of Questions
1. Mathematics	**6–9**
A. Analytic geometry	
B. Calculus	
C. Matrix operations	
D. Vector analysis	
E. Linear algebra	
2. Engineering Sciences	**5–8**
A. Work, energy, and power	
B. Material properties and selection	
C. Charge, energy, current, voltage, and power	
3. Ethics and Professional Practice	**5–8**
A. Codes of ethics and licensure	
B. Agreements and contracts	
C. Professional, ethical, and legal responsibility	
D. Public protection and regulatory issues	
4. Engineering Economics	**10–15**
A. Discounted cash flows (PW, EAC, FW, IRR, amortization)	
B. Types and breakdown of costs (e.g., fixed, variable, direct and indirect labor)	
C. Cost analyses (e.g., benefit-cost, breakeven, minimum cost, overhead)	
D. Accounting (financial statements and overhead cost allocation)	
E. Cost estimation	
F. Depreciation and taxes	
G. Capital budgeting	
5. Probability and Statistics	**10–15**
A. Combinatorics (e.g., combinations, permutations)	
B. Probability distributions (e.g., normal, binomial, empirical)	
C. Conditional probabilities	
D. Sampling distributions, sample sizes, and statistics (e.g., central tendency, dispersion)	
E. Estimation (e.g., point, confidence intervals)	
F. Hypothesis testing	
G. Regression (linear, multiple)	

H. System reliability (e.g., single components, parallel and series systems)
I. Design of experiments (e.g., ANOVA, factorial designs)

6. Modeling and Computations **8–12**

A. Algorithm and logic development (e.g., flow charts, pseudocode)
B. Databases (e.g., types, information content, relational)
C. Decision theory (e.g., uncertainty, risk, utility, decision trees)
D. Optimization modeling (e.g., decision variables, objective functions, and constraints)
E. Linear programming (e.g., formulation, primal, dual, graphical solutions)
F. Mathematical programming (e.g., network, integer, dynamic, transportation, assignment)
G. Stochastic models (e.g., queuing, Markov, reliability)
H. Simulation

7. Industrial Management **8–12**

A. Principles (e.g., planning, organizing, motivational theory)
B. Tools of management (e.g., MBO, reengineering, organizational structure)
C. Project management (e.g., scheduling, PERT, CPM)
D. Productivity measures

8. Manufacturing, Production, and Service Systems **8–12**

A. Manufacturing processes
B. Manufacturing systems (e.g., cellular, group technology, flexible)
C. Process design (e.g., resources, equipment selection, line balancing)
D. Inventory analysis (e.g., EOQ, safety stock)
E. Forecasting
F. Scheduling (e.g., sequencing, cycle time, material control)
G. Aggregate planning
H. Production planning (e.g., JIT, MRP, ERP)
I. Lean enterprises
J. Automation concepts (e.g., robotics, CIM)
K. Sustainable manufacturing (e.g., energy efficiency, waste reduction)
L. Value engineering

9. Facilities and Logistics **8–12**

A. Flow measurements and analysis (e.g., from/to charts, flow planning)
B. Layouts (e.g., types, distance metrics, planning, evaluation)
C. Location analysis (e.g., single- and multiple-facility location, warehouses)
D. Process capacity analysis (e.g., number of machines and people, trade-offs)
E. Material handling capacity analysis
F. Supply chain management and design

10. Human Factors, Ergonomics, and Safety **8–12**

A. Hazard identification and risk assessment
B. Environmental stress assessment (e.g., noise, vibrations, heat)
C. Industrial hygiene
D. Design for usability (e.g., tasks, tools, displays, controls, user interfaces)
E. Anthropometry
F. Biomechanics
G. Cumulative trauma disorders (e.g., low back injuries, carpal tunnel syndrome)

H. Systems safety
I. Cognitive engineering (e.g., information processing, situation awareness, human error, mental models)

11. Work Design **8–12**

A. Methods analysis (e.g., charting, workstation design, motion economy)
B. Time study (e.g., time standards, allowances)
C. Predetermined time standard systems (e.g., MOST, MTM)
D. Work sampling
E. Learning curves

12. Quality **8–12**

A. Six sigma
B. Management and planning tools (e.g., fishbone, Pareto, QFD, TQM)
C. Control charts
D. Process capability and specifications
E. Sampling plans
F. Design of experiments for quality improvement
G. Reliability engineering

13. Systems Engineering **8–12**

A. Requirements analysis
B. System design
C. Human systems integration
D. Functional analysis and allocation
E. Configuration management
F. Risk management
G. Verification and assurance
H. System life-cycle engineering

NCEES
advancing licensure for engineers and surveyors

Fundamentals of Engineering (FE)
MECHANICAL CBT Exam Specifications

Effective Beginning with the January 2014 Examinations

- The FE exam is a computer-based test (CBT). It is closed book with an electronic reference.
- Examinees have 6 hours to complete the exam, which contains 110 multiple-choice questions. The 6-hour time also includes a tutorial and an optional scheduled break.
- The FE exam uses both the International System of Units (SI) and the US Customary System (USCS).

Knowledge	Number of Questions
1. Mathematics	**6–9**
A. Analytic geometry	
B. Calculus	
C. Linear algebra	
D. Vector analysis	
E. Differential equations	
F. Numerical methods	
2. Probability and Statistics	**4–6**
A. Probability distributions	
B. Regression and curve fitting	
3. Computational Tools	**3–5**
A. Spreadsheets	
B. Flow charts	
4. Ethics and Professional Practice	**3–5**
A. Codes of ethics	
B. Agreements and contracts	
C. Ethical and legal considerations	
D. Professional liability	
E. Public health, safety, and welfare	
5. Engineering Economics	**3–5**
A. Time value of money	
B. Cost, including incremental, average, sunk, and estimating	
C. Economic analyses	
D. Depreciation	
6. Electricity and Magnetism	**3–5**
A. Charge, current, voltage, power, and energy	
B. Current and voltage laws (Kirchhoff, Ohm)	
C. Equivalent circuits (series, parallel)	
D. AC circuits	
E. Motors and generators	

7.	**Statics**	**8–12**

A. Resultants of force systems
B. Concurrent force systems
C. Equilibrium of rigid bodies
D. Frames and trusses
E. Centroids
F. Moments of inertia
G. Static friction

8.	**Dynamics, Kinematics, and Vibrations**	**9–14**

A. Kinematics of particles
B. Kinetic friction
C. Newton's second law for particles
D. Work-energy of particles
E. Impulse-momentum of particles
F. Kinematics of rigid bodies
G. Kinematics of mechanisms
H. Newton's second law for rigid bodies
I. Work-energy of rigid bodies
J. Impulse-momentum of rigid bodies
K. Free and forced vibrations

9.	**Mechanics of Materials**	**8–12**

A. Shear and moment diagrams
B. Stress types (axial, bending, torsion, shear)
C. Stress transformations
D. Mohr's circle
E. Stress and strain caused by axial loads
F. Stress and strain caused by bending loads
G. Stress and strain caused by torsion
H. Stress and strain caused by shear
I. Combined loading
J. Deformations
K. Columns

10.	**Material Properties and Processing**	**8–12**

A. Properties, including chemical, electrical, mechanical, physical, and thermal
B. Stress-strain diagrams
C. Engineered materials
D. Ferrous metals
E. Nonferrous metals
F. Manufacturing processes
G. Phase diagrams
H. Phase transformation, equilibrium, and heat treating
I. Materials selection
J. Surface conditions
K. Corrosion mechanisms and control
L. Thermal failure

M. Ductile or brittle behavior
N. Fatigue
O. Crack propagation

11. Fluid Mechanics **9–14**

A. Fluid properties
B. Fluid statics
C. Energy, impulse, and momentum
D. Internal flow
E. External flow
F. Incompressible flow
G. Compressible flow
H. Power and efficiency
I. Performance curves
J. Scaling laws for fans, pumps, and compressors

12. Thermodynamics **13–20**

A. Properties of ideal gases and pure substances
B. Energy transfers
C. Laws of thermodynamics
D. Processes
E. Performance of components
F. Power cycles, thermal efficiency, and enhancements
G. Refrigeration and heat pump cycles and coefficients of performance
H. Nonreacting mixtures of gases
I. Psychrometrics
J. Heating, ventilating, and air-conditioning (HVAC) processes
K. Combustion and combustion products

13. Heat Transfer **9–14**

A. Conduction
B. Convection
C. Radiation
D. Thermal resistance
E. Transient processes
F. Heat exchangers
G. Boiling and condensation

14. Measurements, Instrumentation, and Controls **5–8**

A. Sensors
B. Block diagrams
C. System response
D. Measurement uncertainty

15. Mechanical Design and Analysis **9–14**

A. Stress analysis of machine elements
B. Failure theories and analysis
C. Deformation and stiffness
D. Springs
E. Pressure vessels
F. Beams
G. Piping

H. Bearings
I. Power screws
J. Power transmission
K. Joining methods
L. Manufacturability
M. Quality and reliability
N. Hydraulic components
O. Pneumatic components
P. Electromechanical components

NCEES
advancing licensure for engineers and surveyors

Fundamentals of Engineering (FE)
OTHER DISCIPLINES CBT Exam Specifications

Effective Beginning with the January 2014 Examinations

- The FE exam is a computer-based test (CBT). It is closed book with an electronic reference.
- Examinees have 6 hours to complete the exam, which contains 110 multiple-choice questions. The 6-hour time also includes a tutorial and an optional scheduled break.
- The FE exam uses both the International System of Units (SI) and the US Customary System (USCS).

Knowledge	Number of Questions
1. Mathematics and Advanced Engineering Mathematics	**12–18**
A. Analytic geometry and trigonometry	
B. Calculus	
C. Differential equations (e.g., homogeneous, nonhomogeneous, Laplace transforms)	
D. Numerical methods (e.g., algebraic equations, roots of equations, approximations, precision limits)	
E. Linear algebra (e.g., matrix operations)	
2. Probability and Statistics	**6–9**
A. Measures of central tendencies and dispersions (e.g., mean, mode, variance, standard deviation)	
B. Probability distributions (e.g., discrete, continuous, normal, binomial)	
C. Estimation (e.g., point, confidence intervals)	
D. Expected value (weighted average) in decision making	
E. Sample distributions and sizes	
F. Goodness of fit (e.g., correlation coefficient, least squares)	
3. Chemistry	**7–11**
A. Periodic table (e.g., nomenclature, metals and nonmetals, atomic structure of matter)	
B. Oxidation and reduction	
C. Acids and bases	
D. Equations (e.g., stoichiometry, equilibrium)	
E. Gas laws (e.g., Boyle's and Charles' Laws, molar volume)	
4. Instrumentation and Data Acquisition	**4–6**
A. Sensors (e.g., temperature, pressure, motion, pH, chemical constituents)	
B. Data acquisition (e.g., logging, sampling rate, sampling range, filtering, amplification, signal interface)	
C. Data processing (e.g., flow charts, loops, branches)	
5. Ethics and Professional Practice	**3–5**
A. Codes of ethics	
B. NCEES *Model Law* and *Model Rules*	
C. Public protection issues (e.g., licensing boards)	

6. Safety, Health, and Environment **4–6**

A. Industrial hygiene (e.g., carcinogens, toxicology, MSDS, lower exposure limits)
B. Basic safety equipment (e.g., pressure relief valves, emergency shut-offs, fire prevention and control, personal protective equipment)
C. Gas detection and monitoring (e.g., O_2, CO, CO_2, CH_4, H_2S, Radon)
D. Electrical safety

7. Engineering Economics **7–11**

A. Time value of money (e.g., present worth, annual worth, future worth, rate of return)
B. Cost (e.g., incremental, average, sunk, estimating)
C. Economic analyses (e.g., breakeven, benefit-cost, optimal economic life)
D. Uncertainty (e.g., expected value and risk)
E. Project selection (e.g., comparison of unequal life projects, lease/buy/make, depreciation, discounted cash flow)

8. Statics **8–12**

A. Resultants of force systems and vector analysis
B. Concurrent force systems
C. Force couple systems
D. Equilibrium of rigid bodies
E. Frames and trusses
F. Area properties (e.g., centroids, moments of inertia, radius of gyration)
G. Static friction

9. Dynamics **7–11**

A. Kinematics
B. Linear motion (e.g., force, mass, acceleration)
C. Angular motion (e.g., torque, inertia, acceleration)
D. Mass moment of inertia
E. Impulse and momentum (linear and angular)
F. Work, energy, and power
G. Dynamic friction
H. Vibrations

10. Strength of Materials **8–12**

A. Stress types (e.g., normal, shear, bending, torsion)
B. Combined stresses
C. Stress and strain caused by axial loads, bending loads, torsion, or shear
D. Shear and moment diagrams
E. Analysis of beams, trusses, frames, and columns
F. Deflection and deformations (e.g., axial, bending, torsion)
G. Elastic and plastic deformation
H. Failure theory and analysis (e.g., static/dynamic, creep, fatigue, fracture, buckling)

11. Materials Science **6–9**

A. Physical, mechanical, chemical, and electrical properties of ferrous metals
B. Physical, mechanical, chemical, and electrical properties of nonferrous metals
C. Physical, mechanical, chemical, and electrical properties of engineered materials (e.g., polymers, concrete, composites)
D. Corrosion mechanisms and control

12. Fluid Mechanics and Dynamics of Liquids **8–12**

A. Fluid properties (e.g., Newtonian, non-Newtonian)
B. Dimensionless numbers (e.g., Reynolds number, Froude number)
C. Laminar and turbulent flow
D. Fluid statics
E. Energy, impulse, and momentum equations (e.g., Bernoulli equation)
F. Pipe flow and friction losses (e.g., pipes, valves, fittings, Darcy-Weisbach equation, Hazen-Williams equation)
G. Open-channel flow (e.g., Manning equation, drag)
H. Fluid transport systems (e.g., series and parallel operations)
I. Flow measurement
J. Turbomachinery (e.g., pumps, turbines)

13. Fluid Mechanics and Dynamics of Gases **4–6**

A. Fluid properties (e.g., ideal and non-ideal gases)
B. Dimensionless numbers (e.g., Reynolds number, Mach number)
C. Laminar and turbulent flow
D. Fluid statics
E. Energy, impulse, and momentum equations
F. Duct and pipe flow and friction losses
G. Fluid transport systems (e.g., series and parallel operations)
H. Flow measurement
I. Turbomachinery (e.g., fans, compressors, turbines)

14. Electricity, Power, and Magnetism **7–11**

A. Electrical fundamentals (e.g., charge, current, voltage, resistance, power, energy)
B. Current and voltage laws (Kirchhoff, Ohm)
C. DC circuits
D. Equivalent circuits (series, parallel, Norton's theorem, Thevenin's theorem)
E. Capacitance and inductance
F. AC circuits (e.g., real and imaginary components, complex numbers, power factor, reactance and impedance)
G. Measuring devices (e.g., voltmeter, ammeter, wattmeter)

15. Heat, Mass, and Energy Transfer **9–14**

A. Energy, heat, and work
B. Thermodynamic laws (e.g., 1st law, 2nd law)
C. Thermodynamic equilibrium
D. Thermodynamic properties (e.g., entropy, enthalpy, heat capacity)
E. Thermodynamic processes (e.g., isothermal, adiabatic, reversible, irreversible)
F. Mixtures of nonreactive gases
G. Heat transfer (e.g., conduction, convection, and radiation)
H. Mass and energy balances
I. Property and phase diagrams (e.g., T-s, P-h)
J. Phase equilibrium and phase change
K. Combustion and combustion products (e.g., CO, CO_2, NO_X, ash, particulates)
L. Psychrometrics (e.g., relative humidity, wet-bulb)